I CAN Learn®
EDUCATION SYSTEMS

JRL Enterprises, Inc.
New Orleans, Louisiana

ISBN 978-0-9789017-7-6

JRL Enterprises, Inc.

400 Poydras Street, Suite 1000

New Orleans, LA 70130

www.icanlearn.com

Printed in Canada

Table of Contents

*HA1-861 is an advanced lesson and includes the Division Property for Exponents.

HA1-003: Order of Operations

Suppose you were asked to simplify $4 + 3 \cdot 6$. Because you read from left to right, you would probably add 3 to 4, which is 7, and then multiply 7 and 6 to get the product 42.

This answer, however, is not correct. Expressions that contain more than one operation must be simplified in a particular order. This is called the **order of operations**.

Order of Operations	A series of rules that determine the sequence in which you perform operations. If a particular operation is not included in the problem, go to the next step. Step 1: Perform all operations in grouping symbols, starting with the innermost group of symbols. Examples of grouping symbols are ( ), [ ], { }, and the fraction bar. Step 2: Simplify all exponents. Step 3: Multiply or divide in order from left to right. Step 4: Add or subtract in order from left to right.

Using the steps above, simplify the expression presented earlier, and try more examples.

Example 1 Simplify: $4 + 3 \cdot 6$

Step 1: There are no grouping symbols; move to the next step.

Step 2: There are no exponents; move to the next step.

$4 + 3 \cdot 6$
$= 4 + 18$

Step 3: Perform multiplication and division; multiply 3 and 6.

$= 4 + 18$
$= 22$

Step 4: Add or subtract; add 18 to 4 to get the answer.

Answer: 22

Example 2 Simplify: $2(9 - 4) + 7$

$2(9 - 4) + 7$
$= 2(5) + 7$

Step 1: Perform the operation within the grouping symbol.

Step 2: There are no exponents; move to the next step.

$= 2(5) + 7$
$= 10 + 7$

Step 3: Multiply 2 and 5.

$= 10 + 7$
$= 17$

Step 4: Add 7 to 10 to get the answer.

Answer: 17

Example 3 Simplify: $\frac{(5 \cdot 8) - (9 - 4)}{2 \cdot 4 - 1}$

$\frac{(5 \cdot 8) - (9 - 4)}{2 \cdot 4 - 1}$

$= \frac{(40) - (5)}{2 \cdot 4 - 1}$

Step 1: Perform all operations in the numerator according to the order of operations. In this case, parentheses will come first.

$= \frac{35}{2 \cdot 4 - 1}$

Step 2: Next, subtract the numbers in the numerator.

$= \frac{35}{2 \cdot 4 - 1}$

$= \frac{35}{8 - 1}$

Step 3: Perform all operations in the denominator according to the order of operations. In this case, multiply and then subtract.

$= \frac{35}{7}$

$= 5$

Step 4: Divide the numerator by the denominator.

Answer: 5

Example 4 Simplify: $[6 + 2(8 - 4)] - 5$

$[6 + 2(8 - 4)] - 5$

$= [6 + 2(4)] - 5$

Step 1: Perform the operation in the innermost parentheses first.

$= [6 + 8] - 5$

$= 14 - 5$

Step 2: Perform the operations in the brackets in the correct order: multiplication followed by addition.

$= 14 - 5$

$= 9$

Step 3: Subtract.

Answer: 9

Problem Set

Simplify:

1. $27 - 8 \div 2$

2. $37 + 2 \cdot 3$

3. $25 + 4 \cdot 7$

4. $42 + 6 \cdot 2$

5. $52 + 7 \cdot 5$

6. $6(15 - 12)$

7. $17(10 - 8)$

8. $12(15 + 10)$

9. $10(5 - 3)$

10. $15(10 - 5)$

11. $(100 + 50) \div (15 - 5)$

12. $(13 + 12) \cdot (40 - 17)$

13. $(13 + 12) \cdot (20 + 17)$

14. $\frac{48 + 12}{10 - 4}$

15. $\frac{48 - 12}{10 - 4}$

16. $\frac{48 + 12}{10 - 6}$

17. $\frac{75 + 15}{12 - 9}$

18. $250 - [5(25 + 5) + 17]$

19. $4(24 - 15) - 2(32 - 20)$

20. $\frac{7(75 + 15)}{3(12 - 5)}$

HA1-005: Evaluating Algebraic Expressions

How do $2 + 3$ and $2 + 3 = 5$ differ? What about $a - 5$ and $a - 5 = 9$? The difference is that $2 + 3$ and $a - 5$ do not contain equal signs, whereas $2 + 3 = 5$ and $a - 5 = 9$ do. Math problems such as $2 + 3$ and $a - 5$ are called *expressions*. *Numerical* expressions such as $2 + 3$ contain numbers and operations only. *Algebraic* expressions such as $a - 5$ contain numbers, operations, and variables. Either type of expression may also contain grouping symbols such as parentheses. Math problems such as $2 + 3 = 5$ and $a - 5 = 9$ are called *equations*. This lesson deals with **algebraic expressions**.

Algebraic Expression	A mathematical phrase that contains numbers, variables, one or more operations, and possibly grouping symbols such as parentheses

Variable	A letter or symbol used to represent an unknown number

Look at examples of algebraic expressions and how to read them:

Algebraic expression:	Reads as:
$5 + d$	5 plus d
$n - 7$	n minus 7 or the difference between n and 7
st	s times t
$\frac{p}{8}$	p divided by 8

Notice that you do not know what any of these expressions is equal to because you do not know the value of the variable(s). To evaluate algebraic expressions, you must replace all variables with given values.

If you have the expression $4a + 2$, and if $a = 2$, substitute the number 2 for every a and follow the order of operations to evaluate the expression. Look at the following examples:

Example 1 Evaluate $m - (6 - n)$ for $m = 12$ and $n = 3$.

$12 - (6 - 3)$ **Step 1:** Replace m with 12 and n with 3.

$= 12 - (3)$ **Step 2:** Follow the order of operations. In this case, parentheses followed by subtraction.

$= 9$

Answer: 9

Example 2 Evaluate $5(3a + 2b)$ for $a = 4$ and $b = 1$.

$5[3(4) + 2(1)]$

Step 1: Replace a with 4 and b with 1. Notice parentheses are placed around the numbers that are substituted so that $3a$ does not look like 34, but rather 3(4), which reminds you to multiply.

$= 5[12 + 2]$
$= 5[14]$
$= 70$

Step 2: Follow the order of operations.

Answer: 70

Example 3 Evaluate $\frac{12c}{2d}$ for $c = 3$ and $d = 2$.

$= \frac{12(3)}{2(2)}$

Step 1: Replace c with 3 and d with 2.

$= \frac{36}{4}$
$= 9$

Step 2: Follow the order of operations.

Answer: 9

Example 4 Evaluate $\frac{a+b}{c} + ab$ for $a = 5$, $b = 3$, and $c = 4$.

$= \frac{5+3}{4} + 5(3)$

Step 1: Replace a with 5, b with 3, and c with 4.

$= \frac{8}{4} + 15$
$= 2+15$
$= 17$

Step 2: Follow the order of operations.

Answer: 17

Problem Set

Evaluate for the given values of the variables:

1. $x + 7$ for $x = 11$

2. $m - 8$ for $m = 33$

3. $\frac{a}{b}$ for $a = 20$ and $b = 5$

4. $\frac{mn}{x}$ for $m = 6, n = 4,$ and $x = 8$

5. $2x + 8$ for $x = 5$

6. $5a - 2$ for $a = 3$

7. $8m + 12n$ for $m = 2$ and $n = 2$

8. $4x - 8y$ for $x = 4$ and $y = 2$

9. $\frac{a+b}{2}$ for $a = 20$ and $b = 2$

10. $\frac{a-b}{3}$ for $a = 15$ and $b = 3$

11. $a - (b + 2c)$ for $a = 36, b = 4,$ and $c = 5$

12. $m + (2n - r)$ for $m = 7, n = 5,$ and $r = 8$

13. $k(p - 3d)$ for $k = 4, p = 30,$ and $d = 5$

14. $4h(5e + w)$ for $h = 4, e = 2,$ and $w = 3$

15. $6bh - 2hp$ for $b = 10, h = 5,$ and $p = 6$

16. $cd - \frac{c}{d}$ for $c = 48$ and $d = 2$

17. $\frac{1}{4}xyz$ for $x = 12, y = 2$ and $z = 8$

18. $\frac{7a+4}{2(5+b)}$ for $a = 4$ and $b = 3$

19. $\frac{3(x+y)}{2x-y}$ for $x = 10$ and $y = 2$

20. $\frac{5(m-n)}{2m-3}$ for $m = 2$ and $n = 1$

HA1-015: Graphing Real Numbers Using a Number Line

Do you think that there must be a beginning and an end to numbers? Think about it. What is the greatest number you know? Can you add another number? Of course you can. If you begin counting from zero, it would be impossible for you to name all the numbers that occur before one billion. It is also impossible to name every number that occurs between 0 and 1. This is because just as there is never a number that you can't increase by one, there is no number that cannot be divided into a smaller number or part of a whole. The numbers used in algebra are called **real numbers,** and they include zero, the integers, and all of the fractions and decimals that occur between each integer.

Real Numbers	All positive and negative numbers, including fractions and decimals

All numbers that occur to the right of zero are called **positive** numbers.

Positive Numbers	All numbers greater than zero, to the right of the origin (zero) on the number line

Positive numbers can be written with or without the plus (+) sign, which means 4 and +4 have the same value. Regardless of how you write the number, it is still read as "four" or "positive four."

Now, what is the lowest number you know? You might think it is zero. Actually, our number system includes numbers less than zero; these are called **negative** numbers.

Negative Numbers	All numbers less than zero, to the left of the origin (zero) on the number line

Negative numbers, like –2, are denoted by placing a small hyphen or dash in front of the number. One way to organize real numbers is on a number line. Look at the number line below. Notice the real numbers to the left and right of zero. Also notice the two arrows at the opposite ends of the line. These arrows indicate that numbers continue infinitely in both directions—negative numbers to the left of zero and positive numbers to the right of zero. Zero is neither positive nor negative.

Look at the number line below. Notice the vertical lines that intersect the number line.

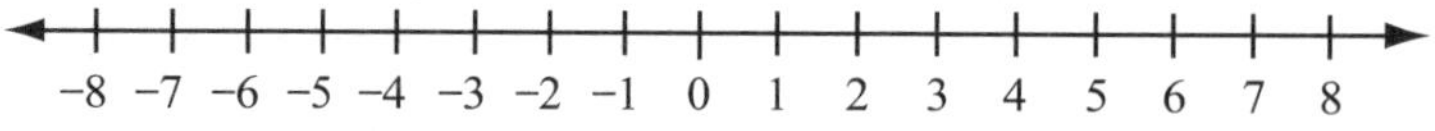

These lines are called **tick marks** and are used to show the unit distance from one vertical line to another. Once the tick marks are made, numbers can be placed on the number line. Each number, called a **coordinate**, can be identified as a point on the number line using a letter.

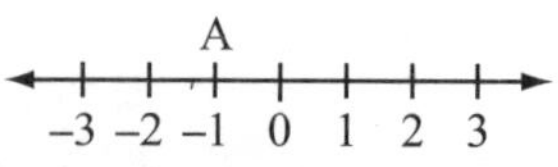

On the number line above, Point A is coordinate –1.

Coordinate	A number that represents a specific point on a number line.

Use this number line for the following examples:

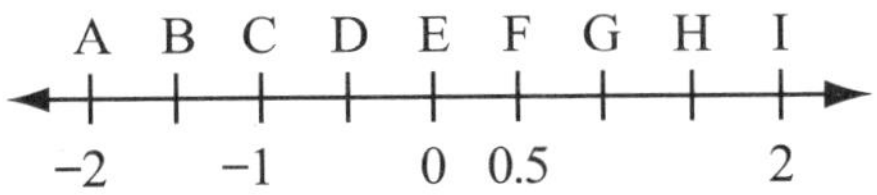

Example 1 The coordinate –1 is associated with which point?

Answer: C

Example 2 State the coordinate of point G.

Step 1: On this number line, each tick mark represents a number that increases by 0.5 more than the previous number. If the coordinate of point F is 0.5, then the coordinate of point G is 1.

Answer: 1

Example 3 State the coordinate of point B.

Answer: Point B is to the left of zero, so it will be negative. Start counting at zero, and decrease the number by 0.5 at each tick mark, and you will find that point B = –1.5.

Answer: –1.5

Problem Set

1. Give the coordinate of point E on the number line below:

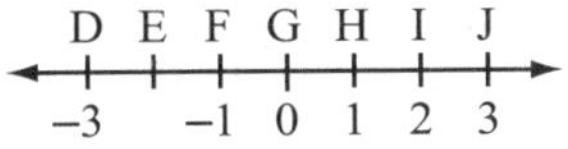

2. Give the coordinate of point M on the number line below:

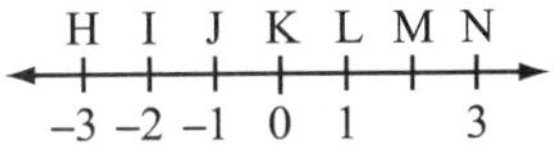

3. Name the letter that corresponds to coordinate –3 on the number line below:

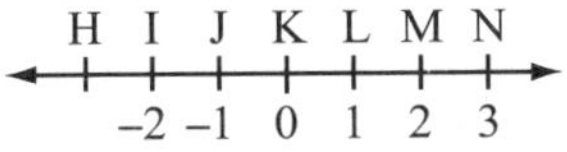

4. Name the letter that corresponds to coordinate 2 on the number line below:

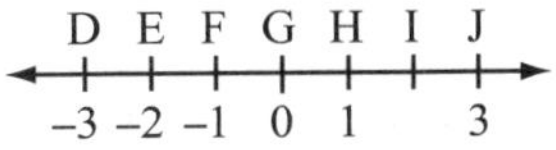

5. Name the letter that corresponds to coordinate 1 on the number line below:

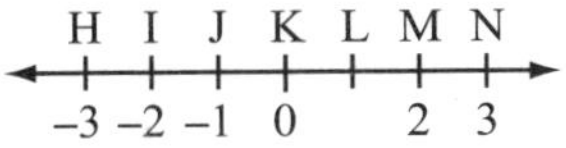

6. Name the letter that corresponds to coordinate 5 on the number line below:

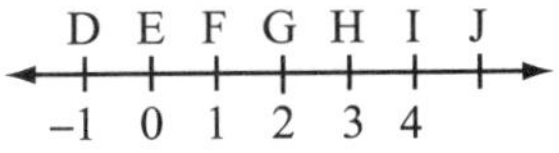

7. Give the coordinate of point F on the number line below:

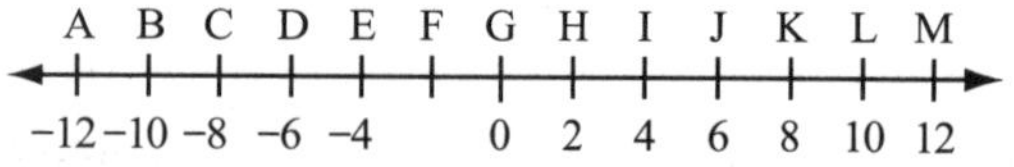

8. Give the coordinate of point J on the number line below:

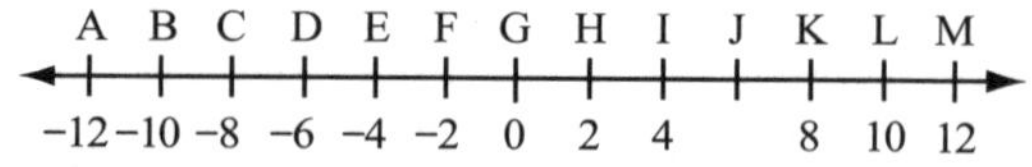

9. Give the coordinate of point A on the number line below:

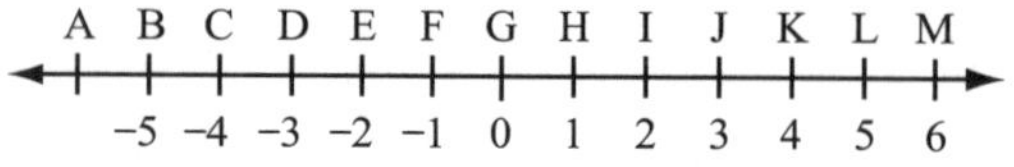

10. Give the coordinate of point C on the number line below:

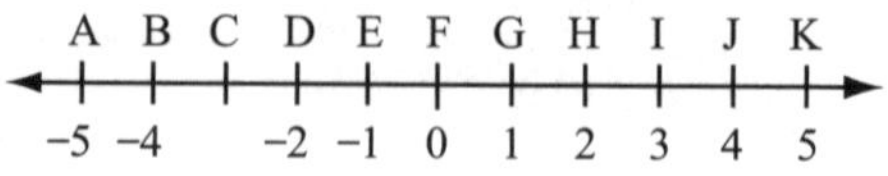

11. Give the coordinate of point J on the number line below:

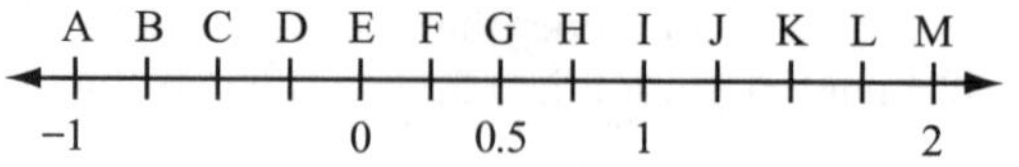

12. Give the coordinate of point M on the number line below:

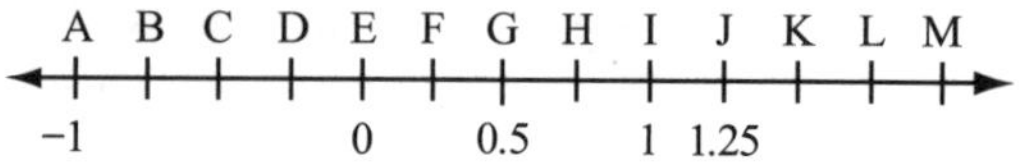

13. Give the coordinate of point G on the number line below:

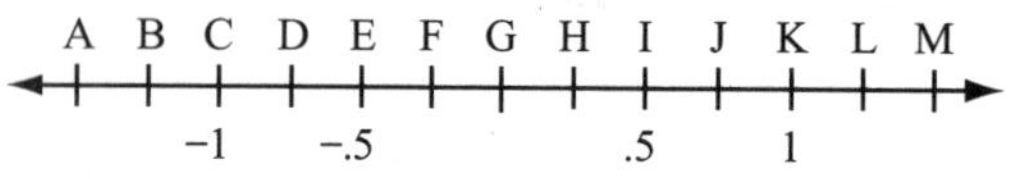

14. Give the coordinate of point B on the number line below:

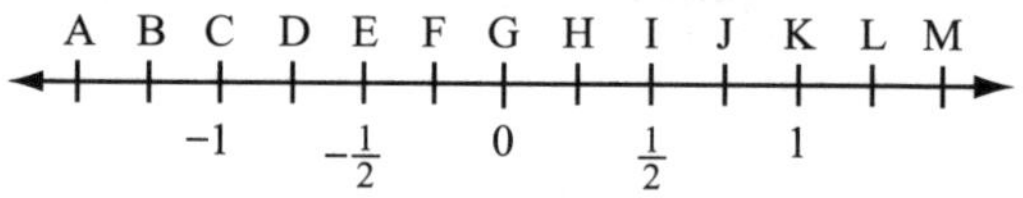

15. Name the letter that corresponds to $-\frac{3}{4}$ on the number line below:

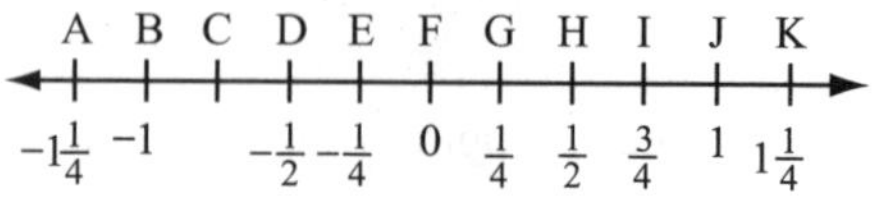

16. Give the coordinate of point D on the number line below:

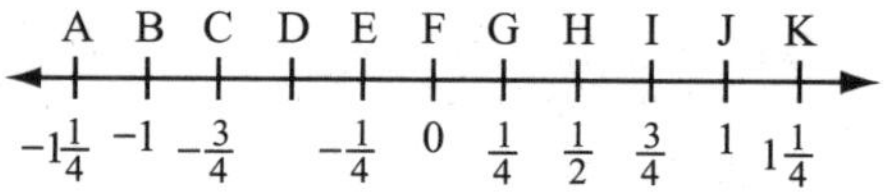

17. Give the coordinate of point H on the number line below:

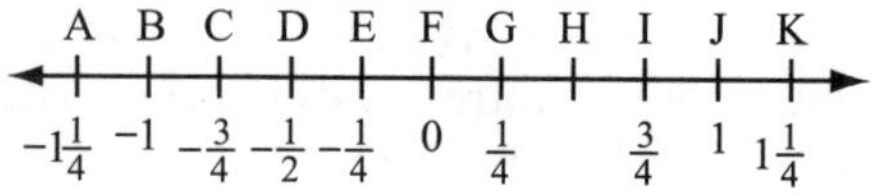

18. A thermometer is also a type of number line. Read the thermometer and tell the temperature that it shows.

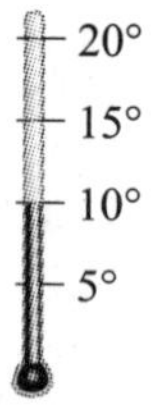

19. If you wished to graph all numbers on a number line greater than zero, which number would be exclude from the list below?

$$0.7,\ \frac{1}{10},\ 0,\ 0.174$$

20. Would −37 be graphed to the left or right of the zero on a number line?

HA1-020: Classifying Numbers into Subsets of Real Numbers

Say you are studying biology, which is the science of all living things. This is a very large group. It might be easier to look at all living things in smaller groups or subsets. For example, all living things could be broken down into plants and animals. Each of these groups could then be broken down even further into different types of plants and animals.

Set	A collection of objects

You can do the same thing with numbers. Mathematicians developed a method that would make numbers easier to understand. To do this, they classified numbers into smaller subsets according to common characteristics. First, numbers can be either real or imaginary. For the purposes of this course, all numbers that are used will be real. Real numbers can then be described as **rational** or **irrational**.

Rational Number	Any number that can be written in the form $\frac{a}{b}$ where a and b are integers and $b \neq 0$

Any number that can be written as a fraction is considered a rational number. However, the denominator can never be zero. Note that even whole numbers can be written as fractions. For example, the whole number 2 can easily be converted to an improper fraction by placing it over the denominator 1, giving you $\frac{2}{1}$. Rational numbers also include repeating and terminating decimals (for example, $0.3333\overline{3}\ldots$ and 0.25).

Irrational Number	Any real number that cannot be written in the form $\frac{a}{b}$ where a and b are integers and $b \neq 0$

Irrational numbers include any infinite, non-terminating, and non-repeating decimals. An example of an irrational number is **pi** (π), a number used to measure the circumference and area of a circle. Although pi is often referred to as 3.14, the actual value of pi is 3.1415920065… You could not possibly memorize it because the digits to the right of the decimal do not repeat or terminate; the number of them is infinite. A good way to determine that a number is irrational is by determining that numbers to the right of the decimal point continue forever without any pattern of repeating digits.

Rational numbers can then be broken down into more specific categories. These categories are natural numbers, whole numbers and integers.

Natural Numbers	All numbers, beginning with one (1) and increased by one. These are also known as **counting numbers.** For example: {1, 2, 3, 4…}

Whole Numbers	Whole numbers are the set of natural numbers and zero. For example: {0, 1, 2, 3, 4, 5…}

Integers	Integers are natural numbers, their opposites, and zero. For example: {…–2, –1, 0, 1, 2, 3…}

Let's look at some numbers and their classifications.

Example 1 The number 15 can be classified into which subset(s) of numbers?

Answer: Real, rational, natural, whole, and integer

Example 2 The number –15 can be classified into which subset(s) of numbers?

Answer: Real, rational, and integer

Example 3 The number $\frac{2}{3}$ can be classified into which subset(s) of numbers?

Answer: When 2 is divided by 3, it equals 0.6666... The 6 repeats; thus, $\frac{2}{3}$ belongs to the sets of real and rational numbers.

Example 4 The number $\sqrt{3}$ can be classified into which subset(s) of numbers?

Answer: When $\sqrt{3}$ is simplified, it equals 1.7320508…, a decimal that does not terminate or repeat; therefore, it belongs to the sets of real and irrational numbers.

Example 5 The number 6.2 can be classified into which subset(s) of numbers?

Answer: This decimal is the equivalent of the mixed number $6\frac{1}{5}$; thus, it belongs to the sets of real and rational numbers.

Example 6 The number $\sqrt{25}$ can be classified into which subset(s) of numbers?

Answer: When $\sqrt{25}$ is simplified, it equals 5. Therefore, it belongs to the sets of real, rational, whole and natural numbers.

Example 7 What is the smallest set of real numbers to which 2 belongs?

Answer: The number 2 is a member of real, rational, integers, whole, and natural numbers. Therefore, the smallest set of real numbers to which 2 belongs is the natural numbers.

Example 8 What is the smallest set of real numbers to which –1 belongs?

Answer: The smallest set is integers.

Problem Set

Solve:

1. To which set(s) of numbers does –18 belong?

2. The number 42 belongs to the sets of rational numbers, natural numbers, and integers. True or False?

3. Which of the following numbers belong to the set of whole numbers?

$-4, 0, 8.7, 4, \frac{3}{5}$

4. What is the smallest set of real numbers to which 3 belongs?

5. What is the smallest set of real numbers to which 0 belongs?

6. What is the smallest set of real numbers to which $\frac{1}{5}$ belongs?

7. What is the smallest set of real numbers to which $\sqrt{5}$ belongs?

8. What is the smallest set of real numbers to which $\sqrt{7}$ belongs?

9. What is the smallest set of real numbers to which –12 belongs?

10. What is the smallest set of numbers to which –4 belongs?

11. Which of the following numbers belong to the set of natural numbers?

$-5, 0, \frac{1}{4}, 2, 6.7, 14$

0, 2, 14	$\frac{1}{4}$, 6.7
2, 14	–5, 0, 2, 14

12. Which of the following is a rational number?

$\sqrt{16}$	$\sqrt{14}$
$\sqrt{26}$	π

13. To which set(s) of numbers does $1.\overline{6}$ belong?

14. Which of the following belong to the set of integers?

$-15, 0, \frac{1}{4}, 5.8, 113$

15. All integers are ____________ numbers.

16. All irrational numbers are ____________ numbers.

17. Name the smallest set of numbers to which $\left\{0.5, \frac{3}{4}, 6\right\}$ belongs.

18. Which number does not belong to the given set of rational numbers?

$\left\{\frac{1}{2}, \frac{1}{4}, \pi, 5, 9\right\}$

19. Which number does not belong to the given set of rational numbers?

$\{-0.75, -0.5, -0.25, 0, \pi\}$

20. Zero is an element of the set of whole numbers and natural numbers. True or False?

HA1-025: Comparing and Ordering Real Numbers

The number line serves more purposes than just showing the position of negative and positive numbers. It can also be used as a tool to help compare and order numbers. For example, if several movie theater owners want to see whose theater is attracting the most moviegoers, they can use the number line to help them. They decide to track and compare ticket sales for a month. The numbers below represent the increase (+) or decrease (–) of ticket sales..

Movie Theater	Change in Ticket Sales
Cinema Seven	+325
ShowTime Theater	–140
Castle Theater	+260
Galaxy Cinema	0
Crescent Cinema	–72

If you put these numbers in order from smallest to largest, the numbers would appear in this order: –140, –72, 0, 260, 325. How can you show this on a number line?

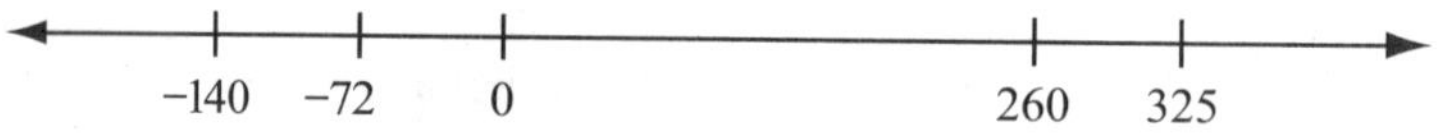

To compare and order numbers on the line, read the number line from left to right. Remember that the values increase as you move to the right and decrease as you move left. The theater owners now know that Cinema Seven is attracting the most moviegoers.

Look at the following set of unordered real numbers {4, –2, 3.5, 0, –5}. If you were to arrange them in order from smallest to largest, you could use the number line below to verify the correct order.

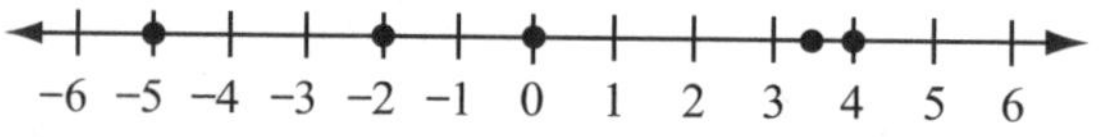

When you ordered the numbers in the set, you arranged them from the smallest number to the largest number. Any of these numbers can be compared using the following symbols:

$<$	less than
$>$	greater than
$=$	equal to

When you use the equal sign, the mathematical sentence is referred to as an **equation**.

Equation	A mathematical sentence stating that two quantities are equal

When you compare numbers using the less than (<) or greater than (>) symbol, the mathematical sentence is referred to as an **inequality**.

Inequality	A mathematical sentence stating that one quantity is greater than or less than another quantity

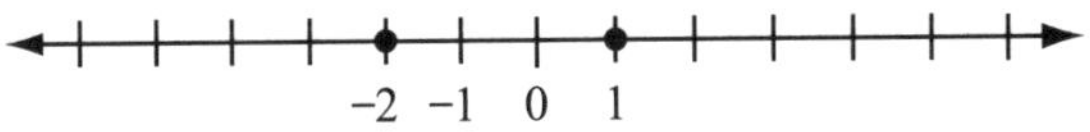

On the number line above, –2 and 1 have been graphed. Of the two numbers, which is greater? Looking at the number line, you would write the mathematical sentence as $1 > -2$, which reads "1 is greater than –2." You could also write $-2 < 1$, which reads "–2 is less than 1."

Note: Check the accuracy of your mathematical sentence by ensuring that the symbol always points to the lesser number.

Try the following examples.

Example 1 Using the number line to compare numbers, fill in the blank with the correct symbol (<, >, or =) to make the mathematical sentence true.

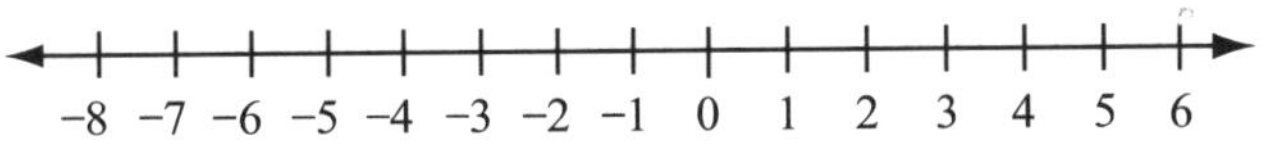

A. $-4 \bigcirc -8$ **Answer:** $-4 > -8$ because –4 is to the right of –8.

B. $-3 \bigcirc 0$ **Answer:** $-3 < 0$ because –3 is to the left of 0.

C. $5 \bigcirc \frac{10}{2}$ **Answer:** $5 = \frac{10}{2}$ Divide 10 by 2; then compare $\frac{10}{2} = 5$; thus, $5 = 5$.

Example 2 List the set of numbers in order from smallest to largest.

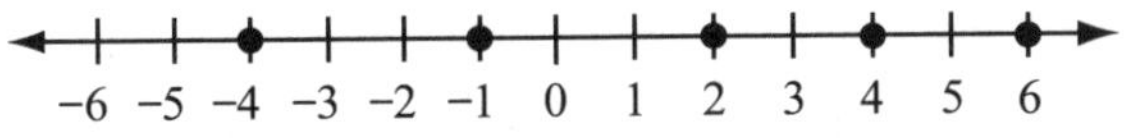

Answer: $\{-4, -1, 2, 4, 6\}$

Example 3 Graph the numbers $\left\{\frac{1}{2}, -3, 1.75, 4, 0, -2.5\right\}$ on the number line and order from smallest to largest.

Answer:

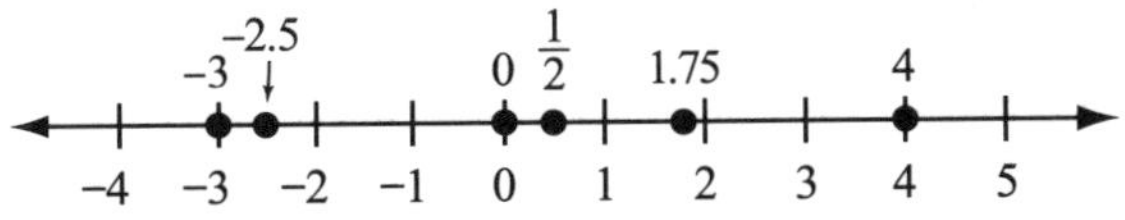

$\left\{-3, -2.5, 0, \frac{1}{2}, 1.75, 4\right\}$

Problem Set

Solve:

1. Fill in the blank with the appropriate symbol to make the statement true:

-1 ______ -3

2. Fill in the blank with the appropriate symbol to make the statement true:

-2 ______ 1

3. Fill in the blank with the appropriate symbol to make the statement true:

3 ______ 3.00

4. Fill in the blank with the appropriate symbol to make the statement true:

$\frac{4}{2}$ ______ 2

5. Fill in the blank with the appropriate symbol to make the statement true:

3.5 ______ -3.5

6. Fill in the blank with the appropriate symbol to make the statement true:

0.50 ______ 1

7. Is the following statement true or false?

$-2 < 0$

8. Is the following statement true or false?

$\frac{15}{5} = 3$

9. Which statement is false?

$-3 < 2$ $\quad 0 < 2$ $\quad -1 > -2$ $\quad 1 > 2$

10. What set of numbers is graphed below?

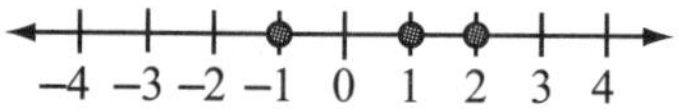

11. Select the number from the set that has the least value:

$-1, 0, 2, 1, 3$

12. Select the number from the set that has the least value.

$1, 4, 3, 0, 2$

13. Select the number from the set that has the greatest value:

$-1, 0, 2, -4, -2$

14. Select the number from the set that has the greatest value:

$5, 0, -1, -4, -5$

15. Fill in the blank with the appropriate symbol to make the statement true:

$3(1)$ ____________ $4(2)$

16. Fill in the blank with the appropriate symbol to make the statement true:

$\frac{1}{4}$ ____________ 0.50

17. Fill in the blank with the appropriate symbol to make the statement true:

$-2\frac{1}{2}$ ____________ -2.5

18. Which number has the least value when placed in order from least to greatest?

$2, -2.5, -1.75, 2\frac{1}{4}, 3$

19. Which number is the fourth term when placed in order from least to greatest?

$3, 2.75, 3.25, 3\frac{1}{2}, 0$

20. The Smith family decides to go to Texas. The family is so large that they must travel in two cars. The two cars leave from the driveway and begin their trip. After one hour, the first car has traveled 40 miles and the second car has traveled 55 miles. Which car has traveled the greater number of miles?

HA1-030: Using Opposites and Absolute Values

A popular game that many young children play is naming opposites. Do you remember playing this game? It went something like this: someone would say "The opposite of up is…", and the other person would say "down"; the opposite of hot is cold; the opposite of short is long, and so on. Understanding opposites is a skill that is also helpful in studying algebra. To fully understand what is meant by opposites, let's use the number line.

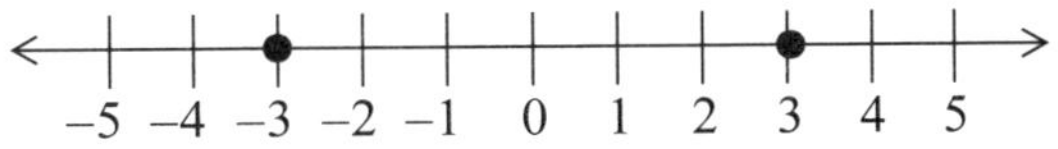

Looking at the number line, notice that –3 and 3 are both graphed. The integers –3 and 3 are opposites. How can you check this? Start at 0, also called the origin, and count the number of units it takes to get to three. You should count three units. Now, count how many units it takes to get to –3 from the origin. Again, you should count three units; therefore, it is true that –3 and 3 are opposites. They are equal distances from zero, but on opposite sides of zero.

Opposite	The number whose distance from zero is the same as the given number, but on the other side of zero

Example 1 What is the opposite of –5?

Answer: From 0 to –5, there are 5 units. Beginning at the origin and counting 5 units to the right, you get +5, or 5. Thus, the opposite of –5 is 5.

Example 2 What is the opposite of 2?

Answer: From 0 to 2, there are two units. Beginning at the origin, count 2 units in the opposite direction. You get –2. The opposite of 2 is –2.

Example 3 What is the opposite of $-4\frac{3}{5}$?

Answer: From 0 to $-4\frac{3}{5}$, there are $4\frac{3}{5}$ units. Beginning at the origin, count $4\frac{3}{5}$ units in the opposite direction. The opposite of $-4\frac{3}{5}$ is $4\frac{3}{5}$.

Refer to the numbers –3 and +3 on the number line above. Although they are on opposite sides of the origin, or zero, they are the same number of units from zero. Therefore, the **absolute value** of both –3 and +3 is simply 3..

Absolute Value	The value of a real number without regard to its sign; the distance on the number line between a real number and zero. The absolute value of a real number n is written as $\lvert n \rvert$.

If you were asked to find the absolute value of -1, you would count the number of units it takes to get from 0 to –1. Even though –1 is a negative number, the absolute value is 1, because its distance is one unit from zero. The symbol for absolute value is two vertical bars, one on each side of the number. For example, the absolute value of negative 1 would be written: $\lvert -1 \rvert$ .The absolute value of a number is always positive because distances are not expressed in negative numbers.

Example 4 SImplify: $-(-4)$

Answer: This can be read as the opposite of -4. Therefore, the answer is 4.

Example 5 Evaluate: $\lvert -12 \rvert$

$\lvert -12 \rvert = 12$

Step 1: The answer is 12, because –12 is 12 units away from the origin.

Answer: 12

Example 6 Evaluate: $\lvert 26 \rvert$

$\lvert 26 \rvert = 26$

Step 1: The answer is 26, because 26 is 26 units away from the origin.

Answer: 26

Example 7 Evaluate: $-\lvert -40 \rvert$

$-\lvert -40 \rvert = -40$

Step 1: Take the absolute value of -40. The absolute value of -40 is 40, because it is 40 units away from the origin. Since there is a negative sign on the outside of the absolute value symbol, it must be placed in front of the number after it is simplified.

Note: Always solve what is inside of the absolute value symbol first.

Answer: –40

Example 8 Simplify: $|-3| + |5|$

$\|3\| = 3$ $\|5\| = 5$	**Step 1:**	First, determine the absolute values of –3 and 5. The absolute value of –3 is 3, because it is 3 units away from the origin. The absolute value of 5 is 5, because it is 5 units away from the origin.
$3 + 5 = 8$	**Step 2:**	Add the absolute values.
	Answer:	8

Example 9 Evaluate the expression $|x - 3|$ if $x = 9$.

$\|x - 3\| = \|9 - 3\|$	**Step 1:**	Substitute 9 for x in the expression.
$\|9 - 3\| = \|6\|$	**Step 2:**	The absolute value of 6 is 6, because it is 6 units away from the origin.
$\|6\| = 6$	**Answer:**	6

Problem Set

Solve:

1. Simplify the following expression: $|-14|$
2. Simplify the following expression: $|33|$
3. Give the opposite of 29.
4. Give the opposite of 31.
5. Give the opposite of 3.65.
6. Give the opposite of 0.125.
7. Give the opposite of –39.
8. Give the opposite of –41.
9. Give the opposite of –84.
10. Give the opposite of –0.56.
11. Simplify: $-(4.5 - 2.7)$
12. Simplify the following expression: $-(-24.9)$
13. Simplify: $-(3.7)$
14. Simplify: $-(82)$
15. Simplify: $-[-(256)]$
16. Simplify: $|2.5|$
17. Simplify: $-|-12.8|$
18. Simplify: $-(12 - 8)$
19. Simplify: $-(4.3 + 2.8)$
20. Simplify: $-[|20| + |-12|]$

HA1-035: Adding Real Numbers Using a Number Line

Now that you know how to graph real numbers on a number line, you can use the number line to add numbers in a number sequence. The key to using the number line is to make sure you start at the origin and move in the correct direction for the specified number of units. Follow the steps below to obtain the correct sum:

Always begin at 0, marking it with a short, vertical line.

Step 1. Using the first number in the expression:

a. if it is *positive*, move *right,* drawing a horizontal arrow for the number of units indicated by the value of the number; or,

b. if it is *negative*, move *left*, drawing a horizontal arrow for the number of units indicated by the value of the number.

Step 2. From that number, move left or right for the number of units indicated by the second number in the expression. Repeat this process for any additional numbers in the expression. Once all the numbers in the expression are graphed, the last number where the arrow points is the correct sum.

Example 1 Add: $(-2) + 3$

Step 1: Because the first number of the expression is –2, move 2 units to the left of the origin.

Note: For each number in the expression, draw the horizontal line above the last graphed line.

Step 2: From there, move three units to the right, because the second number is 3.

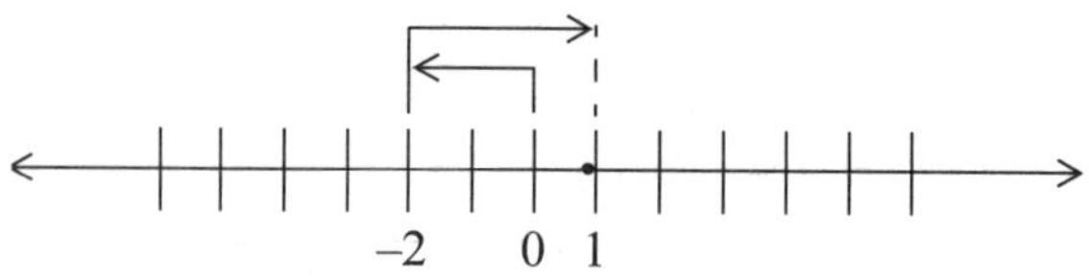

Answer: $(-2) + 3 = 1$

Example 2 Add: $5 + (-8) + 6$

Step 1: Begin at zero and move 5 units right, because 5 is positive.

Step 2: From 5, move 8 units left, because 8 is negative.

Step 3: Finally, move 6 units right.

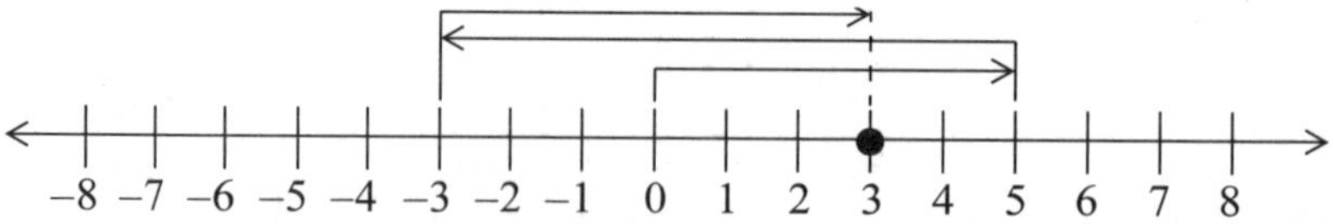

Answer: $5 + (-8) + 6 = 3$

Example 3 Write the correct sentence that the diagram illustrates.

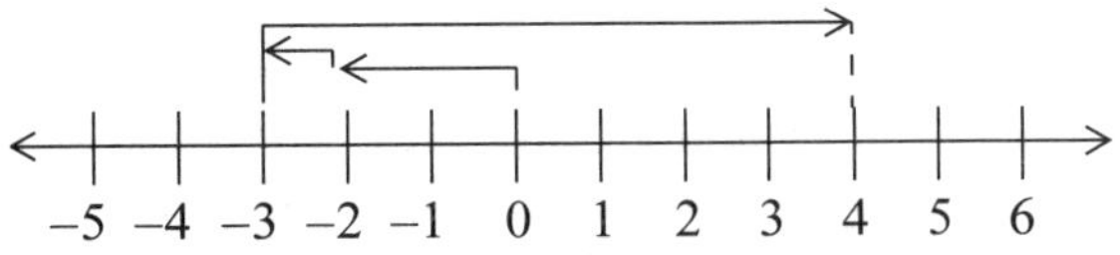

Step 1: The lowest arrow begins at 0, and moves 2 units left (–2).

Step 2: The next arrow moves left one unit (–1).

Step 3: The highest arrow moves 7 units right (7).

Answer: The sentence is: $(-2) + (-1) + 7 = 4$.

Problem Set

1. Choose the addition statement that matches the illustration below.

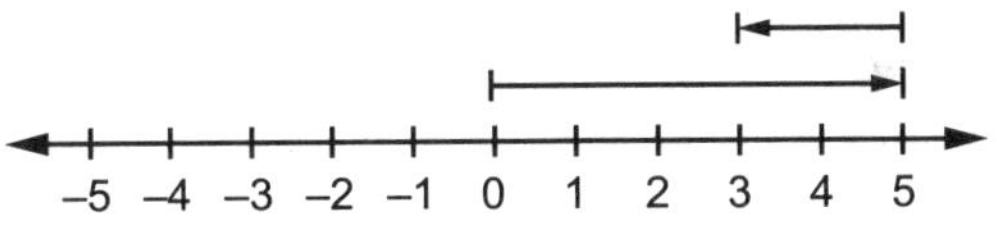

$2 + (-5) = 3$ $\quad$ $5 + (-2) = 3$

$-5 + (-2) = 3$ $\quad$ $-5 + (2) = -3$

2. Choose the addition statement that matches the illustration below.

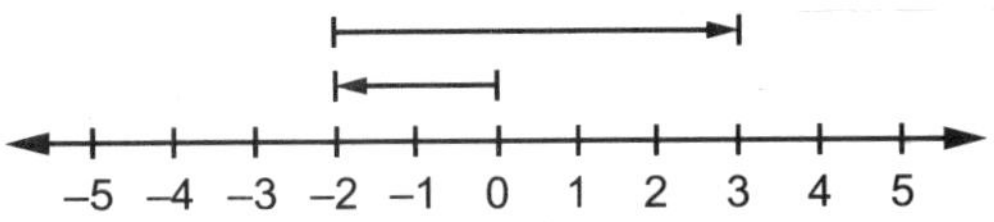

$-5 + (-2) = -3$ $\quad$ $2 + (-5) = 3$

$-5 + (2) = -3$ $\quad$ $-2 + 5 = 3$

3. Choose the addition statement that matches the illustration below.

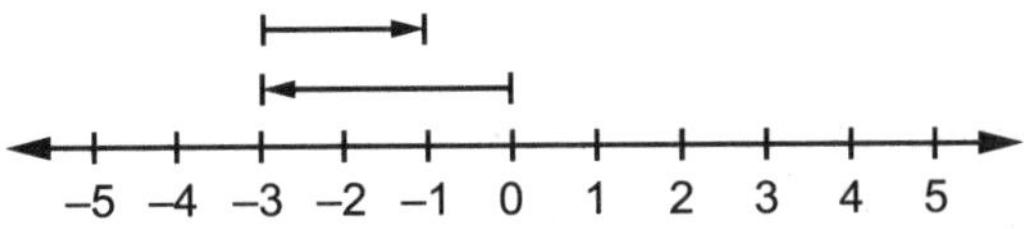

$-1 + 2 = -3$ $\quad$ $-3 + 2 = 1$

$-3 + 2 = -1$ $\quad$ $-3 + 2 = 5$

4. Choose the addition statement that matches the illustration below.

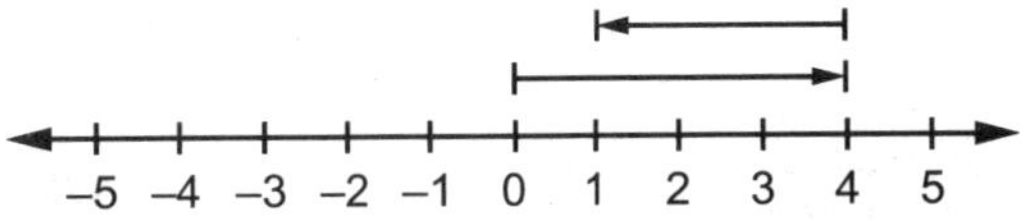

$-3 + 1 = 4$ $\quad$ $4 + (-3) = 1$

$4 + (-3) = -1$ $\quad$ $-3 + 4 = 7$

Find the sum using a number line:

5. $-2 + (-1)$

6. $-2 + 1$

7. $-2 + 4$

8. $-3 + (-4)$

9. $-3 + (-3)$

10. $-3 + 1$

11. $-2 + (-3) + 6$

12. $-2 + (-3) + 5$

13. $-2 + (-3) + (-3)$

14. $-6 + 7 + 5$

15. $-8 + 3 + (-3)$

16. $-7 + 3 + (-3)$

17. $-3 + (-6) + 5$

18. $-1.50 + (-0.75)$

19. $-1.50 + 0.75$

20. $-0.75 + 0.75$

HA1-040: The Addition Rule for Real Numbers

Just as you can use the interstate as a traveling shortcut, you can also use a shortcut for adding real numbers: the number line. A number line helps when adding smaller real numbers such as (–5) + 3 + (–8), but can be cumbersome when adding larger real numbers, such as (–35) + 23 + (–80). In this case, we should use the following rules:

Adding Real Numbers with Like Signs	Add the absolute values of the numbers. The sum takes the common sign.

Adding Real Numbers with Unlike (Different) Signs	Subtract the absolute values of the numbers. The sum takes the sign of the number with the largest absolute value.

Example 1 Add: $4 + 9$

$$4 + 9 = |4| + |9|$$
$$= 4 + 9$$
$$= 13$$

Step 1: Add the absolute values of 4 and 9.

$$= 13$$

Step 2: Place the common sign in front of the sum. Because the sum is already positive, the 13 can remain.

Answer: 13 or +13

Example 2 Add: $-18 + (-5)$

$$-18 + (-5) = |-18| + |-5|$$
$$= 18 + 5$$
$$= 23$$

Step 1: Add the absolute values of –18 and –5.

$$-18 + -5 = -23$$

Step 2: Place the common sign in front of the sum. Both numbers are negative, so the sum is also negative.

Answer: –23

Now try using the second rule with examples where the signs are different.

Example 3 Add: $-3.2 + 2.0$

$-3.2 + 2.0 = |-3.2| - |2.0|$
$= 3.2 - 2.0$
$= 1.2$

Step 1: Subtract the absolute value of 2.0 from the absolute value of –3.2.

$= -1.2$

Step 2: Place the sign of the number with the greater absolute value in front of the sum. The absolute value of –3.2 is greater; therefore, the sum is given the negative sign.

Answer: –1.2

When adding more than two numbers, you should regroup the numbers by their signs. Using the commutative and associative properties of addition, group the numbers with the same sign. Add numbers with the same sign first.

Example 4 Add: $15 + (-6) + 4$

$15 + (-6) + 4 = |15| + |4| + (-6)$
$= 15 + 4 + (-6)$
$= 19 + (-6)$

Step 1: First add the absolute value of any numbers with the same sign.

$= |19| - |-6|$
$= 19 - 6$
$= 13$

Step 2: The partial sum, 19, and –6 have different signs, so subtract their absolute values.

$= 13$

Step 3: Place the sign of the number with the greater absolute value in front of the sum. Thus, the number 13 remains positive.

Answer: 13 or +13

Example 5 Evaluate $x + y$ if $x = -3$ and $y = 1$.

$x + y = -3 + 1$

Step 1: Substitute –3 for x and 1 for y.

$= |-3| - |1|$
$= 3 - 1$
$= 2$

Step 2: Subtract the absolute value of 1 from the absolute value of –3, since the two numbers have different signs.

$= -2$

Step 3: Place the sign of the number with the greater absolute value in front of the sum. The absolute value of –3 is greater, therefore, the sum is given the negative sign.

Answer: –2

Problem Set

1. Find the sum:
$7 + (-15)$

2. Find the sum:
$-9 + (-7)$

3. Find the sum:
$-5 + 18$

4. Find the sum:
$-6 + 14$

5. Use the addition rules to determine whether the sum would be positive or negative.
$-5\frac{7}{8} + 2\frac{8}{9}$

6. Use the addition rules to determine whether the sum would be positive or negative.
$-\frac{9}{10} + \frac{3}{10}$

7. Use the addition rules to determine whether the sum would be positive or negative.
$-25.67 + (-43.42)$

8. Use the addition rules to determine whether the sum would be positive or negative.
$-23.65 + 85.925$

9. Use the addition rules to determine whether the sum would be positive or negative.
$25.67 + (-43.42)$

10. Use the addition rules to determine whether the sum would be positive or negative.
$23.65 + (-85.925)$

11. Find the sum:
$-6 + (-12) + (-10)$

12. Find the sum:
$-9 + 12 + (-10)$

13. Find the sum:
$15 + (-23)$

14. Find the sum:
$-15 + 23$

15. Find the sum:
$48 + (-27)$

16. Find the sum:
$-83 + (-27)$

17. Find the sum:
$15 + (-21) + (-33)$

18. Find the sum:
$|-6| + |15|$

19. Find the sum:
$8 + (-4) + 3$

20. Find the sum:
$|-12| + 7$

HA1-045: Subtracting Real Numbers

You have just been hired for your first job. Now you will have money to buy your own computer games, clothes, and other items. As you decide which items to buy, you must be sure you have enough money to pay for them. Your first paycheck finally arrives. The new stereo system you want to buy costs $250. Your first paycheck is $176. How much additional money do you need to pay for the stereo? In order to determine the amount, you must be able to subtract real numbers.

Subtracting Real Numbers	To subtract two numbers, add the first number to the opposite of the second number, or $a - b = a + (-b)$.

Let's use the rule in the example above. Keep in mind that the word "difference" is often used to mean "subtraction." You will encounter its use in some of the following problems.

Example 1 The new stereo system you want to buy costs $250. Your first paycheck is $176. How much additional money do you need to pay for the stereo?

$250 - 176$	**Step 1:** Set up the subtraction problem.
$= 250 + (-176)$	**Step 2:** Now add the opposite of 176.
$= 74$	**Step 3:** Use the rule for adding real numbers with different signs.
$74	**Answer:** In order to purchase the stereo system, you will need an additional $74.

As you continue to work, you decide to open a checking account. The balance in your account increases when you deposit your check, and decreases when you make a purchase. Using the subtraction rule, you can keep an accurate account of your money. Let's look at the example below.

Example 2 You have $13.75 in your checking account. A CD is on sale for $17.99. How much money is left in your account after you purchase the CD?

$13.75 – $17.99 = 13.75 + (–17.99)	**Step 1:** Change 17.99 to its opposite, –17.99, and add it to 13.75.
= –4.24	**Step 2:** Add 13.75 to –17.99 using the rule for adding real numbers with different signs.
–$4.24	**Answer:** The account has a balance of –$4.24.

Actually, you did not have enough money in your checking account to buy the CD. Because you have already purchased it, your account has a negative balance of $4.24, which means your account is overdrawn. You must get to the bank quickly and make a deposit.

Example 3 You are going on a scuba diving trip. For your first dive, you reach a depth of 28 feet below sea level. The second dive is to a depth of 57 feet below sea level. What is the difference between the two water depths? Remember, when you refer to "below sea level," a negative sign must precede the measurement.

$-28-(-57)$	**Step 1:** Change –57 to the opposite +57, and add it to –28.
$=-28+(+57)$	**Step 2:** Use the rule for adding real numbers with different signs.
$=+29$ feet	**Answer:** The difference between the two water depths is 29 feet.

Example 4 The temperature in Fairbanks, Alaska, is –18° F on February 3. The temperature in Chicago, Illinois is +25° F on the same day. What is the difference in temperature between the two cities?

$-18°\text{ F}-25°\text{ F}$	**Step 1:** Change 25 to the opposite –25.
$=-18+(-25)$ $=-43$	**Step 2:** Add –18 to –25 because the signs are the same.
43°	**Answer:** The difference in temperature between the two cities is 43° F.

Example 5 You own stock in a company, and yesterday, each share increased by $\frac{3}{8}$. Today, each share decreased by $\frac{5}{8}$. Over this two-day period, what was the net increase or decrease?

$\frac{3}{8}-\frac{5}{8}$	**Step 1:** Change $\frac{5}{8}$ to its opposite, $-\frac{5}{8}$, and add it to $\frac{3}{8}$.
$\frac{3}{8}+\left(-\frac{5}{8}\right)=-\frac{2}{8}$	**Step 2:** Use the rule for adding real numbers with different signs.
$-\frac{2}{8}=-\frac{1}{4}$	**Answer:** The net decrease of each share was $\frac{1}{4}$.

Problem Set

Simplify:

1. $3-8$
2. $-4-7$
3. $12-16$
4. $10-13$
5. $9-(-6)$
6. $-24-(-15)$
7. $-43-54$
8. $-37-56$
9. $50-105$
10. $60-140$
11. $-1.8-4.2$
12. $-2.4-6.8$
13. $-11.3-4.6$
14. $12.3-15.1$
15. $58-62.5$
16. $9-11.4$
17. $-10.5-(-11)$
18. $10-(-8-3)$
19. $\frac{2}{3}-\frac{5}{6}$
20. Evaluate: $x-(y-z)$ if $x=13$, $y=-12$, and $z=7$

HA1-050: Multiplying Real Numbers

What is the sum of 2 + 2 + 2 + 2? The answer, of course, is 8. You can use a shortcut any time the number you are adding is repeated over and over again. This shortcut is called multiplication. Another way to write 2 + 2 + 2 + 2 is 4 · 2. For example, you already know that 3 · 5 means adding five three times (5 + 5 + 5) or adding three five times (3 + 3 + 3 + 3 + 3). Either way, both expressions equal fifteen. When you multiply two or more numbers, you can refer to those numbers as **factors**.

Factor	A quantity that can be evenly divided into a given number, with no remainder

Product	The answer to a multiplication problem

Notice in the expression 3 · 5, both factors are positive, so the product is positive. But in algebra, an expression may contain both positive and negative numbers. In this case, how would you determine if the answer is negative or positive? When multiplying signed numbers, you multiply as you normally would. Then depending on the combination of positive and negative numbers, you determine the sign of the product. Let's consider the possible combinations using the following numbers: –3, + 3, –5, + 5.

The expression:	Is interpreted as:
(+3) · (+5)	(+5) + (+5) + (+5) = 15 or (+3) · (+5) = +15
(+3) · (–5)	(–5) + (–5) + (–5) = –15 or (+3) · (–5) = –15
(–3) · (+5)	(+5) is subtracted 3 times, so that – (+5) – (+5) – (+5) = –15 or (–3) · (+5) = – 15
(–3) · (–5)	(–5) is subtracted 3 times, so that – (–5) – (–5) – (–5) = 15 or (–3) · (–5) = + 15

From these examples, you can conclude:

$+a \cdot +b = +ab$	positive · positive = positive
$+a \cdot -b = -ab$	positive · negative = negative
$-a \cdot +b = -ab$	negative · positive = negative
$-a \cdot -b = +ab$	negative · negative = positive

Therefore, you can state the following rule concerning the **multiplication of real numbers**:

Multiplying Real Numbers	The product of two factors having like signs is positive and the product of two factors having unlike signs is negative.

Using the rule for multiplying real numbers, try the examples below:

Example 1 Find the product of $(+8) \cdot (+2)$.

$8 \cdot 2 = 16$	**Step 1:**	Multiply as usual.
$= (+16)$	**Step 2:**	Make the product positive because the numbers have like signs.
	Answer:	+16 or 16

Example 2 What is the product of $(-2.3) \cdot (+1.5)$?

$2.3 \cdot 1.5 = 3.45$	**Step 1:**	Multiply as usual.
$= (-3.45)$	**Step 2:**	Make the product negative because the numbers have unlike signs.
	Answer:	–3.45

Example 3 What is the product of $(-11) \cdot (-10)$?

$11 \cdot 10 = 110$	**Step 1:**	Multiply as usual.
$= (+110)$	**Step 2:**	Make the product positive because the numbers have like signs.
	Answer:	110

Now consider multiplying several factors, such as $(+2)(-3)(-2)(+1)$. You know from the rule of signs that the product of two negative factors is positive. So what sign would you use if you multiplied three, four, or five negative factors?

If $(-a) \cdot (-b) = (+ab)$	2 negatives
then $(-a) \cdot (-b) \cdot (-c) = (+ab) \cdot (-c) = (-abc)$	3 negatives
and $(-a) \cdot (-b) \cdot (-c) \cdot (-d) = (-abc) \cdot (-d) = (+abcd)$	4 negatives
and $(-a) \cdot (-b) \cdot (-c) \cdot (-d) \cdot (-e) = (+abcd) \cdot (-e) = (-abcde)$	5 negatives

Consider these examples:

Example 4 What is the product of $(+2)(-3)(-2)(+1)$?

$2 \cdot 3 \cdot 2 \cdot 1 = 12$	**Step 1:**	Multiply the numbers as usual.
$= (+12)$	**Step 2:**	Count the negative factors. There are two negative factors, so the product is positive.
	Answer:	12

Example 5 What is the product of (+5)(–2)(+6)?

(5)(2)(6)	**Step 1:**	Multiply the numbers as usual.
= (–60)	**Step 2:**	Count the negative factors. There is one negative factor. One is odd, so the product is negative.
	Answer:	–60

Example 6 What is the product of (–2)(–2)(–2)(–2)?

(2)(2)(2)(2)	**Step 1:**	Multiply the numbers as usual.
= (+16)	**Step 2:**	Count the negative factors. Four is even, so the product is positive.
	Answer:	16

Example 7 What is the product of (–1)(+3)(–5)(–2)(+4)?

(1)(3)(5)(2)(4)	**Step 1:**	Multiply the numbers as usual.
= (–120)	**Step 2:**	Count the negative factors. Three is odd, so the product is negative.
	Answer:	–120

What does $5 \cdot 0$ mean? $0 + 0 + 0 + 0 + 0$. The sum is 0. What about $10 \cdot 0$? $0 + 0 + 0 + 0 + 0 + 0 + 0 + 0 + 0 + 0$. The sum, of course, is 0. If you multiply zero by other numbers, you will see that the product will always equal zero.

Multiplication by Zero Property	The product of a number n and zero is zero. $n \cdot 0 = 0$

What does $6 \cdot 1$ mean? $1 + 1 + 1 + 1 + 1 + 1$. The sum is 6. What does $3 \cdot 1$ mean? $1 + 1 + 1$. The sum is 3. Try this with other numbers. You will notice that the product of any number multiplied by 1 equals that number (itself). This is called the Identity Property of Multiplication.

Identity Property of Multiplication	The product of a number n and 1 is n. $n \cdot 1 = n$

Example 8 What is the product of $(-7.3) \cdot 0$?

$(-7.3) \cdot 0 = 0$	**Step 1:** Because one factor is zero, the product is 0.
	Answer: 0

Example 9 What is the product of $(2.7) \cdot 1$?

$(2.7) \cdot 1 = 2.7$	**Step 1:** Because 1 is a factor, the product is the other factor, 2.7.
	Answer: 2.7

Example 10 Evaluate $a(-b)$ when $a = -5$ and $b = 4$.

$a(-b) = (-5)(-4)$	**Step 1:** Substitute –5 for a and 4 for b.
$= 20$	**Step 2:** Multiply. The product is positive because the numbers have like signs.
	Answer: 20

Problem Set

Solve:

1. Multiply: $(-8)(-15)$
2. Multiply: $(10)(3)(-6)$
3. Would the following product be positive, negative, or zero? $-5.4(3.6)$
4. Would the following product be positive, negative, or zero? $4.5(-2.3)$
5. Would the following product be positive, negative, or zero? $-6.7(3.6)$
6. Would the following product be positive, negative, or zero? $2.9(-2.3)$
7. Would the following product be positive, negative, or zero? $(-5)(0)$
8. Would the following product be positive, negative, or zero? $(0)(-4)$
9. Would the following product be positive, negative, or zero? $(-8)(7)(5)$
10. Would the following product be positive, negative, or zero? $(-8)(7)(0)$
11. Multiply: $-\frac{3}{4}(-32)\left(3\frac{3}{7}\right)(0)\left(\frac{1}{6}\right)$
12. Multiply: $\frac{5}{9}(-36)\left(-2\frac{3}{4}\right)(-12)\left(-\frac{7}{11}\right)(0)$
13. Multiply: $(-9)(0)$
14. Multiply: $(-4)(-12)$
15. Multiply: $(-11)(0)$
16. Multiply: $(-11)(-1)$
17. Multiply: $(20)(-8)$
18. Multiply: $-2.36(0.7)$
19. Evaluate: $a(-b)(-c)$ if $a = -4$, $b = -\frac{5}{6}$, and $c = 8$.
20. Multiply: $\left(-\frac{5}{7}\right)\left(\frac{2}{3}\right)$

HA1-055: Dividing Real Numbers

You have learned that the operation of division can be expressed in three ways. For example, 15 divided by 5 can be represented by:

$$5\overline{)15}^{\,3} \quad or \quad \frac{15}{5} \quad or \quad 15 \div 5 = 3$$

A dividend (15), a divisor (5), and a quotient (3) are in each representation.

Division and multiplication are called inverse operations. For example, $6 \cdot 7 = 42$ illustrates multiplication, but $42 \div 7 = 6$ illustrates division, its inverse. Since division and multiplication are inverse operations, you can use what you have already learned about multiplying signed numbers to determine what happens when you divide signed numbers.

Dividing Real Numbers	When the dividend and divisor have like signs, the quotient is positive; when they have unlike signs, the quotient is negative.

If	Then
$(+a) \cdot (+b) = (+ab)$	$(+ab) \div (+a) = (+b)$
$(+a) \cdot (-b) = (-ab)$	$(-ab) \div (+a) = (-b)$
$(-a) \cdot (+b) = (-ab)$	$(-ab) \div (-a) = (+b)$
$(-a) \cdot (-b) = (+ab)$	$(+ab) \div (-a) = (-b)$

Try the following examples using the rule of signed numbers. To check your answers, multiply the quotient and the divisor. If the product equals the dividend, then it is correct.

Example 1 Find the quotient of $+11\overline{)+99}$.

$$+11\overline{)+99}^{\,+9}$$

Step 1: Divide the numbers. Because the dividend and the divisor have like signs, the quotient is positive.

Answer: The quotient is +9 or 9. Check: $(+9) \cdot (+11) = 99$.

Example 2 Find the quotient of $\frac{+63}{-9}$.

$\frac{+63}{-9} = -7$	**Step 1:**	Divide the numbers. Because the dividend and the divisor have unlike signs, the quotient is negative.
$(-7) \cdot (-9) = (+63)$	**Step 2:**	The quotient is –7. Check your answer.

Answer: –7

Example 3 Find the quotient of $\frac{4.9}{0.7}$.

$\frac{4.9}{0.7} = 7$	**Step 1:**	Divide the numbers. Because the dividend and the divisor have like signs, the quotient is positive.
$(7) \cdot (0.7) = (4.9)$	**Step 2:**	The quotient is 7. Check your answer.

Answer: 7

Reciprocal	
	If a is a nonzero real number, then there is one number $\frac{1}{a}$ that is the reciprocal of a. Zero has no reciprocal. Two numbers are reciprocals of each other if their product is one. The reciprocal of 2 is $\frac{1}{2}$ because $2 \cdot \frac{1}{2} = \frac{2}{1} \cdot \frac{1}{2} = 1$. Reciprocals are used to define division: $\frac{a}{b} = \frac{a}{1} \cdot \frac{1}{b}$. (Dividing by a number is the same thing as multiplying by the reciprocal.)

Example 4 Find the quotient of $\frac{2}{3} \div \left(-\frac{1}{4}\right)$.

$$\frac{2}{3} \div \left(-\frac{1}{4}\right) = \frac{2}{3} \cdot \left(-\frac{4}{1}\right)$$
$$= -\frac{8}{3}$$
$$= -2\frac{2}{3}$$

Step 1: Divide the numbers. Because the dividend and the divisor have unlike signs, the quotient is negative.

$\left(-2\frac{2}{3}\right)\left(-\frac{1}{4}\right)$

$= \left(-\frac{8}{3}\right)\left(-\frac{1}{4}\right)$

$= +\frac{8}{12}$

$= +\frac{2}{3}$

Step 2: The quotient is $-\frac{8}{3}$ or $-2\frac{2}{3}$.

Answer: $-\frac{8}{3}$ or $-2\frac{2}{3}$

Example 5 What is the quotient of $\frac{-5.6}{+0.8}$?

$\frac{-5.6}{+0.8} = -7$

Step 1: Divide the numbers. Because the dividend and the divisor have unlike signs, the quotient is negative.

$(-7) \cdot (+0.8) = -5.6$

Step 2: The quotient is –7. Check your answer.

Answer: –7

Division Involving Zero	
	Zero divided by any nonzero number is zero. For example, $0 \div a = 0$.
	Division by zero is undefined. For example, $\frac{a}{0}$ is undefined.

Example 6 What is the quotient of $\frac{8}{0}$?

Step 1: Remember that multiplying the quotient by the divisor should equal the dividend. There is no number that can be multiplied by zero to give an answer of 8.

Answer: No solution; division by 0 is undefined.

Example 7 Evaluate $a \div b$ when $a = -\frac{2}{5}$ and $b = \frac{8}{15}$.

$a \div b = -\frac{2}{5} \div \frac{8}{15}$

Step 1: Substitute $-\frac{2}{5}$ for a and $\frac{8}{15}$ for b.

$= -\frac{2}{5} \cdot \frac{15}{8}$

$= -\frac{\cancel{2}^{1}}{{}_{1}\cancel{5}} \cdot \frac{\cancel{15}^{3}}{\cancel{8}_{4}}$

$= -\frac{3}{4}$

Step 2: Take the reciprocal of $\frac{8}{15}$ and multiply, because division by a number is the same as multiplication by the reciprocal.

Answer: $-\frac{3}{4}$

Problem Set

1. Divide, if possible: $0 \div (-0.5)$

2. Divide, if possible: $-56 \div 14$

3. Determine if the quotient is positive, negative, zero, or undefined: $48 \div (-6)$

4. Determine if the quotient is positive, negative, zero, or undefined: $0 \div 8$

5. Determine if the quotient is positive, negative, zero, or undefined: $-24 \div 3$.

6. Determine if the quotient is positive, negative, zero, or undefined: $0 \div (-24)$

7. Determine if the quotient is positive, negative, zero, or undefined: $18 \div (-6)$.

8. Determine if the quotient is positive, negative, zero, or undefined: $0 \div 18$

9. What is the reciprocal of $\frac{1}{3}$?

10. What is the reciprocal of -3?

11. Divide, if possible: $-2.345 \div 0$

12. Divide, if possible: $-4\frac{1}{2} \div 1\frac{1}{8}$

13. Divide: $-42 \div (-7)$

14. Divide, if possible: $-24 \div 4$

15. Divide, if possible: $-28 \div (-2)$

16. Divide, if possible: $-28 \div 2$

17. Divide, if possible: $\frac{40}{2}$

18. Evaluate $m \div n$ if $m = -\frac{2}{7}$ and $n = 6$.

19. Divide, if possible: $\frac{2}{3} \div \frac{3}{4}$

20. Evaluate $y \div x$ if $y = -124$ and $x = 2$.

HA1-060: Evaluating Numerical Expressions Using the Order of Operations

We have already seen how important order of operations is to mathematics. Whether we begin evaluating a problem with either addition or multiplication, order of operations can make all the difference. Using order of operations allows us to organize our approach to mathematical expressions and ensures a correct answer. Remember to perform the operations in grouping symbols, starting with the innermost set. When there are no grouping symbols, or when working inside a grouping symbol, first simplify exponents from left to right. Then do all multiplications and divisions in order from left to right. Next, do all additions and subtractions, again, in order from left to right.

Example 1 Simplify: $8 - 9 \cdot 7$

Step 1: There are no grouping symbols; move to the next step.

Step 2: There are no exponents; move to the next step.

$8 - 9 \cdot 7$
$= 8 - 63$

Step 3: Perform multiplication and division, so multiply 9 and 7.

$= 8 - 63$
$= -55$

Step 4: Add or subtract, so subtract 63 from 8 to get the answer.

Answer: -55

Example 2 Simplify: $7(2 + 5) + 22$

$7(2 + 5) + 22$
$= 7(7) + 22$

Step 1: Perform the operation in the grouping symbol.

Step 2: There are no exponents; move to the next step.

$= 7(7) + 22$
$= 49 + 22$

Step 3: Multiply 7 and 7.

$= 49 + 22$
$= 71$

Step 4: Add 49 + 22 to get the answer.

Answer: 71

Example 3 Simplify: $\dfrac{(5 \cdot 6) + (7 - 4)}{4 \cdot 3 - 9}$

$= \dfrac{30 + 3}{4 \cdot 3 - 9}$

Step 1: Perform all operations in the numerator according to the order of operations. In this case, parentheses will come first.

$= \dfrac{33}{4 \cdot 3 - 9}$

Step 2: Next, add the numbers in the numerator

$= \dfrac{33}{12-9}$

Step 3: Perform all operations in the denominator according to the order of operations. In this case, multiply then subtract.

$= \dfrac{33}{3}$

$= 11$

Step 4: Divide the numerator by the denominator.

Answer: 11

Example 4 Evaluate the expression $6(5a+3b)$ if $a = 3$, $b = 2$

$6(5a+3b)$ if $a=3$, $b=2$

$= 6[5(3)+3(2)]$

Step 1: Replace a with 3 and b with 2. Notice parentheses are placed around the numbers that are substituted so that $5a$ does not look like 53, but rather 5(3), which reminds you to multiply.

$=6[5(3)+3(2)]$

$=6(15+6)$

$=6(21)$

$=126$

Step 2: Follow the order of operations.

Answer: 126

Example 5 Evaluate the expression $\dfrac{2w-k}{b}+kb$ if $k = 4$, $w = 5$, $b = 2$

$\dfrac{2w-k}{b}+kb$ if $k = 4$, $w = 5$, $b = 2$

$= \dfrac{2(5)-4}{2}+4(2)$

Step 1: Replace k with 4, w with 5, b with 2.

$= \dfrac{2(5)-4}{2}+4(2)$

$= \dfrac{10-4}{2}+8$

$= \dfrac{6}{2}+8$

$= 3+8$

$= 11$

Step 2: Follow the order of operations.

Answer: 11

Problem Set

Simplify the expressions:

1. $2 + 4 \cdot 6$
2. $28 - 7 \cdot 3$
3. $(10 - 8) \cdot (-2)$
4. $8(-3) - 4(-2)$
5. $5(-7) - 3(-4)$
6. $-9(-1) + 4(-20)$
7. $-5(-7) - 3(-4)$
8. $10 \div 5(-3)$
9. $10(3) \div (-2)$
10. $\frac{-24}{-8} + 6(2)$
11. $8 - (-3)(4 + 7)$
12. $8 - (-3)(4 - 7)$
13. $6 - (-3)(4 - 7)$
14. $24 \div 3 \cdot (-2 + 1)$

Evaluate the expressions:

15. Evaluate the expression if $x = -2$, $y = 3$ and $z = 4$:

$$xz - xy$$

16. Evaluate the expression if $a = -1$, $b = -2$ and $c = 0$:

$$30 - 3abc$$

17. Evaluate the expression if $g = 4$ and $h = -5$:

$$3g + 2h$$

18. Evaluate the expression if $m = -2$ and $n = 9$:

$$\frac{m + 2n}{6 - m}$$

19. Evaluate the expression if $m = 2$ and $n = 9$:

$$\frac{m + 2n}{6 - m}$$

20. Evaluate the expression if $m = 4$ and $n = -3$:

$$\frac{m + 2n}{6 - m}$$

HA1-062: Adding, Subtracting, Multiplying, and Dividing Real Numbers

Addition:

- To find the sum of two real numbers with like signs, add their absolute values. The sum has the same sign as the given numbers.
- To find the sum of two real numbers with unlike signs, subtract their absolute values. The sum has the same sign as the number with the larger absolute value.

Subtraction:

- To find the difference of two real numbers, add the opposite of the number being subtracted: $a - b = a + (-b)$
- Use the addition rule for adding two real numbers with unlike signs.

Multiplication and Division:

To multiply or divide two real numbers, multiply or divide the absolute value of the two numbers and give the product or quotient the appropriate sign:

- The product of two numbers with like signs is always positive.
- The product of two numbers with unlike signs is always negative.
- The quotient of two numbers with like signs is always positive.
- The quotient of two numbers with unlike signs is always negative.

Note: The rules for the signs of multiplication and division of two real numbers are the same.

Example 1 Find the sum of $-5 + (-4)$.

$|-5| + |-4| = 5 + 4 = 9$ **Step 1:** Since both numbers are negative, we add their absolute values.

$-5 + (-4) = -9$ **Step 2:** Since both numbers are negative, we make the sum negative.

Answer: -9

Example 2 Find the sum of $-17 + 12$.

$|-17| - |12| = 17 - 12 = 5$ **Step 1:** Since the numbers have opposite signs, we take the difference of their absolute values and give the result the sign of the number with the larger absolute value.

$-17 + 12 = -5$ **Step 2:** Since -17 has the larger absolute value, the answer is negative.

Answer: (-5)

Example 3 Find the difference of $3 - 9$.

$3 - 9 = 3 + (-9)$ **Step 1:** This is a subtraction problem, so we add the opposite of the number being subtracted.

$|-9| - |3| = 9 - 3 = 6$ **Step 2:** Since we created an addition problem with numbers of opposite signs in Step 1, we find the difference of their absolute values and give the sum the sign of the number with the larger absolute value.

$3 - 9 = -6$	**Step 3:**	Since -9 has the larger absolute value, the answer is negative.
	Answer:	-6

Example 4 Find the product of $-4(-8)$.

$\|-4\| = 4,\ \|-8\| = 8$ $4(8) = 32$	**Step 1:**	Multiply the absolute values of the numbers.
$-4(-8) = 32$	**Step 2:**	Since the numbers have the same sign, the product is positive.
	Answer:	32

Example 5 Find the quotient of $-40 \div 5$.

$\|-40\| = 40,\ \|5\| = 5$ $40 \div 5 = 8$	**Step 1:**	Divide the absolute values of the two numbers.
$-40 \div 5 = -8$	**Step 2:**	Since the numbers have opposite signs, the result is negative.
	Answer:	-8

Problem Set

Perform the indicated operation:

1. $-12 + (-3)$
2. $7(-6)$
3. $-9 + (-4)$
4. $\frac{-27}{-3}$
5. $-8 + (-7)$
6. $\frac{0}{-4}$
7. $-6 + (-9)$
8. $\frac{4}{0}$
9. $-5 + (-15)$
10. $(-8)(-10)$
11. $-10 + (-8)$
12. $\frac{50}{-10}$
13. $-9 + (-3)$
14. $9(-11)$
15. $-7 + (-13)$
16. $6 + (-9)$
17. $10 - (-6)$
18. $-8 + 2$
19. $5 + (-2)$
20. $11 + (-3)$
21. $0 - (-4)$
22. $-3 + 6$
23. $13 - (-5)$
24. $7 + (-10)$
25. $1 - 8$
26. $8 + (-3)$
27. $3 - 10$
28. $-2 + 5$
29. $5 - (-6)$
30. $-9 + 6$
31. $-4 - (-2)$
32. $-16 - (-10)$
33. $-3 - 11$
34. $-13 - 3$
35. $-4 - (-5)$
36. $-2 - 6$
37. $-5 - (-9)$
38. $-3 - 7$
39. $-2 - (-8)$
40. $-6 - 8$

HA1-065: Evaluating Expressions Containing Exponents

Imagine that you are cleaning up after a dinner party. The plates and glasses must be put onto the proper shelves. You want to know the number of glasses you can place on each shelf. You know that you can place three rows of three glasses stacked three glasses high. You could count each glass separately, for a total of 27 glasses, but this would be time-consuming. Instead, you can get the same answer by multiplying 3 glasses times 3 rows times 3 glasses per stack $(3 \cdot 3 \cdot 3)$. Just as you can use multiplication as a shortcut to addition, you can use *exponents* as a shortcut to multiplication.

Power

3^3 — Exponent

Base

An **exponent** is a shorthand notation of repeated multiplication. Numbers expressed using exponents are called **powers**. The exponent 3 tells you to multiply the base three times.

Exponent	A number or variable that represents how many times the base is multiplied. In an expression that involves exponents, such as a^n, the exponent is n.
Base	A number or variable that has an exponent attached to it. In an expression that involves exponents, such as a^n, the base is a.
Power	An expression with a base and an exponent, such as 5^2

In the example above, the number 3 is being multiplied 3 times. Instead of writing this in the expanded factored form $(3 \cdot 3 \cdot 3)$, you can write 3^3 (read "three to the third power"), and still get the total of 27.

Note: Be careful not to multiply the base by the exponent. Doing so will result in an incorrect answer. For example: $3^3 = 27$, not $3^3 = 9$.

Example 1 Simplify: 10^2

$= 10 \cdot 10$

$= 100$

Answer: The base is 10. The exponent 2 tells you to multiply 10 two times. Therefore, $10^2 = 100$.

Example 2 Simplify: $(-5)^2$

$= (-5) \cdot (-5)$
$= 25$

Answer: The base is –5. The exponent 2 tells you to multiply –5 twice. Therefore, $(-5)^2 = 25$. Because there is an even number of negatives, the product will be positive.

Example 3 Simplify: -5^2

Answer: The base is 5. The exponent only applies to what is immediately in front of it: in this case, 5. Notice the difference between this problem and the one in Example 2. In Example 2, the parentheses are immediately before the exponent, so the entire contents of the parentheses are multiplied. In this example, multiply 5 twice, then make the product negative. Therefore, $-5^2 = -25$.

Example 4 Simplify: $\left(\frac{1}{2}\right)^3$

$\left(\frac{1}{2}\right)^3 = \frac{1}{2} \cdot \frac{1}{2} \cdot \frac{1}{2}$
$= \frac{1}{8}$

Answer: The base is $\frac{1}{2}$. The exponent tells you to multiply $\frac{1}{2}$ three times. Therefore, $\left(\frac{1}{2}\right)^3 = \frac{1}{8}$.

Example 5 Simplify: $3(2-6)^2$

$3(2-6)^2 = 3(-4)^2$

Step 1: Follow the order of operations–perform the operation in the parentheses, $2-6$.

$= 3 \cdot (-4) \cdot (-4)$
$= 3 \cdot 16$

Step 2: Simplify the exponent.

$= 48$

Step 3: Multiply 3 and 16 to get the answer.

Answer: 48

Example 6 Simplify: $3(2-6)^2$

$3(2-6)^2 = 3(-4)^2$	**Step 1:**	Follow the order of operations–perform the operation in the parentheses,
$= 3 \cdot (-4)(-4)$ $= 3 \cdot 16$	**Step 2:**	Simplify the exponent.
$= 48$	**Step 3:**	Multiply 3 and 16 to get the answer.
	Answer:	48

The instruction to "evaluate" simply means to substitute the given value of a variable into an expression and follow the order of operations. These examples include exponents, the second step in the order of operations.

Example 7 Evaluate 4^x if $x = 3$.

$4^x = 4^3$	**Step 1:**	Substitute 3 for x.
$= 4 \cdot 4 \cdot 4$ $= 64$	**Step 2:**	Simplify 4^3.
	Answer:	64

Example 7 Evaluate $z^{.5t}$ if $z = 8$ and $t = 4$.

$z^{.5t} = 8^{(.5 \cdot 4)}$	**Step 1:**	Substitute 8 for z and 4 for t.
$= 8^2$	**Step 2:**	Simplify $8^{(.5 \cdot 4)}$.
$= 8 \cdot 8$ $= 64$	**Step 3:**	Simplify 8^2.
	Answer:	64

Example 8 Evaluate $(x+3)^{y+1}$ if $x = 5$ and $y = 0$.

$(x+3)^{y+1} = (5+3)^{0+1}$	**Step 1:**	Substitute 5 for x and 0 for y.

$= (8)^1$ **Step 2:** Simplify 8^1.

Note: Any real number is assumed to have an exponent of 1, whether the exponent is written or not. Therefore, 8 and 8^1 are the same.

Answer: 8

Problem Set

1. Simplify: $-(5^2)$

2. Simplify: $(-2)^3$

3. Simplify: -10^2

4. What is the base in the expression 8^3?

5. What is the exponent in the expression 2^7?

6. What is the base in the expression 4^{10}?

7. Rewrite this expression using exponents: $a \cdot a \cdot b \cdot b$

8. Rewrite this expression using exponents: $r \cdot s \cdot s \cdot s$

9. Rewrite this expression in the expanded form: $2^3 \cdot 5^4$

10. Rewrite this expression in expanded form: $6^2 \cdot 3^4$

11. Simplify: $2(4+3)^2$

12. Simplify: $4(6+3)^2$

13. Evaluate the following expression if $x = -3$:

$$4x^2$$

14. Simplify: 2^4

15. Simplify: -8^2

16. Simplify: 9^2

17. Simplify: $\left(-\frac{2}{3}\right)^2$

18. Evaluate the expression if $x = -3$, $y = 2$, and $z = -5$.:

$$xy^2 - z^3$$

19. Simplify: $-(-3)^3$

20. Evaluate: $(-3a)^2$ if $a = 2$

HA1-070: Evaluating Formulas for Given Values of the Variables

A carpet layer needs to determine how much carpet she needs to cover the floor of a room. To do this, she will need to calculate the total square footage of the room.

An investigator at an accident scene needs to determine how fast each vehicle was traveling when the accident occurred. To do this, he will take certain measurements and calculate the speed.

A teacher needs to determine a student's semester grade.

In each of these cases, the person will use a **formula** to find the answer.

Formula	An equation that shows a relationship between two or more quantities

Formulas generally contain variables, numbers, and operations. In other words, formulas are algebraic expressions.

Formulas can be evaluated with given values for the variable. There are too many formulas to name, but some common formulas are listed below.

Perimeter of a rectangle	$2l + 2w$	l = length; w = width
Interest formula	$p \cdot r \cdot t$	p = principal; r = rate (%); t = time
Volume of a rectangular solid	$l \cdot w \cdot h$	l = length; w = width; h = height
Area of a circle	πr^2	$\pi \approx 3.14$; r = radius
Distance formula	$r \cdot t$	r = rate (speed); t = time
Area of a triangle	$\frac{1}{2}bh$	b = base; h = height

Example 1 The formula for distance is $r \cdot t$. Find the distance if $r = 85$ mph and $t = 12$ hours. Note that 85 mph = 85 miles divided by 1 hour.

$d = r \cdot t$	**Step 1:** Start with the formula.
$= 85 \cdot 12$	**Step 2:** Substitute the known values and multiply.
$= 1{,}020$	**Answer:** 1,020 miles

Example 2 The formula for the area of a circle is $\pi \cdot r^2$. Find the approximate area if $r = 5$ inches and $\pi = 3.14$.

$A \approx \pi \cdot r^2$ — **Step 1:** Start with the formula.

$\approx 3.14 \cdot 5^2$ — **Step 2:** Substitute the known values and evaluate using order of operations.

$\approx 3.14 \cdot 25$

≈ 78.5

Answer: 78.50 sq in. or 78.50 in^2

Example 3 The formula for volume is $l \cdot w \cdot h$. Find the volume if $l = 12$ft, $w = 10$ft, and $h = 14ft$.

$v = l \cdot w \cdot h$ — **Step 1:** Start with the formula.

$= 12 \cdot 10 \cdot 14$ — **Step 2:** Substitute the known values and multiply.

$= 1{,}680$

Answer: 1,680 cubic feet or 1,680 ft^3

Example 4 The formula for the area of a triangle is $\frac{1}{2}bh$. Find the area if $b = 4$ inches and $h = 10$ inches.

$A = \frac{1}{2}bh$ — **Step 1:** Start with the formula.

$= \frac{1}{2}(4 \cdot 10)$ — **Step 2:** Substitute the known values and multiply.

$= \frac{1}{2}(40)$

$= 20$ — **Answer:** 20 sq in. or 20 in^2

Problem Set

1. The total surface area, S, of a cube is found using the formula $S = 6a^2$, where a is the length of one side of the cube. Find S when $a = 1.2$.

2. The volume, V, of a cube is found using the formula $V = a^3$, where a is the length of one side of the cube. Find V when $a = 7$.

3. Evaluate the formula for the given value of the variables:

$A = bh \div 2$ when $b = 15$ and $h = 6$.

4. Evaluate the formula for the given value of the variables:

$A = bh \div 2$ when $b = 8$ and $h = 12$.

5. Evaluate the formula for the given value of the variables:

$I = prt$ when $p = 250$, $r = 7$, and $t = 4$.

6. Evaluate the formula for the given value of the variables:

$I = prt$ when $p = 200$, $r = 7$, and $t = 3$.

7. Evaluate the formula for the given value of the variable: $K = -2n + 4$ when $n = 25$.

8. Evaluate the formula for the given value of the variable: $K = -2n + 4$ when $n = 30$.

9. Evaluate the formula for the given value of the variables: $j = -3(h - k)$ when $h = 12$ and $k = -5$.

10. Evaluate the formula for the given value of the variables: $d = (w + b) \div (-2)$ when $w = 10$ and $b = 8$.

11. The volume, *V*, of a sphere is found by the formula $V = \frac{1}{6}\pi d^3$, where π is a constant and *d* is the diameter of the sphere. Find *V* when $\pi = 3.14$ and $d = 3$.

12. Kinetic energy, *K*, of an object is found using the formula $K = \frac{1}{2}mv^2$, where *m* is the mass of the object and *v* is the velocity that the object is traveling. Find *K* when $m = 3.5$ and $v = 12$.

13. Evaluate the formula for the given value of the variables: $h = vt - 5t^2$ when $v = 70$ and $t = 8$.

14. Evaluate the formula for the given value of the variables: $h = vt - 5t^2$ when $v = 120$ and $t = 12$.

15. Evaluate the formula for the given value of the variable: $C = 0.6n + 4$ when $n = 5$.

16. Evaluate the formula for the given value of the variable: $C = 0.6n + 4$ when $n = 30$.

17. Evaluate the formula for the given value of the variable: $A = 4\pi r^2$ when $\pi = 3.14$ and $r = 20$.

18. The volume, *V*, of a cone is found by the formula $V = \frac{1}{3}\pi r^2 h$, where *r* is the radius of the base and *h* is the height of the cone. Find *V* when $\pi = 3.14$, $r = 6$, and $h = 2$.

19. The diameter, *d*, of the floor of a stadium is 710 feet. Use $\pi = 3.14$ and the formula $A = \pi\left(\frac{d}{2}\right)^2$ to approximate the floor area for the given arena.

20. In constructing a building, a bricklayer must estimate the number of bricks that will be required to build a wall. The formula that he uses is $N = 7LH$. The variable *N* represents the number of bricks, *L* is the length of the wall, and *H* is the height of the wall. Estimate the number of bricks that would be needed to construct a wall with a length of 14 feet and a height of 10 feet.

HA1-075: Simplifying Algebraic Expressions by Combining Like Terms

Take a look at $2x^2 - 3x + 5x^2 - 8x + 4$ or $7x^2 - 11x + 4$. These two expressions are equivalent, but the second expression is a simplified version of the first. In math, when we say simplified we mean that the expression is written in simplest terms. This expression was simplified by combining like terms.

Term	A number, variable, or a product or quotient of numbers and variables

For example, x, $2y$, and $-5xyz$ are terms. The expression $x + y$ is not one term but actually two terms combined with an addition sign.

Note: Addition and subtraction signs separate terms in an expression.

Numerical Coefficient (Coefficient)	The numerical part of the term—also referred to simply as coefficient

When a term consists of a number and a variable, such as $-2x$, the numeric part is called a **coefficient**. Like terms may or may not have common coefficients. **Any variable not preceded by a number is assumed to have a coefficient of 1.**

Identify the coefficient of each term.

Term	Coefficient
$4x$	4
$-8x$	-8
$3xy$	3
y	1

Like Terms	Terms that contain the same variable, raised to the same power

Some examples of like terms are listed here.

$3x, 5x$ $\qquad y^2, -4y^2$ $\qquad z, 6z$ $\qquad -4xy, 4xy$

Suppose you were asked to simplify the following algebraic expression: $3x + 4y + 6 - 2x$.

Step 1: Determine how many terms you have. In this case, there are four terms: $3x$, $4y$, 6, and $-2x$.
Step 2: See if there are any like terms. Because $3x$ and $-2x$ have the same variable and exponent, they are like terms.
Step 3: Combine like terms using addition rules for integers and real numbers.
Solution: $x + 4y + 6$

When writing terms, begin with the term with the highest power and then write remaining terms in descending order.

Example 1 Simplify: $3d^2 + 5g - d^2 + 6g$

$3d^2 + 5g - d^2 + 6g.$	**Step 1:**	There are four terms: $3d^2$, $5g$, $-d^2$, and $6g$.
$= 2d^2 + 5g + 6g$	**Step 2:**	$3d^2$ and $-d^2$ are like terms, so they can be combined. The difference is $2d^2$, leaving three terms: $2d^2$, $5g$, and $6g$.
$= 2d^2 + 11g$	**Step 3:**	Because the terms $5g$ and $6g$ are alike, you can combine them. The sum is $11g$.
	Answer:	$2d^2 + 11g$

Example 2 Simplify: $2x^2 - 3x + 5x^2 - 8x + 4$

$2x^2 - 3x + 5x^2 - 8x + 4$	**Step 1:**	There are five terms: $2x^2$, $-3x$, $5x^2$, $-8x$, and 4.
$2x^2 + 5x^2 - 3x - 8x + 4$	**Step 2:**	Find the terms that are alike. The terms $2x^2$ and $5x^2$ are alike, and the terms $-3x$ and $-8x$ are alike.
$7x^2 - 11x + 4$	**Step 3:**	Combine the like terms.
	Answer:	$7x^2 - 11x + 4$

Problem Set

Solve the following:

1. Is y a term?

2. Is $3n + 7$ a term?

3. Is $12n$ a term?

4. What is the coefficient of $15m^3$?

5. What is the coefficient of d^2?

6. Are $2a$ and $3a$ like terms?

7. Are $3rs$ and $3st$ like terms?

Simplify the following expressions:

8. $6d + 9d$

9. $-6a + 4a$

10. $-2x - 7x$

11. $7d + 4 - 2d$

12. $7d - 4 + 2d$

13. $-3r - 8 + 6r$

14. $3 + 2h - 8$

15. $3 - 2h + 8$

16. $5 + 7k - 3$

17. $-5 + 7k - 3$

18. $2 + 4a + 12 - 5a$

19. $8x + 12y - 7 + 14x - 4y$

Solve the following:

20. Perimeter is found by adding the lengths of the sides. Find the perimeter of the square:

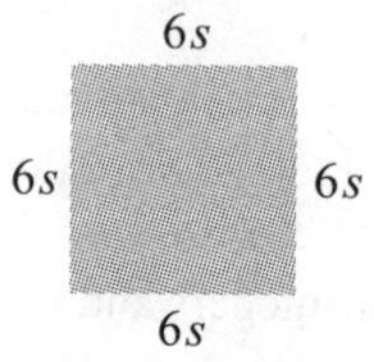

HA1-076: Basic Distributive Property

Multiplying a number and a sum can be done two ways. For example, $3(4+1)$ can be solved by first adding within the parentheses and then multiplying. We get:

$$\begin{aligned} 3(4+1) &= 3(5) \\ &= 15 \end{aligned}$$

We can also "distribute" the factor of 3 to each of the two summands. We get:

$$\begin{aligned} 3(4+1) &= 3(4)+3(1) \\ &= 12+3 \\ &= 15 \end{aligned}$$

Distributive Property of Multiplication	For real numbers a, b and c: $a(b+c) = ab+ac$; $(b+c)a = ba+ca$

Example 1 Multiply using the distributive property $4(x+2)$.

$4(x+2) = 4(x)+4(2)$	**Step 1:** Multiply each of the terms, x and 2, by 4.
$= 4x+8$	**Step 2:** Simplify.
	Answer: $4x+8$

Example 2 Multiply using the distributive property $5(2x-3)$.

$5(2x-3) = 5(2x)-5(3)$	**Step 1:** Multiply each of the terms, $2x$ and 3, by 5.
$= 10x-15$	**Step 2:** Simplify.
	Answer: $10x-15$

Example 3 Multiply using the distributive property $6\left(3x+\frac{2}{3}\right)$.

$6\left(3x+\frac{2}{3}\right) = 6(3x)+6\left(\frac{2}{3}\right)$	**Step 1:** Multiply each of the terms, $3x$ and $\frac{2}{3}$, by 6.
$= 18x+\frac{12}{3}$	**Step 2:** Simplify.

$= 18x + 4$ **Step 3:** Since $\frac{12}{3} = 4$, we can further simplify.

Answer: $18x + 4$

Example 4 Multiply using the distributive property $(2 - x)(-9)$.

$(2-x)(-9) = 2(-9) - x(-9)$	**Step 1:**	Multiply each of the terms, 2 and x, by -9.
$= -18 - (-9x)$	**Step 2:**	Simplify.
$= -18 + 9x$	**Step 3:**	To find the difference, add the opposite of the number being subtracted, $a - b = a + (-b)$.

Answer: $-18 + 9x$

Example 5 Multiply using the distributive property $\frac{x}{2}(6y + 10)$.

$\frac{x}{2}(6y+10) = \frac{x}{2}(6y) + \frac{x}{2}(10)$	**Step 1:**	Multiply each of the terms, $6y$ and 10, by $\frac{x}{2}$.
$= \frac{6xy}{2} + \frac{10x}{2}$	**Step 2:**	Simplify.
$= 3xy + 5x$	**Step 3:**	Since $\frac{6}{2} = 3$ and $\frac{10}{2} = 5$, we can further simplify.

Answer: $3xy + 5x$

Problem Set

Simplify:

1. $5(a+4)$ **2.** $6(c+3)$ **3.** $4(c+7)$ **4.** $12(d+8)$ **5.** $7(g+11)$

6. $9(6+m)$ **7.** $10(c+a)$ **8.** $8(p-5)$ **9.** $15(c-3)$ **10.** $20(m-8)$

11. $10(3-p)$ **12.** $11(h-6)$ **13.** $12(n-6)$ **14.** $3(9-x)$ **15.** $7(a-b)$

16. $6\left(x+\frac{1}{3}\right)$ **17.** $15\left(x-\frac{1}{5}\right)$ **18.** $18\left(x-\frac{5}{9}\right)$ **19.** $20\left(x+\frac{3}{4}\right)$ **20.** $12\left(x+\frac{5}{6}\right)$

21. $4\left(x-\frac{3}{4}\right)$ **22.** $14\left(x+\frac{3}{7}\right)$ **23.** $(x+8)2$ **24.** $(x-9)5$ **25.** $(x+7)3$

26. $(x+4)7$ **27.** $(x+5)8$ **28.** $(x-12)3$ **29.** $(x+4)(-8)$ **30.** $(x-7)(-8)$

31. $\frac{2}{3}(x+12)$ **32.** $\frac{3}{8}(x-24)$ **33.** $\frac{4}{5}(x+25)$ **34.** $\frac{2}{9}(x-18)$ **35.** $\frac{3}{7}(x+28)$

36. $\frac{4}{9}(x-36)$ **37.** $\frac{5}{8}(x+48)$ **38.** $\frac{5}{12}(x+60)$ **39.** $\frac{3}{4}(x+44)$ **40.** $\frac{7}{8}(x-16)$

HA1-079: Using a Concrete Model to Simplify Algebraic Expressions

This lesson describes how to use algebra tiles to simplify an algebraic expression into fewer terms, which is usually easier to evaluate than the original expression.

Algebra Tiles	Manipulatives that model algebraic expressions

The basic algebra tiles are as follows:

Expression	Tile
x	[x]
$-x$	[$-x$] (shaded)
1	[+]
-1	[−] (shaded)

Two properties of addition are used to simplify expressions. The first property is the **additive inverse property**.

Additive Inverse Property	The property that states that for any number x, $x+(-x) = 0$. The additive inverse of x is $-x$..

The second property is the **additive identity property.**

Additive Identity Property	The property that states that for any number x, $x+0 = x$. The additive identity of x is 0.

Zero Pair	The result of pairing one positive algebra tile with one negative algebra tile of the same size.

The practical value of the additive identity property is that it allows one to remove zero pairs when simplifying expressions with algebra tiles.

$x \quad + \quad -x \quad = \quad 0$

zero pair ----→ 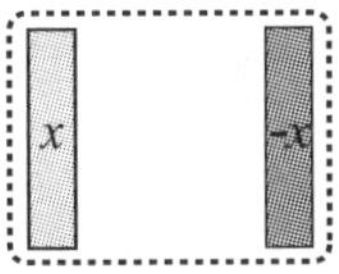 $=\ 0$

$1 \quad + \quad -1 \quad = \quad 0$

zero pair ----→  $=\ 0$

Example 1 Write an algebraic expression to match the model below.

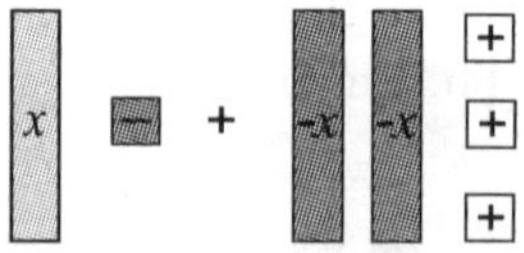

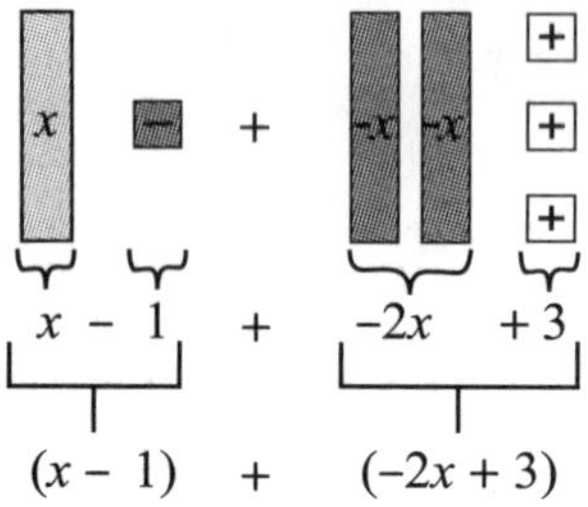

Step 1: On the left side of the plus sign, there is one green rectangle, which represents *x*, and one red square, which represents –1. Because these tiles are in the same group, they represent the expression $x - 1$.

On the right side of the plus sign, there are two red rectangles, which represent –2*x*, and three yellow squares, which represent 3. Because these tiles are in the same group, they represent the expression –2*x* + 3.

Answer: Therefore, the model represents the algebraic expression $(x - 1) + (-2x + 3)$.

Example 2 Model (4*x* – 3) + (–2*x* + 4) with algebra tiles.

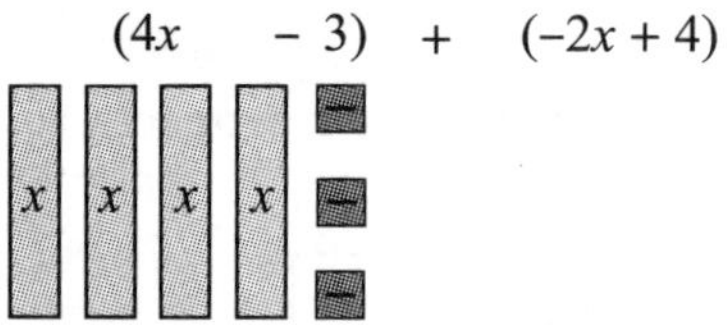

Step 1: Model the expression (4*x* – 3) on the left of the plus sign with four green rectangles, which represent 4*x*, and three red squares, which represent –3.

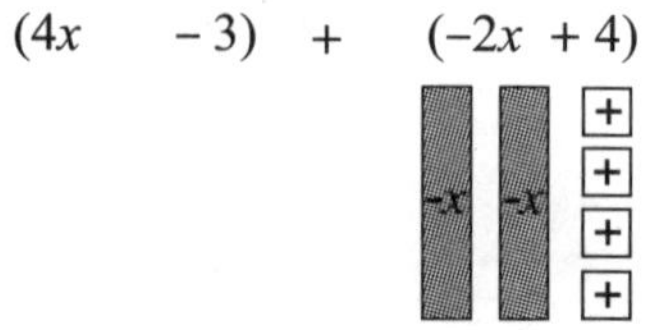

Step 2: Model the expression (–2*x* + 4) on the right of the plus sign with two red rectangles, which represent –2*x*, and four yellow squares, which represent 4.

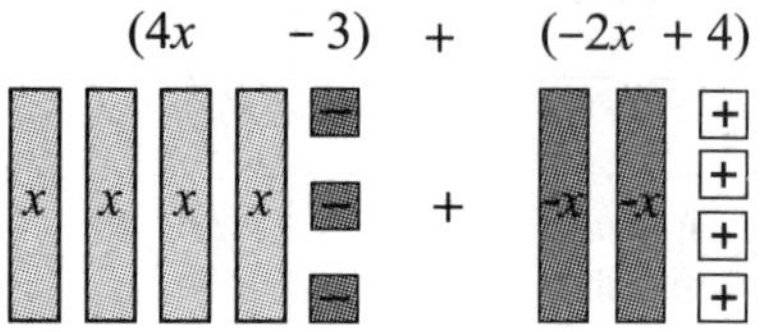

Step 3: Add the two models together to get a model of the sum.

Answer: Therefore, the model below represents the expression (4*x* – 3) + (–2*x* + 4).

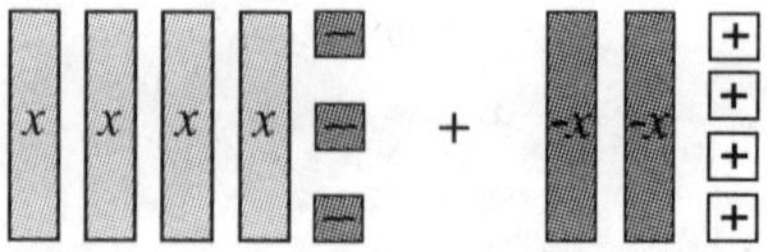

Example 3 Write an expression modeled by the algebra tiles and then simplify.

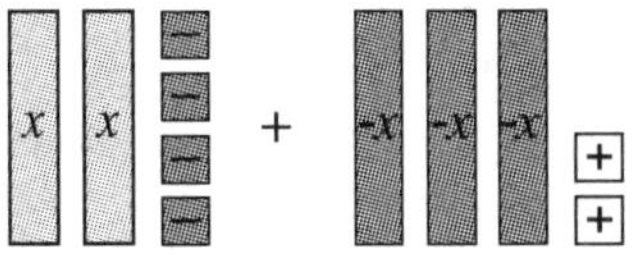

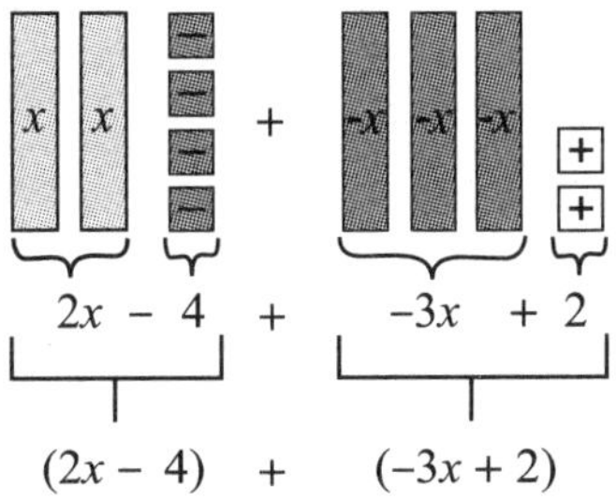

Step 4: On the left side of the plus sign, there are two green rectangles, which represent $2x$, and four red squares, which represent -4. Because these tiles are in the same group, they represent the expression $2x - 4$.

On the right side of the plus sign, there are three red rectangles, which represent $-3x$, and two yellow squares, which represent 2. Because these tiles are in the same group, they represent the expression $-3x + 2$.

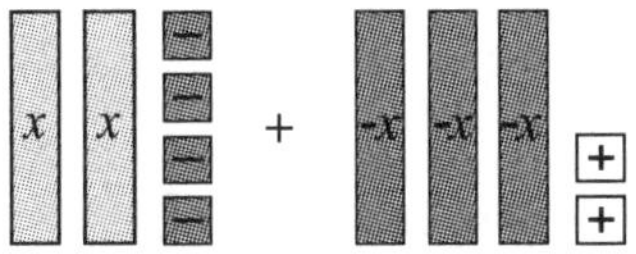

Step 5: Therefore, the model represents the algebraic expression $(2x - 4) + (-3x + 2)$.

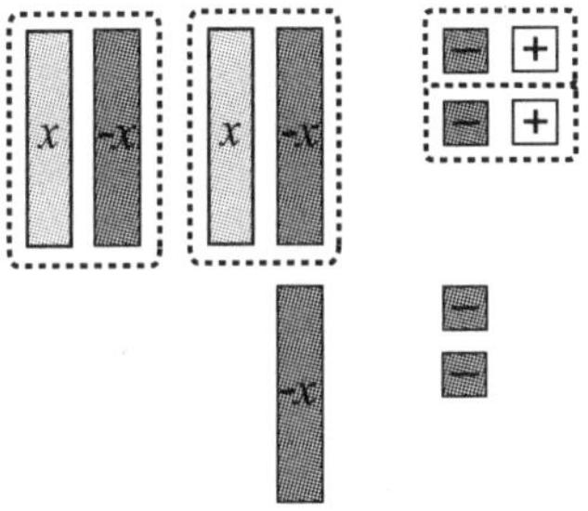

Step 6: Simplify the model by grouping zero pairs together.

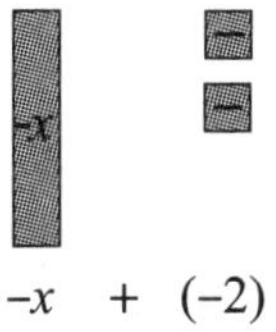

Step 7: Remove the zero pairs. The red rectangle represents $-x$, and the two red squares represent -2.

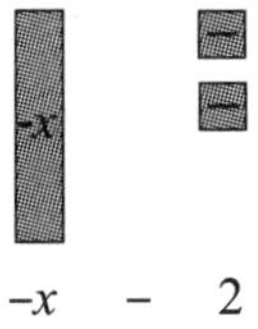

Step 8: Change addition of a negative to subtraction.

Answer: Therefore, the simplified model represents $(2x - 4) + (-3x + 2) = -x - 2$.

Example 4 Simplify $4(x - 2) + 3(-x + 1)$ using algebra tiles.

$4[x + (-2)] + 3(-x + 1)$

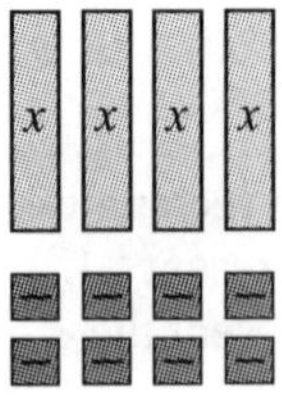

Step 1: Model $4(x - 2)$ with four groups of one green rectangle and two red squares.

$4[x + (-2)] + 3(-x + 1)$

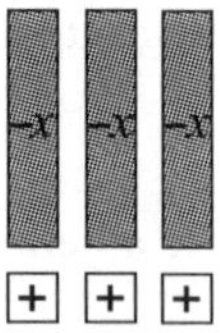

Step 2: Model $3(-x + 1)$ with three groups of one red rectangle and one yellow square.

$4[x + (-2)] + 3(-x + 1)$

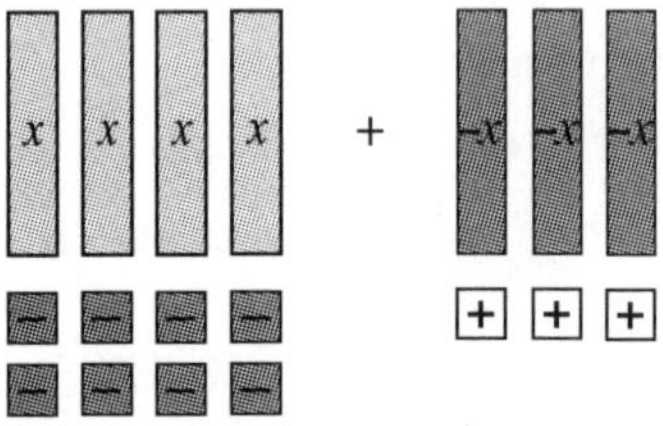

Step 3: Add the two models to get a model for the sum.

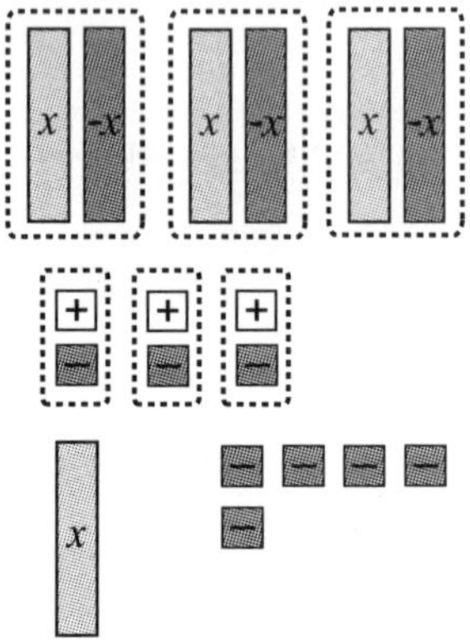

Step 4: Simplify the model by grouping zero pairs together.

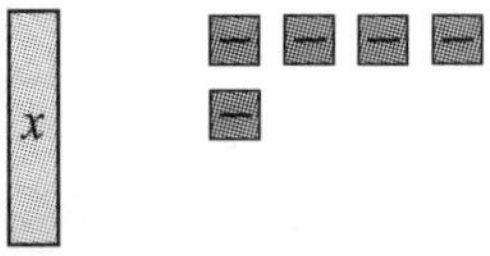

$x + (-5)$

Step 5: Remove the zero pairs. The green rectangle represents x, and the five red rectangles represent -5.

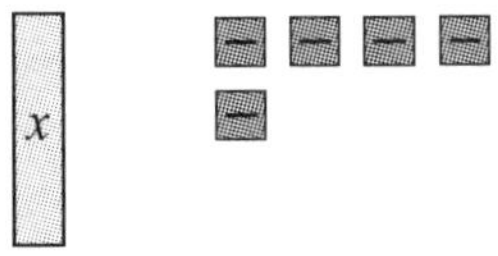

$x \quad - \quad 5$

Step 6: Change addition of a negative to subtraction.

Answer:

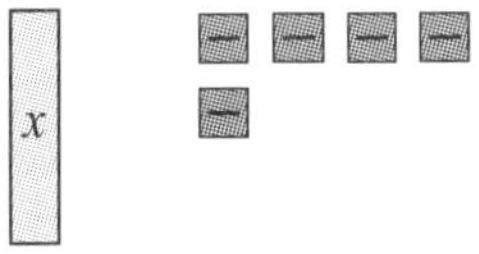

Example 5

Model the phrase below using algebra tiles and simplify the resulting expression. Let x represent the number.

"Twice the difference of one and a number added to four times the number decreased by three."

$2(1 - x)$

Step 1: Translate "twice the difference of one and the number."

$2(1 - x) + (4x - 3)$

Step 2: Translate "added to four times the number decreased by three."

$2[1 + (-x)] \quad + \quad (4x \quad - \quad 3)$

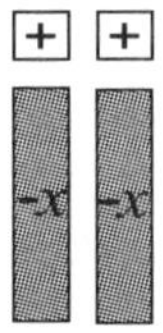

Step 3: Model $2(1 - x)$ with two groups of one yellow square and one red rectangle.

$2[1 + (-x)] \quad + \quad (4x \quad - \quad 3)$

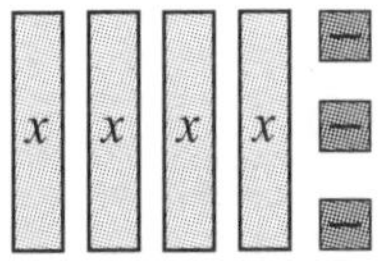

Step 4: Model $(4x - 3)$ with four green rectangles and three red squares.

$2[1 + (-x)] \quad + \quad (4x \quad - \quad 3)$

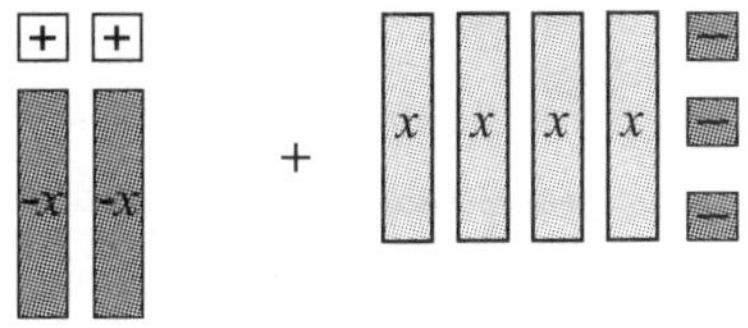

Step 5: Add the two models to get a model for the sum.

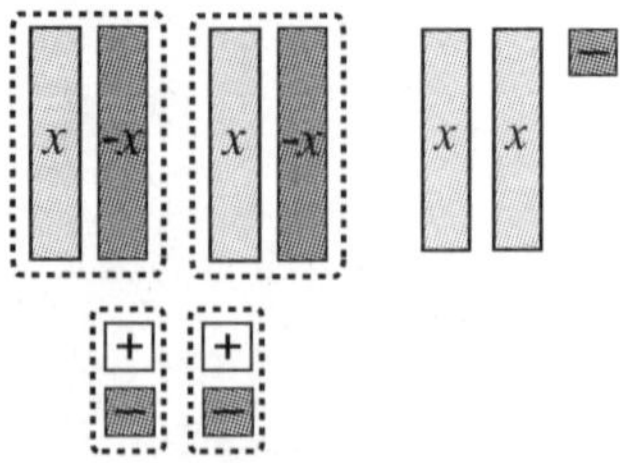

Step 6: To simplify, group zero pairs together.

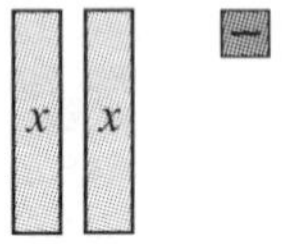

$2x \quad + (-1)$

Step 7: Remove the zero pairs. The two green rectangles represent $2x$, and the red square represents -1.

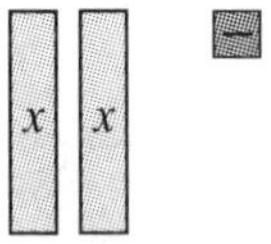

$2x \quad - \ 1$

Step 8: Change addition of a negative to subtraction.

Answer: Therefore, the model represents the simplified algebraic expression $2x - 1$.

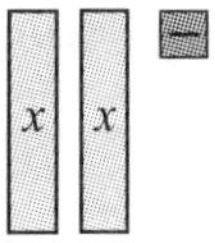

Problem Set

Write an algebraic expression modeled by the algebra tiles below:

1. 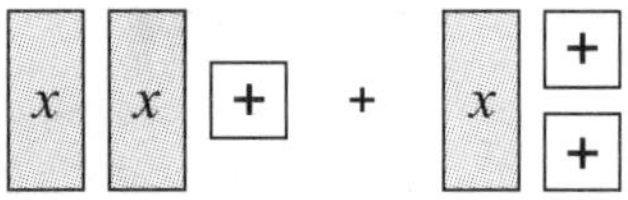

2. 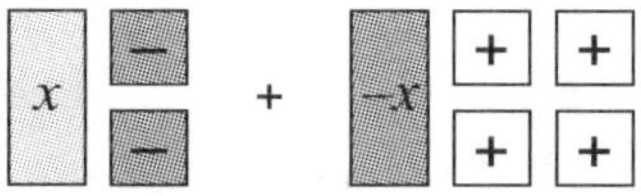

3. 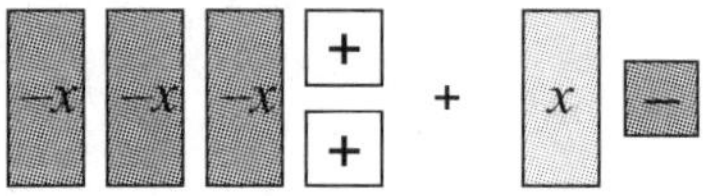

4. 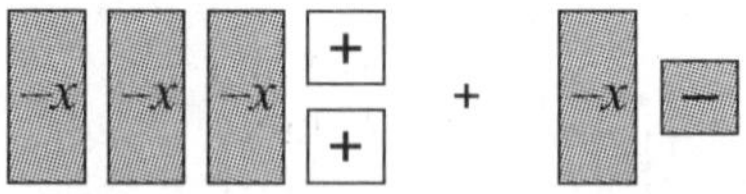

5. 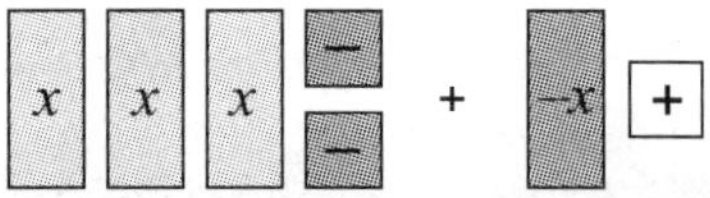

6. 

7. 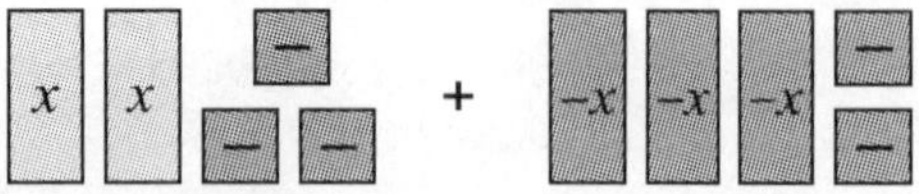

8. 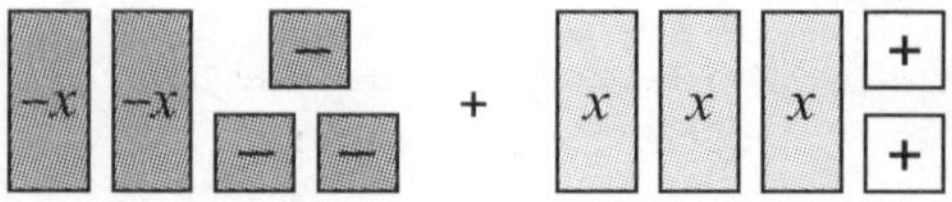

Create an algebra tile model that represents each of the following expressions:

9. $(2x-3)+(x+4)$

10. $(2x+3)+(-x+1)$

11. $(x-2)+(-x+4)$

12. $(x-3)+(-2x-1)$

13. $(3x+1)+(-x+3)$

14. $(-2x+2)+(3x-2)$

15. $(x+3)+(-2x-2)$

Write an algebraic expression modeled by the algebra tiles below and then simplify each.

16.

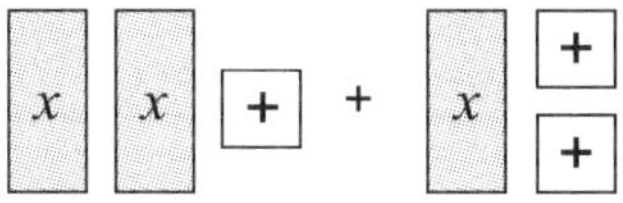

17.

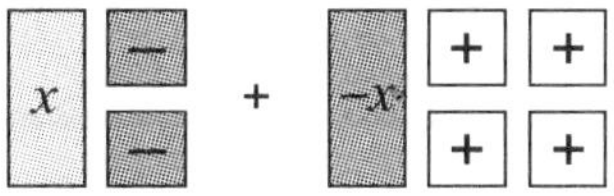

18.

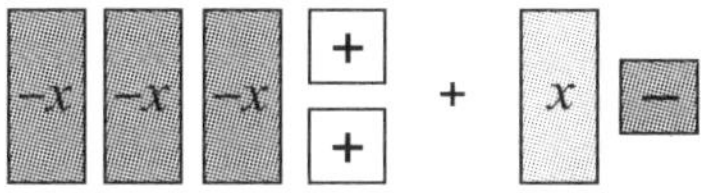

19.

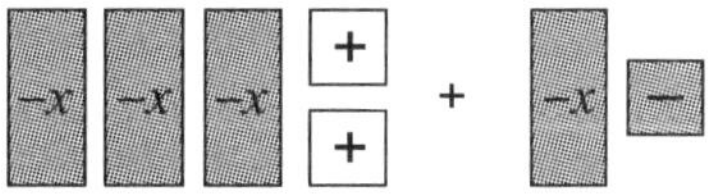

20.

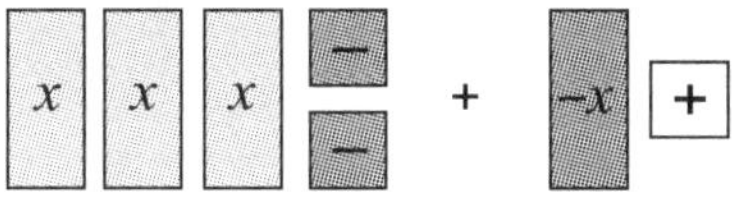

21.

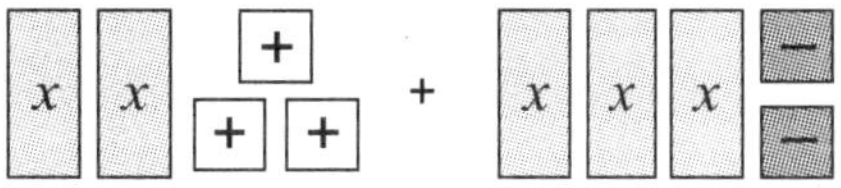

22.

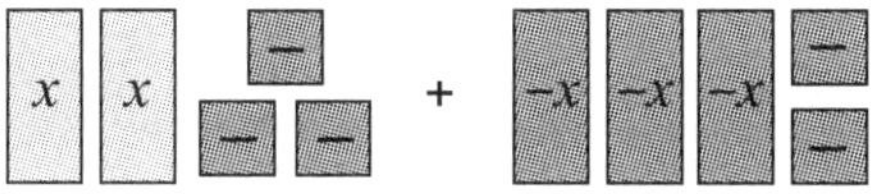

23.

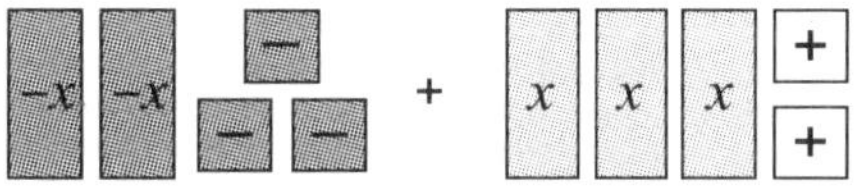

Simplify the following using algebra tiles:

24. $3(x+1)+2(-x+2)$

25. $2(x+1)+2(-x-2)$

26. $3(x+2)+(x-5)$

27. $2(-x+1)+3(x-1)$

28. $3(x+1)+2(x-2)$

29. $(-x-3)+2(-x+2)$

30. $(-x-3)+3(x-1)$

For each of the following, model the phrase using algebra tiles and simplify the resulting expression. Let x represent the number.

31. The sum of a number increased by one and twice the number increased by three.

32. The sum of a number increased by two and the difference of the number and three

33. The sum of the difference of four and a number and twice the sum of the number and three.

34. The sum of twice a number decreased by three and the difference of five and the number.

35. The sum of twice a number decreased by two and the differnce of one and three times the number.

36. The sum of three more than a number and twice the number decreased by three.

37. The sum of three times a number decreased by one and the number increased by three.

38. The sum of the difference of three and twice a number and twice the difference of the number and four.

39. The sum of one less than a number and the difference of three and twice the number.

40. The sum of twice the difference of two and twice the number and three times the difference of the number and one.

HA1-080: Simplifying and Evaluating Algebraic Expressions Containing Grouping Symbols

In this lesson you will combine what you have already learned about substituting values for variables in an algebraic expression with what you have learned about simplifying expressions by following the rules for order of operations. In algebra, the order in which you simplify an expression is crucial. Use the examples below to practice evaluating expressions that contain exponents and grouping symbols.

Example 1 Evaluate $2x^2 + 4y$, if $x = 3$ and $y = 6$.

$= 2(3)^2 + 4(6)$ **Step 1:** Substitute values for x and y.

$= 2(9) + 4(6)$ **Step 2:** Follow rules for order of operations. In this expression, you will simplify 3^2 first.

$= 18 + 24$ **Step 3:** Multiply $2 \cdot 9$ and $4 \cdot 6$.

$= 42$ **Step 4:** Add.

Answer: 42

Example 2 Simplify: $4x - 5 - (2x + 3)$

$= 4x - 5 - 2x - 3$ **Step 1:** Distribute –1 inside the parentheses.

$= 4x - 2x - 5 - 3$ **Step 2:** Combine like terms.
$= 2x - 5 - 3$
$= 2x - 8$

Answer: $2x - 8$

Example 3 Simplify: $-5(y + 2) + 3(y - 4)$

$= -5y - 10 + 3y - 12$ **Step 1:** Use the distributive property to eliminate the parentheses.

$= -5y + 3y - 10 - 12$ **Step 2:** Combine like terms.
$= -2y - 10 - 12$
$= -2y - 22$

Answer: $-2y - 22$

Example 4 Simplify $2+3[-6(y+1)]$, then evaluate for $y=4$.

$= 2+3[-6y-6]$	**Step 1:**	Distribute –6 to the terms in parentheses.
$= 2-18y-18$	**Step 2:**	Distribute 3 to the terms in the brackets.
$= -18y-16$	**Step 3:**	Combine like terms.
$= -18(4)-16$	**Step 4:**	Substitute 4 for *y*.
$= -72-16$ $= -88$	**Step 5:**	Follow the order of operations.
	Answer:	–88

Problem Set

Simplify the following expressions:

1. $-(-h-6)$

2. $-(h-6)$

3. $-(5+h)$

4. $-(-d^2+6)$

5. $-(w^2+8)$

6. $-(-k^2-1)$

7. $-(-5-7h)$

8. $-(-4d^2+6)$

9. $-(-3-9v^2)$

10. $-(6x-3y+5)$

11. $(3h^2-2)-(4h^2+5)$

12. $(5h+5)+(4h-3)$

13. $4f-5-(-f+2)$

14. Evaluate the given value of the variables:
$-(x-y)$ for $x = 8$ and $y = -2$

15. Evaluate the given value of the variables:
$-(x-y^2)$ for $x = 8$ and $y = -2$

16. Evaluate the given value of the variables:
$-(x^2+y)$ for $x = 2$ and $y = -8$

17. Evaluate the given value of the variable:
$2+4[-3(x-6)+7x]$ for $x = 1$

18. To find the perimeter of a figure, you must add the lengths of the sides. Find the expression that represents the perimeter of the given figure:

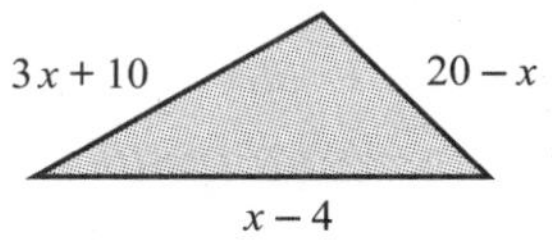

19. To find the perimeter of a figure, you must add the lengths of the sides. Find the expression that represents the perimeter of the given figure:

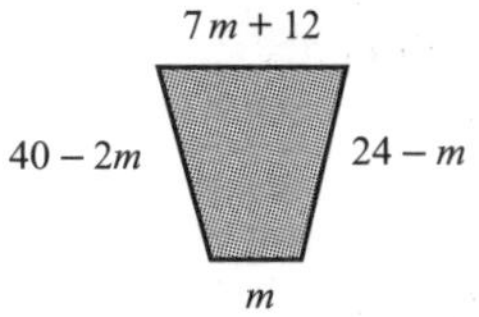

20. To find the perimeter of a figure, you must add the lengths of the sides. Find the expression that represents the perimeter of the given figure:

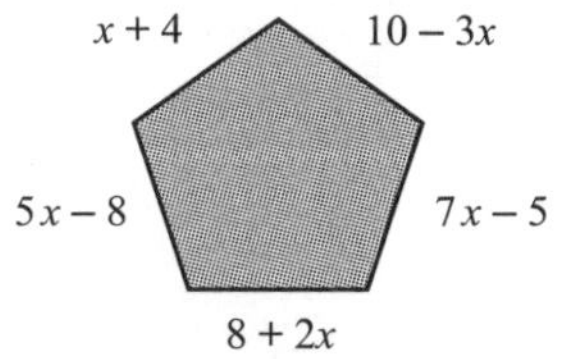

HA1-085: Simplifying Expressions Using the Properties of Real Numbers

Knowing certain properties of real numbers can help you simplify expressions. The **Commutative Property** states that the order in which you add or multiply numbers does not affect the sum or product. You can use this property when adding or multiplying several numbers.

Commutative Property of Addition	For all real numbers a and b, $a + b = b + a$.

Commutative Property of Multiplication	For all real numbers a and b, $a \cdot b = b \cdot a$.

Consider the expression $6 + 12 + 4 + 8$. Because the combinations of $6 + 4$ and $12 + 8$ are easier to add, change the order of the numbers. The addition of $6 + 4 = 10$ and $12 + 8 = 20$ makes it easier to evaluate the expression and find its solution, which is 30. Let's look at some examples.

Example 1 Add: $7 + 5 + 3 + 15$

$= \underbrace{7 + 3} + \underbrace{5 + 15}$

Step 1: Change the order of the numbers, because $7 + 3$ and $15 + 5$ are easier to combine and add.

$= 10 + 20$

$= 30$

Step 2: Add.

Answer: 30

Example 2 Multiply: $4 \cdot 3 \cdot 5 \cdot 2$

$= \underbrace{4 \cdot 5} \cdot \underbrace{3 \cdot 2}$

Step 1: Change the order of the numbers.

$= 20 \cdot 6$

$= 120$

Step 2: Multiply.

Answer: 120

Example 3 Add: $2 + 7 + 8 + 3 + 6 + 4$

$= \underbrace{2+8} + \underbrace{7+3} + \underbrace{6+4}$	**Step 1:**	Change the order of the numbers.
$= 10 + 10 + 10$ $= 30$	**Step 2:**	Add *Note: It is easier to add numbers whose sums are multiples of 5 or 10.*
	Answer:	30

The **Associative Property** states that when adding or multiplying more than two numbers, changing the grouping of the addends or factors does not change the sum or product. The results are the same, but regrouping allows you to use easier combinations to solve problems more efficiently.

Associative Property of Addition	For all real numbers a, b, and c, $a + (b + c) = (a + b) + c$

Associative Property of Multiplication	For all real numbers a, b, and c, $a \cdot (b \cdot c) = (a \cdot b) \cdot c$

Notice with the associative property that parentheses are used; however, the terms are separated by arithmetic symbols. How can you use this strategy to make solving problems easier?

Consider the following problems.

Regroup		
$12 + (48 + 17)$ $= 12 + 65$ $= 77$	Both 12 and 48 are multiples of 12.	$(12 + 48) + 17$ $= 60 + 17$ $= 77$
$15 + (105 + 23)$ $= 15 + 128$ $= 143$	Both 15 and 105 are multiples of 5.	$(15 + 105) + 23$ $= 120 + 23$ $= 143$
$10 + (43 + 30)$ $= 10 + 73$ $= 83$	Both 10 and 30 are multiples of 10.	$(10 + 30) + 43$ $= 40 + 43$ $= 83$

The **Distributive Property of Multiplication** states that the sum of two addends multiplied by a number is the sum of the product of each addend and the number. To use this property, multiply the number or variable on the outside of the parentheses by each term inside the parentheses. This property can also be used to simplify algebraic expressions.

Distributive Property of Multiplication	For all real numbers a, b, and c, $a(b-c) = (a \cdot b) - (a \cdot c)$ $a(b+c) = (a \cdot b) + (a \cdot c)$

Example 4 Simplify: $3(10 + 2)$

$= 30 + 6$ **Step 1:** Distribute the 3 by multiplying it by all numbers in the parentheses.

$= 36$ **Step 2:** Add.

OR OR

$= 3(12)$
$= 36$ **Step 3:** Follow rules for the order of operations; add the terms inside the parentheses first, then multiply.

Answer: 36

Example 5 Simplify: $5(6a + 12b)$

$= 30a + 60b$ **Step 1:** Distribute the 5 by multiplying it by all terms in the parentheses.

$= 30a + 60b$ **Step 2:** Combine any like terms; if there are no like terms, leave the expression as it is.

Answer: $30a + 60b$

Example 6 Simplify: $x(4x + 2) + 8x$

$= 4x^2 + 2x + 8x$ **Step 1:** Distribute x by multiplying it by all the terms in the parentheses.

$= 4x^2 + 10x$ **Step 2:** Combine any like terms; if there are no terms, leave the expression as it is.

Answer: $4x^2 + 10x$

Example 7 Simplify: $8(y-4)-2y$

$= 8y - 32 - 2y$	**Step 1:**	Distribute the 8 by multiplying it by all the terms in the parentheses.
$= 8y - 2y - 32$ $= 6y - 32$	**Step 2:**	Combine like terms.

Answer: $6y - 32$

The **Reflexive Property of Equality** states that a number is equal to itself.

Reflexive Property of Equality	$a = a$

The **Transitive Property of Equality** states that if one quantity, a, is equal to another quantity, b, and quantity b is equal to quantity c, then quantities a and c must also be equal.

Transitive Property of Equality	If $a = b$, and $b = c$, then $a = c$.

The **Symmetric Property of Equality** states that two quantities are equal.

Symmetric Property of Equality	If $a = b$, then $b = a$.

Examples of the Reflexive, Transitive, and Symmetric Properties of Equalities are listed in the table below. Note that each property given could be used to justify that the statement is true.

Statements	Properties
$2(c+3) = 2(c+3)$	Reflexive Property of Equality
If $n+3 = m+5$ and $m+5 = 6$, then $n+3 = 6$.	Transitive Property of Equality
If $12 = 2+10$, then $2+10 = 12$	Symmetric Property of Equality

Problem Set

Simplify the following expressions using the properties of real numbers:

1. $(12 + 14) + 45$

2. $25(4 + 2)$

3. $11 + (37 + 29)$

Answer "Yes" or "No" to the following:

4. Is the following expression equivalent to $3n + 11$?

$11 + 3n$

5. Is the following expression equivalent to $(8 - 5) - x$?

$8 - (5 - x)$

6. Is the following expression equivalent to $(5 + x)8$?

$40 + 8x$

7. Is the following expression equivalent to $7 \cdot (a \cdot 10)$?

$(7 \cdot a) \cdot 10$

8. Is the following expression equivalent to $(7 \div a) \div 10$?

$7 \div (a \div 10)$

9. Is the following expression equivalent to $6n - 15$?

$15 - 6n$

10. Is the following expression equivalent to $6n - 15$?

$n \cdot 6 - 15$

11. Is the following expression equivalent to $6(5a - 7c)$?

$6(-7c + 5a)$

12. Is the following expression equivalent to $6(5a - 7c)$?

$6(5a) - 6(7c)$

Answer the following:

13. Which property is illustrated by the following equation?
If $5 = n + 4$ and $n + 4 = y$, then $5 = y$

14. Which property is illustrated by the following equation?

$$2(n + 18) = 2n + 36$$

15. Which property is illustrated by the following equation?

$$n + 4 = 4 + n$$

16. Which property is illustrated by the following equation?

$$5(c + 7) = 5(c + 7)$$

17. Rewrite the following expression using the Distributive Property:

$$5(-2 + 6)$$

18. Is the following expression equivalent to $6(5a - 7c)$?

$$6(7a - 5c)$$

No

Yes

19. Rewrite this expression using the symmetric property:

$$3n + 9 = -12$$

20. If $34 = 2n + 9$ and $2n + 9 = x - 3$, then:

$34 = x$

$34 = x - 3$

$2n + 9 = 34 - x$

$2n + 9 = 34x - 3$

HA1-090: Simplifying Expressions Using the Property of –1

The principle called the **Identity Property of Multiplication** states that when you multiply a number by 1, the product is that number. For example, $1 \cdot 8 = 8$; $1 \cdot 800 = 800$; $1 \cdot 8{,}000 = 8{,}000$.

You can also use the **Multiplication Property of –1** to simplify some expressions.

Multiplication Property of –1	For every real number a, it is true that $-a = (-1)(a)$.

Look at the following examples using this property: $-20 = -1 \cdot 20$; $-10 = -1 \cdot 10$.

When multiplying numbers, remember that if the factors have the same signs, the product is positive. If the factors have different signs, the product is negative.

Evaluate the following examples using the Multiplication Property of –1.

Example 1 Multiply: $-5 \cdot 9$

$= (-1)(5)(9)$ **Step 1:** Multiply 5 and 9.

$= (-1)(45)$ **Step 2:** Multiply the partial product by –1 to get the final product.
$= -45$

Answer: -45

Example 2 Multiply: $x \cdot -4$

$= x(-1)(4)$ **Step 1:** Multiply x and 4.

$= (-1)(4x)$ **Step 2:** Multiply the partial product by –1 to get the final product.
$= -4x$

Answer: $-4x$

Example 3 Simplify: $-(x-5)$

$= -1(x-5)$ **Step 1:** Rewrite the problem.

$= -1(x) + -1(-5)$ **Step 2:** Distribute the –1 by multiplying it by each term in parentheses.

$= -x + 5$ **Step 3:** Simplify.

Answer: $-x + 5$

Problem Set

1. Is the following statement True or False?

$-a = -1 \cdot a$

2. Is the following statement True or False?

$-x = -1 \cdot x$

3. Is the following statement True or False?

$-(-5) = -1(5)$

4. Is the following statement True or False?

$-(5) = -1(5)$

5. Is the following statement True or False?

$-(7) = -1(-7)$

6. Is the following statement True or False?

$-(-2.1) = -1(-2.1)$

7. Is the following statement True or False?

$-(-m) = -1(m)$

8. Is the following statement True or False?

$-(m) = -1(m)$

9. Is the following statement True or False?

$-1(-j) = -(-j)$

10. Is the following statement True or False?

$-1(6) = -6$

11. Is the following statement True or False?

$-(m-n) = -1(m-n)$

12. Is the following statement True or False?

$-y = -1(y)$

13. Is the following statement True or False?

$-1[-1(-b)] = -b$

14. Is the following statement True or False?

$-(-2w) = 2w$

15. Is the following statement True or False?

$-(-3.1n) = -3.1n$

16. Is the following statement True or False?

$-1[-1(-3.1n)] = 3.1n$

17. Is the following statement True or False?

$-1[-(c+d)] = -(c+d)$

18. Is the following statement True or False?

$-1(-n) = n$

19. Choose the expression that is equivalent to the given expression.

$-(-8+c)$

$8-c$ $\quad$ $-8+c$

$-8-c$ $\quad$ $8+c$

20. Choose the expression that is equivalent to the given expression.

$-(4+a+c)$

$-4+a+c$ $\quad$ $-4-a-c$

$4+a+c$ $\quad$ $4-a-c$

HA1-095: Translating Word Phrases into Algebraic Expressions

Have you ever noticed that there are often many words that mean the same thing? If you wanted to say that two people looked the same, you could say that they looked alike or that they looked similar or that they resembled each other. All of these words translate to the same meaning.

Likewise, algebra has its own *translation system*. Certain words and phrases are used to indicate that a specific arithmetic operation should be used. Those words and phrases mean the same thing as the algebraic operation they stand for.

Add	Subtract	Multiply	Divide
Sum Plus Added to More than Increased by Total	Difference Minus Subtracted from Less than Decreased by	Product of Times Multiplied by Twice	Quotient Divided by

Now that you are familiar with the types of phrases that indicate certain operations, let's try interpreting some examples.

Phrase	Expression
Ten increased by s	$10 + s$ or $s + 10$
Eleven less than w	$w - 11$
Fifteen times n	$15n$
Twelve divided by p	$\frac{12}{p}$

Review the examples below.

Example 1 Write an algebraic equivalent for each of the following expressions. When you see the phrase "the quotient of," the first number is the numerator and the second number is the denominator.

The sum of 7 and y.	The word *sum* refers to addition.	$7 + y$
The product of d and $39s$.	The word *product* refers to multiplication.	$d \cdot 39s$
The difference between y and $8z$.	The word *difference* refers to subtraction.	$y - 8z$
The quotient of 2 and $3x$.	The word *quotient* refers to division.	$\frac{2}{3x}$

Example 2 Translate the following word phrases into algebraic expressions.

Word Phrase	Algebraic Expression
The sum of 4 and h 4 added to h h more than 4	$4 + h$
The quotient of 4 and y 4 divided by y	$\frac{4}{y}$
The difference of 9 and $3g$ 9 decreased by $3g$ 3 times g less than 9 (The phrase "less than" reverses the order of the terms.)	$9 - 3g$
The product of 10 and n 10 times n 10 multiplied by n	$10n$

Example 3 Let y represent Ann's age. What will her age be 10 years from now?

$y + 10$ **Step 1:** Start with y, which represents Ann's age now. Add 10 to represent 10 years.

Answer: $y + 10$

Example 4 Last month, your electric bill was x dollars. This month, your electric bill is \$10 less. How much is this month's bill?

$x - 10$ **Step 1:** Start with x, which represents last month's bill. Subtract 10, because this month's bill is \$10 less.

Answer: $x - 10$

Example 5 You are charged \$0.25 for each call you make from your cellular phone. You make c calls. How much will the calls cost?

c **Step 1:** Start with c, which represents the number of calls you make.

$0.25 \cdot c$

$0.25 \cdot c = 0.25c$

Step 2: Because you will be charged \$0.25 for each call, you can multiply the number of calls by the cost of each call. You can then rewrite $0.25 \cdot c$ as $0.25c$.

Answer: $0.25c$

Problem Set

1. Translate the following word phrase into an algebraic expression:
   k increased by m

2. Translate the following word phrase into an algebraic expression:
   The difference between t and q

3. Translate the following word phrase into an algebraic expression:
   The product of b and d

4. Translate the following word phrase into an algebraic expression:
   The sum of f and h

5. Translate the following word phrase into an algebraic expression:
   k minus t

6. Translate the following word phrase into an algebraic expression:
   Twenty-five more than a number

7. Translate the following word phrase into an algebraic expression:
   A number decreased by 25

8. Translate the following word phrase into an algebraic expression:
   The quotient of a number and 3

9. Translate the following word phrase into an algebraic expression:
   The product of 10 and a number squared.

10. Translate the following word phrase into an algebraic expression:
    The difference of eight and a number

11. Translate the following algebraic expression into the appropriate word phrase:
    xy

12. Translate the following algebraic expression into the appropriate word phrase:
    $z + x$

13. Translate the following algebraic expression into the appropriate word phrase:
    $k + c$

14. Translate the following algebraic expression into the appropriate word phrase:
    $m \div 8$

15. Translate the following word phrase into an algebraic expression:
    Eight times the sum of a number and twelve

16. Translate the following word phrase into an algebraic expression:
    Twice the difference of a number and eleven

17. Translate the following word phrase into an algebraic expression:
    Twenty-eight more than two times a number

18. Let f equal Frank's age. Jerry is seven years older than Frank. What expression represents Jerry's age?

19. Let p equal the number of points Derrick scored in a game. Kevin scored twenty-one fewer points than Derrick. What expression represents Kevin's points?

20. Rosita makes $6.20 an hour at a clothing store. If y equals the number of hours Rosita worked last week, what expression represents the amount of her paycheck?

HA1-100: Finding Solution Sets of Open Sentences from Given Replacement Sets

You have used different properties to determine the value of the unknown variable. Normally, once you have found one solution, you check to make sure the solution is correct.

Equation	A mathematical sentence that uses the symbol "=" to connect two algebraic expressions

Inequality	A mathematical sentence that uses the symbols $<$, $>$, $\leq$, $\geq$, or $\neq$ to connect two algebraic expressions

What you probably do not know is that equations can have more than one solution. An equation that contains one or more variables is called an **open sentence**.

Open Sentence	An equation or inequality that contains one or more variables

Mathematicians use **replacement sets** to limit the values that can be used for the variable.

Replacement Set	The set of numbers that may be substituted for a variable

When you are given a replacement set, you must replace the variable in the equation with a number from the replacement set. Numbers belonging to the replacement set are enclosed in braces. A replacement set may be $\{1, -3, 0, 12\}$. Any number from the replacement set that makes the equation or inequality true is a solution, and is a member of the **solution set.**

Solution Set	The set of all the numbers from the replacement set that makes the equation true

If you do not find a number from the replacement set that makes the equation true, then the solution set is empty, which is denoted by empty braces, { }, or the empty set symbol, $\varnothing$.

Example 1 Find the solution set for $2x + 3 = 7$. The replacement set is $\{-1, 0, 1, 2\}$.

Check –1

$$2(-1) + 3 = 7$$
$$-2 + 3 = 7$$
$$1 \neq 7$$

False

Step 1: Substitute –1 into the equation.

Check 0

$$\begin{aligned} 2(0)+3 &= 7 \\ 0+3 &= 7 \\ 3 &\neq 7 \end{aligned}$$

False

Step 2: Substitute 0 into the equation.

Check 1

$$\begin{aligned} 2(1)+3 &= 7 \\ 2+3 &= 7 \\ 5 &\neq 7 \end{aligned}$$

False

Step 3: Substitute 1 into the equation.

Check 2

$$\begin{aligned} 2(2)+3 &= 7 \\ 4+3 &= 7 \\ 7 &= 7 \end{aligned}$$

True

Step 4: Substitute 2 into the equation.

Step 5: Look at the equations again. The substitution of 2 into the equation makes the equation true. Thus, the solution set is $\{2\}$.

Answer: The solution set is $\{2\}$.

Example 2 Find the solution set for $3(x+1) = 3x+3$. The replacement set is $\{-1, 1, 2\}$.

Check –1

$$\begin{aligned} 3(-1+1) &= 3(-1)+3 \\ 3(0) &= -3+3 \\ 0 &= 0 \end{aligned}$$

True

Step 1: Substitute –1 into the equation.

Check 1

$$\begin{aligned} 3(1+1) &= 3(1)+3 \\ 3(2) &= 3+3 \\ 6 &= 6 \end{aligned}$$

True

Step 2: Substitute 1 into the equation.

Check 2

$$\begin{aligned} 3(2+1) &= 3(2)+3 \\ 3(3) &= 6+3 \\ 9 &= 9 \end{aligned}$$

True

Step 3: Substitute 2 into the equation.

Step 4: Look at the equations again. Substituting each of the numbers in the replacement set makes the equation true. Thus, the solution set is the same as the replacement set, $\{-1, 1, 2\}$.

Answer: The solution set is $\{-1, 1, 2\}$.

Example 3 Find the solution set for $3x - 2 \neq 3$. The replacement set is $\{1, 2\}$.

Check 1

$3(1) - 2 \neq 3$

$3 - 2 \neq 3$

$1 \neq 3$

True

Step 1: Substitute 1 into the equation.

Check 2

$3(2) - 2 \neq 3$

$6 - 2 \neq 3$

$4 \neq 3$

True

Step 2: Substitute 2 into the equation.

Step 3: Look at the equations again. The substitution of 1 or 2 into the equation makes the equation true. Thus, the solution set is $\{1, 2\}$.

Answer: The solution set is $\{1, 2\}$.

Example 4 Find the solution set for $2x + 1 < 4$. The replacement set is $\{-1, 0, 2\}$.

Check –1

$2(-1) + 1 < 4$

$-2 + 1 < 4$

$-1 < 4$

True

Step 1: Substitute –1 into the equation.

Check 0

$2(0) + 1 < 4$

$0 + 1 < 4$

$1 < 4$

True

Step 2: Substitute 0 into the equation.

Check 2

$2(2) + 1 < 4$

$4 + 1 < 4$

$5 < 4$

False

Step 3: Substitute 2 into the equation.

Step 4: Look at the equations again. The substitution of –1 and 0 into the equation makes the equation true. Thus, the solution set is $\{-1, 0\}$.

Answer: The solution set is $\{-1, 0\}$.

Example 5 Find the solution set for $7 - 4x < 2$. The replacement set is $\{-1, 0, 1\}$.

Check –1

$7 - 4(-1) < 2$

$7 - (-4) < 2$

$11 < 2$

False

Step 1: Substitute –1 into the equation.

Check 0

$7 - 4(0) < 2$

$7 - 0 < 2$

$7 < 2$

False

Step 2: Substitute 0 into the equation.

Check 1

$7 - 4(1) < 2$

$7 - 4 < 2$

$3 < 2$

False

Step 3: Substitute 1 into the equation.

Step 4: Look at the equations again. Since none of the values in the replacement set made the inequality true, the solution set is the empty set, $\varnothing$.

Answer: The solution set is $\varnothing$.

Example 6 Find the solution set for $4x - 2 \geq 2$. The replacement set is $\{0, 1, 2\}$.

Check 0

$4(0) - 2 \geq 2$

$0 - 2 \geq 2$

$-2 \geq 2$

False

Step 1: Substitute 0 into the equation.

Check 1

$4(1) - 2 \geq 2$

$4 - 2 \geq 2$

$2 \geq 2$

True

Step 2: Substitute 1 into the equation.

Check 2

$4(2) - 2 \geq 2$

$8 - 2 \geq 2$

$6 \geq 2$

True

Step 3: Substitute 2 into the equation.

Step 4: Look at the equations again. The substitution of 1 and 2 into the equation makes the equation true. Thus, the solution set is $\{1, 2\}$.

Answer: The solution set is $\{1, 2\}$.

Problem Set

1. Find the solution set for the inequality. The replacement set is $\{-2, -1, 0, 1\}$.

$$3x - 1 \geq 0$$

2. Find the solution set for the inequality. The replacement set is $\{-2, -1, 0, 1\}$.

$$3x - 1 \neq -7$$

3. Find the solution set for the inequality. The replacement set is $\{-1, 0, 1, 2\}$.

$$x + 2 > -1$$

4. Find the solution set if the replacement set is $\{-4, 0, 4\}$.

$$r < 2$$

5. Find the solution set if the replacement set is $\{-4, 0, 4\}$.

$$g \geq 0$$

6. Find the solution set if the replacement set is $\{-7, 0, 7\}$.

$$r < 2$$

7. Find the solution set if the replacement set is $\{-7, 0, 7\}$.

$$k \geq 0$$

8. Find the solution set if the replacement set is $\{-1.5, 0, 1.5\}$.

$$y < 2$$

9. Find the solution set if the replacement set is $\{-1.5, 0, 1.5\}$.

$$p \geq 0$$

10. Find the solution set if the replacement set is $\{-3, 0, 3\}$.

$$c \leq 0$$

11. Find the solution set for the inequality. The replacement set is $\{-2, -1, 0, 1\}$.

$$2x + 1 \leq 0$$

12. Choose the solution set for the inequality. The replacement set is $\{-2, -1, 0, 1\}$.

$$6 \geq 5a + 1$$

13. If the replacement set is $\{1, 2, 3, 4\}$, what is the solution set?

$$r + 1 > 4$$

14. If the replacement set is $\{1, 2, 3, 4\}$, what is the solution set?

$$r - 1 < 4$$

15. If the replacement set is $\{1, 2, 3, 4\}$, what is the solution set?

$$2r + 1 > 5$$

16. If the replacement set is $\{1, 2, 3, 4\}$, what is the solution set?

$$3r - 1 \geq 8$$

17. Find the solution set of $x - 8 > 5$ if the replacement set is $\{10, 15, 20\}$.

18. Choose the solution set for the inequality. The replacement set is $\{-2, -1, 0, 1\}$.

$$4x - 5 \geq -5$$

19. Find the solution set of $7x - 3 = -10$ if the replacement set is $\{-1, 0, 2\}$.

20. Find the solution set of $-4 + 6g \leq 8$ if the replacement set is $\{-2, 0, 2\}$.

HA1-104: Translating Word Statements into Equations

Mathematical expressions like $3x + 4$ contain constant and variable elements along with operators like: +, –, ×, and ÷. A **constant** is a number like 3 or 4 or –296. These quantities have an unchanging value.

Constant	A quantity with a fixed and unchanging value. Numbers are constants.

A **variable** represents an unspecified numerical value. We use a symbol like "x" to represent a variable. For example, in the expression $3x + 4$, x is a variable. If $x = 1$ this expression has a value of $3(1) + 4 = 3 + 4 = 7$. However, if $x = 2$ this expression has a value of $3(2) + 4 = 6 + 4 = 10$.

Variable	A quantity, usually represented by a letter of the alphabet, with a value that is unspecified or may change

Variables are useful when you need to write a mathematical expression but you don't know the value of a certain quantity.

Example 1 Shani has 3 more coins than John has. Write an algebraic expression for the number of coins Shani has.

x — **Step 1:** Since we don't know how many coins John has, let's denote this number of coins by the variable x.

$x + 3$ — **Step 2:** The quantity that is 3 more than x is $x + 3$.

Answer: If John has x coins, then Shani has $x + 3$ coins.

When translating words into algebraic or mathematical expressions, it is important to note that there are many words that can lead to the same expression. For example, when talking about addition we might have the expression $n + 9$. This can result from the words:

- A number plus nine
- The sum of a number and nine
- A number increased by nine
- Nine more than a number
- Nine added to a number

Example 2 Write an algebraic expression that represents the statement: The sum of a number and 24.

n — **Step 1:** Let the unknown number be represented by n.

$n + 24$ — **Step 2:** "The sum of the number and 24" can be expressed as $n + 24$.

Answer: "The sum of a number and 24" can be written as $n + 24$.

When talking about subtraction, we have many words that can give rise to the same algebraic or mathematical expression. For example, the expression $x - 2$ can result from:

- A number minus two
- The difference of a number and two
- A number decreased by two
- Two less than a number
- Two subtracted from a number

Multiplication and division can also be expressed using a variety of words.

The expression $12p$ can arise from:

- Twelve times a number
- The product of twelve and a number
- A number multiplied by 12

The expression $\frac{r}{6}$ can arise from the words:

- A number divided by six
- The quotient of a number and six
- The fraction of a number over six
- Six divides into a number
- Ratio of a number and six

Example 3 Write an algebraic expression that represents the statement: Nine times a number divided by four.

n	**Step 1:**	Let the unknown number be represented by n.
$9n$	**Step 2:**	The term "nine times a number" can be expressed as $9n$.
$\frac{9n}{4}$	**Step 3:**	Since this quantity is "divided by four", we have $\frac{9n}{4}$.

Answer: "Nine times a number divided by four" can be written as $\frac{9n}{4}$.

•

Example 4 Write an algebraic expression that represents the statement: Five increased by six times a number.

x	**Step 1:**	Let the unknown number be represented by the variable x.
$6x$	**Step 2:**	"Six times a number" translates into the expression $6x$.
$5 + 6x$	**Step 3:**	"Five increased by six times a number" can now be represented as $5 + 6x$.

Answer: "Five increased by six times a number" can be written as $5 + 6x$.

Sometimes, statements translated into mathematical ideas are not expressions alone, but a relationship between expressions. These types of statements can be written as an equation: that is, two expressions that have the same value.

A variety of words can lead to the same equation. For example, the equation $2x + 3 = 9$ can arise from the following statements:

- Twice a number plus three equals nine
- Twice a number plus three is nine
- The sum of twice a number and three gives a result of nine
- Three more than twice a number is equal to nine

Example 5 Write an equation that represents the statement: Three times the number of cans plus five gives twenty-six.

c	**Step 1:**	Let the unknown number of cans be represented by the variable c.
$3c$	**Step 2:**	"Three times the number of cans" can be translated into the expression $3c$.
$3c + 5$	**Step 3:**	"Plus five" gives us the expression $3c + 5$.
$3c + 5 = 26$	**Step 4:**	"Gives twenty six" leads us to the equation $3c + 5 = 26$.

Answer: "Three times the number of cans plus five gives twenty-six" can be written as the equation: $3c + 5 = 26$.

Problem Set

Write an algebraic expression to represent the following:

1. A number decreased by nine.

2. Eighteen less than a number.

3. The sum of a number and five.

4. Twelve more than a number.

5. The product of six and a number.

6. Three times a number.

7. The quotient of 75 and a number.

8. The sum of a number and 14 is 20.

9. The product of a number and 3 is equal to 45.

10. The quotient of 18 and a number is six.

11. Seven decreased by a number equals thirty.

12. Twelve more than a number equals 23.

13. Seventeen less than a number is fifty-two.

14. A number increased by 11 is 90.

15. Twice a number is equal to 42.

16. Five times a number decreased by three.

17. The product of four and a number increased by seven.

18. The quotient of 12 and a number decreased by eight.

19. The quotient of a number and 15 minus 3.

20. The sum of a number and 40 divided by 6.

21. Nineteen less than 3 times a number.

22. The difference of six times a number and 17.

23. Seventy-seven decreased by four times a number equals 22.

24. The quotient of 20 and a number increased by 12 is 16.

25. Forty-two is equal to the difference of 8 times a number and 12.

26. Sixteen decreased by the quotient of 9 and a number is 15.

27. Five times the sum of a number and 4 is 25.

28. Twelve less than twice a number is equal to 48.

29. Twenty-one increased by the quotient of 4 and a number equals 65.

30. Three multiplied by 12 less than a number is 36.

31. Three times seven more than the quotient of 4 and a number is 50.

32. The product of seven and the difference of 20 and 12 increased by a number is 90.

33. The quotient of a number and 12 minus the product of 7 and 3 is 34.

34. The difference of twice the product of 9 and a number and 15 is 40.

35. Twelve more than the quotient of 7 times a number and 10 is 19.

36. The difference of the product of a number and 5 and the quotient of 12 and 2 equals 20.

37. Twenty-two increased by the difference of 12 and 4 times a number is 30.

38. The product of sixteen and a number decreased by the sum of 15 and 4 is 28.

39. Thirty-five increased by the quotient of 8 and 3 times a number equals 48.

40. The sum of eight times a number and the quotient of 22 and the same number is 50.

HA1-105: Translating Word Statements into Inequalities

Translating word statements into algebraic expressions is a key to solving problems. In previous lessons, you had to translate word statements into algebraic equations. However, many problems can be solved using algebraic inequalities. So, it is important to know how to translate word statements into algebraic inequalities.

Inequalities	Inequalities are mathematical statements that compare quantities. The following symbols are used to make the comparisons: $>$ Greater than $\geq$ Greater than or equal to $<$ Less than $\leq$ Less than or equal to

Some additional expressions that may be useful in translating word statements to algebraic inequalities are:

$>$ More than

$\geq$ No less than, at least, minimum

$<$ Fewer than

$\leq$ No more than, at most, maximum

Example 1 Translate the following statement into an algebraic inequality: "The sum of a number and ten is less than twenty."

Let n be the unknown number. **Step 1:** Define the unknown quantity in terms of a variable.

The sum of a number and ten is less than twenty. **Step 2:** Translate the word statements into an algebraic inequality.

$n+10$ $<$ 20

Answer: $n + 10 < 20$

Example 2 Translate the following statement into an algebraic inequality. "A number divided by three is at most thirty."

Let n be the unknown number. **Step 1:** Define the unknown quantity in terms of a variable.

A number divided by three is at most thirty.

$$\frac{n}{3} \quad \le \quad 30$$

Step 2: Translate the word statements into an algebraic inequality.

Answer: $\frac{n}{3} \le 30$

Example 3 On the last English test, 30 students earned a C and at least 70 students earned a B or a C. Write an inequality to express the fact that at least 70 students earned a B or a C.

Let x be the number of students who earned a B.

Number of students who earned a C: 30

Number of students who earned a B or C: $x + 30$

Step 1: Define the unknown quantity in terms of a variable and state the given information.

Number of students who earned a B or C is at least seventy.

$$x+30 \quad \ge \quad 70$$

Step 2: Translate the word statements into an algebraic inequality.

Answer: $x + 30 \ge 70$

Example 4 Angela window-shopped last weekend and realized that she would have to spend at least \$550 to purchase a new TV. She has saved \$300 and is currently working at a rate of \$8 per hour to earn additional money. Write an inequality that will represent the amount of hours she needs to work to be able to purchase a TV.

Let x be the number of hours Angela needs to work.

We are given that:

- The minimum price of the TV is \$550.
- She saved \$300.
- She gets paid \$8 per hour.

Step 1: Define the unknown quantity in terms of a variable and state the given information.

The total amount of money that Angela will have is equal to the amount of money she has saved, plus the amount of money that she will earn working x number of hours, which is 8 times x.

Therefore: the amount of money $= 300 + 8x$.

Step 2: Use the given information to express algebraically the number of hours that Angela will have to work.

Total amount of money she needs is at least \$500.

$$300+8x \quad \ge \quad 500$$

Step 3: Translate the word statements into an algebraic inequality.

Answer: $300 + 8x \ge 500$

Example 5 Maria plans to put lace around a rectangular table cloth. The lace is sold in rolls and each roll contains nine feet of lace. If the table cloth is eight feet long and 4 feet wide, write the inequality that represents the minimum number of rolls of lace Maria should buy.

Step 1: Define the unknown quantity in terms of a variable and state the given information.

The amount of lace Maria needs is equal to the perimeter of the tablecloth.

The perimeter of a rectangle $= 2l + 2w$.

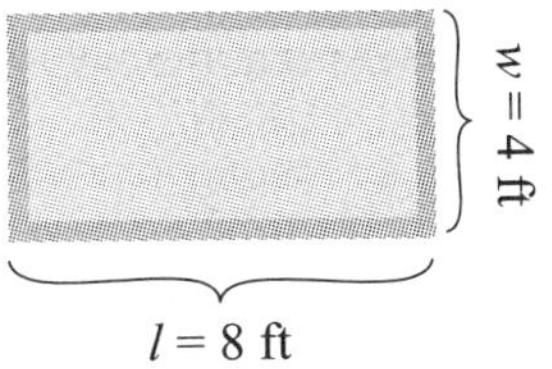

$$\begin{aligned}\text{Amount of lace} &= \text{Perimeter of the tablecloth}\\ &= 2l + 2w\\ &= 2(8) + 2(4)\\ &= 16 + 8\\ &= 24 \text{ ft}\end{aligned}$$

Step 2: Use the given information to express algebraically the number of rolls of lace Maria should buy.

Length of a single roll = 9 feet

Amount of lace in x rolls = $9x$ feet

Recall that "minimum" is represented by " ≥ ".

Step 3: Translate the word statements into an algebraic inequality.

Amount of lace in x rolls ≥ amount of lace for the tablecloth

Answer: $9x \geq 24$

Problem Set

Write each of the following statements as an inequality:

1. One more than a number is greater than ten.

2. The product of six and a number is more than twenty.

3. The quotient of a number and two is less than five.

4. The difference of a number and twelve is less than six.

5. A number increased by nine is less than or equal to twenty.

6. Five less than a number is greater than or equal to nine.

7. Twice a number is at most seven.

8. A number decreased by two is greater than or equal to ten.

9. Thirteen decreased by a number is no more than three.

10. Nine increased by a number is greater than or equal to fifteen.

11. The sum of a number and two is greater than six.

12. The product of a number plus one and three is no more than fourteen.

13. The quotient of a number plus five and seven is at most thirteen.

14. Eight more than twice a number is no less than fifty.

15. The quotient of a number and five increased by one is no more than ten.

16. A football team has won 3 games. In order to be in first place, it must win more than 9 games. Write an inequality to determine how many additional games the team must win to be in first place. Use x for the number of additional games the team must win.

17. Frank weighs 230 pounds and is on a diet. His goal is to weigh less than 210 pounds one month from now. Write an inequality to determine the number of pounds he must lose to meet his goal. Use x for the number of pounds he must lose.

18. A bag of candy is to be divided up among 4 people. Each person will get more than 10 pieces of candy. Write an inequality to determine how many pieces of candy are in the bag. Use x to represent the number of pieces of candy in the bag.

19. Jose earns $7 per hour. He needs to save an amount greater than $150 to buy a new bicycle. Write an inequality to determine how many hours he must work to earn enough money to buy a bicycle. Use x for the number of hours worked.

20. A farmer wants to plant rows of lettuce with 12 plants in each row. He must plant fewer than 300 lettuce plants. Write an inequality to determine the number of rows he can plant. Use x to represent the number of rows.

21. Ada has $425 in a savings account and is going shopping for some clothes. She wants to have more than $300 left in her savings account after her shopping spree. Write an inequality to determine how much money she may spend on the clothes. Use x for the number of dollars she may spend.

22. Javier is a wrestler and wishes to wrestle in a lighter weight class. In order to wrestle in the lighter class, he must lose more than 8 pounds so that his weight will be 180 pounds. Write an inequality to determine his present weight. Use x to represent the number of pounds Javier weighs.

23. Marinda is planning to buy some socks that cost $4 per pair. She must spend less than $55. Write an inequality to determine the number of pairs of socks Marinda can buy. Use x for the number of pairs of socks she buys.

24. Carlos has a savings account that contains $520. He plans to save $100 per month until he has enough money to purchase a motorcycle. Carlos has learned that he needs a minimum of $1,000 to buy a motorcycle. Write an inequality to determine the number of months he must save his money. Use x for the number of months.

25. Maria is purchasing burgers and sodas for her friends. She plans to purchase twice as many burgers as sodas. Each burger costs $2.50 and each soda costs $0.50. She can spend a maximum of $35. Write an inequality to determine how many sodas she can buy. Use x to represent the number of sodas.

26. The base price of Lynn's cell phone bill is $30 per month which includes a certain number of minutes. But she pays $0.40 a minute for each additional minute used. Lynn wants her cell phone bill to be no more than $45. Write an inequality to determine how many additional minutes she can use, if x represents the number of additional minutes.

27. Ken's internet provider charges a base price of $8 per month; for this price, he is allowed to use the service for a certain number of hours. He is charged $0.35 per hour for any additional hours he uses. Ken can afford a maximum of $12 per month. Write an inequality to determine how many additional hours Ken can use the Internet. Use x for the number of additional hours used.

28. Twan has already spent $180 on school clothes. He wants to purchase some jeans that cost $34 per pair. Write an inequality to determine how many pairs of jeans he can buy if he plans to spend no more than $250 in total on school clothes. Let x represent the number of jeans he can buy.

29. A basketball team scored 34 points during the first half. Their goal is to score at least 90 points in the game. If the baskets made in the second half are each worth 2 points, how many of these 2-point baskets do they need to score to meet their goal? Write an inequality to answer the question and use x to represent the number of baskets they need.

30. Josie was shopping for a new watch and determined that she would need to spend at least $125. She already has $35 saved that she can use. She is working and earning $7 per hour. How many hours must she work to have enough money to purchase a new watch? Write an inequality to answer the question and use x to represent the number of hours she must work.

31. Arlo wants to put a fence around his rectangular garden. The garden is 35 feet long and 20 feet wide. The fence comes in 8-foot sections. Let x represent the number of sections of fence he needs. Write an inequality to determine the minimum number of sections he should buy.

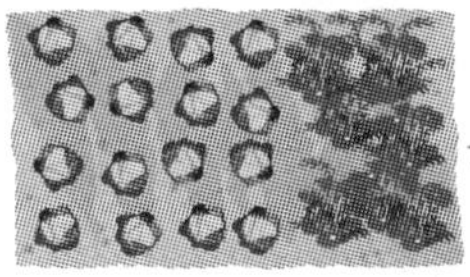

$w = 20$ ft

$l = 35$ ft

32. Leah wants to put a border around the walls in her rectangular bedroom. The room is 12 feet long and 11 feet wide. The border comes in 3-foot rolls. Let x represent the number of rolls of border she needs. Write an inequality to determine the minimum number of rolls she should buy.

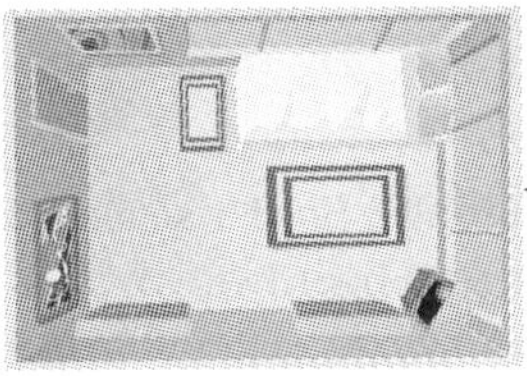

$w = 11$ ft

$l = 12$ ft

33. A builder needs to put baseboards around the floors of a rectangular family room. The room is 20 feet long and 17 feet wide. The baseboard comes in 8-foot pieces. Write an inequality to find the minimum number of pieces of baseboard he must purchase. Use x for the number of pieces of baseboard he needs.

$w = 17$ ft

$l = 20$ ft

34. Martha is making a rectangular rug and she plans to put fringe around the entire rug. The rug is 10 feet long and 8 feet wide, and the fringe comes in 5-foot sections. Write an inequality to determine the minimum number of sections of fringe she should buy. Use x to stand for the number of sections needed.

$w = 8$ ft

$l = 10$ ft

35. A rectangular flower bed is 12 feet long and 5 feet wide. It is going to be bordered with blocks that are 0.8 feet in length. Write an inequality to determine the minimum number of blocks needed to enclose the bed. Use x to stand for the number of blocks.

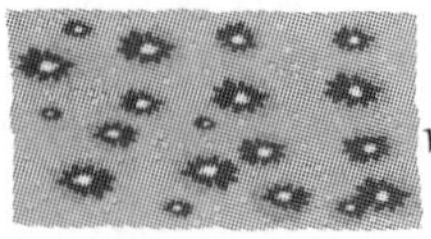

$w = 5$ ft

$l = 12$ ft

36. Cara is decorating the border around a rectangular table. The table is 6 feet long and 3.5 feet wide. The decorations come in strips that are 2 feet in length. Write an inequality to determine the minimum number of strips she should purchase. Let x stand for the number of strips needed.

$w = 3.5$ ft

$l = 6$ ft

37. Hillary is making a rectangular bedspread and plans to put lace around the outer edge. The bedspread is 15 feet long and 9 feet wide. The lace comes in 2.5-foot sections. Write an inequality to determine the minimum number of sections of lace Hillary should buy. Use x to stand for the number of sections.

$w = 9$ ft

$l = 15$ ft

38. Herman is building a rectangular table. He is using a decorative wooden strip around the edges of the table. The table is 6 feet long and 4 feet wide. The decorative wooden strips come in 3-foot sections. Write an inequality to determine the minimum number of sections he should buy. Let x represent the number of sections.

$w = 4$ ft

$l = 6$ ft

39. A teacher is putting a border around her rectangular bulletin board which is 22 feet long and 5 feet wide. The border comes in strips that are 6 feet in length. Write an inequality to determine the minimum number of strips of border needed. Use x to represent the number of strips.

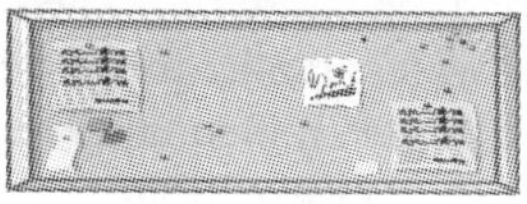

$w = 5$ ft

$l = 22$ ft

40. A farmer is fencing in a rectangular grazing area for his sheep. The field is 240 feet long and 190 feet wide. The fence comes in 8-foot long sections. Write an inequality to determine the minimum number of fence sections the farmer should purchase. Use x for the number of fence sections needed.

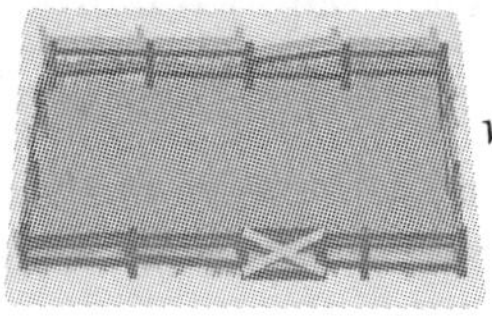

$w = 190$ ft

$l = 240$ft

HA1-115: Using the Addition and Subtraction Properties for Equations

In this lesson, you will continue working with equations. An equation is a mathematical sentence stating that two quantities are equal. Remember that an equation is similar to a balanced scale. Throughout the process of solving for the unknown, you must keep the scale balanced by performing the same operation on both sides of the equation. The same number may be added to both sides of an equation without changing the equality, or the same number may be subtracted from both sides of an equation without changing the equality. To solve an equation, use inverse operations to isolate the variable. Addition and subtraction are inverse operations. When solving algebraic equations, we often express the solution as a set of numbers.

Addition Property for Equations	For all real numbers, a, b, and c, if $a = b$, then $a + c = b + c$.

Subtraction Property for Equations	For all real numbers, a, b, and c, if $a = b$, then $a - c = b - c$.

Solution Set	The set of numbers that makes the equation true

Let's look at a sample equation.

Example 1 $n + 3 = 19$

$n + 3 - 3 = 19 - 3$ **Step 1:** Isolate the n by subtracting 3 from both sides of the equation.

$n = 16$ **Step 2:** Use your knowledge of signed numbers to evaluate the expression.

Step 3: Check your answer.

$$n + 13 = 19$$
$$16 + 3 = 19$$
$$19 = 19$$

Answer: The equation is true; the solution set is $\{16\}$.

Let's solve some equations using addition and subtraction properties for equations.

Example 2 $x + 6 = 9$

$x + 6 - 6 = 9 - 6$ **Step 1:** Isolate the x by subtracting 6 from both sides of the equation.

$x = 3$ **Step 2:** Use your knowledge of signed numbers to evaluate the expression.

$x + 6 = 9$
$3 + 6 = 9$
$9 = 9$

Step 3: Check your answer.

Answer: The equation is true; the solution set is {3}.

Example 3 $x - 4 = -25$

$x - 4 + 4 = -25 + 4$ **Step 1:** Isolate the x by adding 4 to both sides of the equation.

$x = -21$ **Step 2:** Use your knowledge of signed numbers to evaluate the expression.

$x - 4 = -25$
$-21 - 4 = -25$
$-25 = -25$

Step 3: Check your answer.

Answer: The equation is true; the solution set is {–21}.

Example 4 $7 = -6 + x$

$7 + 6 = -6 + x + 6$ **Step 1:** Isolate the x by adding 6 to both sides of the equation.

$13 = x$ **Step 2:** Use your knowledge of signed numbers to evaluate the expression.

$7 = -6 + x$
$7 = -6 + 13$
$7 = 7$

Step 3: Check your answer.

Answer: The equation is true; the solution set is {13}.

Example 5 $15 = n - (-8)$

$15 = n + 8$ **Step 1:** Simplify.

$15 - 8 = n + 8 - 8$ **Step 2:** Isolate the n by subtracting 8 from both sides of the equation.

$7 = n$ **Step 3:** Use your knowledge of signed numbers to evaluate the expression.

$15 = n - (-8)$
$15 = 7 - (-8)$
$15 = 15$

Step 4: Check your answer.

Answer: The equation is true; the solution set is {7}.

Example 6 $18 + y = -13$

$18 + y - 18 = -13 - 18$	**Step 1:**	Isolate y by subtracting 18 from both sides of the equation.
$y = -31$	**Step 2:**	Use your knowledge of signed numbers to evaluate the expression.
$18 + y = -13$ $18 + (-31) = -13$ $-13 = -13$	**Step 3:**	Check your answer.
	Answer:	The equation is true; the solution set is $\{-31\}$.

Example 7 A car salesman sold 8 cars this month. In the last two months, he sold a total of 15 cars. How many cars did he sell last month?

x = number of cars sold last month	**Step 1:**	Define the variable.
$x + 8 = 15$	**Step 2:**	Write an equation to represent the word problem.
$x + 8 - 8 = 15 - 8$	**Step 3:**	Isolate the x by subtracting 8 from both sides of the equation.
$x = 7$	**Step 4:**	Use your knowledge of signed numbers to evaluate the expression.
$7 + 8 = 15$	**Step 5:**	Check your answer. Is the total number of cars sold 15, if last month 7 cars were sold and this month 8 cars were sold?
	Answer:	The salesman sold 7 cars last month.

Problem Set

Solve:

1. $25 = x + 19$

2. $16 = x + 4$

3. $n + 7 = -15$

4. $n - 7 = 15$

5. $n + 52 = -27$

6. $n - 52 = 27$

7. $k - 89 = -13$

8. $j + 87 = 27$

9. $y + 65 = -50$

10. $y - 65 = 50$

11. $-45 = a + 14$

12. $a - (-12) = 24$

13. $34 = h + (-8)$

14. $15 = g + 27$

15. $13 = f - 30$

16. $-13 = 30 + f$

17. $78 + c = -57$

18. $0.9 = h + 1.2$

19. $b - (-7.2) = -8.3$

20. $c + (-5.9) = -6.0$

HA1-120: Using the Multiplication and Division Properties for Equations

In a previous lesson, you learned that an equation must remain balanced by using inverse operations on both sides of the equation to isolate the variable. Just as addition and subtraction are inverse operations, multiplication and division are inverse operations. Use the following two rules to isolate variables.

Multiplication Property for Equations	For all real numbers, a, b, and c, if $a = b$, then $a \cdot c = b \cdot c$.

Division Property for Equations	For all real numbers, a, b, and c, if $a = b$ and $c \neq 0$, then $\frac{a}{c} = \frac{b}{c}$.

Let's review examples of how to solve equations using these properties.

Example 1 Solve $\frac{x}{5} = 4$ using the multiplication property.

$5 \cdot \frac{x}{5} = 4 \cdot 5$

Step 1: Multiply both sides of the equation by 5 to "undo" the division and to isolate the x.

$x = 20$

Step 2: Use your knowledge of signed numbers to evaluate the expression.

Step 3: Check your answer.

$\frac{x}{5} = 4$

$\frac{20}{5} = 4$

$4 = 4$

Answer: The equation is true; the solution set is $\{20\}$.

Let's try a problem that uses the division property.

Example 2 Solve: $5n = 105$

$\frac{5n}{5} = \frac{105}{5}$

Step 1: Isolate the n by dividing both sides of the equation by 5.

$n = 21$

Step 2: Use your knowledge of signed numbers to evaluate the expression.

$5n = 105$ $5 \cdot 21 = 105$ $105 = 105$	**Step 3:**	Check your answer.
	Answer:	The equation is true; the solution set is {21}.

The goal of solving equations is to isolate the variable. We want to find the most efficient and straightforward way possible to do that. Sometimes, using inverse operations seems to make solving the problem more complicated. For example, if the coefficient of the variable is a fraction, multiplying both sides of the equation by the reciprocal is a more efficient approach. However, you may remember from what you've learned about fractions that dividing by a fraction and multiplying by its reciprocal are equivalent operations. So we can say that the inverse operation of multiplying by a fraction can be to multiply by the reciprocal of that fraction.

Example 3 Solve: $\frac{2}{3}n = 6$

$\frac{3}{2} \cdot \frac{2}{3}n = \frac{6}{1} \cdot \frac{3}{2}$	**Step 1:**	To isolate the *n*, multiply both sides of the equation by the reciprocal of $\frac{2}{3}$.
$n = 9$	**Step 2:**	Use your knowledge of signed numbers to evaluate the expression.
$\frac{2}{3}n = 6$ $\frac{2}{3} \cdot \frac{9}{1} = 6$ $6 = 6$	**Step 3:**	Check your answer.
	Answer:	The equation is true; the solution set is {9}.

Try solving the following examples.

Example 4 Solve: $36 = -4x$

$\frac{36}{-4} = \frac{-4x}{-4}$	**Step 1:**	Divide both sides of the equation by –4 to isolate the *x*.
$-9 = x$	**Step 2:**	Use your knowledge of signed numbers to evaluate the expression.
$36 = -4x$ $36 = -4 \cdot -9$ $36 = 36$	**Step 3:**	Check your answer.
	Answer:	The equation is true; the solution set is {–9}.

Example 5 Solve: $\frac{3}{7}n = -6$

$\frac{7}{3} \cdot \frac{3}{7}n = \frac{-6}{1} \cdot \frac{7}{3}$	**Step 1:**	To isolate the n, multiply both sides of the equation by the reciprocal of $\frac{3}{7}$.
$n = -14$	**Step 2:**	Use your knowledge of signed numbers to evaluate the expression.
$\frac{3}{7}n = -6$ $\frac{3}{7} \cdot \frac{-14}{1} = -6$ $-6 = -6$	**Step 3:**	Check your answer.

Answer: The equation is true; the solution set is $\{-14\}$.

Example 6 Solve: $21 = 0.3x$

$\frac{21}{0.3} = \frac{0.3x}{0.3}$	**Step 1:**	Isolate the x by dividing both sides of the equation by 0.3.
$70 = x$	**Step 2:**	Use your knowledge of signed numbers to evaluate the expression.
$21 = 0.3x$ $21 = 0.3 \cdot 70$ $21 = 21$	**Step 3:**	Check your answer.

Answer: The equation is true; the solution set is $\{70\}$.

Example 7 Kayla spent $720 on carpet. If the carpet cost $36 per square yard, how many square yards did Kayla buy?

x = number of sq yd of carpet	**Step 1:**	Define the variable.
$36x = 720$	**Step 2:**	Write an equation to represent the word problem. (cost of carpet per sq yd)(number of sq yd) = cost
$\frac{36x}{36} = \frac{720}{36}$	**Step 3:**	Isolate the x by dividing both sides of the equation by 36.
$x = 20$	**Step 4:**	Use your knowledge of signed numbers to evaluate the expression.
$\$36(20) = \$720$	**Step 5:**	Check your answer. When you multiply the price of the carpet by the number of yards, does the total equal $720? Yes.

Answer: Kayla bought 20 square yards of carpet.

Problem Set

1. Solve: $14 = 7x$

2. Solve: $-45 = -a$

3. Tell what must be done to each side of the equation so that the variable will be alone on one side.

$$\frac{2a}{5} = 28$$

4. Tell what must be done to each side of the equation so that the variable will be alone on one side.

$$-a = 28$$

5. Tell what must be done to each side of the equation so that the variable will be alone on one side.

$$\frac{-5d}{2} = 25$$

6. Tell what must be done to each side of the equation so that the variable will be alone on one side.

$$-d = -25$$

7. Tell what must be done to each side of the equation so that the variable will be alone on one side.

$$\frac{9j}{11} = 18$$

8. Tell what must be done to each side of the equation so that the variable will be alone on one side.

$$-j = 18$$

9. Tell what must be done to each side of the equation so that the variable will be alone on one side.

$$\frac{3p}{5} = 60$$

10. Tell what must be done to each side of the equation so that the variable will be alone on one side.

$$-p = 60$$

11. Solve: $\frac{y}{4} = 7$

12. Solve: $\frac{c}{6} = 18$

13. Solve: $\frac{-4g}{7} = -84$

14. Solve: $-g = -8$

15. Solve: $\frac{k}{2} = 28$

16. Solve: $-k = 13$

17. Solve: $5r = -20$

18. Solve: $-5.2m = -26$

19. Solve: $-28 = -x$

20. Solve: $-72 = -w$

HA1-124: Using a Concrete Model to Solve One- and Two-Step Equations

Algebra tiles can be used to solve one- and two-step equations. Different expressions in algebra are represented by the color and the shape of a tile. A green rectangle represents the variable x. A red rectangle represents the opposite of x, or $-x$. A yellow square represents 1, and a red square represents -1.

Expression	Tile
x	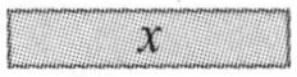
$-x$	
1	
-1	

When solving equations with algebra tiles, zero pairs, which occur on the same side of the equation, can be removed. A green rectangle, which represents x, and a red rectangle, which represents $-x$, form a zero pair because $x + -x = 0$. Also, a yellow square, which represents 1, and a red square, which represents -1, form a zero pair because $1 + -1 = 0$.

Zero Pairs

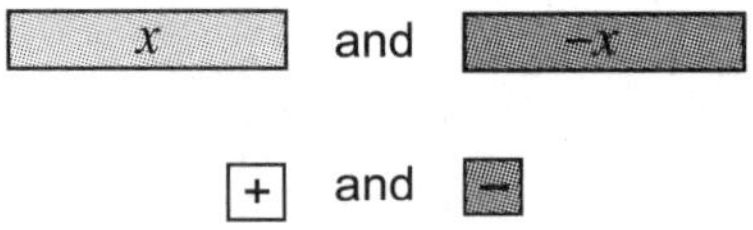

It is also important to note that when using algebra tiles to solve equations, whatever is done to one side of the equation must be done to the other side. This must be done to maintain the balance of the equation.

Example 1 Model and solve the equation using algebra tiles: $x + 3 = -1$.

Step 1: Model the equation. Place one green rectangle and three yellow squares on the left side of the equation to represent $x + 3$. Place one red square on the right side of the equation to represent -1.

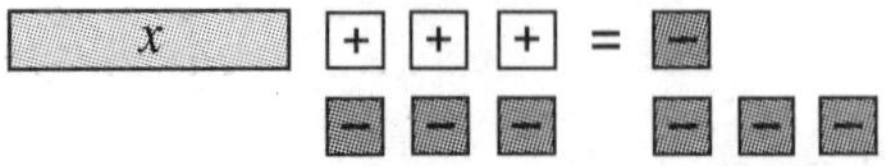

Step 2: Place three red squares on each side to isolate the green rectangle.

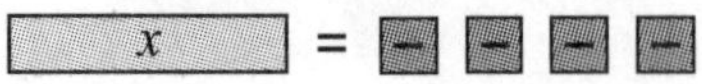

Step 3: Remove the zero pairs. The solution is $x = -4$.

Answer: 

$x = -4$

Example 2 Model and solve the equation using algebra tiles: $3x = -6$.

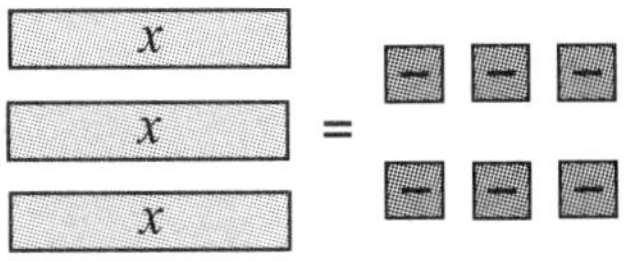

Step 1: Model the equation. Place three green rectangles on the left side of the equation to represent $3x$. Place six red squares on the right side of the equation to represent -6.

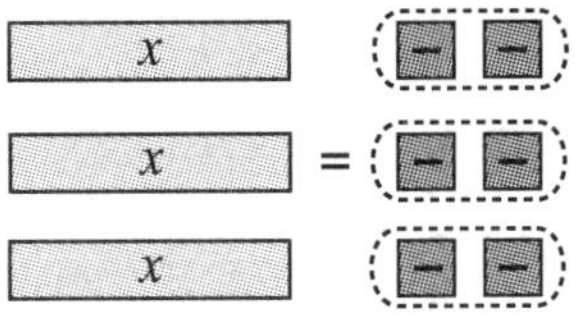

Step 2: Divide the square tiles into three equal groups to find
the value of one green rectangle, which represents x.
The solution is $x = -2$.

Answer:

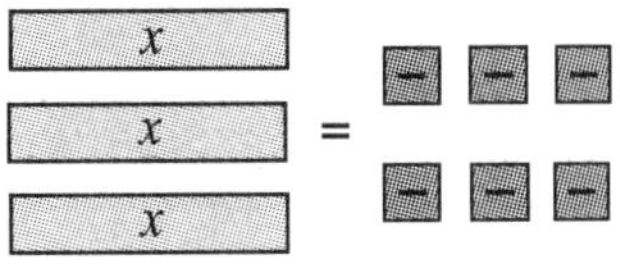

$x = -2$

Example 3 Use algebra tiles to model and solve the problem: Monique rented two movies at the video store for seven dollars. If she has six dollars left, how much money did she originally have?

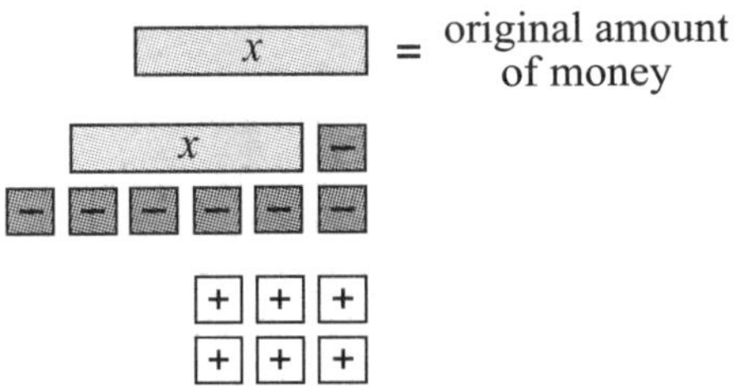

Step 1: Let one green rectangle equal the original amount of money, which represents x. Let one green rectangle and seven red squares represent the spent seven dollars from the original amount, and let six yellow squares represent the six dollars that is left.

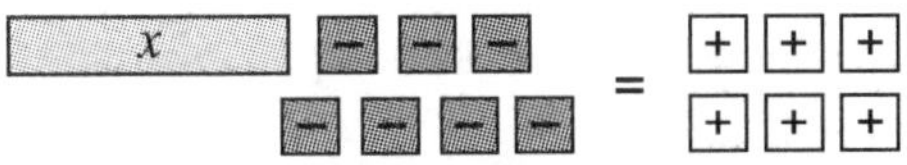

Step 2: Model the equation. Place the one green rectangle and seven red squares on the left side of the equal sign and the six yellow squares on the right side.

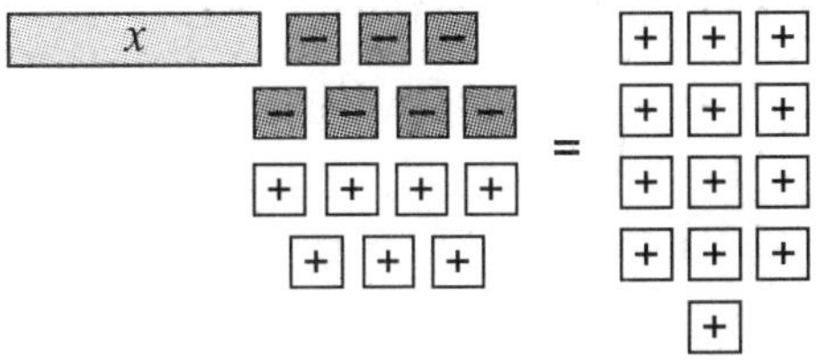

Step 3: Place seven yellow squares on each side to isolate the green rectangle.

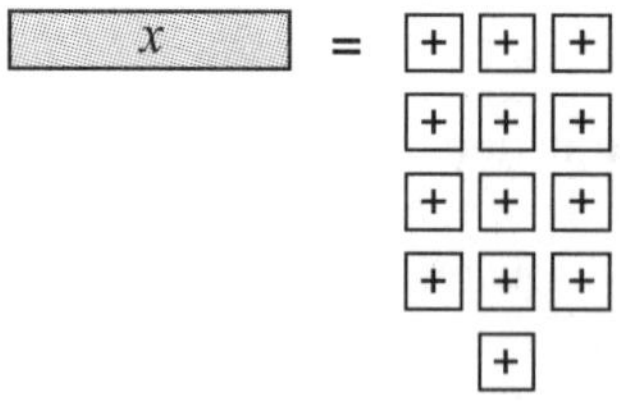

Step 4: Remove the zero pairs. The solution is $x = 13$.

Answer:

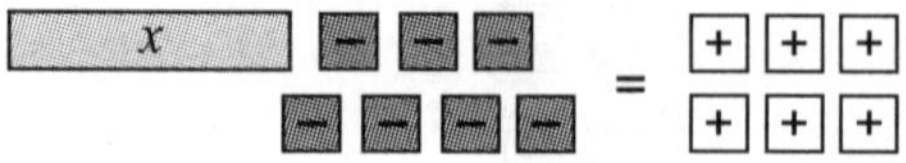

$x = 13$

Therefore, Monique originally had $13.00.

Example 4 Model and solve the equation with algebra tiles: $3(x-1) = 6$.

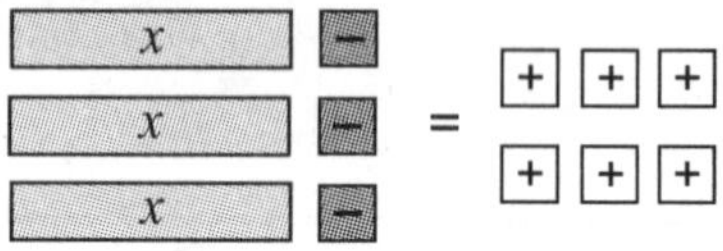

Step 1: Model the equation. Place three sets of one green rectangle and one red square on the left side of the equation to represent $3(x-1)$. Place six yellow squares on the right side of the equation to represent 6.

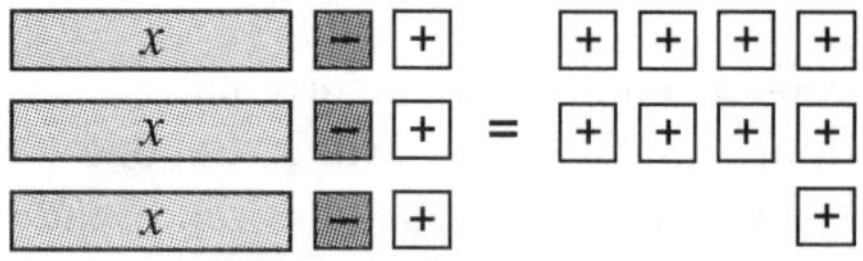

Step 2: Place three yellow squares on each side to isolate the green rectangles.

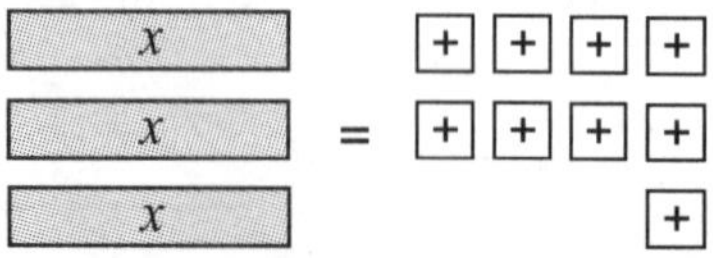

Step 3: Remove the zero pairs.

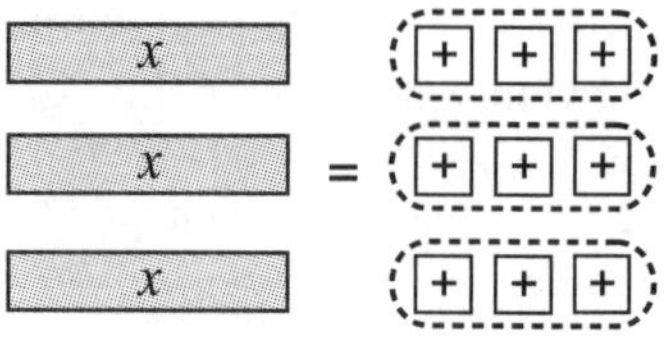

Step 4: Divide the square tiles into three equal groups. The solution is $x = 3$.

Answer:

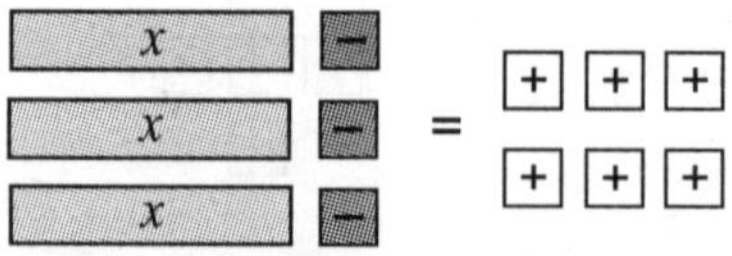

$x = 3$

Example 5 Model and solve the equation using algebra tiles: $-2x+3 = -1$.

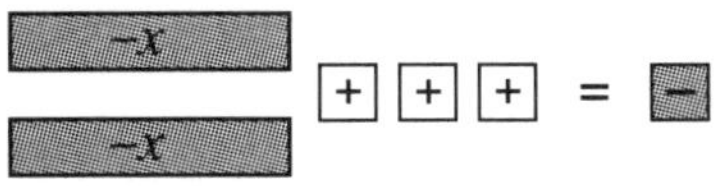

Step 1: Model the equation. Place two red rectangles and three yellow squares on the left side of the equation to represent $-2x+3$. Place one red square on the right side of the equation to represent -1.

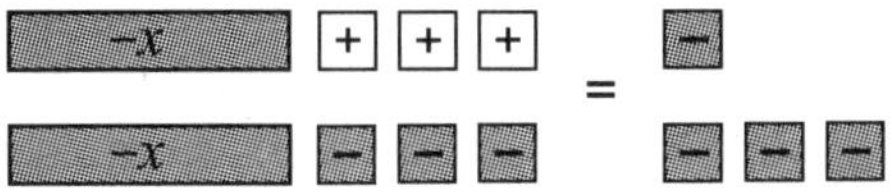

Step 2: Place three red squares on each side to isolate the red rectangles.

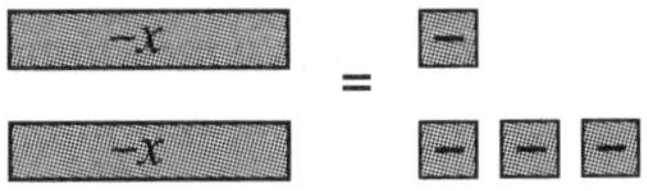

Step 3: Remove the zero pairs.

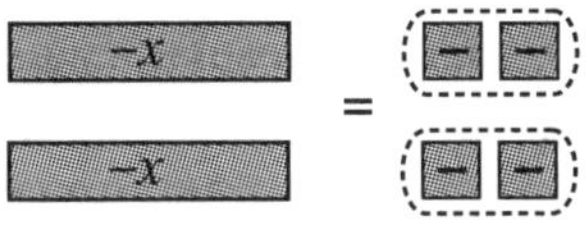

Step 4: Divide the square tiles into two equal groups. The model represents $-x = -2$.

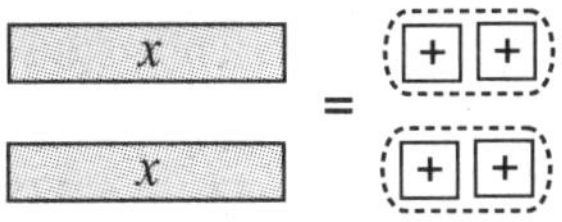

Step 5: Solve for x. Change the two red rectangles to green rectangles. Also, change the red squares to yellow squares. The solution is $x = 2$.

Answer:

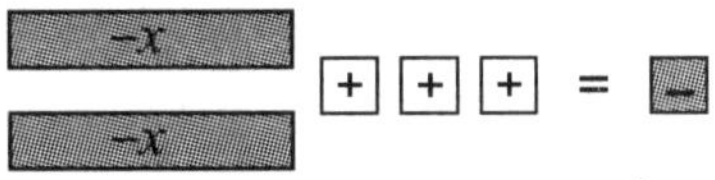

$x = 2$

Problem Set

Model and solve the given equations using algebra tiles:

1. $x+3 = -1$

2. $x+4 = -3$

3. $x+7 = 5$

4. $x+5 = 2$

5. $x-3 = 4$

6. $x-7 = -3$

7. $x-2 = -5$

8. $x-1 = -3$

9. $2x = 6$

10. $2x = -10$

11. $3x = 9$

12. $4x = 4$

13. $4x = -4$

14. $2x = -12$

15. $4x = 12$

Use algebra tiles to model and solve the problems stated below.

16. Brandon is four years older than his sister. If Brandon is seven, how old is his sister?

17. Danielle is twice as old as her brother. If Danielle is twelve, how old is her brother?

18. Mark worked 6 more hours this week than he did last week. Last week he worked 10 hours. How many hours did he work this week?

19. Brenton worked 3 less hours this week than he did last week. Last week he worked eleven hours. How many hours did he work this week?

20. Jared has three times as much money as Terrence. If Jared has $18.00, how much money does Terrence have?

21. Monique spent $9.00 renting videos. If she has $7.00 left, how muchg did she originally have?

22. Taylor is three years younger than her sister. If Taylor is ten, how old is her sister?

Model and solve the given equations using algebra tiles:

23. $2(x-1) = 4$

24. $2(x-2) = 6$

25. $2(x+1) = 6$

26. $2(x+2) = 6$

27. $2(x-1) = -6$

28. $3(x-1) = -6$

29. $3(x+2) = 3$

30. $3(x+3) = -3$

31. $-2x+1 = 3$

32. $-2x-1 = 5$

33. $-2x-2 = 6$

34. $-2x+3 = 5$

35. $-2x+1 = -1$

36. $-2x+3 = -3$

37. $-3x+5 = 2$

38. $-3x+1 = -5$

39. $-3x-2 = -5$

40. $-3x-1 = 2$

HA1-125: Solving Equations Using More Than One Property

You remember from order of operations that we often have to carry out more than one step to complete a problem, and that the order in which we perform those steps is important. The same is true for solving algebraic equations that require more than one step. There are certain procedures that you must follow, and following them in order will ensure a correct answer.

Let's look at an example. Imagine you are planning a trip to the art museum. The school has agreed to contribute \$125 to the cost of the trip. The total cost is \$375. If a total of 20 students plan to go to the museum, how much money must each student contribute? Let m represent the amount of money each student needs to pay. First, let's set up the equation.

125	+	$20m$	=	375
Amount the school will give		Amount students pay		Total cost of trip

Now solve the equation.

Example 1 Solve $125 + 20m = 375$ to determine how much money each student must contribute towards the trip to the art museum.

$$125 - 125 + 20m = 375 - 125$$
$$20m = 250$$

Step 1: Since neither side can be simplified, subtract 125 from both sides of the equation to isolate the term containing the variable.

$$\frac{20m}{20} = \frac{250}{20}$$

Step 2: The variable has not been isolated. Divide both sides of the equation by the coefficient of the variable.

$$m = \$12.50$$

Step 3: Use your knowledge of signed numbers to evaluate the expression.

$$\$125 + 20m = \$375$$
$$\$125 + 20(\$12.50) = \$375$$
$$\$125 + \$250 = \$375$$
$$\$375 = \$375$$

Step 4: Each student will need to pay \$12.50. To verify your answer, substitute \$12.50 for m in the original equation. Be sure to follow the rules for the order of operations.

Answer: \$12.50

Let's look at an equation that contains parentheses.

Example 2 Solve: $5(h + 2) = 30$

$$5h + 10 = 30$$

Step 1: Using the distributive property, multiply 5 times h and 5 times 2.

$$5h + 10 - 10 = 30 - 10$$
$$5h = 20$$

Step 2: Subtract 10 from both sides of the equation to isolate the term containing the variable.

$$\frac{5h}{5} = \frac{20}{5}$$

Step 3: The variable has not been isolated. Divide both sides of the equation by the coefficient of the variable.

$$h = 4$$

Step 4: Use your knowledge of signed numbers to evaluate the expression.

$$5(h+2) = 30$$
$$5(4+2) = 30$$
$$5(6) = 30$$
$$30 = 30$$

Step 5: Check your answer.

Answer: The equation is true; the solution set is $\{4\}$.

Steps for Solving Equations	1. Use the Distributive Property to remove grouping symbols. 2. Use the Addition and Subtraction Properties for Equations to isolate the term that contains the variable. 3. Use the Multiplication and Division Properties for Equations to isolate the variable. 4. Check your answer.

Let's try a few more examples.

Example 3 Solve: $-12 = -3 - r$

$$-12 + 3 = -3 - r + 3$$
$$-9 = -r$$

Step 1: The distributive property is not needed, so the first step is to add 3 to both sides of the equation to isolate the term with the variable.

$$\frac{-9}{-1} = \frac{-r}{-1}$$

Step 2: Divide both sides by -1 to isolate the variable.

$$9 = r$$

Step 3: Use your knowledge of signed numbers to evaluate the expression.

$$-12 = -3 - r$$
$$-12 = -3 - 9$$
$$-12 = -12$$

Step 4: Check your answer.

Answer: The equation is true; the solution set is $\{9\}$.

Example 4 Solve: $\frac{d}{6} - 3 = 7$

$$\frac{d}{6} - 3 + 3 = 7 + 3$$
$$\frac{d}{6} = 10$$

Step 1: The distributive property is not needed, so the first step will be to add 3 to both sides of the equation to isolate the term with the variable.

$$\frac{6}{1} \cdot \frac{d}{6} = 10 \cdot 6$$

Step 2: Multiply both sides by 6 to isolate the variable.

$$d = 60$$

Step 3: Use your knowledge of signed numbers to evaluate the expression.

$$\frac{d}{6} - 3 = 7$$

$$\frac{60}{6} - 3 = 7$$

$$10 - 3 = 7$$

$$7 = 7$$

Step 4: Check your answer.

Answer: The equation is true; the solution set is {60}.

Example 5 Solve: $45 = 5(2b + 3) + 10$

$45 = 10b + 15 + 10$

Step 1: Using the distributive property, multiply 5 times 2*b* and 5 times 3 to remove the parentheses.

$45 = 10b + 25$

Step 2: An extra step is needed to simplify the right side by combining 15 and 10.

$45 - 25 = 10b + 25 - 25$

$20 = 10b$

Step 3: Subtract 25 from both sides of the equation to isolate the term with the variable.

$$\frac{20}{10} = \frac{10b}{10}$$

Step 4: Divide both sides by 10 to isolate the variable.

$2 = b$

Step 5: Use your knowledge of signed numbers to evaluate the expression.

$45 = 5(2b + 3) + 10$

$45 = 5(2 \cdot 2 + 3) + 10$

$45 = 5(4 + 3) + 10$

$45 = 5(7) + 10$

$45 = 35 + 10$

$45 = 45$

Step 6: Check your answer.

Answer: The equation is true; the solution set is {2}.

Problem Set

Solve:

1. $7 - h = -6$

2. $3x - 2 = 16$

3. $-3x - 6 = 21$

4. $8 = -1 - x$

5. $-4 = 6 - c$

6. $\frac{1}{2}h + 7 = 15$

7. $2h - 7 = 15$

8. $2g - 11 = 47$

9. $3f - 15 = 51$

10. $-4x - 26 = 54$

11. $4(x + 8) = 20$

12. $70 = 5(c - 4)$

13. $3(r - 4) = 27$

14. $3(r + 4) = 21$

15. $9(6 - r) = 54$

16. $9(r - 6) = 54$

17. $\frac{1}{4}(b + 6) = 12$

18. $2.4 = 0.8 - 0.4b$

19. $1.5(2.4 - m) = 7.5$

20. $10 - \frac{2}{5}y = 30$

HA1-130: Identifying Postulates, Theorems, and Properties

Every field of study has its own language or jargon. To be successful in a particular field or subject, you must understand the terminology. Since algebra is a system of statements represented as numbers, you must be able to comprehend the terminology used to present these statements as well as the rules governing these statements. These statements and rules are the basic processes you will use to solve equations and inequalities.

Many problems in algebra can be solved using **postulates** or **theorems**.

Postulate	A statement that we assume, without proof, to be true

An example of a postulate is "a number is equal to itself." You do not need to prove this; it is assumed to be true.

Theorem	A statement that must be proven true using definitions, postulates, rules of logic, and other theorems

Once a theorem is proven, it can be used to prove other theorems. Several methods can be used to prove theorems. One method is called the **direct proof**.

Direct Proof	The steps we take to prove a true conclusion based on one or more true **hypotheses** (assumptions)

A direct proof begins with a hypothesis. In a direct proof, hypotheses are called "given statements." The hypotheses are followed by a series of logical steps, which lead to the **conclusion**, or the final statement of what you are trying to prove.

$a+b = b+a$	The **Commutative Property of Addition** states that changing the order in which numbers are added does not change the sum.
$ab = ba$	The **Commutative Property of Multiplication** states that changing the order in which numbers are multiplied does not change the product.
$y+0 = y$	The **Identity Property of Addition**, also known as the **Additive Identity**, states that when zero is added to any real number, the result is that same real number.
$y \cdot 1 = y$	The **Identity Property of Multiplication**, also known as the **Multiplicative Identity**, states that when a real number is multiplied by one, the result is that same real number.
$y+(-y) = 0$	The **Inverse Property of Addition**, also known as the **Additive Inverse**, states that if a number is added to its opposite, the result is zero.
$y \cdot \frac{1}{y} = 1$	The **Inverse Property of Multiplication**, also known as the **Multiplicative Inverse**, states that a real number multiplied by its reciprocal is equal to one.

$(a+b)+c = a+(b+c)$	The **Associative Property of Addition** states that numbers may be grouped in different ways and then added without changing the sum.
$(ab)c = a(bc)$	The **Associative Property of Multiplication** states that numbers may be grouped in different ways and then multiplied without changing the product.
$a(b+c) = ab+ac$	The **Distributive Property** states that a number being multiplied by the sum of two addends will result in the sum of the product of each addend and the number.

In the examples below, note the property you would use to justify the equation.

$(-2)+(2) = 0$	Inverse Property of Addition
$\frac{2}{3}\cdot\frac{3}{2} = 1$	Inverse Property of Multiplication
$15+0 = 15$	Identity Property of Addition
$2(5a+3) = 10a+6$	Distributive Property
$(3+6)+7 = 16$ $3+(6+7) = 16$	Associative Property of Addition
$4+5 = 9$ $5+4 = 9$	Commutative Property of Addition

Problem Set

Solve:

1. Which property states that the sum of zero and any real number is equal to that number?

   Commutative Property of Addition

   Additive Inverse

   Additive Identity

   Associative Property of Addition

2. Which property states that the product of one and any real number is equal to that number?

   Commutative Property of Multiplication

   Associative Property of Multiplication

   Multiplicative Identity

   Multiplicative Inverse

3. Name the property of addition illustrated in the given statement.

$$a + b + c = c + b + a$$

Associative Property of Addition

Commutative Property of Addition

Additive Identity

Additive Inverse

4. Name the property of addition illustrated in the given statement.

$$a + b + c = b + c + a$$

Associative Property of Addition

Commutative Property of Addition

Additive Identity

Additive Inverse

5. Name the property of addition illustrated in the given statement.

$$(k + j) + m = k + (j + m)$$

Commutative Property of Addition

Associative Property of Addition

Additive Identity

Additive Inverse

6. Name the property of addition illustrated in the given statement.

$$(m + n) + o + p = m + (n + o) + p$$

Commutative Property of Addition

Associative Property of Addition

Additive Inverse

Additive Identity

7. Name the property illustrated in the given statement.

$$9 \cdot \frac{1}{9} = 1$$

Multiplicative Identity

Commutative Property of Multiplication

Associative Property of Multiplication

Multiplicative Inverse

8. Name the property illustrated in the given statement.

$$16 \cdot \frac{1}{16} = 1$$

Associative Property of Multiplication

Commutative Property of Multiplication

Multiplicative Inverse

Multiplicative Identity

9. Name the property illustrated in the given statement.

$$8 \cdot 1 = 8$$

Multiplicative Inverse

Commutative Property of Multiplication

Multiplicative Identity

Associative Property of Multiplication

10. Name the property illustrated in the given statement.

$$\frac{1}{5} \cdot 1 = \frac{1}{5}$$

Multiplicative Inverse

Associative Property of Multiplication

Multiplicative Identity

Commutative Property of Multiplication

11. A statement that can be proven to be true by using definitions, postulates, proved theorems and rules of logic is a hypothesis.

False True

12. True or False:
The given statement in a proof is called the hypothesis.

13. Is the given statement true or false?

$$58 + 3 = 3 + 58$$

14. Is the given statement true or false?

$$n - m = m - n$$

15. Is the given statement true or false?

$$(k + j) + m = k + (j + m)$$

16. Is the given statement true or false?

$$m + n - o + p = m + o - n + p$$

17. Is the given statement true or false?

$$16 \cdot \frac{6}{16} = 1$$

18. The statement which we wish to verify in a proof is called the ________________.

Hypothesis Theorem

Postulate Conclusion

19. Name the property illustrated in the given statement:

$$n + mx = mx + n$$

Commutative Property of Multiplication

Associative Property of Multiplication

Associative Property of Addition

Commutative Property of Addition

20. Name the property illustrated in the given statement:

$$\left[6\left(\frac{1}{3}\right)\right]2 = 6\left[\frac{1}{3}(2)\right]$$

Multiplicative Identity

Multiplicative Inverse

Associative Property of Multiplication

Commutative Property of Multiplication

HA1-135: Evaluating Formulas

By now, you are familiar with formulas. Recall that a formula is a mathematical statement that expresses a relationship among quantities. Each variable represents a specific quantity in the formula.

Example 1 Mrs. Martinez wants to enclose a rectangular garden with a white fence. The garden is 14 feet long and 10 feet wide. How many feet of fencing does she need? Use the formula for perimeter of a rectangle: $P = 2l + 2w$.

$\mathrm{P} = 2l + 2w$ **Step 1:** Start with the formula.

$\mathrm{P} = 2(14) + 2(10)$ **Step 2:** Substitute the values for the variables: $l = 14$ and $w = 10$.

$\mathrm{P} = 28 + 20$ **Step 3:** Solve for the remaining variable.

$\mathrm{P} = 48$

Answer: Mrs. Martinez needs 48 feet of fencing.

Note: Include the unit of measure in your final answer if units are given.

Example 2 On a rainy day, Jaime drove home from college in three hours. If she lives 135 miles from school, what was her average speed? Use the formula $d = r \cdot t$.

$d = r \cdot t$ **Step 1:** Start with the formula.

$135 = r \cdot 3$ **Step 2:** Substitute the values for the variables: $d = 135$ and $t = 3$.

$135 = 3r$ **Step 3:** Solve for the remaining variable.

$\frac{135}{3} = \frac{3r}{3}$

$45 = r$

Answer: Jaime averaged 45 mph.

Example 3 Given the formula $\mathrm{P} = 2l + 2w$, find w if $p = 200$ and $l = 54$.

$p = 2l + 2w$ **Step 1:** Start with the formula.

$200 = 2(54) + 2w$ **Step 2:** Substitute the values for the variables: $p = 200$ and $l = 54$.

$200 = 108 + 2w$ **Step 3:** Solve for the remaining variable.

$200 - 108 = 108 + 2w - 108$

$92 = 2w$

$\frac{92}{2} = \frac{2w}{2}$

$46 = w$

Answer: $w = 46$

Example 4 Given the formula for the area of a circle, $A = \pi r^2$, find the radius, r, if $A = 16\pi$ square feet.

$A = \pi r^2$

Step 1: Start with the formula.

$16\pi = \pi r^2$

Step 2: Substitute the values for the variables.

$\frac{16\pi}{\pi} = \frac{\pi r^2}{\pi}$

$16 = r^2$

$\sqrt{16} = \sqrt{r^2}$

Step 3: Solve for the remaining variable.

$4 = r$

Step 4: What number squared equals 16?

Answer: The radius is 4 feet.

Example 5 Given the formula $b = \frac{a+c}{5}$, find c if $a = 9$ and $b = 6$.

$b = \frac{a+c}{5}$

Step 1: Start with the formula.

$6 = \frac{9+c}{5}$

Step 2: Substitute the values for the variables: $a = 9$ and $b = 6$.

$5 \cdot 6 = \frac{9+c}{5} \cdot \frac{5}{1}$

$30 = 9 + c$

$30 - 9 = 9 + c - 9$

$21 = c$

Step 3: Solve for the remaining variable.

Note: Since the variable is part of the numerator, you must first isolate the numerator. This is done by multiplying both sides of the equation by the denominator.

Answer: $c = 21$

Problem Set

1. The formula for finding the volume of a pyramid is $V = \frac{1}{3}bh$. Find b if $V = 144$ and $h = 12$.

2. The distance formula is $d = rt$. Find d if $r = 56$ and $t = 7$.

3. The formula for finding the area of a triangle is $A = \frac{1}{2}bh$. Find A if $b = 27$ and $h = 10$.

4. The formula for finding the volume of a rectangular prism or box is $V = lwh$.
Find w if $V = 1,200$, $l = 10$, and $h = 15$.

5. Given the formula $C = \frac{a+b}{2}$, find the value of a if $C = 40$ and $b = 10$.

6. Given the formula $A = lw$, find the value of w if $l = 4.4$ and $A = 550$.

7. Given the formula $P = 2b + 2h$, find the value of h if $P = 64$ and $b = 12$.

8. The distance formula is given by $d = rt$. Find r if $d = 550$ and $t = 5.5$.

9. Given the formula $P = 2L + 2W$, find the value of W if $P = 232$ and $L = 70$.

10. Solve the formula $A = \frac{1}{2}h(a + b)$ for h if $A = 120$, $a = 7$, and $b = 13$.

11. The formula for total receipts is $T = 7A + 3C$, where T is total receipts, A is the number of adult tickets sold, and C is the number of children tickets sold. Find A when $T = 208$ and $C = 32$.

12. The formula for wages is $W = S + 2C$, where W is wages, S is salary, and C is commission. Find S when $W = \$1,175$ and $C = \$325$.

13. The formula for finding the number of bricks needed to cover a wall is $N = 7LH$, where N is the number of bricks needed, L is the length of the wall in feet, and H is the height of the wall in feet. Find H when $L = 12$ feet and $N = 672$.

14. The formula for finding the number of bricks needed to cover a wall is $N = 7LH$, where N is the number of bricks needed, L is the length of the wall in feet, and H is the height of the wall in feet. Find H when $L = 12$ feet and $N = 840$.

15. Given the formula $A = P(1 + r)^4$, find the value of P if $A = 16,200$ and $r = 2$.

16. The formula for finding the area of a right circular cylinder is $T = 2\pi rh + 2\pi r^2$. Find h if $T = 156\pi$ and $r = 6$.

17. Given the formula $A = \frac{24f}{b(p + 1)}$, find the value of f if $A = 0.1$, $b = 1,920$, and $p = 24$.

18. The distance required for a moving automobile to stop is affected by its speed in miles per hour, M, and a drag factor, f. The drag factor, f, is determined by the type of road and its condition. The drag factor, f, of each type of road and condition is based on the stopping distance in feet, D, and the speed of the car, M. The drag factor, f, can be calculated by using the formula $30fD = M^2$. Find the drag factor of an icy road if a car traveling 60 miles per hour requires 1,000 feet of stopping distance.

19. Mr. Johnson pays a base price, b, of \$19.95 per month to use his cell phone. He also pays \$0.10 per minute, m. He wants to determine the number of minutes he used last month. His charge, c, for the month was \$49.45. Use the formula $c = b + 0.10m$ to determine the number of minutes he used last month.

20. A rental car agency charges a fee, c, to rent a car. This fee is based on the number of days the car is rented, d; a daily rate, r; and the number of miles driven, m. The mileage charge is \$0.25 per mile driven. Using the formula $c = dr + 0.25m$, determine the maximum number of miles Shenita can drive if she has \$150 to rent a car from an agency that charges a daily rate of \$20 for 5 days.

HA1-140: Solving Equations by Combining Like Terms

You may have realized that to solve equations, you must combine several algebra skills. As you have solved equations, you have applied your knowledge of postulates and basic arithmetic functions to solve for the variable. Now, another skill is required – combining like terms. Recall from a previous lesson that like terms may or may not always have common coefficients, but they do contain the same variables, raised to the same powers. To combine like terms, add or subtract the coefficients as indicated.

Some questions ask you to solve for the variable in an equation, while others only require that you simplify an expression. The first two examples are simplification-only questions.

Example 1 Combine like terms and simplify: $6x + 2(x + 7)$

$= 6x + 2x + 14$ **Step 1:** Distribute the 2 to all terms in the parentheses.

$= 8x + 14$ **Step 2:** Combine the like terms $6x$ and $2x$. The expression cannot be simplified any further; thus, $8x + 14$ is in simplest form.

Answer: $8x + 14$

Example 2 Combine like terms and simplify: $5x + 2(9x + 12y)$

$= 5x + 18x + 24y$ **Step 1:** Distribute the 2 to all terms in the parentheses.

$= 23x + 24y$ **Step 2:** Combine the like terms $5x$ and $18x$. The expression cannot be simplified any further; thus, $23x + 24y$ is in simplest form.

Answer: $23x + 24y$

As you solve equations in this lesson, you must look for opportunities to simplify the equation by combining like terms. Sometimes you may be required to combine like terms before you begin applying any properties; at other times you may combine like terms later, during the simplification process. In other words, you must determine the steps necessary to simplify the equation. The following steps are useful when solving equations with like terms:

Step 1: Simplify both sides of the equation if possible.
Step 2: Use the Addition and Subtraction Properties for Equations to isolate the term that contains the variable.
Step 3: Use the Multiplication and Division Properties for Equations to isolate the variable.
Step 4: Check the answer.

Now use these steps to solve the following equations.

Example 3 Solve for x: $12x - 17x = 20$

$-5x = 20$ **Step 1:** Combine like terms $12x$ and $17x$.

$$\frac{-5x}{-5} = \frac{20}{-5}$$
$$x = -4$$

Step 2: Divide both sides of the equation by –5.

$$
\begin{aligned}
12x - 17x &= 20 \\
12(-4) - 17(-4) &= 20 \\
-48 + 68 &= 20 \\
20 &= 20
\end{aligned}
$$

Step 3: Check your answer.

Answer: $x = -4$

Example 4 Solve for m: $-15 = (8m + 2) - (6m - 7)$

$-15 = 8m + 2 - 6m + 7$

Step 1: Distribute the –1 to all terms in the second parentheses.

$-15 = 2m + 9$

Step 2: Combine like terms $8m$ and $-6m$, and 2 and 7.

$$
\begin{aligned}
-15 - 9 &= 2m + 9 - 9 \\
-24 &= 2m
\end{aligned}
$$

Step 3: Subtract 9 from both sides of the equation to isolate the term containing the variable.

$-\frac{24}{2} = \frac{2m}{2}$

Step 4: The variable has not been isolated. Divide both sides of the equation by the coefficient of the variable.

$-12 = m$

Step 5: Use your knowledge of signed numbers to evaluate the expression.

$$
\begin{aligned}
-15 &= (8m + 2) - (6m - 7) \\
-15 &= (8(-12) + 2) - (6(-12) - 7) \\
-15 &= (-96 + 2) - (-72 - 7) \\
-15 &= -94 - (-79) \\
-15 &= -94 + 79 \\
-15 &= -15
\end{aligned}
$$

Step 6: Check your answer.

Answer: $m = -12$

Example 5 Solve for m: $12 = m - (3m + 8)$

$12 = m - 3m - 8$

Step 1: Distribute the –1 to all terms in the parentheses.

$12 = -2m - 8$

Step 2: Combine like terms m and $-3m$.

$$
\begin{aligned}
12 + 8 &= -2m - 8 + 8 \\
20 &= -2m
\end{aligned}
$$

Step 3: Add 8 to both sides of the equation to isolate the term containing the variable.

$\frac{20}{-2} = \frac{-2m}{-2}$

Step 4: The variable has not been isolated. Divide both sides of the equation by the coefficient of the variable.

$-10 = m$

Step 5: Use your knowledge of signed numbers to evaluate the expression.

$12 = m - (3m + 8)$ $12 = -10 - (3(-10) + 8)$ $12 = -10 - (-30 + 8)$ $12 = -10 - (-22)$ $12 = 12$	**Step 6:** Check your answer.

Answer: $m = -10$

Example 6 John's electric bill for the past two years was $1,500. If the bill for last year was $300 more than the bill for the previous year, what was the cost of electricity the previous year?

x = cost of electricity two years ago $x + 300$ = cost of electricity last year	**Step 1:** Define the variable.
$x + (x + 300) = 1{,}500$	**Step 2:** Write an equation to represent the word problem.
$x + x + 300 = 1{,}500$ $2x + 300 = 1{,}500$	**Step 3:** Simplify the left side of the equation by combining like terms $x + x$.
$2x + 300 - 300 = 1{,}500 - 300$ $2x = 1{,}200$	**Step 4:** Subtract 300 from both sides of the equation to isolate the term with the variable.
$\frac{2x}{2} = \frac{1{,}200}{2}$	**Step 5:** Divide both sides by 2 to isolate the variable.
$x = 600$	**Step 6:** Use your knowledge of signed numbers to evaluate the expression.
$x + 300 =$ $= 600 + 300$ $= 900$ $\$600 + \$900 = \$1{,}500$	**Step 7:** Check your answer. Was last year's bill $300 more than the previous year? Yes. Does the cost of electricity for the two years equal $1,500? Yes.

Answer: The cost of electricity two years ago was $600.

Problem Set

Solve the following equations:

1. $2x + 4x = 24$
2. $16 = 12x - 4x$
3. $28 = 5d - 9d$
4. $3c + 4c = -84$
5. $d + d - 9 = 5$
6. $4k - 2k - 8 = 48$
7. $m + 5m = -60$
8. $m + 5m - 24 = 60$
9. $3c - 6c = 36$
10. $3d - 6d - 9 = 36$
11. $3(x + 5) = 15$
12. $16 = 2(w - 7)$
13. $(x + 2) + (x + 4) = 34$
14. $3(2n + 4) = 8$
15. $(2x + 6) + (3x - 9) = 18$
16. $-10 = (a - 1) + (a - 3)$
17. $2x + (6x - 16) = -24$
18. $(n - 2) + (3n + 5) - 9 = 2$
19. $5b - 3(b - 2) = 34$
20. $18 = 6f - 2(f + 3)$

HA1-144: Using a Concrete Model to Solve Equations with Variables on Both Sides

Algebra tiles can be used to solve equations with variables on both sides. Various expressions in algebra are represented by tiles of particular color and shapes. A green rectangle represents the variable x. A red rectangle represents the opposite of the variable x, or $-x$. A yellow square represents 1, and a red square represents -1.

Expression	Tile
x	x
$-x$	$-x$
1	+
-1	−

When solving equations with algebra tiles, we can remove zero pairs, which occur on the same side of the equation. A green rectangle, which represents x, and a red rectangle, which represents $-x$, form a zero pair because $x + (-x) = 0$. Also, a yellow square, which represents 1, and a red square, which represents -1, form a zero pair because $1 + (-1) = 0$.

Zero Pairs	x and $-x$ + and −

Example 1 Is $x = -2$ a solution to the equation that is modeled by algebra tiles?

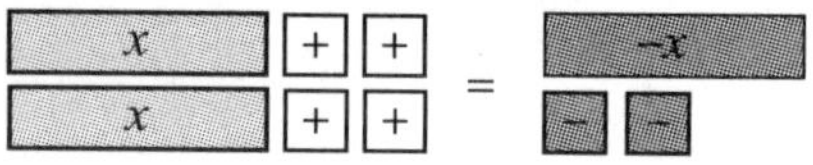

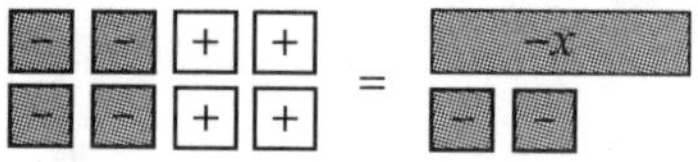

Step 1: Because $x = -2$ means a green rectangle equals two red squares, substitute two red squares for each green rectangle in the equation.

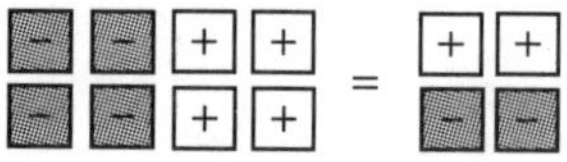

Step 2: Since $x = -2$, $-x = 2$. This means that a red rectangle equals two yellow squares, so substitute two yellow squares for the red rectangle in the equation.

$$-4 + 4 = 2 + (-2)$$
$$0 = 0$$

Step 3: Evaluate each side of the equation based on the tile values. Recall that each red square represents -1 and each yellow square represents 1.

Answer: $x = -2$ is a solution to the equation.

Example 2 Model and solve the equation using algebra tiles: $-3x + 2 = -2x - 1$.

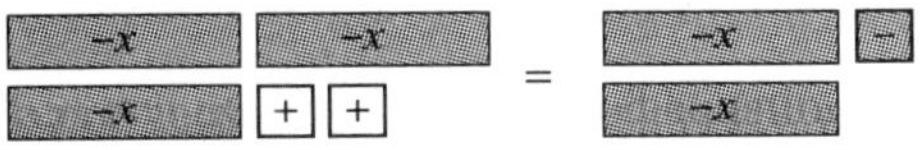

Step 1: Model the equation. Place three red rectangles and two yellow squares on the left of the equation to represent $-3x + 2$. Place two red rectangles and one red square on the right of the equation to represent $-2x - 1$.

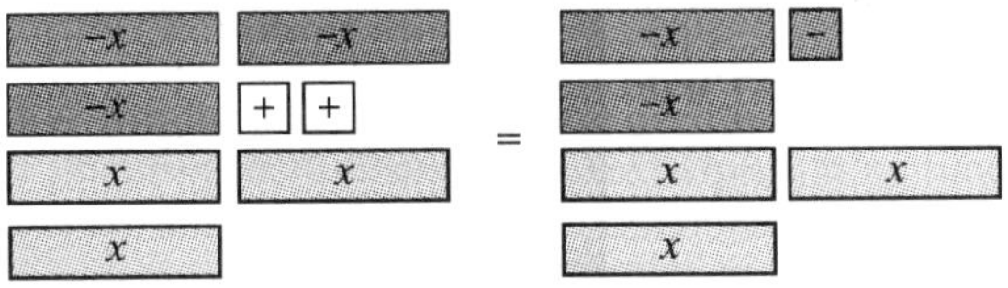

Step 2: Place three green rectangles on each side to eliminate the red rectangles by removing zero pairs.

Step 3: Remove the zero pairs.

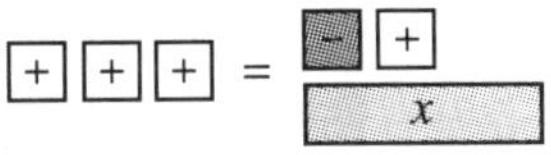

Step 4: Place one yellow square on each side to eliminate the red square by removing the zero pair. This will isolate the green rectangle, which represents x.

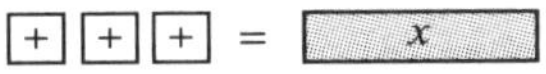

Step 5: Remove the zero pair. We are left with $x = 3$.

Answer:

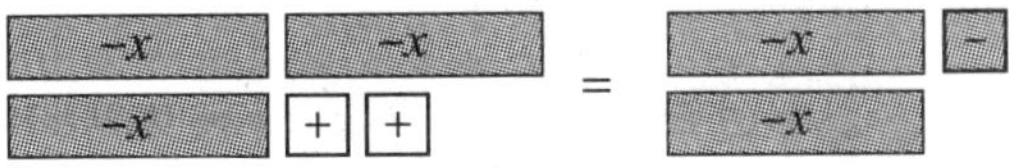

$x = 3$

Example 3 Model and solve the equation using algebra tiles: $3x + 6 = 5x$.

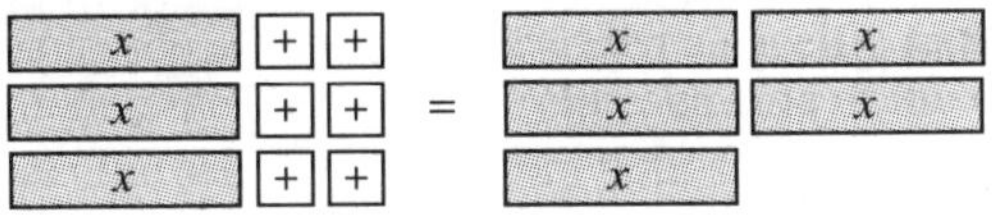

Step 1: Model the equation. Place three green rectangles and six yellow squares on the left side of the equation to represent $3x + 6$. Place five green rectangles on the right side of the equation to represent $5x$.

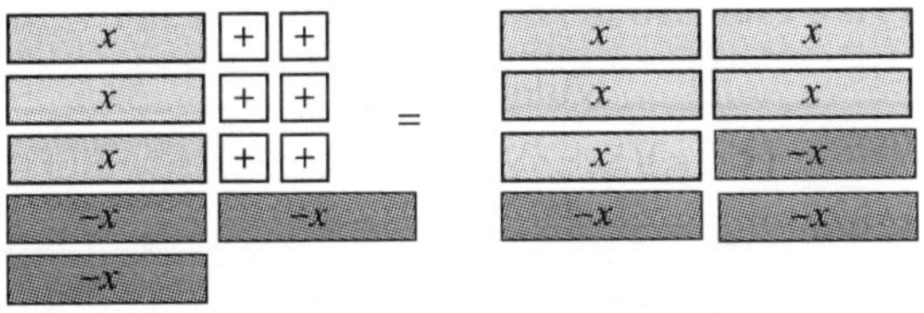

Step 2: Place three red rectangles on each side of the equation to get all of the green rectangles, which represent x, on one side.

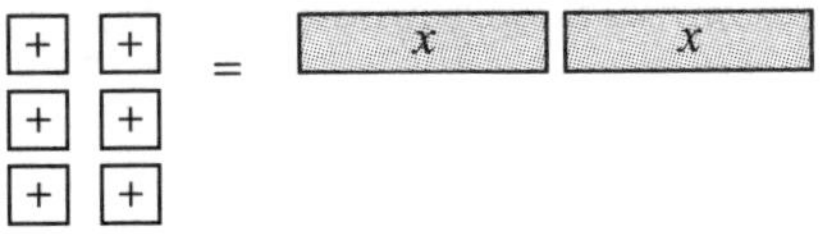

Step 3: Remove the zero pairs.

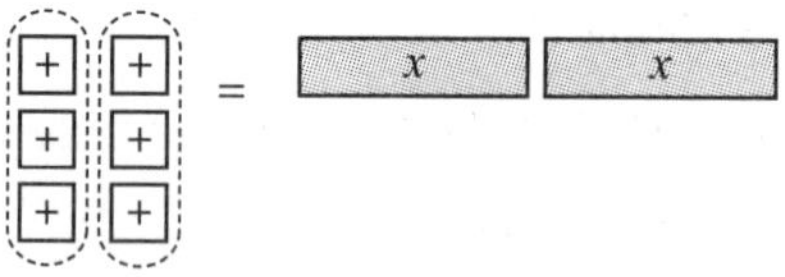

Step 4: Divide the square tiles into two equal groups to find the value of one green rectangle, which represents x. One green rectangle represents 3, or $x = 3$.

Answer:

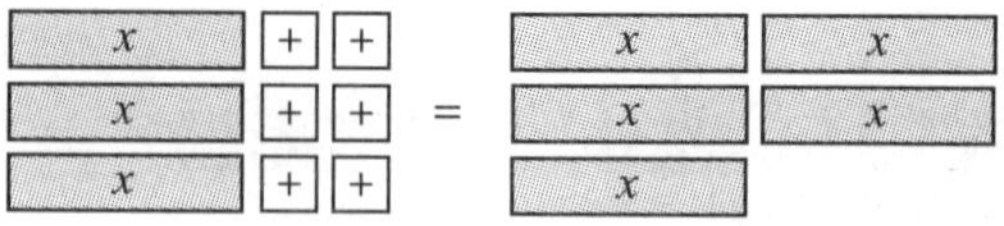

$x = 3$

Example 4 Model and solve the equation using algebra tiles: $2(x - 3) = 3x - 2x$.

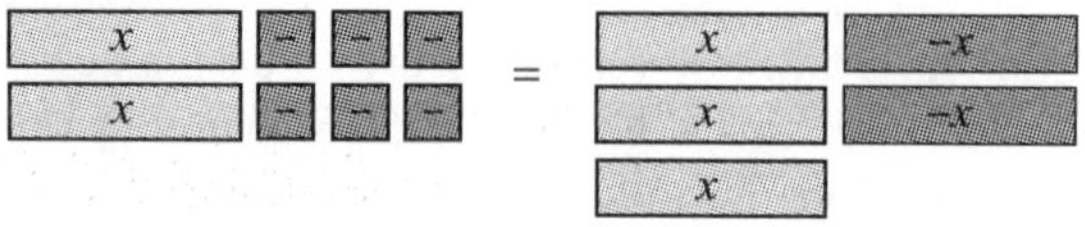

Step 1: Model the equation. Place two sets of tiles, consisting of a green rectangle and three red squares, on the left to represent $2(x - 3)$. Place three green rectangles and two red rectangles on the right to represent $3x - 2x$.

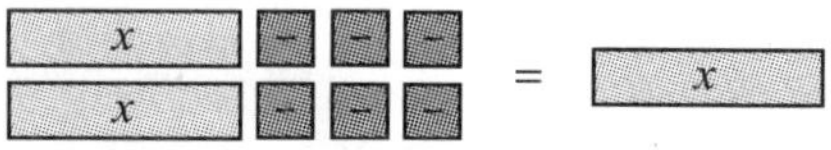

Step 2: Remove the zero pairs on the right side of the equation in order to simplify.

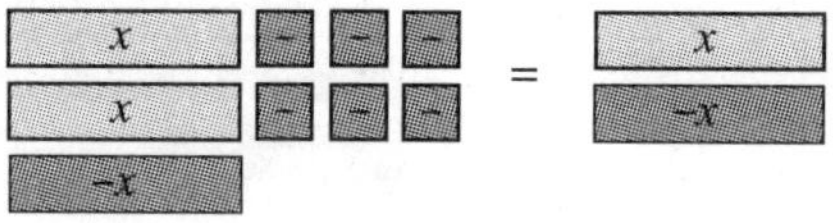

Step 3: Place one red rectangle on each side of the equation to get all of the green rectangles, which represent x, on one side.

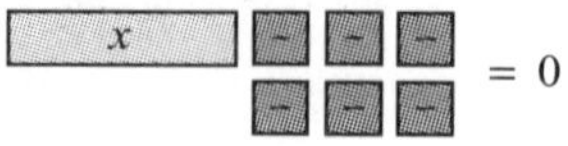

Step 4: Remove the zero pairs.

Note: If all tiles are removed from one side of the equation, the value is 0.

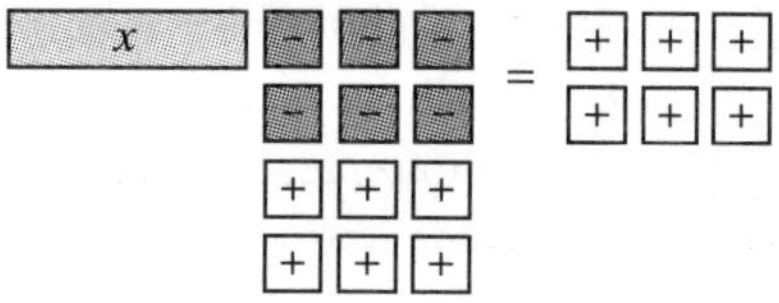

Step 5: Place six yellow squares on each side of the equation in order to isolate the green rectangle, which represents x.

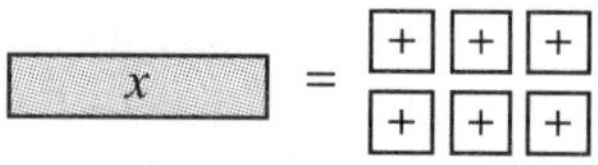

Step 6: Remove the zero pairs. We are left with $x = 6$.

Answer:

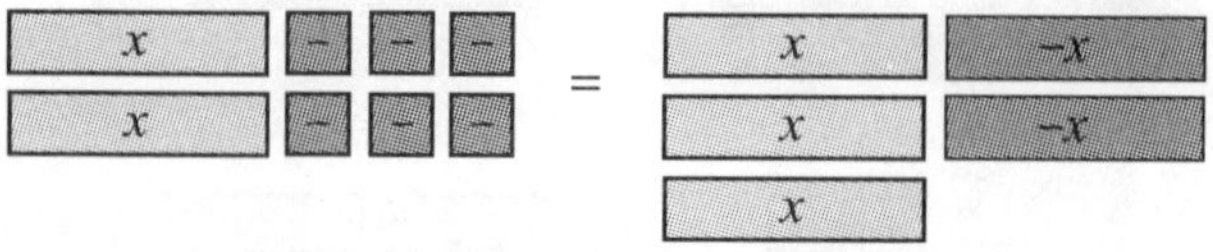

$x = 6$

Example 5 Use algebra tiles to model and solve the given word problem. Let x represent the number.

Two decreased by a number is equal to the number increased by four. Find the number.

Step 1: Translate the phrase "two decreased by a number" into the algebraic expression $2 - x$. Model it with algebra tiles using two yellow squares and one red rectangle.

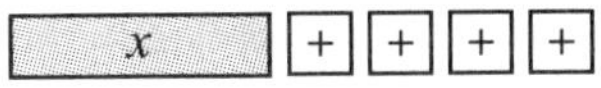

Step 2: Translate the phrase "the number increased by four" into the algebraic expression $x + 4$. Model it with algebra tiles using one green rectangle and four yellow squares.

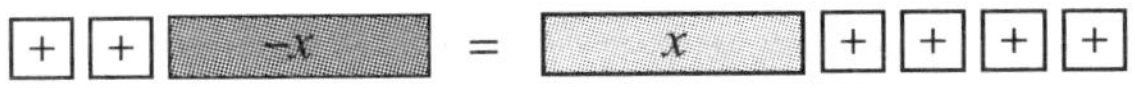

Step 3: The word problem says that the two expressions are equal, so set the two sets of tiles equal to each other. Then solve using algebra tiles. This is the model.

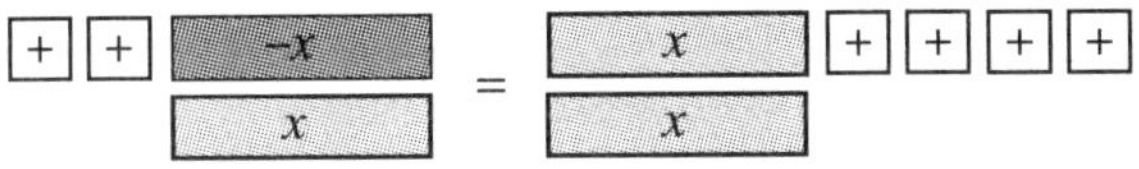

Step 4: Place one green rectangle on each side of the equation to eliminate the red rectangle.

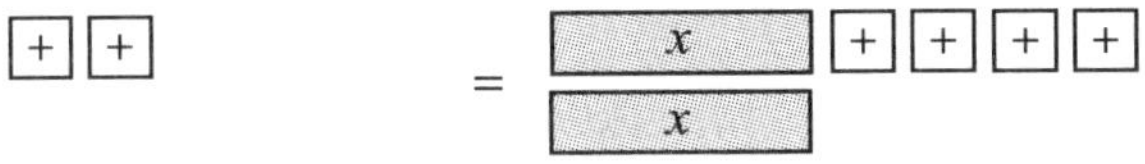

Step 5: Remove the zero pair on the left side.

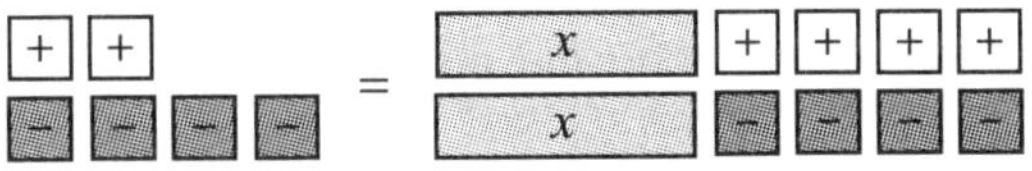

Step 6: Place four red squares on each side in order to isolate the green rectangles, which represent x.

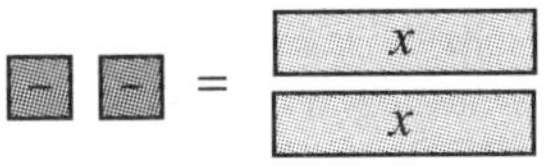

Step 7: Remove the zero pairs.

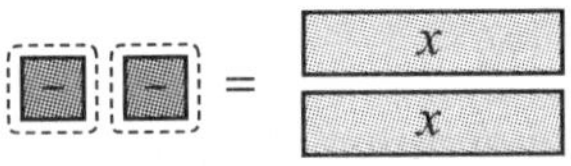

Step 8: Divide the red squares into two equal groups. We can see that one green rectangle is equal to one red square, so we are left with $x = -1$.

Answer:

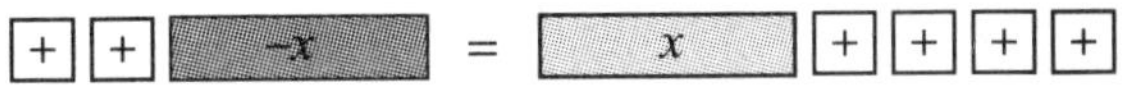

$x = -1$

Problem Set

Find the value of x that is the solution to the equation that is modeled by the algebra tiles below:

1. 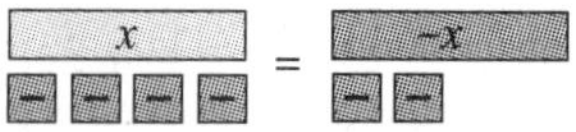

2. 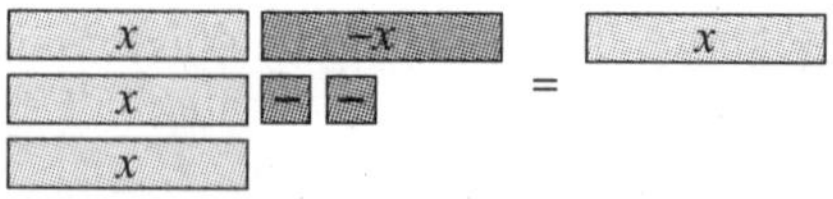

3. 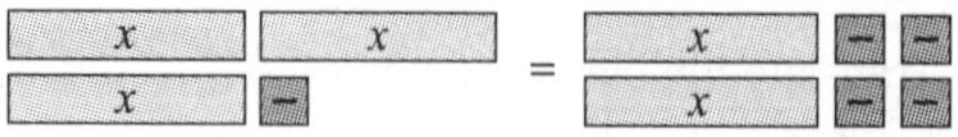

4. 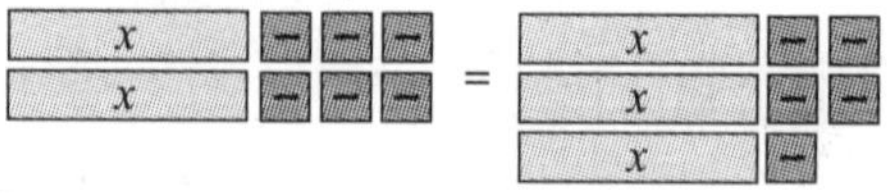

5. 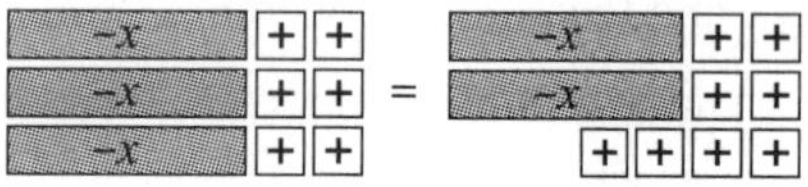

6. 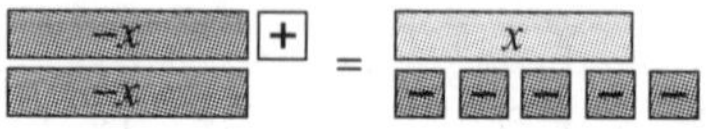

7. 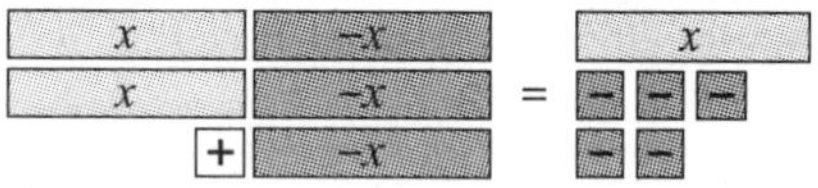

8. 

Model and solve the given equations using algebra tiles:

9. $2x - 4 = x + 2$

10. $-x - 1 = -2x + 2$

11. $3x - 5 = 2x + 2$

12. $x - 3 = 2x - 5$

13. $4 - 2x = 3 - x$

14. $-3x = -2x - 2$

15. $2x - 1 = 3x$

16. $2x + 2 = -2x + 6$

17. $5x - 3 = -x + 9$

18. $4x + 2 = -2x - 10$

19. $6x - 2 = x + 8$

20. $-2x - 2 = 4 - 4x$

21. $5x = 2x - 9$

22. $-3x + 6 = -5x - 8$

23. $4x + 1 = x - 5$

24. $2(x - 3) = 3x + 1$

25. $3(x + 2) = -x - 2$

26. $4(-x - 1) = -3x + 5$

27. $x - 4x - 2 = -x - 6$

28. $6 + 2x - 4 = 3x - 6$

29. $2 + 3x - 6 = 2x - 3$

30. $2(2 - 2x) = -x + 2x - 6$

Use algebra tiles to model and solve the problems below. Let x represent the number.

31. Twice a number decreased by one is equal to four more than the number. Find the number.

32. Two more than three times a number is equal to twice the number increased by one. Find the number

33. Twice the difference of a number and two is three less than the number. Find the number.

34. Three times the sum of a number and one is the same as twice the number. Find the number.

35. If a number is decreased by two and then the difference is multiplied by three, the result is the same as they number itself. Find the number.

36. "If a number is decreased by four, the result is the same as multiplying the number by three. Find the number.

37. If a number is multiplied by five, the result is the same as the number increased by eight. Find the number.

38. The sum of a number and twice the same number is the same as the number increased by six. Find the number.

39. Four times a number increased by four is the same as negative six decreased by the number. Find the number.

40. Four decreased by twice a number is equal to the number decreased by eight. Find the number.

HA1-145: Solving Equations with Variables on Both Sides

Remember that your goal when solving equations is to move the variables to one side of the equation. However, the key is to make sure that whatever you do on one side of the equation, you do on the other side. If you perform an operation on only one side of the equation, the two sides will no longer be equal. We call this balancing the equation. Let's look at a simple equation that contains variables on both sides:

Example 1 Solve for x: $2x = x + 7$

$2x - x = x + 7 - x$

Step 1: The first step is to use the Addition and Subtraction Properties to get all the terms that contain the variable, x, on one side of the equation. Since it is easier to work with positive numbers, subtract x from both sides of the equation.

$x = 7$

Step 2: Then combine like terms and simplify. Because the variable is isolated, no other work is needed.

$$\begin{aligned} 2x &= x + 7 \\ 2(7) &= (7) + 7 \\ 14 &= 14 \end{aligned}$$

Step 3: Check your answer.

Answer: The solution set is $\{7\}$.

Example 2 Solve for y: $25 - 3y = 7y + 5$

$$\begin{aligned} 25 - 3y + 3y &= 7y + 5 + 3y \\ 25 &= 10y + 5 \end{aligned}$$

Step 1: Use the Addition and Subtraction Properties to get the terms that contain the variables on the same side of the equation. Try to move the term with the smaller coefficient to the term with the larger one, using the rules of signed numbers. This will often keep the coefficient of your variable positive, thereby making the entire equation easier to solve.

$$\begin{aligned} 25 &= 10y + 5 \\ 25 - 5 &= 10y + 5 - 5 \\ 20 &= 10y \end{aligned}$$

Step 2: To isolate the term containing the variable, subtract 5 from each side of the equation.

$$\begin{aligned} \frac{20}{10} &= \frac{10y}{10} \\ 2 &= y \end{aligned}$$

Step 3: The variable y is still not isolated; it is being multiplied by 10. Divide both sides by 10 to *undo* the multiplication.

$$\begin{aligned} 25 - 3y &= 7y + 5 \\ 25 - 3(2) &= 7(2) + 5 \\ 25 - 6 &= 14 + 5 \\ 19 &= 19 \end{aligned}$$

Step 4: Check your answer.

Answer: The answer checks, so the solution set is $\{2\}$.

Example 3 Solve for n: $10n + 3 = 6n - 5$.

$$10n + 3 - 6n = 6n - 5 - 6n$$
$$4n + 3 = -5$$

Step 1: Use the Addition and Subtraction Properties to get the variables on one side of the equation by subtracting $6n$ from both sides.

$$4n + 3 - 3 = -5 - 3$$

Step 2: Subtract 3 from both sides of the equation.

$$\frac{4n}{4} = \frac{-8}{4}$$
$$n = -2$$

Step 3: Isolate the variable by dividing both sides by 4.

$$10n + 3 = 6n - 5$$
$$10(-2) + 3 = 6(-2) - 5$$
$$-20 + 3 = -12 - 5$$
$$-17 = -17$$

Step 4: Check your answer

Answer: The equation is true; the solution set is $\{-2\}$.

Example 4 Solve for h: $-4(h + 2) + 3 = 5(-2h + 3) - 2$

$$-4h - 8 + 3 = -10h + 15 - 2$$

Step 1: Distribute -4 and 5 to remove the parentheses.

$$-4h - 5 = -10h + 13$$

Step 2: Combine like terms: –8 and 3, 15 and –2.

$$-4h - 5 + 10h = -10h + 13 + 10h$$
$$6h - 5 = 13$$

Step 3: Add $10h$ to both sides of the equation.

$$6h - 5 + 5 = 13 + 5$$
$$6h = 18$$

Step 4: Add 5 to both sides of the equation to isolate the term containing the variable.

$$\frac{6h}{6} = \frac{18}{6}$$

Step 5: The variable has not been isolated. Divide both sides of the equation by the coefficient of the variable.

$$h = 3$$

Step 6: Use your knowledge of signed numbers to evaluate the expression.

$$-4(h + 2) + 3 = 5(-2h + 3) - 2$$
$$-4(3 + 2) + 3 = 5(-2 \cdot 3 + 3) - 2$$
$$-4(5) + 3 = 5(-6 + 3) - 2$$
$$-20 + 3 = 5(-3) - 2$$
$$-17 = -15 - 2$$
$$-17 = -17$$

Step 7: Check your answer.

Answer: The solution set is $\{3\}$.

Example 5 The Rainbow Paint Store has 700 gallons of paint in its inventory. Four times the number of gallons of outdoor paint is equal to 3 times the number of gallons of indoor paint. How many gallons of outdoor paint are in the inventory?

x = number of gallons of outdoor paint $700 - x$ = number of gallons of indoor paint	**Step 1:** Define the variable.
4 (gallons of outdoor paint) = 3 (gallons of indoor paint)	**Step 2:** Write an equation to represent the word problem.
$4x = 3(700 - x)$ $4x = 2{,}100 - 3x$	**Step 3:** Use the distributive property to simplify the right side of the equation.
$4x + 3x = 2{,}100 - 3x + 3x$ $7x = 2{,}100$	**Step 4:** Add $3x$ to both sides of the equation.
$\frac{7x}{7} = \frac{2{,}100}{7}$	**Step 5:** The variable has not been isolated. Divide both sides of the equation by the coefficient of the variable.
$x = 300$	**Step 6:** Use your knowledge of signed numbers to evaluate the expression.
$4(300) = 3(700 - 300)$ $4(300) = 3(400)$ $1{,}200 = 1{,}200$ $300 + 400 = 700$	**Step 7:** Check your answer. Is 4 times the number of gallons of outdoor paint equal to 3 times the number of gallons of indoor paint? Yes. Is their sum 700? Yes.
	Answer: There are 300 gallons of outdoor paint in the inventory.

Problem Set

Solve the equations:

1. $3h - 12 = 5h$ **2.** $2h - 12 = 5h$ **3.** $2d - 20 = 6d$ **4.** $4j + 36 = 7j$

5. $2b + 18 = 5b$ **6.** $7w + 60 = -5w$ **7.** $2w - 56 = -5w$ **8.** $2y = -5y - 63$

9. $2x = -8x - 80$ **10.** $-r = -9r - 24$ **11.** $4s + 12 = 6s - 18$ **12.** $3n - 6 = 6n - 8$

13. $5s + 2 = 8(s - 2)$ **14.** $8(x + 2) = 7x$ **15.** $5x - 26 = 2x - 5$ **16.** $2a - 40 = 3a - 17$

17. $10f + 4 = 3f - 17$ **18.** $6(r - 3) - 2 = -4 + 2r$

19. $7m + 3 + 2m = 7m - 4 + m$ **20.** $6k + 7 + 2k = 5k - 9 - k$

HA1-150: Writing an Equation to Solve Word Problems

Math is everywhere! Simple activities such as reading the newspaper, going shopping, or eating at a restaurant sometimes require problem-solving skills. Being able to understand certain math words or phrases will help you write equations to solve such problems. As you read word problems, be sure you understand what you are being asked to find before you proceed. This will be your variable, or the unknown information. Then, using key terms, translate the word problem into an equation and begin solving for the variable.

Example 1 Many football teams contribute both time and money to charitable organizations. Suppose a team contributes $25 to an organization for every yard gained on punt returns. They have also contributed $5,000 to another organization. So far this year, they have contributed a total of $10,000 to these two organizations. How many yards has the punt return team gained this year?

y = number of yards gained on punt returns — **Step 1:** Define the variable.

$$\$25y + \$5,000 = \$10,000$$

Step 2: Set up the equation.

$$\$25y + \$5,000 - \$5,000 = \$10,000 - \$5,000$$

Step 3: Isolate the term with the variable by subtracting 5,000 from both sides.

$$\frac{\$25y}{\$25} = \frac{\$5,000}{\$25}$$

$$y = 200$$

Step 4: Isolate the y by dividing both sides of the equation by the coefficient of y, $25.

Step 5: Check your answer. Is the number of yards gained times $25 plus $5,000 equal to $10,000? Yes.

Answer: The team gained a total of 200 yards.

Example 1 resembles the type of equation that may be used in real-world applications. Some problems may be easier to solve than others. However, all problems can be solved by carefully reading the information and determining what you are being asked to find. To set up equations, remember the key words and phrases that indicate the arithmetic operations you use to write the algebraic expressions. You can be sure a statement is an equation if it uses the word "is," which usually means the same as "equals."

Example 2 Write the following phrase as an equation and solve: "The product of two and three times a number, x, is twelve.".

x = unknown number — **Step 1:** Define the variable.

$$(2)3x = 12$$

Step 2: Set up the equation.

$$6x = 12$$

Step 3: Simplify the left side of the equation.

$$\frac{6x}{6} = \frac{12}{6}$$

Step 4: Isolate the variable by dividing both sides of the equation by 6.

$$x = 2$$

Step 5: Solve for x.

$2(3 \cdot 2) = 2(6) = 12$

Step 6: Check: Is the product of two and three times two equal to twelve? Yes.

Answer: $(2)3x = 12$

$x = 2$

Example 3 Write the following phrase as an equation and solve: "A number decreased by the quotient of fifteen and three is ten."

x = unknown number

Step 1: Define the variable.

$$x - \frac{15}{3} = 10$$

Step 2: Set up the equation.

$$x - 5 = 10$$

Step 3: Simplify the left side of the equation.

$$x - 5 + 5 = 10 + 5$$

Step 4: Isolate the variable by adding 5 to both sides.

$$x = 15$$

Step 5: Solve for x.

$$15 - \frac{15}{3} = 10$$

$$15 - 5 = 10$$

$$10 = 10$$

Step 6: Check. Is fifteen decreased by the quotient of fifteen and three ten? Yes.

Answer: $x - \frac{15}{3} = 10$

$x = 15$

Example 4 A triangle has two sides of equal measure, and one side measuring 15 cm. If the perimeter of the triangle is 47 cm, find the lengths of the sides of equal measure. Remember, the perimeter is the sum of the lengths of the sides.

x = length of one side

Step 1: Define the variable.

$$x + x + 15 = 47$$

Step 2: Write an equation to represent the word problem.

$$2x + 15 = 47$$

Step 3: Combine like terms: x and x.

$$2x + 15 - 15 = 47 - 15$$

$$2x = 32$$

Step 4: Subtract 15 from both sides of the equations.

$$\frac{2x}{2} = \frac{32}{2}$$

Step 5: Divide both sides of the equation by 2, the coefficient of the variable.

$$x = 16$$

Step 6: Solve for x.

$16 + 16 + 15 = 47$	**Step 7:** Check: Is the perimeter 47 cm, if the 2 equal sides are 16 cm and the unequal side is 15 cm? Yes.

Answer: The triangle has two sides that measure 16 cm, and one side that measures 15 cm.

Problem Set

Solve the following:

1. Twice a number is 48. Find the number.
2. Three times a number is 45. Find the number.
3. Nineteen is 4 less than a number. Find the number.
4. Ten is the quotient of 5 and a number. Find the number.
5. Five more than twice a number is –17. Find the number.
6. Ten more than a number is 41. Find the number.
7. The difference of 4 times a number and 8 is 36. Find the number.
8. Twenty is the quotient of 100 and a number. Find the number.
9. The difference of 2 times a number and 6 is –24. Find the number.
10. Two multiplied by the difference of a number and 15 is 38. Find the number.
11. Kevin weighs 10 pounds more than Tim. The sum of their weights is 470 pounds. How much does Tim weigh?
12. Kevin weighs 10 pounds more than Tim. The sum of their weights is 470 pounds. How much does Kevin weigh?
13. Bryson's new computer costs \$178.50 less than his old one. The sum of the costs of the two computers is \$3,500. How much did his old computer cost?
14. Bryson's new computer costs \$178.50 less than his old one. The sum of the costs of the two computers is \$3,500. How much does his new computer cost?
15. Kelsey is 4 years older than Kierra. Five years ago the sum of their ages was 14. How old is Kelsey?
16. In one week, Tanya earned \$7 an hour babysitting and \$15 running errands. She earned a total of \$85 during that week. How many hours of babysitting did she work?
17. Louis put \$3,000 down on a new car. He will pay \$450 a month. The total cost of the car was \$30,000. How many months will he pay on the car?
18. A quadrilateral has two sides of equal measure. The remaining two sides are 4 and 12 meters. Given that the perimeter is 52 meters, find the length of each of the two missing sides.
19. A rectangle has length three times its width. The perimeter of the rectangle is 56 inches. Find the length of the rectangle.
20. A pentagon has three sides of equal measure. The remaining two sides are 10 and 8 centimeters. Given that the perimeter is 63 centimeters, find the length of each of the three missing sides.

HA1-155: Writing an Equation to Solve Consecutive Integer Problems

The set of integers consists of all of the whole numbers and their opposites. The word "consecutive" means to follow in order, one by one, without skipping a number. The numbers 1, 2, 3 and 4 are consecutive integers. So are the numbers –105, –104 and –103.

Consecutive Integers	Consecutive integers increase by one and can be represented by x, $x + 1$, $x + 2$, $x + 3$...

The numbers 6, 8 and 10 are consecutive even integers.

- To what number must you add six to get eight? → Two, because $8 = 6 + 2$.
- To what number must you add six to get ten? → Four, because $10 = 6 + 4$.

If you replace each six with x, you will have algebraic expressions for consecutive even integers.

$$6 \rightarrow x$$
$$8 = 6 + 2 \rightarrow x + 2$$
$$10 = 6 + 4 \rightarrow x + 4$$

Consecutive Even Integers	Consecutive even integers increase by two and can be represented by x, $x + 2$, $x + 4$...

The numbers 11, 13 and 15 are consecutive odd integers.

- To what number must you add eleven to get thirteen? → Two, because $13 = 11 + 2$.
- To what number must you add eleven to get fifteen? → Four, because $15 = 11 + 4$.

If you replace each eleven with x, you will have algebraic expressions for consecutive odd integers.

$$11 \rightarrow x$$
$$13 = 11 + 2 \rightarrow x + 2$$
$$15 = 11 + 4 \rightarrow x + 4$$

Consecutive Odd Integers	Consecutive odd integers increase by two and can be represented by x, $x + 2$, $x + 4$...

Note: It is common mistake to use x, x+1, x+3... for consecutive odd integers. All consecutive odd integers increase by two.

To solve consecutive integer word problems:

1. Choose a variable to represent the first consecutive integer.
2. Represent the other integers using the same variable in the expression.
3. Write an equation that represents the conditions stated in the problem.
4. Solve the equation.

5. Evaluate your expressions to determine the other integers.
6. Check your answer with the conditions stated in the problem.
7. Answer the question.

Example 1 Find the second of three consecutive integers whose sum is 48.

x = first consecutive integer	**Step 1:** Choose a variable to represent the first integer.
$x + 1$ = second consecutive integer $x + 2$ = third consecutive integer	**Step 2:** Represent the other integers.
$x + (x + 1) + (x + 2) = 48$	**Step 3:** Write an equation.
$3x + 3 = 48$ $3x + 3 - 3 = 48 - 3$ $3x = 45$ $\frac{3x}{3} = \frac{45}{3}$ $x = 15$	**Step 4:** Solve the equation.
$x + 1 = 15 + 1$ $x + 1 = 16$ $x + 2 = 15 + 2$ $x + 2 = 17$	**Step 5:** Determine the other integers.
$15 + 16 + 17 = 48$	**Step 6:** Check your answer. Do you have three consecutive integers? Yes, 15, 16, and 17. Is their sum 48? Yes.
The second consecutive integer is 16.	**Step 7:** Answer the question.
	Answer: 16

Example 2 Find three consecutive odd integers whose sum is 39.

x = first consecutive odd integer	**Step 1:** Choose a variable to represent the first integer.
$x + 2$ = second consecutive odd integer $x + 4$ = third consecutive odd integer	**Step 2:** Represent the other integers.
$x + (x + 2) + (x + 4) = 39$	**Step 3:** Write an equation.
$3x + 6 = 39$ $3x + 6 - 6 = 39 - 6$ $3x = 33$ $\frac{3x}{3} = \frac{33}{3}$ $x = 11$	**Step 4:** Solve the equation.

$x + 2 = 11 + 2$ $x + 2 = 13$ $x + 4 = 11 + 4$ $x + 4 = 15$	**Step 5:** Determine the other integers.
$11 + 13 + 15 = 39$	**Step 6:** Check your answer. Do you have three consecutive odd integers? Yes, 11, 13, and 15. Is their sum 39? Yes.
The consecutive odd integers are 11, 13 and 15.	**Step 7:** Answer the question.
	Answer: 11, 13 and 15

Numbers sometimes exist in patterns as multiples. Using your knowledge of multiplication and consecutive patterns, you can set up equations for these as well. Here is an example.

Example 3 The sum of three consecutive multiples of 7 is 105. What are the three multiples of 7?

$x =$ first multiple of 7	**Step 1:** Choose a variable to represent the first number.
$x + 7 =$ second multiple of 7 $x + 14 =$ third multiple of 7	**Step 2:** Multiples of 7 would increase each time by 7; for example, x, $x + 7$, $x + 14$, and so on.
$x + (x + 7) + (x + 14) = 105$	**Step 3:** Write an equation.
$3x + 21 = 105$ $3x + 21 - 21 = 105 - 21$ $3x = 84$ $\frac{3x}{3} = \frac{84}{3}$ $x = 28$	**Step 4:** Solve the equation.
$x = 28$ $x + 7 = 35$ $x + 14 = 42$	**Step 5:** Determine the other integers.
$28 + 35 + 42 = 105$	**Step 6:** Check your answer. Do you have three multiples of 7? 28, 35, 42 $\rightarrow$ Yes Is their sum 105? $\rightarrow$ Yes
The three consecutive multiples of 7 are 28, 35, and 42.	**Step 7:** Answer the question.
	Answer: 28, 35, and 42

Problem Set

1. Find two consecutive integers whose sum is equal to 25. List answers in order from least to greatest with a comma separating the numbers.

2. Find two consecutive integers whose sum is equal to 3. List answers in order from least to greatest with a comma separating the numbers.

3. Find two consecutive integers whose sum is equal to 7. List answers in order from least to greatest with a comma separating the numbers.

4. Find two consecutive integers whose sum is equal to 11. List answers in order from least to greatest with a comma separating the numbers.

5. Find two consecutive integers whose sum is equal to –9. List answers in order from least to greatest with a comma separating the numbers.

6. Find two consecutive integers whose sum is equal to –13. List answers in order from least to greatest with a comma separating the numbers.

7. Find three consecutive integers whose sum is equal to 6. List answers in order from least to greatest with a comma separating the numbers.

8. Find three consecutive integers whose sum is equal to 21. List answers in order from least to greatest with a comma separating the numbers.

9. Find three consecutive integers whose sum is equal to –12. List answers in order from least to greatest with a comma separating the numbers.

10. Find three consecutive integers whose sum is equal to –30. List answers in order from least to greatest with a comma separating the numbers.

11. Find four consecutive integers whose sum is equal to 14. List answers in order from least to greatest with a comma separating the numbers.

12. Find four consecutive even integers whose sum is equal to 68. List answers in order from least to greatest with a comma separating the numbers.

13. Find three consecutive even integers whose sum is equal to 24. List answers in order from least to greatest with a comma separating the numbers.

14. Find two consecutive even integers whose sum is equal to 106. List answers in order from least to greatest with a comma separating the numbers.

15. Find two consecutive odd integers whose sum is equal to 84. List answers in order from least to greatest with a comma separating the numbers.

16. Find three consecutive odd integers whose sum is equal to 9. List answers in order from least to greatest with a comma separating the numbers.

17. Find four consecutive odd integers whose sum is equal to 16. List answers in order from least to greatest with a comma separating the numbers.

18. Find the largest of three consecutive integers if the sum of the two smaller integers equals the largest integer.

19. Find the largest of three consecutive even integers if the sum of the two smaller integers is 16 more than the largest integer.

20. Find the smallest of four consecutive even integers if twice the sum of the second and third integers is equal to the sum of the first and the fourth integers increased by ten.

HA1-160: Writing an Equation to Solve Distance, Rate, and Time Problems

In an earlier lesson, you learned that the formula for distance is $d = r \cdot t$.

d	$=$	r	$\cdot$	t
distance		rate or speed		time
how far away?		how fast?		how long did it take?

Example 1 How far can you travel in 4 hours if you are driving 65 mph?

$d = r \cdot t$	**Step 1:** Write the formula.
$d = 65 \cdot 4$	**Step 2:** Substitute the given values for the variables.
$d = 260$ miles	**Step 3:** Solve for d. Be sure to include units.
You can travel 260 miles in 4 hours at a speed of 65 mph.	**Step 4:** Answer the question.
	Answer: 260 miles

Example 2 If you travel 125 miles in 5 hours, how fast are you going?

$d = r \cdot t$	**Step 1:** Write the formula.
$125 = r \cdot 5$	**Step 2:** Substitute the given values for the variables.
$\frac{125}{5} = \frac{r \cdot 5}{5}$ $25 = r$	**Step 3:** Solve for r.
You are driving 25 mph if you travel 125 miles in 5 hours.	**Step 4:** Answer the question.
	Answer: 25 mph

Example 3 How long would it take to travel 650 miles at 50 mph?

$d = r \cdot t$	**Step 1:** Write the formula.
$650 = 50 \cdot t$	**Step 2:** Substitute the given values for the variables.
$\frac{650}{50} = \frac{50t}{50}$ $13 = t$	**Step 3:** Solve for t.

It would take 13 hours to travel 650 miles at 50 mph.

Step 4: Answer the question.

Answer: 13 hours

When solving more difficult problems, you may need to set up a table for the formula $d = r \cdot t$. Use the information given to fill in the table and then set up an equation to solve for the correct variable.

Example 4 Two trains leave a station going in opposite directions. One train is traveling at a speed of 80 mph, and the other train is traveling at 95 mph. How long will it take both trains to be a distance of 1,750 miles apart?

Step 1: Fill in a chart.

	d	r	t
train one	$80x$	80	x
train two	$95x$	95	x

Step 2: Write an equation.

$$80x + 95x = 1{,}750$$

Step 3: Solve the equation.

$$175x = 1{,}750$$

$$\frac{175x}{175} = \frac{1{,}750}{175}$$

$$x = 10$$

Step 4: Answer the question.

It will take 10 hours for the trains to be exactly 1,750 miles apart.

Answer: 10 hours

Example 5 A moving van left the Pitres' house averaging 51 mph. One hour later, the Pitres left, averaging 68 mph. How long will it take them to catch up to the moving van?

Step 1: Fill in a chart.

	d	r	t
moving van	$51(h+1)$	51	$h+1$
Pitres	$68h$	68	h

Step 2: Write an equation.

$$51(h+1) = 68h$$

Step 3: Solve the equation.

$$51h + 51 = 68h$$

$$51h + 51 - 51h = 68h - 51h$$

$$51 = 17h$$

$$\frac{51}{17} = \frac{17h}{17}$$

$$3 = h$$

The Pitres will catch up to the moving van in 3 hours.

Step 4: Answer the question.

Answer: 3 hours

Example 6 Rose drove to visit her parents at an average speed of 60 km per hour. She returned home in heavy traffic, along the same route, at an average speed of 45 km per hour. It took her an hour longer to return home than it did to get to her parent's house. How long did it take her to get home?

Step 1: Fill in a chart.

	d	r	t
To	$60(x-1)$	60	$x-1$
Return	$45x$	45	x

Step 2: Write an equation.

$$60(x-1) = 45x$$

Step 3: Solve the equation.

$$\begin{aligned} 60x - 60 &= 45x \\ 60x - 60 - 60x &= 45x - 60x \\ -60 &= -15x \\ \frac{-60}{-15} &= \frac{-15x}{-15} \\ 4 &= x \end{aligned}$$

Rose drove home in four hours.

Step 4: Answer the question.

Answer: 4 hours

Problem Set

1. Two planes, traveling in opposite directions, leave an airport at the same time. If their average speeds are 375 miles per hour and 325 miles per hour, respectively, in how many hours will they be 1,050 miles apart?

2. Two planes, traveling in opposite directions, leave an airport at the same time. If their average speeds are 480 miles per hour and 540 miles per hour, respectively, in how many hours will they be 2,040 miles apart?

3. How many miles can you travel in 6 hours at an average rate of 66 miles per hour?

4. How many miles can you travel in 6 hours at an average rate of 72 miles per hour?

5. How many hours will it take to travel 200 miles at an average rate of 80 miles per hour?

6. How many hours will it take to travel 300 miles at an average rate of 60 miles per hour?

7. How many hours will it take to travel 3,600 miles at an average rate of 400 miles per hour?

8. How many hours will it take to travel 3,600 miles at an average rate of 300 miles per hour?

9. If you travel 480 miles in 4 hours, what is your average rate?

10. If you travel 480 miles in 10 hours, what is your average rate?

11. While traveling to her parents' house, Rose drove at an average speed of 60 kilometers per hour. She returned home along the same route at an average speed of 45 kilometers per hour. Due to heavy traffic, it took her an hour longer to return home than it did to get to her parents' house. How long did it take her to get home?

12. Two planes leave at the same time from cities 1,980 miles apart and fly toward each other. One travels at 540 miles per hour and the other at 560 miles per hour. How long will it take for the two planes to meet?

13. Two airplanes leave the same airport at the same time and travel in opposite directions. One airplane averages 450 miles per hour, and the other airplane averages 475 miles per hour. Let y equal the number of hours the airplanes will have flown by the time they are 1,850 miles apart. Write an equation in terms of y that represents the best first step to solve this problem for y.

14. Two cars leave the same hotel at the same time and travel in opposite directions. One car averages 63 miles per hour, and the other car averages 72 miles per hour. Let n equal the number of hours the cars will have traveled by the time they are 270 miles apart. Write an equation in terms of n that represents the best first step to solve this problem for n.

15. Mr. and Mrs. Carlton will each be driving one of the moving vans they have rented. Mr. Carlton will drive at an average rate of 60 miles per hour. Mrs. Carlton, who will be leaving one hour after Mr. Carlton, will average 65 miles per hour. Let c equal the number of hours that will pass before Mrs. Carlton catches up with Mr. Carlton. Write an equation in terms of c that represents the best first step to solve this problem for c.

16. Mr. and Mrs. Carlton will each be driving one of the moving vans they have rented. Mr. Carlton will drive at an average rate of 60 miles per hour. Mrs. Carlton, who will be leaving two hours after Mr. Carlton, will average 65 miles per hour. Let c equal the number of hours that will pass before Mrs. Carlton catches up with Mr. Carlton. Write an equation in terms of c that represents the best first step to solve this problem for c.

17. Two buses, traveling in opposite directions, leave the same depot at the same time. In 2 hours they are 250 miles apart. One bus travels 5 miles per hour faster than the other bus. If t equals the rate of the faster bus, write an equation in terms of t that represents the best first step to solve this problem for t.

18. Two cars on a highway passed one another going in opposite directions. The eastbound car was traveling 5 mph faster than the westbound car. One hour later they were 125 miles apart. At what speed was the westbound car traveling?

19. One hiker starts hiking at 7:00 am, walking at an average speed of 6 miles per hour. Another hiker starts on the same trail at 9:00 am, walking at an average speed of 7 miles per hour. How long will it take for the second hiker to catch up to the first hiker?

20. Two buses, traveling in opposite directions, leave the same depot at the same time. In 3 hours they are 405 miles apart. If one bus travels 5 miles per hour faster than the other bus, find the rate of the faster bus.

HA1-165: Using Equations to Solve Percent Problems

Juanita has been saving her money to buy a sweater that costs $50.00. As she approaches the display, she sees a salesperson putting a 30% off sale sign on the rack holding the sweater she wants. How much money will she save by buying the sweater on sale?

To solve a percent problem, you must be able to change a percent to a decimal or fraction, and you must be able to change a decimal to a percent. Use the following rules:

Percent Rules	1. To change a percent to a decimal or fraction, divide the percent by 100. 2. To change a decimal to a percent, multiply the decimal by 100.

Example 1 Change 50% to a decimal.

$50\% = \frac{50}{100}$ — **Step 1:** Divide the percent by 100.

$50\% = 0.50$ — **Step 2:** Dividing a whole number or decimal by 100 is the same as moving the decimal point two places to the left.

Answer: 0.50

Example 2 Change 42.5% to a decimal.

$42.5\% = \frac{42.5}{100}$

$42.5\% = 0.425$

Step 1: Dividing a whole number or decimal by 100 is the same as moving the decimal point two places to the left.

Answer: 0.425

Example 3 Change $33\frac{1}{3}\%$ to a fraction.

Step 1: Divide the percent by 100.

$$33\frac{1}{3} \div 100 =$$
$$= \frac{100}{3} \cdot \frac{1}{100}$$
$$= \frac{1}{3}$$

Answer: $\frac{1}{3}$

Example 4 Change 0.3 to a percent.

$0.3(100) = 30.0$	**Step 1:** Multiply the decimal by 100.
$0.3 = 30\%$	**Step 2:** Multiplying a whole number or decimal by 100 is the same as moving the decimal point two places to the right.
	Answer: 30%

You can use equations to solve problems involving percents. A percent question consists of key words and phrases.

what, what number, what percent ⟷ indicates an unknown value, represented by a variable

of ⟷ indicates multiplication

is ⟷ indicates an equal sign

Now, let's determine how much money Juanita will save by buying the $50.00 sweater on sale at 30% off the regular price.

Example 5 What is 30% of $50?

↓ ↓ ↓ ↓ ↓

$n = 30\% \cdot 50$	**Step 1:** Write an equation.
$n = 0.3(50)$	**Step 2:** Change the percent to a decimal.
$n = 15.0$	**Step 3:** Solve for *n*.
	Answer: Juanita will save $15.

Example 6 3% of what number is 15?

↓ ↓ ↓ ↓ ↓

$3\% \cdot n = 15$	**Step 1:** Write an equation.
$0.03n = 15$	**Step 2:** Change the percent to a decimal.
$\frac{0.03n}{0.03} = \frac{15}{0.03}$ $n = 500$	**Step 3:** Solve for *n*.
	Answer: 3% of 500 is 15.

Example 7 What is $\frac{2}{3}\%$ of 600?

↓ ↓ ↓ ↓ ↓

$x = \frac{2}{3}\% \cdot 600$	**Step 1:** Write an equation.

$x = \left(\frac{2}{3} \div 100\right) \cdot 600$

Step 2: Change the percent to a decimal.

$x = \left(\frac{2}{3} \cdot \frac{1}{100}\right) \cdot 600$

$x = \left(\frac{1}{150}\right) \cdot \frac{600}{1}$

$x = 4$

Step 3: Solve for x.

Answer: $\frac{2}{3}$% of 600 is 4.

Example 8 There are 50 students trying out for the dance team. Only 20 students will be selected. What percent of the total number of students trying out will be selected for the dance team?

What percent	of	50	is	20?
↓	↓	↓	↓	↓
p	•	50	=	20

Step 1: Write a single statement for the problem.

Step 2: Write an equation.

$50p = 20$

$\frac{50p}{50} = \frac{20}{50}$

$p = 0.4$

Step 3: Solve for p.

$0.4(100) = 40.0$

$0.4 = 40\%$

Step 4: Change the decimal to a percent.

Answer: Forty percent of the students trying out will be selected for the dance team.

Problem Set

1. What number is 150% of 2?

2. What number is 20% of 25?

3. What number is 50% of 6,000?

4. What number is 50% of 570?

5. What number is 75% of 3,484?

6. What number is 75% of 4,572?

7. What number is 1% of 230?

8. What number is 1% of 930?

9. What number is 33% of 33?

10. What number is 33% of 162?

11. 195 is 65% of what number?

12. 225 is 15% of what number?

13. 14 is 25% of what number?

14. 21 is 25% of what number?

15. 195 is 3% of what number?

16. 195 is 30% of what number?

17. 328 is 40% of what number?

18. 24.5 is 35% of what number?

19. 31 is what percent of 62? (Be sure to include the % sign at the end of your answer.)

20. 15 is what percent of 100? (Be sure to include the % sign at the end of your answer.)

HA1-170: Solving Percent of Change Problems

Each day we experience problems involving percent of change. These problems occur when we buy items with sales tax added, items that have been reduced by a percentage, or when we want to find the difference in the amount of two numbers. Most people use a calculator to find the percentages for these items; however, you should be able to calculate percentages for your own purchases to ensure that the calculations are correct.

There are other examples in which you might need to find percent of change. For example, if your employer gives all employees a yearly standard-of-living raise, you should be able to calculate your new salary and compare it to your old salary. If your landlord tells you that your rent will increase by 10%, you must be able to calculate the new rent. In order to solve these problems, you must first decide whether there is an increase or decrease in the original amount if the problem does not provide the information. Then you will know whether to add to the original cost or subtract from it.

First, let's review how to find a percent of a number.

Example 1 What number is 60% of 250?

$\downarrow \quad \downarrow \quad \downarrow \quad \downarrow \quad \downarrow$

$n = 60\% \bullet 250$ — **Step 1:** Write an equation.

$n = 0.6(250)$ — **Step 2:** Change the percent to a decimal.

$n = 150$ — **Step 3:** Solve for n.

Answer: 60% of 250 is 150.

Example 2 A case of sodas costs $8.00 in a city that charges 5% tax. How much will the case cost with the tax included?

5% of $8.00 is the tax — **Step 1:** Write a percent statement.

$\downarrow \quad \downarrow \quad \downarrow \quad \downarrow \quad \downarrow$

$5\% \bullet 8 = x$ — **Step 2:** Write an equation.

$0.05 \cdot 8 = x$ — **Step 3:** Change the percent to a decimal.

$0.40 = x$ — **Step 4:** Solve for x.

$\$8.00 + \$0.40 = \$8.40$ — **Step 5:** Since a tax increases the cost, add the tax to the original price.

Answer: The case of sodas will cost $8.40.

Example 3 The advertising department was notified that their budget for next year was cut by 20%. If their budget was $750, what will it be next year?

20% of $750 is the budget cut	**Step 1:** Write a percent statement.
↓ ↓ ↓ ↓ ↓	
$20\% \cdot \$750 = c$	**Step 2:** Write an equation.
$0.2 \cdot 750 = c$	**Step 3:** Change the percent to a decimal.
$150 = c$	**Step 4:** Solve for *c*.
$\$750 - \$150 = \$600$	**Step 5:** Since a budget cut is a decrease, subtract it from the original budget.
	Answer: The budget for advertising will be $600 next year.

You'll find there are many daily situations that require you to calculate percentage of change. A merchant often indicates a sale by posting signs that state what percentage of the price the customer will save. Changes in a city's population and changes in a school's scores on standardized tests are often written as a percent of increase or decrease. Use the following formulas to determine the percent of a change.

Percent of Increase	$\% \text{ Increase} = \dfrac{\text{amount of increase}}{\text{original amount}}$
Percent of Decrease	$\% \text{ Decrease} = \dfrac{\text{amount of decrease}}{\text{original amount}}$

Example 4 The population of Trickling Creek increased from 3,500 last year to 3,650 this year. Find the percent of increase to the nearest tenth of a percent.

$3{,}650 - 3{,}500 = 150$	**Step 1:** Determine the amount of increase.
$\% \text{ increase} = \dfrac{\text{amount of increase}}{\text{original amount}}$	**Step 2:** Write the formula.
$\% \text{ increase} = \dfrac{150}{3{,}500}$	**Step 3:** Substitute the values into the formula.
$\dfrac{150}{3{,}500} = 0.0428$	**Step 4:** Simplify.

% increase = 0.0428 % increase = 4.3%	**Step 5:** Change the decimal to a percent and round to the nearest tenth of a percent
	Answer: The percent of increase is 4.3%.

Example 5 The price of a graphing calculator has fallen from $150 to $90. What is the percent of decrease in price?

$150 - 90 = 60$	**Step 1:** Determine the amount of decrease.
$\% \text{ decrease} = \dfrac{\text{amount of decrease}}{\text{original amount}}$	**Step 2:** Write the formula.
$\% \text{ decrease} = \dfrac{60}{150}$	**Step 3:** Substitute the values into the formula.
$\dfrac{60}{150} = 0.4$	**Step 4:** Simplify.
% decrease = 0.4 % decrease = 40%	**Step 5:** Change the decimal to a percent.
	Answer: The percent of decrease is 40%.

Problem Set

1. Maria earns $320 each month. Her supervisor told her that she will give her a 20% increase in pay. How much will Maria earn?

2. Maria earns $320 each month. Her supervisor told her that she will give her a 5% increase in pay. How much will Maria earn?

3. The price of a shirt that regularly sells for $26 has been decreased by 15%. What is the new price of the shirt?

4. The price of a shirt that regularly sells for $26 has been decreased by 40%. What is the new price of the shirt?

5. The price of a skirt that regularly sells for $18 has been decreased by 10%. What is the new price of the skirt?

6. The price of a hat that regularly sells for $18 has been decreased by 40%. What is the new price of the hat?

7. The Everlasting Paint Company notified all of their distributors that the price of their paint will increase by 20%. If the old price was $12, what will be the new price?

8. Dr. Jackson decided to decrease the dosage of Lakeisha's medication by 5%. If her current dose is 120 cc, what will be her next dose?

9. Jabari plans to sell his stock when the price per share increases 50%. If the price per share is $12.50, at what price will he sell it?

10. The price of cereal has gone up 24%. If the price of a box of cereal was $2.25, what would the price be now?

11. In 10 years, the population of a city increased from 2,000 to 2,200. Find the percent of increase.

12. In 10 years, the population of a city increased from 10,000 to 10,500. Find the percent of increase.

13. In 10 years, the population of a city increased from 10,000 to 13,000. Find the percent of increase.

14. In 5 years, the population of a city decreased from 4,000 to 3,800. Find the percent of decrease.

15. The enrollment in Career Explorations increased from 500 to 520. Find the percent of increase.

16. The enrollment in Psychology I dropped from 300 to 270. Find the percent of decrease.

17. The manager of Sound Systems wants to advertise the percent of the price a customer will save. If a CD player regularly sells for $120 and the sale price will be $96, determine the percent of decrease in the price.

18. Due to inflation, the price of a stereo speaker increased from $300 to $315. Determine the percent of increase in the price.

19. In 2 years, the population of a city decreased from 4,000 to 3,200. Find the percent of decrease.

20. Miguel's monthly salary increased from $1,200 to $1,380. Find the percent of increase.

HA1-175: Solving Literal Equations

In previous lessons, you learned to solve equations such as:

$x + 4 = 9$	$x - 6 = 13$	$4x = 20$	$\frac{x}{9} = 5$
$x + 4 - 4 = 9 - 4$	$x - 6 + 6 = 13 + 6$	$\frac{4x}{4} = \frac{20}{4}$	$\frac{x}{9} \cdot 9 = 5 \cdot 9$
$x = 5$	$x = 19$	$x = 5$	$x = 45$

Do you notice anything that these equations have in common? In each case, the equation had only one variable. Some equations (like many of the formulas you use) contain more than one variable. These types of equations are called **literal equations**.

Literal Equation	An equation that contains more than one variable is a literal equation.

Formulas are examples of literal equations:

Perimeter of a rectangle:	$p = 2l + 2w$
Area of a rectangle:	$A = l \cdot w$
Volume of a box:	$v = l \cdot w \cdot h$
Circumference:	$c = 2\pi r$

You may be asked to solve for any variable that is a part of the literal equation. To do this, you continue to use inverse operations to isolate the variable. Solve $p = 2l + 2w$ for l. This means you will isolate l.

Example 1 Solve for l: $p = 2l + 2w$

$p = 2l + 2w$ — **Step 1:** Start with the formula.

$p - 2w = 2l + 2w - 2w$

$p - 2w = 2l$ — **Step 2:** Isolate the term containing l by subtracting $2w$ from both sides of the equation.

$\frac{p-2w}{2} = \frac{2l}{2}$

$\frac{p-2w}{2} = l$ — **Step 3:** Isolate the l by dividing both sides of the equation by its coefficient, 2.

Answer: $\frac{p-2w}{2} = l$

Example 2 Solve for w: $v = l \cdot w \cdot h$

$v = l \cdot w \cdot h$

Step 1: Start with the formula.

$\frac{v}{lh} = \frac{l \cdot w \cdot h}{lh}$

$\frac{v}{lh} = w$

Step 2: The w is being multiplied by lh, so divide both sides of the equation by lh.

Answer: $\frac{v}{lh} = w$

Example 3 Solve for h: $A = \frac{1}{2}bh$

$A = \frac{1}{2}bh$

Step 1: Start with the formula.

$2A = \frac{1}{2}bh \cdot 2$

$2A = bh$

Step 2: Eliminate the fraction on the right side of the equation by multiplying both sides by 2.

$\frac{2A}{b} = \frac{bh}{b}$

$\frac{2A}{b} = h$

Step 3: Since h is being multiplied by b, divide both sides of the equation by b.

Answer: $\frac{2A}{b} = h$

Example 4 Solve for g: $a + b + cd = gk + h$

$a + b + cd = gk + h$

Step 1: Start with the formula.

$a + b + cd - h = gk + h - h$

$a + b + cd - h = gk$

Step 2: Isolate the term containing g by subtracting h from both sides of the equation.

$\frac{a + b + cd - h}{k} = \frac{gk}{k}$

$\frac{a + b + cd - h}{k} = g$

Step 3: Isolate g by dividing both sides of the equation by k.

Answer: $\frac{a + b + cd - h}{k} = g$

Example 5 Solve for y: $3x + 5y = 8$

$3x + 5y = 8$	**Step 1:**	Start with the formula.
$3x + 5y - 3x = 8 - 3x$ $5y = 8 - 3x$	**Step 2:**	Isolate the term containing y by subtracting $3x$ from both sides of the equation.
$\frac{5y}{5} = \frac{8-3x}{5}$ $y = \frac{8-3x}{5}$	**Step 3:**	Isolate y by dividing both sides of the equation by its coefficient, 5.

Answer: $y = \frac{8-3x}{5}$

Example 6 The area of a trapezoid is expressed by the formula $A = \frac{1}{2}h(b_1 + b_2)$. Rewrite the formula for h.

$A = \frac{1}{2}h(b_1 + b_2)$	**Step 1:**	Start with the formula.
$2A = \left[\frac{1}{2}h(b_1 + b_2)\right] \cdot 2$ $2A = h(b_1 + b_2)$	**Step 2:**	Eliminate the fraction on the right side of the equation by multiplying both sides of the equation by 2.
$\frac{2A}{(b_1 + b_2)} = \frac{h(b_1 + b_2)}{(b_1 + b_2)}$ $\frac{2A}{b_1 + b_2} = h$	**Step 3:**	Isolate h by dividing both sides of the equation by the quantity $b_1 + b_2$.

Answer: $\frac{2A}{b_1 + b_2} = h$

Problem Set

1. Solve the following literal equation for d:
$$d + 12 = e$$

2. Solve the following literal equation for d:
$$\frac{d}{12} = e$$

3. Solve the following literal equation for b:
$$a = b + h$$

4. Solve the following literal equation for b:
$$a = b - h$$

5. Solve the following literal equation for m:

$$m + 7 = g$$

6. Solve the following literal equation for m:

$$7m = g$$

7. Solve the following literal equation for w:

$$lwh = V$$

8. Solve the following literal equation for h:

$$lwh = V$$

9. Solve the following literal equation for m:

$$j = m + n - c$$

10. Solve the following literal equation for c:

$$j = m + n - c$$

11. Solve the following literal equation for v:

$$w = a + v - d$$

12. Solve the following literal equation for b:

$$a = \frac{3}{4}bh$$

13. Solve the following literal equation for d:

$$9e + 8d = 15e$$

14. Given $m = 2(l + w)$, find w if $m = 40$ and $l = 4$.

15. Solve for a:

$$f = np + a$$

16. Solve for n:

$$f = np + a$$

17. Solve for b:

$$A = \frac{1}{2}bh$$

18. Solve the following literal equation for s:

$$rs + p = -5p + 8a$$

19. Solve the following literal equation for h:

$$a = \frac{2}{5}(b + h)$$

20. The formula for the area of a trapezoid is $A = \frac{1}{2}h(b_1 + b_2)$ where h is the height, b_1 is the length of one base, and b_2 is the length of the other base. Given $A = 140$, $h = 10$, and $b_2 = 7$, find b_1.

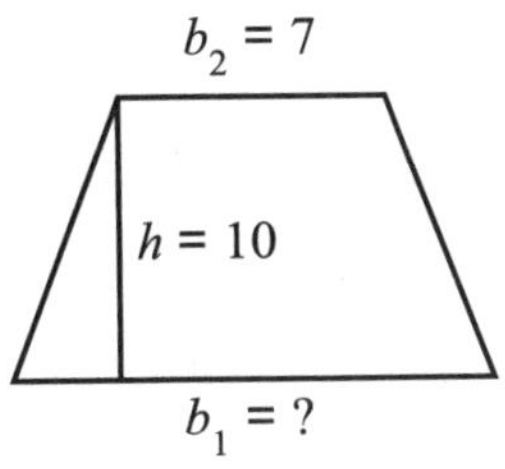

HA1-180: Graphing Equations and Inequalities on the Number Line

There are pictorial representations all around us; we use them to express ideas that we don't want to write out. Bathrooms have symbols for "girls" and "boys." Crosswalk signals flash a hand to say "stop" and a human figure to say "walk". These kinds of pictures, ones we use to convey information, are often referred to as "symbols" or "symbolic representations". We use them in algebra not only to indicate operations, but to represent spatial or quantity relationships. The number line is a symbolic representation that can be used to show which numbers are part of a solution.

Before you begin graphing some equations and inequalities, review the inequality signs:

$<$	is less than
$>$	is greater than
$\leq$	is less than or equal to
$\geq$	is greater than or equal to

For any two numbers graphed on a number line, the number to the right is greater than the number to the left.

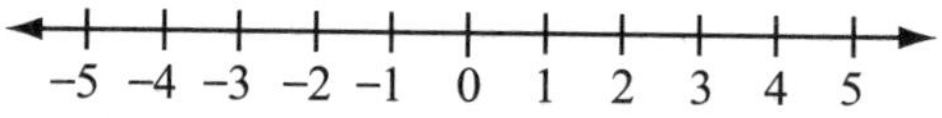

Look at the following examples:

$5 > 3$	Five is greater than three; therefore, five is to the right of three on the number line.
$1 > -1$	One is greater than negative one; therefore one is to the right of negative one on the number line.
$-2 > -5$	Negative two is greater than negative five; therefore negative two is to the right of negative five on the number line.

Example 1 Solve and graph: $x + 4 = 6 - 3$

$$x + 4 = 6 - 3$$
$$x + 4 = 3$$
$$x + 4 - 4 = 3 - 4$$
$$x = -1$$

Step 1: Solve the equation for the given variable.

Answer: To graph $x = -1$, simply put a closed circle on -1 on the number line as in the example below.

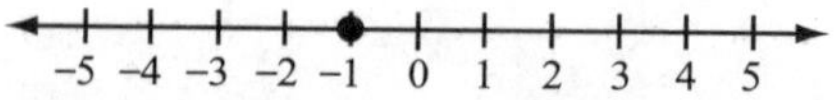

Graphing the solution to an equation is not difficult. To graph inequalities, you must determine when to use an open circle rather than a closed circle. Use the chart below to help you.

Symbol	Type of circle	Circle
$<$ $>$ $\neq$	Open circle	○
$\leq$ $\geq$ $=$	Closed circle	●

Example 2 Graph: $x \geq 2$

Answer: The "greater than or equal to" symbol is used, so you will put a point on the number two, and shade all of the numbers greater than two. The point on two is called a **closed circle**.

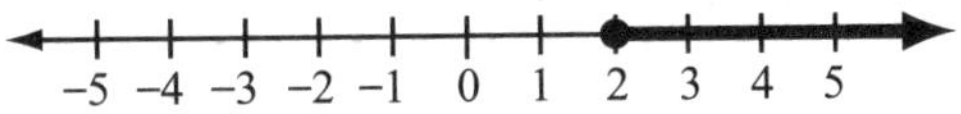

Example 3 Graph the inequality: $y < -1$

Answer: Since y is less than negative one, you will shade all of the numbers to the left of negative one. To indicate that the shading is to be as close to negative one as possible without including negative one, place an **open circle** on negative one.

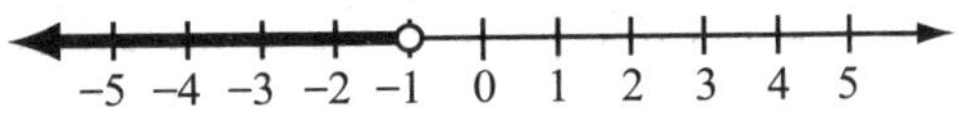

Example 4 Graph: $n \neq 2$

Answer: Since n does not equal two, you will shade every number except two. This means you will place an open circle on two and shade the rest of the number line.

Example 5 Graph: $x > 3$

Answer: The "greater than" symbol means to place an open circle on three and shade all of the numbers to the right of three.

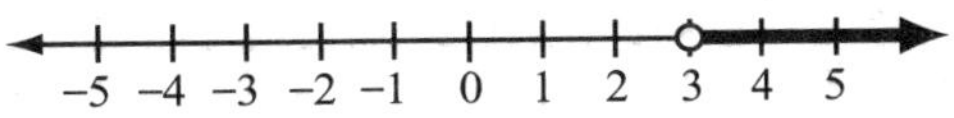

Example 6 Graph: $x \leq 0$

Answer: The "less than or equal to" symbol means to place a closed circle on zero and shade all of the numbers to the left of zero.

Problem Set

1. State the inequality graphed below:

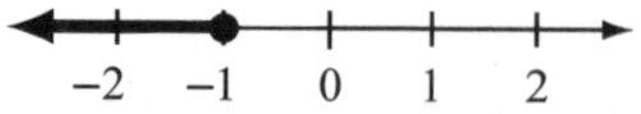

2. State the inequality graphed below:

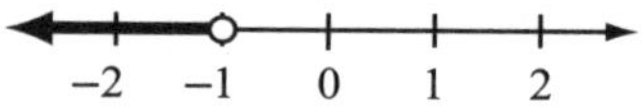

3. State the inequality graphed below:

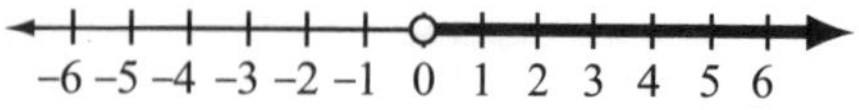

4. State the inequality graphed below:

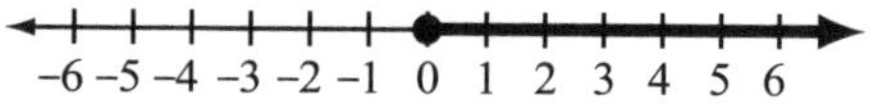

5. State the solution set of the following graph:

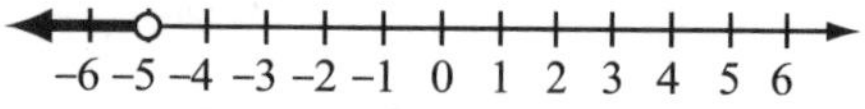

6. State the solution set of the following graph:

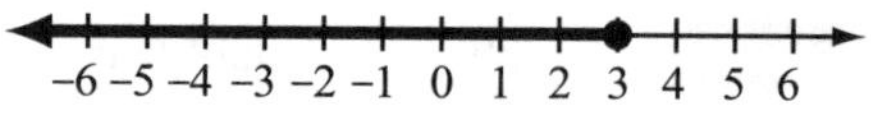

7. State the solution set of the following graph:

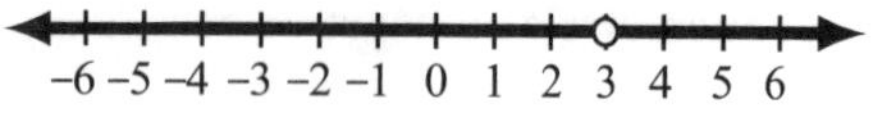

8. Solve the equation and graph the solution:

$$3x + 4 = 5x + 4$$

9. Graph the following inequality:

$$b > -\frac{1}{2}$$

10. Solve the equation and graph the solution:

$$3x + 2 = 7x - 6$$

11. Graph the following inequality:

$$b \leq -\frac{1}{2}$$

12. State the inequality graphed below:

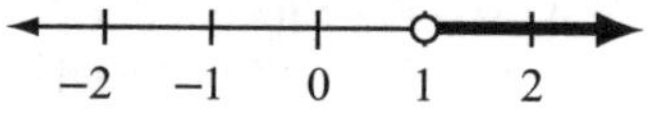

13. State the inequality graphed below:

14. State the inequality graphed below:

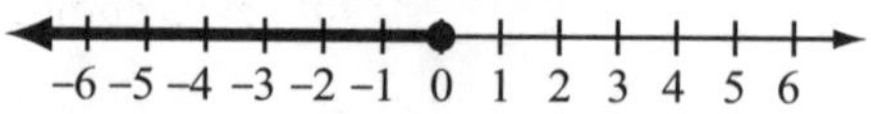

15. State the inequality graphed below:

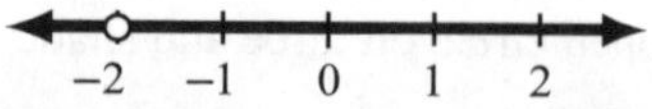

16. Graph: $x \geq -1$

17. Graph: $m \leq -3$

18. Graph: $b \neq -1$

19. Graph: $-1 \leq b$

20. Graph: $-3 \leq d$

HA1-185: Solving Inequalities Using the Addition and Subtraction Properties

In a previous lesson, you solved equations and inequalities given a replacement set. Remember that you must substitute each element of the replacement set into the equation or inequality to determine which ones make the inequality true. The solution set consists of the elements that make the equation or inequality true.

Example 1 Find the solution set for $n + 2 < 3$ The replacement set is $\{-1, 0, 1\}$.

Check –1

$n + 2 < 3$

$(-1) + 2 < 3$

$1 < 3$

True

Step 1: Substitute –1 into the equation.

Check 0

$n + 2 < 3$

$(0) + 2 < 3$

$2 < 3$

True

Step 2: Substitute 0 into the equation.

Check 1

$n + 2 < 3$

$(1) + 2 < 3$

$3 < 3$

False

Step 3: Substitute 1 into the equation.

Step 4: Look at the equations again. Because only (–1) and (0) make the inequality true, the solution set is $\{-1, 0\}$.

Note: The statement $3 < 3$ is false because both are equal to each other. The symbol $\leq$ must be used to make this statement true.

Answer: The solution set is $\{-1, 0\}$.

In this lesson you will not be given a specific replacement set. You learned to solve equations by adding or subtracting the same value from each side of the equation. If you start with a true inequality, can you add or subtract the same number from both sides and get another true inequality?

Start with a true inequality:	$6 > -4$
Add the same number to both sides:	$6 + 9 > -4 + 9$
Simplify:	$15 > 5$
True or false?	True

Start with a true inequality:	$6 > -4$
Subtract the same number from both sides:	$6 - 3 > -4 - 3$
Simplify:	$3 > -7$
True or false?	True

Addition Property for Inequalities	$a < b$ and $a + c < b + c$ are equivalent inequalities.

Subtraction Property for Inequalities	$a < b$ and $a - c < b - c$ are equivalent inequalities.

Example 2 Solve and graph the solution set for $n - 3 < 4$.

$n - 3 + 3 < 4 + 3$
$n < 7$

Step 1: Add 3 to both sides to isolate the variable.

Step 2: Graph $n < 7$.

Answer: The solution set is {all real numbers less than 7}.

Example 3 Solve and graph the solution set for $12 < y + 8$.

$12 - 8 < y + 8 - 8$
$4 < y$
or
$y > 4$

Step 1: Subtract 8 from both sides to isolate the variable. The resulting inequality, $4 < y$, can be rewritten as $y > 4$.

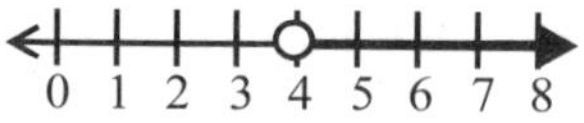

Step 2: Graph $y > 4$.

Answer: The solution set is {all real numbers greater than 4}.

Example 4 Solve $3(n - 3) \geq 2(n - 8)$

$3n - 9 \geq 2n - 16$

Step 1: Use the distributive property to simplify both sides of the equation.

$3n - 9 - 2n \geq 2n - 16 - 2n$
$n - 9 \geq -16$

Step 2: Subtract $2n$ from both sides to get the variable terms on the same side.

$n - 9 + 9 \geq -16 + 9$
$n \geq (-7)$

Step 3: Isolate the variable by adding 9 to both sides.

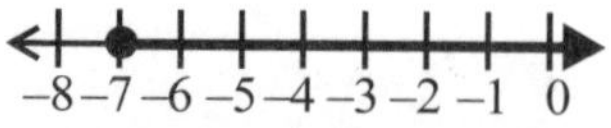

Step 4: Graph $n \geq -7$.

Answer: The solution set is {all real numbers greater than or equal to –7}.

Some word problems can be solved using inequalities. Certain phrases indicate that an inequality should be used instead of an equation.

Inequality Phrases	
$<$ is less than is fewer than	$>$ is greater than is more than
$\leq$ is less than or equal to is at most is no more than	$\geq$ is greater than or equal to is at least is no less than

Example 5 Raquel is planning a party, and her budget for the refreshments is at most \$50. She has ordered a cake that will cost \$22. How much can she spend on the other refreshments?

x = cost of other refreshments — **Step 1:** Define your variable.

$22 + x \leq 50$ — **Step 2:** Write an inequality.

Step 3: Solve the inequality.

$$22 + x \leq 50$$
$$22 + x - 22 \leq 50 - 22$$
$$x \leq 28$$

Answer: Raquel can spend \$28 or less on the other refreshments.

Problem Set

Solve the inequality:

1. $-11 \leq m + 21$
2. $m - 2 > 2$
3. $a + (-13) > -5$
4. $a - (-7) < 5$
5. $18 + f < -14$
6. $12 > e - 5$
7. $-24 < a - (-16)$
8. $-34 < h - 16$
9. $-43 > y - 32$
10. $-43 < y + 32$

Solve the inequality and graph the solution set:

11. $y + 15 \leq 13$
12. $y - 22 \leq -23$
13. $h + 5 \geq 7$
14. $f - 4 \leq 5$
15. $g + (-2) > -6$
16. $g - (-2) < -6$
17. $-4 + x > -1$

Solve the inequality:

18. $15h + 18 - 14h < -29$
19. $-15y + 23 + 16y > -52$
20. $11h + 18 - 10h + 17 \geq 32$

HA1-190: Solving Inequalities Using the Multiplication and Division Properties

In a previous lesson you learned to solve equations by multiplying or dividing both sides by the same number. If you start with a true inequality, can you multiply or divide both sides by the same number and get another true inequality?

Start with a true inequality:	$6 > -4$
Multiply both sides by the same number:	$6(9) > -4(9)$
Simplify:	$54 > -36$
True or false?	True

Start with a true inequality:	$6 > -4$
Multiply both sides by the same number:	$6(-3) > -4(-3)$
Simplify:	$-18 > 12$
True or false?	False

Note: Notice that the opposite is true; $-18 < 12$.

Multiplication and Division Properties for Inequalities	
If c is a positive non-zero number:	The inequalities $a < b$ and $a \cdot c < b \cdot c$ are equivalent. The inequalities $a < b$ and $\frac{a}{c} < \frac{b}{c}$ are equivalent.
If c is a negative non-zero number:	The inequalities $a < b$ and $a \cdot c > b \cdot c$ are equivalent. The inequalities $a < b$ and $\frac{a}{c} > \frac{b}{c}$ are equivalent.

Example 1 Solve and graph the solution set for $3x > -12$.

$$\frac{3x}{3} > \frac{-12}{3}$$

$$x > -4$$

Step 1: Divide both sides by 3 to isolate the variable. The inequality sign remains the same because both sides of the inequality are being divided by a positive number.

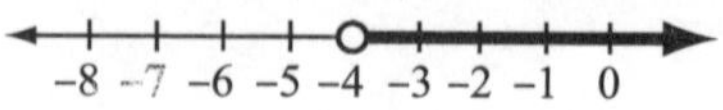

Step 2: Graph $x > -4$.

Answer: The solution set is {all real numbers greater than -4}.

Example 2 Solve and graph the solution set for $-4x > 28$.

$\frac{-4x}{-4} > \frac{28}{-4}$

$x < -7$

Step 1: Divide both sides by –4 to isolate the variable. Reverse the inequality sign because both sides of the inequality are being divided by a negative number.

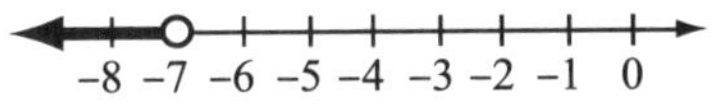

Step 2: Graph $x < -7$.

Answer: The solution set is {all real numbers less than –7}.

Example 3 Solve and graph the solution set for $\frac{x}{-2} \leq 2$.

$-2 \cdot \frac{x}{-2} \leq 2(-2)$

$x \geq -4$

Step 1: Multiply both sides by –2 to isolate the variable. Reverse the inequality sign because both sides of the inequality are being multiplied by a negative number.

Step 2: Graph $x \geq -4$.

Answer: The solution set is {all real numbers greater than or equal to –4}.

Example 4 Solve and graph the solution set for $\frac{2}{3}x \leq -6$.

$\frac{3}{2} \cdot \frac{2}{3}x \leq -6 \cdot \frac{3}{2}$

$x \leq -9$

Step 1: Multiply both sides by $\frac{3}{2}$ to isolate the variable.

The inequality sign remains the same because both sides of the inequality are being multiplied by a positive number.

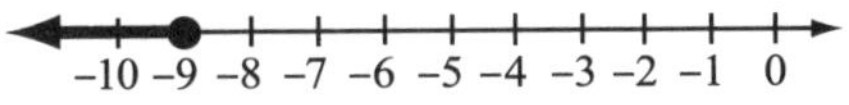

Step 2: Graph $x \leq -9$.

Answer: The solution set is {all real numbers less than or equal to –9}.

Example 5 Jeremy pays \$0.15 per minute for his cell phone. How many minutes can he use to ensure that his bill is no greater than \$24?

x = number of minutes he can use his phone

$0.15x$ = amount of his bill

Step 1: Define the variables in this situation.

Jeremy's bill is not greater than \$24. ↓ ↓ ↓ $0.15x$ $\leq$ 24	**Step 2:** Write an inequality.
$0.15x \leq 24$ $\frac{0.15x}{0.15} \leq \frac{24}{0.15}$ $x \leq 160$	**Step 3:** Solve the inequality.
	Answer: Jeremy can use his phone for 160 minutes or fewer.

Problem Set

1. Solve the inequality and graph the solution set: $\frac{x}{2} > -1$

2. Solve the inequality and graph the solution set: $5y \leq 15$

3. Solve: $7a > 28$

4. Solve: $3b < -60$

5. Solve: $\frac{m}{8} > 8$

6. Solve: $4b < -52$

7. Solve: $2b > -28$

8. Solve: $\frac{m}{10} > 20$

9. Solve: $\frac{n}{8} > -8$

10. Solve: $\frac{n}{2} < -20$

11. Solve: $-3x < 39$

12. Solve: $-8x \geq 32$

13. Solve: $\frac{n}{-3} \leq -4$

14. Solve: $52 > -13x$

15. Solve the inequality and graph the solution set: $-2x < 6$

16. Solve the inequality and graph the solution set: $-7x \geq -21$

17. Solve the inequality and graph the solution set: $-m \geq 3$

18. Solve: $\frac{2}{-3}p > 6$

19. Solve the inequality and graph the solution set: $-27 \geq -9y$

20. Solve: $-2 \leq \frac{a}{-3}$

HA1-195: Solving Inequalities Using More Than One Property

After learning to use addition, subtraction, multiplication and division to solve inequalities, you are ready to solve ones requiring you to use more than one operation. You will solve these inequalities the same way you solve equations using more than one property.

Example 1 Solve and graph the solution set for $22 - 6 \cdot y \leq 34$.

$22 - 6y - 22 \leq 34 - 22$
$-6y \leq 12$

Step 1: Isolate the term containing the variable by subtracting 22 from each side of the inequality.

$\frac{-6y}{-6} \leq \frac{12}{-6}$
$y \geq -2$

Step 2: Isolate the variable by dividing both sides of the inequality by negative six.The inequality sign is reversed because both sides are being divided by a negative number.

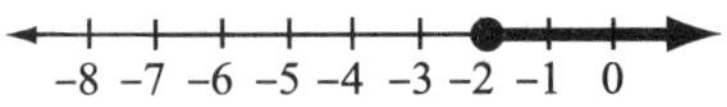

Step 3: Graph $y \geq -2$.

Answer: The solution set is {all real numbers greater than or equal to –2}.

Example 2 Solve and graph the solution set for $5n + 16 > 3n + 6$.

$5n + 16 - 3n > 3n + 6 - 3n$
$2n + 16 > 6$

Step 1: Eliminate one of the variable terms by subtracting $3n$ from both sides.

$2n + 16 - 16 > 6 - 16$
$2n > -10$

Step 2: Isolate the term containing the variable by subtracting 16 from both sides.

$\frac{2n}{2} > \frac{-10}{2}$
$n > -5$

Step 3: Divide both sides by 2 to isolate the variable. The inequality sign remains the same because both sides are being divided by a positive number.

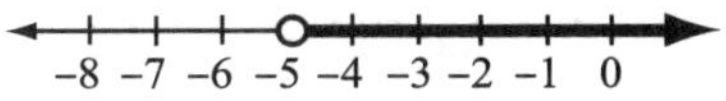

Step 4: Graph $n > -5$.

Answer: The solution set is {all real numbers greater than –5}.

Example 3 Solve and graph the solution set for $7(x + 3) < 3(2x + 8)$.

$7x + 21 < 6x + 24$

Step 1: Simplify each side of the inequality using the distributive property.

$7x + 21 - 6x < 6x - 6x + 24$
$x + 21 < 24$

Step 2: Eliminate one of the variable terms by subtracting $6x$ from each side.

$x + 21 - 21 < 24 - 21$
$x < 3$

Step 3: Isolate the variable by subtracting 21 from each side.

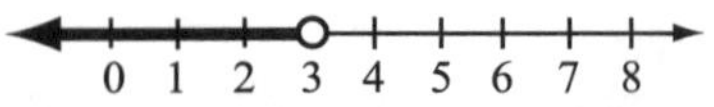

Step 4: Graph $x < 3$.

Answer: The solution set is {all real numbers less than 3}.

Example 4 Mike Walters, a place kicker for the Uptown Bulldogs, has pledged to give his favorite charity \$2,500 plus an additional \$250 for every field goal he makes during the football season. How many field goals would Mike have to make for his charity to receive at least \$5,000?

Step 1: Define the variables in this situation.

x = number of field goals he must make

$250x$ = donation for making field goals

Step 2: Write an inequality.

Donation for making field goals + \$2,500 is at least \$5,000.

$$\underset{250x}{\downarrow} \quad \underset{+\$2{,}500}{\downarrow} \quad \underset{\geq}{\downarrow} \quad \underset{\$5{,}000}{\downarrow}$$

Step 3: Subtract 2,500 from both sides to isolate the term containing the variable.

$$250x + 2{,}500 \geq 5{,}000$$
$$250x + 2{,}500 - 2{,}500 \geq 5{,}000 - 2{,}500$$
$$250x \geq 2{,}500$$

Step 4: Divide both sides by 250 to isolate the variable.

$$\frac{250x}{250} \geq \frac{2{,}500}{250}$$
$$x \geq 10$$

Answer: Mike must make at least ten field goals for his charity to receive a minimum of \$5,000.

Problem Set

Solve the inequality:

1. $3x + 12 < 48$

2. $3x - 12 > 48$

3. $6x + 18 < 42$

4. $6x - 18 > 42$

5. $-18 < 8x + 22$

6. $-18 > 8x - 2$

7. $-45 \leq 9x - 27$

8. $-45 \geq 9x + 27$

9. $-36 \leq 3x + 24$

10. $-36 \geq 3x - 24$

11. $-3x - 12 < 48$

12. $-6x + 18 > 42$

13. $-3x + 24 \leq -36$

Solve the inequality and graph the solution set:

14. $2x + 5 > 9$

15. $2x + 5 \leq 9$

16. $6x + 4 \leq 10$

17. $6x + 4 > 10$

18. Solve the inequality: $3(x - 4) < 5(x - 8)$

19. Solve the inequality and graph the solution set:
$$2(4 + x) \leq -(x + 7)$$

20. Solve the inequality and graph the solution set:
$$2(4 + x) > -(x + 7)$$

HA1-200: Combined Inequalities

There are a number of ways to express the combination of inequalities, and different ways of combining inequalities are represented differently on the number line. When the word "and" appears between two inequalities, the mathematical sentence is called a **conjunction**.

Conjunction	A mathematical sentence that consists of two sentences connected by the word "and." A conjunction is true only if *both* sentences are true.

To find the solution of a conjunction, graph both inequalities. Consider the following: $x < 3$ and $x > -1$.

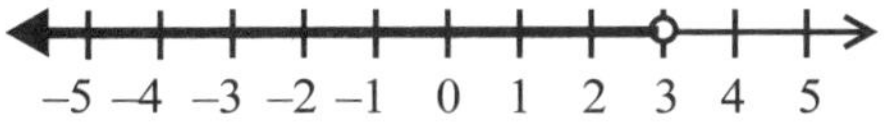

Graph $x < 3$. Place an open circle on 3 and shade all numbers to the left.

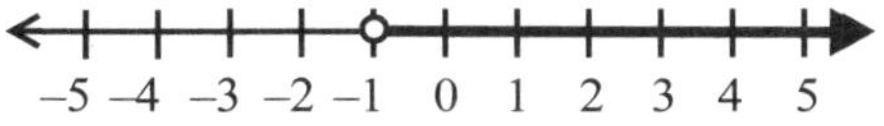

Graph $x > -1$. Place an open circle on –1 and shade all numbers to the right.

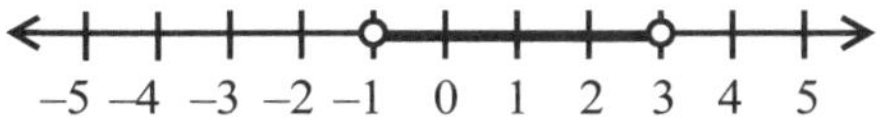

The solution is where the two graphs overlap.

This overlapping is called the **intersection**—the points that the two graphs share or have in common.

Intersection	The points that the two graphs share or have in common

This intersection is the solution of the conjunction. The actual solution can be written as $x > -1$ and $x < 3$, or in compact form, $-1 < x < 3$ (read "x is greater than –1 and less than 3").
The solution set is {all real numbers greater than –1 and less than 3}.

Any number between –1 and 3 makes both inequalities true. Remember that –1 and 3 are not included as part of the solution, because x is not equal to –1 or 3.

When the word "or" appears between two inequalities, the inequality is called a **disjunction**.

Disjunction	A mathematical sentence that consists of two sentences connected by the word "or." A disjunction is true if either or both sentences are true.

When finding the solution of a disjunction, you graph both inequalities. View the following: $x > 1$ or $x \leq -2$.

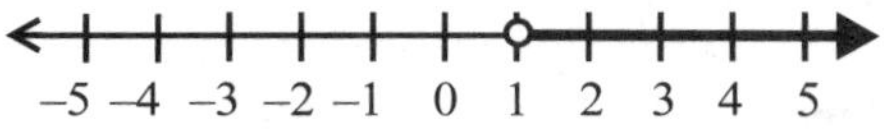

Graph $x > 1$. Place an open circle on 1 and shade all numbers to the right.

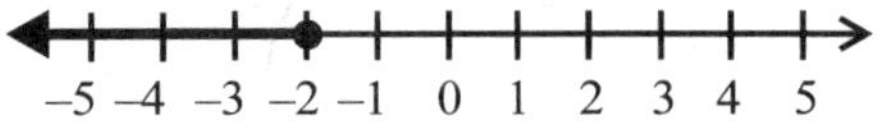

Graph $x \le -2$. Since x may be equal to –2, place a closed circle on –2 and shade all numbers to the left.

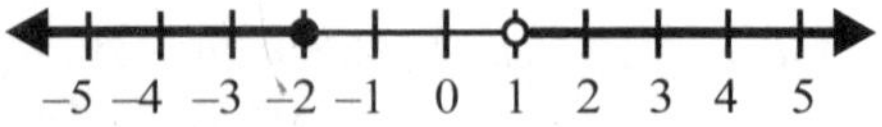

The solution is all points on both graphs.

When you combine the two graphs you have a **union**—all points on both graphs.

Union	All points on both graphs

The union is the solution of the disjunction. Thus, the solution to $x \le -2$ or $x > 1$ is {all real numbers less than or equal to –2 or greater than 1}.

When a conjunction or disjunction uses certain symbols, it is called a **combined inequality.**

Combined Inequality	A conjunction or disjunction expressed using $<$, $>$, $\ge$, and/or $\le$

Graph and state the solutions to the following examples.

Example 1 Graph and state the solution to $x < 0$ or $x \ge 1$.

$x < 0$

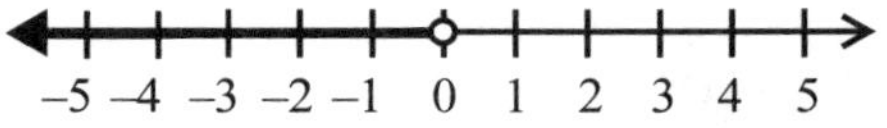

$x \ge 1$

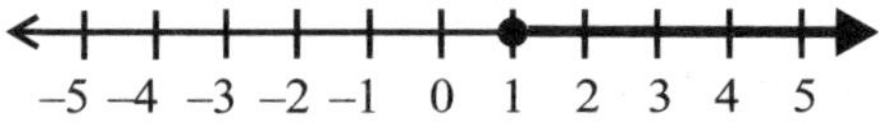

$x < 0$ or $x \ge 1$

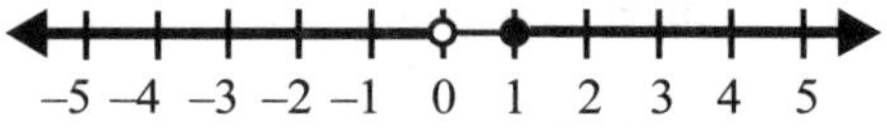

Disjunction (Union)

Answer: {All real numbers less than 0 or greater than or equal to 1}

Example 2 Graph and state the solution to $x > -2$ and $x < 3$.

$x > -2$

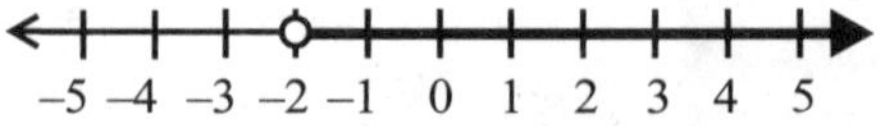

$x < 3$

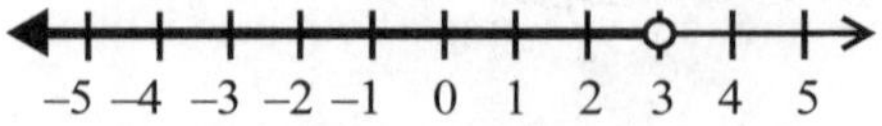

$-2 < x < 3$

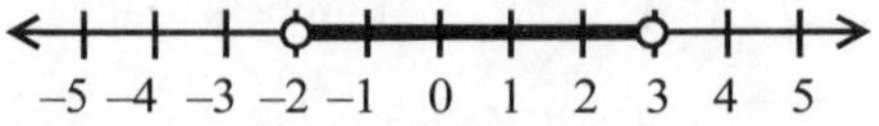

Answer: {All real numbers greater than –2 and less than 3}

Example 3 Graph and state the solution to $x < 1$ and $x > 2$.

$x < 1$

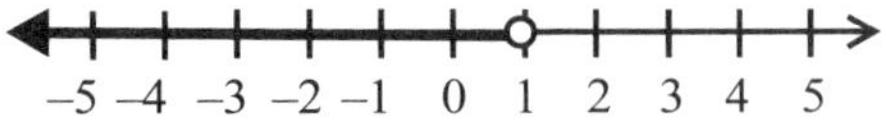

$x > 2$

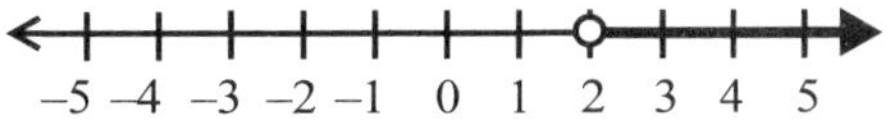

$x < 1$ and $x > 2$

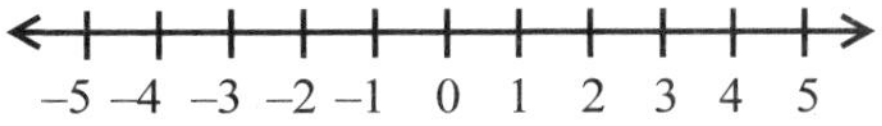

Conjunction; no intersection.

Answer: ∅ Null set; in this example, there is no solution because the two graphs do not intersect.

Problem Set

Graph the following:

1. $n > 4$ or $n < 3$

2. $x > 3$ or $x < -4$

3. $x > 2$ or $x < -2$

4. $n \leq 3$ and $n \geq 2$

5. $n < 3$ and $n < -1$

6. $n \leq -3$ or $n \geq 1$

7. $n \leq -4$ or $n \geq 1$

8. $x \geq 2$ and $x \leq 4$

9. $n < 4$ or $n > -4$

10. $n > 4$ and $n < 0$

11. Select the inequality that describes the graph below:

$x \geq 1$ or $x \leq 4$ $\quad$ $1 < x < 4$

$x > 1$ or $x < 4$ $\quad$ $-1 \leq x \leq 4$

12. Select the inequality that describes the graph below:

$-3 < x < 1$ $\quad$ $x \geq -3$ or $x \leq 1$

$x > -3$ or $x < 1$ $\quad$ $-3 \leq x \leq 1$

13. Select the inequality that describes the graph below:

$-2 < x < 1$ $\quad$ $x \geq -2$ or $x \leq 1$

$x > -2$ or $x < 1$ $\quad$ $-2 \leq x \leq 1$

14. Select the inequality that describes the graph below:

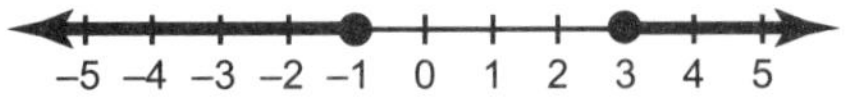

$x < -1$ or $x > 3$ $\quad$ $-1 \leq x \leq 3$

$x \leq -1$ or $x \geq 3$ $\quad$ $-1 < x < 3$

15. Select the inequality that describes the graph below:

$-3 \leq x \leq 2$ $\quad$ $-3 < x < 2$

$x \geq -3$ or $x \leq 2$ $\quad$ $x > -3$ or $x < 2$

16. Select the inequality that describes the graph below:

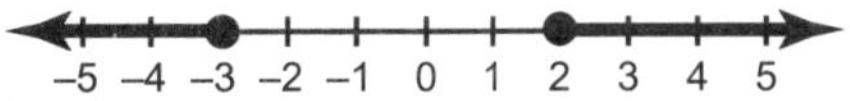

$x \leq -3$ or $x \geq 2$ $\quad$ $-3 \leq x \leq 2$

$x < -3$ or $x > 2$ $\quad$ $-3 < x < 2$

17. Select the inequality that describes the graph below:

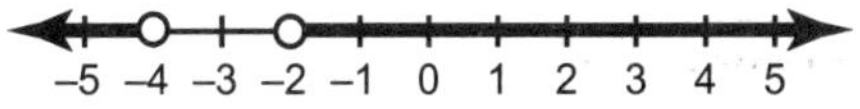

$-4 \leq x \leq -2$ $\quad$ $x < -4$ or $x > -2$

$-4 < x < -2$ $\quad$ $x \leq -4$ or $x \geq -2$

18. Graph: $4 \geq x \geq 1$

19. Graph: $-4 < x < -1$

20. Graph: $0 > x > -3$

HA1-205: Solving Combined Inequalities

This lesson requires you to do a bit of mixing and combining as you continue working with inequalities. In this lesson, you will identify graphed inequalities, solve inequalities using the four properties of real numbers, and graph the inequalities.

Before you actually begin solving and graphing inequalities, make sure you can identify graphed inequalities as conjunctions or disjunctions. Look at the examples below.

Example 1 $x < -2$ or $x \geq 1$

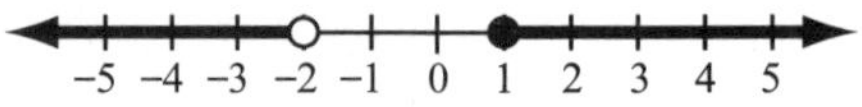

Answer: Disjunction

Example 2 $x \geq -4$ and $x \leq 2$ or written in compacted form $-4 \leq x \leq 2$

Answer: Conjunction

Example 3 $x < 21$ or $x > 28$

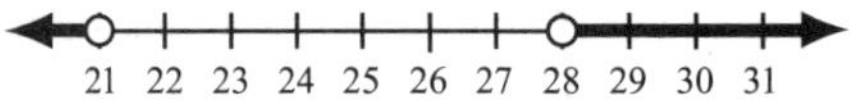

Answer: Disjunction

Example 4 $x > 102$ and $x \leq 104$ or $102 < x \leq 104$

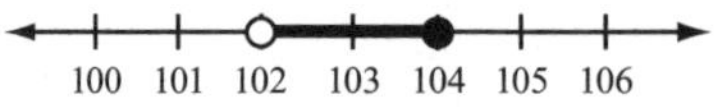

Answer: Conjunction

Now look at examples in which you need to solve the inequalities first.

Example 5 Solve and graph: $x + 4 > 8$ or $x + 3 \leq 4$

$$x + 4 > 8 \qquad x + 3 \leq 4$$
$$x + 4 - 4 > 8 - 4 \qquad x + 3 - 3 \leq 4 - 3$$
$$x > 4 \qquad x \leq 1$$

Step 1: Solve each inequality separately using the properties of real numbers.

Step 2: Next, graph the solution by determining whether the inequality contains the word "and" or "or." This example contains the word "or"; therefore, each inequality would be graphed independently of the other.

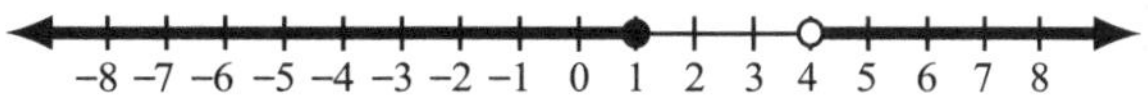

Answer: {All real numbers less than or equal to 1 or greater than 4}

Example 6 Solve and graph: $15 < 5x < 35$

$\frac{15}{5} < \frac{5x}{5} < \frac{35}{5}$

Step 1: Solve both parts of the combined inequality at the same time by dividing each part of the combined inequality by 5.

$3 < x < 7$

Step 2: Graph the solution.

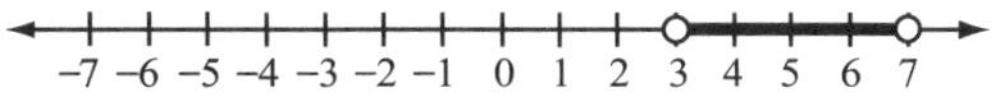

Answer: This solution is a **compact conjunction** and tells you that these two inequalities share common points called an **intersection**. The solution tells you that x is greater than 3 but less than 7, so all the points between 3 and 7 are considered part of the solution. {All real numbers greater than 3 but less than 7}

Example 7 Solve and graph: $27 < 11x + 5 \leq 60$

$27 - 5 < 11x + 5 - 5 \leq 60 - 5$

Step 1: Solve both parts of the combined inequality at the same time. Isolate the term with the variable by subtracting 5 from each part.

$\frac{22}{11} < \frac{11x}{11} \leq \frac{55}{11}$

Step 2: Divide each part by 11.

$2 < x \leq 5$

Answer: The solution is an intersection. Graph the solution.

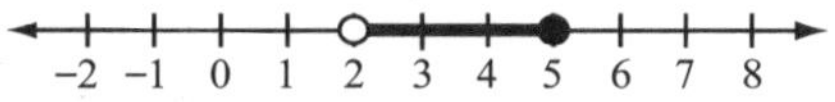

{All real numbers greater than 2 but less than or equal to 5}

Example 8 Solve and graph: $\frac{x}{2} > 6$ or $-3x > 12$

$\frac{x}{2} > 6$ $\quad -3x > 12$

$2 \cdot \frac{x}{2} > 6 \cdot 2$ $\quad \frac{-3x}{-3} > \frac{12}{-3}$

Step 1: Solve each inequality separately. Use the properties of real numbers to isolate the variable in each inequality.

$x > 12$ $x < -4$ **Step 2:** Remember to reverse the inequality sign in the second inequality because you divided by a negative number.

Answer: This is a disjunction. Graph the union of both inequalities.

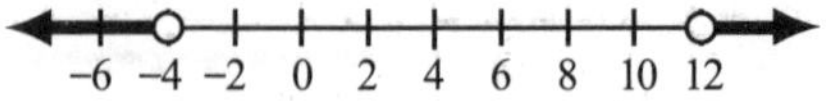

{All real numbers less than –4 or greater than 12}

Problem Set

Solve the following compound inequalities:

1. $2 - x < -2$ or $2 > 2x$

2. $6x \geq -6$ and $x - 5 \leq -3$

3. $-4x > -8$ or $x - 7 > -4$

4. $x + 8 > 11$ or $-2x > 12$

5. $63 > -7x > -14$

6. $8 < 11 + x < 13$

7. $2x > 8$ or $-3 + x < 7$

8. $2x > 16$ or $-5 + x < 4$

9. $-8 \leq x - 5 \leq -4$

10. $-7 < 7x < 42$

11. Solve and graph the compound inequality:
$x < 10$ or $x > 5$

12. Solve and graph the compound inequality:
$-3 + x > 8$ or $-x > -2$

13. Solve and graph the compound inequality:
$-4 + x > -10$ or $-x > 1$

14. Solve the compound inequality:
$-3x + 1 < 4$ and $-2x - 7 < 17$

15. Solve the compound inequality:
$-3x + 4 < 10$ and $-4x - 7 < 17$

16. Solve the compound inequality:
$-6x - 15 > -15$ and $-4x - 8 > 12$

17. Solve the compound inequality:
$-2x + 8 > 10$ and $-6x + 8 > 50$

18. Solve the compound inequality:
$17 < -3x + 11 < 23$

19. Solve and graph the compound inequality:
$7 < 7x + 14 < 56$

20. Solve and graph the compound inequality:
$-3x + 4 < 10$ and $-4x - 7 < 17$

HA1-220: Identifying and Multiplying Monomials

You have learned how an exponent is used as a shortcut for multiplication. An exponent is written a little to the right and raised above its base, and indicates the number of times to use the base as a factor. Consider the following example:

In 2^4, the exponent, 4, tells you to use the base, 2, as a factor four times.

$$\begin{aligned} 2^4 &= 2 \cdot 2 \cdot 2 \cdot 2 \\ &= 4 \cdot 4 \\ &= 16 \end{aligned}$$

Therefore, $2^4 = 16$.

The same concept is applied when a letter is used as a base.

$$\begin{aligned} a^3 &= a \cdot a \cdot a \\ b^5 &= b \cdot b \cdot b \cdot b \cdot b \end{aligned}$$

Recall that an algebraic expression consists of numbers, letters, or combinations of numbers and letters called terms. Terms are numbers, variables, or combinations of numbers and variables that are separated by the addition (+) and/or subtraction (–) symbols. Exponents can be a part of an algebraic expression. When an algebraic expression has only one term, it is called a **monomial.**

Monomial	A number, variable, or the product of numbers and variables containing non-negative integer exponents

Observe the sample monomials below.

$2a$ $\quad$ $4ab^2$ $\quad$ x^3 $\quad$ m^4n^5 $\quad$ 18 $\quad$ $3x \cdot 2y$

Note: An expression that has a variable in the denominator is not considered a monomial, such as the expression $\frac{7}{4x^2}$.

Refer to the expression $4ab^2$. Even though it consists of a coefficient and two variables being multiplied together, the term is still a monomial. Can you determine the total number of terms in the expression $3ab + 2a^4 - 6b$? The expression has three monomials that are separated by either an addition or subtraction operation symbol. Therefore, the expression $3ab + 2a^4 - 6b$ contains three terms. You should remember that numbers or variables that are multiplied or divided but not separated by addition or subtraction signs are considered one term. For example, there are only two terms in the expression $3d^2 + 5b(f) \cdot 4d$.

Suppose you are asked to multiply two monomials containing exponents. Given $2^4 \cdot 2^2$, you would probably write out four 2's to represent 2^4 and then write two more 2's to represent 2^2. This is called the **factored form** and looks like this:

$$2^4 \cdot 2^2 = 2 \cdot 2 \cdot 2 \cdot 2 \cdot 2 \cdot 2 = 2^6$$

When you are required to multiply, an easier way to simplify the problem is to keep the base and simply add the exponents, as shown in the examples below:

$$2^4 \cdot 2^2 = 2^{4+2} = 2^6$$
$$a^2 \cdot a^3 = a^{2+3} = a^5$$
$$b^3 \cdot b^3 = b^{3+3} = b^6$$
$$x^a \cdot x^b = x^{a+b}$$

Notice that the terms being multiplied must have the same base. As you multiply monomials, you must be sure that you multiply only those terms that have the same base. When terms have the same base, you can multiply them regardless of the number of terms..

Product Rule	If a is a real number and m and n are positive integers, then $a^m a^n = a^{m+n}$.

To multiply monomials, use the following steps:

Step 1: Multiply the coefficients.
Step 2: Group like variables together.
Step 3: Add the exponents for each variable

Try the following examples.

Example 1 Multiply: $(3ab^2c)(2a^2bc^3)$

$= 6 \cdot (ab^2c)(a^2bc^3)$	**Step 1:** Multiply the coefficients 3 and 2.
$= 6 \cdot (a \cdot a^2)(b^2 \cdot b)(c \cdot c^3)$	**Step 2:** Multiply like variables (in other words, variables that have the same base).
$= 6a^3b^3c^4$	**Step 3:** Add the exponents of like variables.
	Answer: $6a^3b^3c^4$

Example 2 Multiply: $(-4)(3x^2y^2)(3x^2y^3)$

$= -36\,(x^2y^2)(x^2y^3)$	**Step 1:** Multiply the coefficients.
$= -36 \cdot (x^2 \cdot x^2) \cdot (y^2 \cdot y^3)$	**Step 2:** Multiply like variables.
$= -36x^4y^5$	**Step 3:** Add the exponents of like variables.
	Answer: $-36x^4y^5$

Example 3 Multiply: $(5x^4y^6) \cdot (6xy^7z)$

$= 30 \cdot (x^4y^6) \cdot (xy^7z)$	**Step 1:** Multiply the coefficients.
$= 30 \cdot (x^4 \cdot x) \cdot (y^6 \cdot y^7) \cdot z$	**Step 2:** Multiply like variables.

$= 30x^5y^{13}z$ **Step 3:** Add the exponents of like variables.

Answer: $30x^5y^{13}z$

Problem Set

1. Is this expression a monomial?

$-10y$

No Yes

2. Is this expression a monomial?

$n - 2m$

No Yes

3. Is this expression a monomial?

$4ab$

No Yes

4. Is this expression a monomial?

s^4

No Yes

5. Is this expression a monomial?

-61

Yes No

6. Simplfy: $x^3 \cdot x^6$

7. Simplfy: $x^5 \cdot x^6$

8. Simplfy: $(x^4y^7)(x^8y^2)$

9. Simplfy: $(x^4y^6)(x^8y^2)$

10. Simplfy : $(ab^2)(a^3b)$

11. Simplfy: $(7c^2d^3)(-2c^4d^2)$

12. Simplfy: $(a^4b^3)(-9ab^4)$

13. Simplfy: $(6x^2y^3)(2x^3y^5)$

14. Simplfy: $(5x^4y^9)(3x^8y^2)$

15. Simplfy: $(-3x^4y^6)(5x^8y^2)$

16. Simplfy: $(-x^6y^7)(6x^9y^2)$

17. Simplfy: $(7x^{12}y^8)(-3x^8y^3)$

18. Simplfy: $-6(6x^4y^5)(10x^8y^{10})$

19. Simplfy: $-3(x^4y^{11})(8x^9y^2)$

20. Find the volume of the given figure.

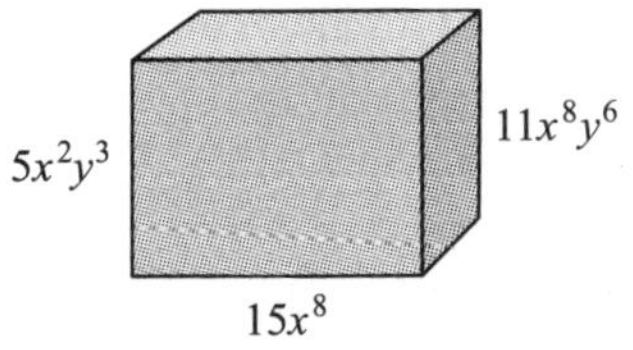

21. The length of a rectangle measures $3x$ inches and the width measures $7y$ inches. Find the expression that represents the area of the rectangle in square inches.

22. A car moves at $5x^2y^3$ miles per hour for $3x^2y^4$ hours. How many miles does it travel?

HA1-225: Dividing Monomials and Simplifying Expressions Having an Exponent of Zero

Recall that a monomial has only one term that consists of numbers, variables, and/or exponents.To multiply a monomial, add the exponents of the like variables. The opposite is true when you divide monomials. When dividing monomials, combine the exponents that have the same base, then subtract the exponents. Then determine whether the exponent is part of the numerator or the denominator.

Quotient Rule	For x, a non-zero real number, m and n are positive integers: If $m > n$, then $\frac{x^m}{x^n} = x^{m-n}$ If $n > m$, then $\frac{x^m}{x^n} = \frac{1}{x^{n-m}}$ If $m = n$, then $\frac{x^m}{x^n} = x^{m-n} = x^0 = 1$

Let's simplify $\frac{y^6}{y^2}$ using two methods, the factored form and the rule for dividing monomials. First, try the factored form. Beginning with the numerator, you must replace y^6 with six y's: $y \cdot y \cdot y \cdot y \cdot y \cdot y$. Next, replace y^2 in the denominator with two y's: $\frac{y \cdot y \cdot y \cdot y \cdot y \cdot y}{y \cdot y}$. You can cross out each y that appears in both the numerator and denominator, because $\frac{y}{y} = 1$. This is sometimes called cancelling. The y in the numerator is cancelled by the y in the denominator.

$$\frac{\cancel{y} \cdot \cancel{y} \cdot y \cdot y \cdot y \cdot y}{\cancel{y} \cdot \cancel{y}} = \frac{y^4}{1} = y^4$$

After all cancellations, multiply the numerator, and then multiply the denominator. The answer written in simplest form is y^4. Let's apply the rule for dividing monomials to the same problem. The rule states you can subtract exponents that share a common base. Therefore, you would have the following:

$$\frac{y^6}{y^2} = y^{6-2} = y^4$$

When using the rule to divide monomials, you must make sure that you correctly place the difference. In the above example, the larger exponent for base y is in the numerator; the difference, y^4, has to be placed in the numerator.

Now look at an example in which the larger exponent is in the denominator:

$$\frac{y^2}{y^4}$$

The rule for dividing monomials says to subtract the smaller exponent from the larger exponent; the difference is placed wherever the larger exponent was originally. In this case, $4 > 2$, so the difference is placed in the denominator.

$$\frac{1}{y^{4-2}} = \frac{1}{y^2}$$

Another way to evaluate the expression is to leave the y in the numerator and subtract the exponents. In this case you would have $y^{2-4} = y^{-2}$. You may recall that

$$y^{-2} = \frac{1}{y \cdot y} = \frac{1}{y^2}$$

Either method yields the same answer: $\frac{1}{y^2}$

Sometimes a problem will include coefficients. When it does, recall how you solved such divisions in arithmetic. Consider the next example.

$$\frac{5a^2}{10a}$$

The first step is to separate the coefficients from the variables, being careful not to change any part of the original problem.

$$\frac{5a^2}{10a} = \frac{5}{10} \cdot \frac{a^2}{a}$$

It should now be easy to see any divisible coefficients or terms. In this problem, divide 5 by 10 to get $\frac{1}{2}$. Then divide a^2 by a, which equals a. The final answer is $\frac{a}{2}$.

$$\frac{5a^2}{10a} = \frac{5}{10} \cdot \frac{a^2}{a} = \frac{1}{2} \cdot \frac{a^{2-1}}{1} = \frac{a}{2}$$

In some instances, a base and exponent may be found in only the numerator or denominator. If this is the case, the term still must be included in the answer. Look at the following example:

A coefficient and the variables a and b (to some power) appear both in the numerator and the denominator. However, c^3 only appears in the numerator. Therefore, c^3 will not be divided by any other term and will remain in the numerator as part of the final answer.

Example 1 Simplify: $\frac{4a^2bc^3}{2a^5b^3}$

$$\frac{4a^2bc^3}{2a^5b^3} = \frac{4}{2} \cdot \frac{a^2}{a^5} \cdot \frac{b}{b^3} \cdot \frac{c^3}{1}$$

Step 1: Separate the coefficients from the variables.

$$= \frac{2}{1} \cdot \frac{1}{a^{5-2}} \cdot \frac{1}{b^{3-1}} \cdot \frac{c^3}{1}$$

Step 2: Divide like variables by subtracting their exponents.

$$= \frac{2}{1} \cdot \frac{1}{a^3} \cdot \frac{1}{b^2} \cdot \frac{c^3}{1}$$

Step 3: Simplify the equation.

Answer: $\frac{2c^3}{a^3b^2}$

Example 2 Simplify: $\frac{x^2y^3}{x^3y^2}$

$\frac{1}{x^{3-2}} \cdot \frac{y^{3-2}}{1}$	**Step 1:** Divide like variables by subtracting their exponents.
$\frac{1}{x} \cdot \frac{y}{1} = \frac{y}{x}$	**Step 2:** Multiply numerators and denominators.
	Answer: $\frac{y}{x}$

Example 3 Simplify: $\frac{7a^6b^4}{21a^3b^9}$

$\frac{7}{21} \cdot \frac{a^6}{a^3} \cdot \frac{b^4}{b^9}$	**Step 1:** Divide the coefficients.
$\frac{1}{3} \cdot \frac{a^{6-3}}{1} \cdot \frac{1}{b^{9-4}}$	**Step 2:** Divide like variables by subtracting their exponents.
$\frac{1}{3} \cdot \frac{a^3}{1} \cdot \frac{1}{b^5} = \frac{a^3}{3b^5}$	**Step 3:** Multiply numerators and denominators.
	Answer: $\frac{a^3}{3b^5}$

Zero Exponent Rule	Let a be a non-zero real number $a^0 = 1$.

Recall what you learned in basic arithmetic about any fraction with the same numerator and denominator: that this type of fraction always equals 1. For example:

$$\frac{5}{5} = 1 \qquad \frac{18}{18} = 1 \qquad \frac{101}{101} = 1$$

When dividing monomials, you subtract exponents. When subtracting, the only way to get a zero result is for the subtrahend and the minuend to be the same number, such as in 5 – 5 = 0. Therefore, whenever you see a common base and an identical exponent in both the numerator and denominator, the exponent becomes zero and any variable with a zero exponent is always 1. Look at the example below.

$$\frac{a^5}{a^5} = a^{5-5} = a^0 = 1$$

Because a^5 appears in both the numerator and denominator, the answer is 1 regardless of what the base is. Try the following examples.

Example 4 Simplify: $\frac{c^{15}}{c^{15}}$

$c^{15-15} = c^0 = 1$

Answer: 1

Example 5 Simplify: $\frac{-4h^{23}}{12h^{23}n}$

$-\frac{4}{12} \cdot \frac{h^{23-23}}{1} \cdot \frac{1}{n}$

Step 1: Divide the coefficients and each variable.

$= -\frac{1}{3} \cdot \frac{h^0}{1} \cdot \frac{1}{n} = -\frac{1}{3} \cdot \frac{1}{1} \cdot \frac{1}{n} = \frac{-1}{3n}$

Notice that $\frac{h^{23}}{h^{23}} = h^0 = 1$.

Answer: $\frac{-1}{3n}$

Problem Set

Simplify the following:

1. $\frac{x^3}{x^5}$
2. $\frac{x^2}{x^8}$
3. $\frac{y^3}{y^7}$
4. $\frac{y^3}{y^{17}}$
5. $\frac{x^4}{x^5}$
6. $\frac{a^5}{a^2}$
7. $\frac{a^{15}}{a^{12}}$
8. $(3x)^0$
9. 7^0
10. $\frac{x^8}{x^8}$
11. $\frac{a^7b^9}{a^8b^5}$
12. $\frac{a^4b^5}{a^{10}b^2}$
13. $\frac{x^4y^5}{x^{10}y^2}$
14. $\frac{a^3b^5}{a^3b^2}$
15. $-5y^0$
16. $\frac{a^0b}{3}$
17. $(2x)(x^0)$
18. $\frac{-40a^4b^8}{5a^4b^7}$
19. $\frac{-30a^8b^7}{5a^4b^7}$
20. $\frac{21x^3y^2z^5}{14x^{10}y^{12}z^2}$

HA1-230: Raising a Monomial or Quotient of Monomials to a Power

In this lesson, you will learn to raise monomials or quotients of monomials to a power. First, let's look at the **Power of a Power** rule.

Power of a Power Rule	If a is a real number and m and n are positive integers, then $(a^m)^n = a^{mn}$.

In simpler terms, this rule means that the two exponents are multiplied to get a new exponent. What does $(x^2)^4$ mean? It would look like this: $(x^2)(x^2)(x^2)(x^2)$. Since you add exponents of common bases when you multiply, the product would be x^8. The simpler way is to multiply the exponents.

Example 1 Simplify: $(c^5)^4$

$(c^5)^4$
$= c^{5 \times 4}$
$= c^{20}$

Step 1: Follow the rules for exponents.

Answer: c^{20}

Example 2 Simplify: $(k^{20})^2$

$(k^{20})^2$
$= k^{20 \times 2}$
$= k^{40}$

Step 1: Follow the rules for exponents.

Answer: k^{40}

Next, view the rule for taking the power of a product.

Power of a Product Rule	If a and b are real numbers and n is a positive integer, $(ab)^n = a^n b^n$.

The **Power of a Product** rule states that when two or more factors are within parentheses and are raised to a power, each number or variable is independently raised to the same power. For example, if you simplify $(2c)^3$, the answer is derived by raising the number 2 and the variable c to the third power as in 2^3c^3, thereby obtaining an answer of $8c^3$.

Consider the following examples:

Example 3 Simplify $(-5xy^2z^3)^3$

$= (-5)^3(x)^3(y^2)^3(z^3)^3$ — **Step 1:** Raise each coefficient and variable to the third power.

$= -125x^3y^6z^9$ — **Step 2:** Multiply, following the rules for multiplying exponents.

Answer: $-125x^3y^6z^9$

Example 4 Simplify: $7m(4m)^2$

$= 7 \cdot m \cdot 4^2 \cdot m^2$ — **Step 1:** Raise both 4 and m to the second power.

$= 7 \cdot 4^2 \cdot m \cdot m^2$ — **Step 2:** Regroup coefficients together and variables together.

$= 7 \cdot 16 \cdot m \cdot m^2$
$= 112m^3$ — **Step 3:** Simplify 4^2 and multiply, following the rules for multiplying exponents.

Answer: $112m^3$

The final rule that helps govern this process is the **Power of a Quotient Rule**.

Power of a Quotient Rule	If a and b are real numbers, $b \neq 0$, and n is a positive integer, $\left(\frac{a}{b}\right)^n = \left(\frac{a^n}{b^n}\right)$.

According to the rule of Power of a Quotient Rule, when a monomial being divided by another monomial is raised to a power, each number or variable in the numerator and denominator is raised to that power, as shown in the next examples.

Example 5 Simplify: $\left(\frac{4t}{u^4}\right)^2$

Step 1: Follow the rules for exponents.

$$\left(\frac{4t}{u^4}\right)^2$$

$$= \frac{(4)^2(t)^2}{(u^4)^2}$$

$$= \frac{16t^2}{u^8}$$

Answer: $\frac{16t^2}{u^8}$

Example 6 Simplify: $\left(\frac{j^3}{5k^2}\right)^3$

$\left(\frac{j^3}{5k^2}\right)^3$

Step 1: Follow the rules for exponents.

$= \frac{(j^3)^3}{5^3(k^2)^3}$

$\frac{j^9}{125k^6}$

Answer: $\frac{j^9}{125k^6}$

Problem Set

Simplify each of the following:

1. $(y^8)^9$

2. $(x^5)^6$

3. $(m^6)^8$

4. $(m^2)^{11}$

5. $\left(\frac{x^2}{y^3}\right)^5$

6. $\left(\frac{x^8}{y^9}\right)^2$

7. $\left(\frac{4}{x}\right)^2$

8. $\left(\frac{5}{x}\right)^2$

9. $\left(\frac{y}{6}\right)^2$

10. $\left(\frac{y}{7}\right)^2$

11. $(9x^2y^3)^2$

12. $(2x^2y^3z^4)^2$

13. $(2x^3y^4z^5)^3$

14. $(4a^8b^9c^{10})^2$

15. $(-3a^6b^{12}c)^2$

16. $2x^4(2x)^4$

17. $3x^5(2x)^4$

18. $\left(\frac{3}{x^9y^8z^5}\right)^4$

19. $\left(\frac{-3a}{2b^3c^2}\right)^3$

20. $(10x^4y^{11})(2x^3y^5)^3$

HA1-235: Writing, Multiplying, and Dividing Numbers Written in Scientific Notation

Before the technological age, all forms of note-taking and dictation were done by hand. It did not matter how fast an individual spoke or how quickly the events in a meeting unfolded, you had to rapidly write the information presented so that no vital material was lost. One of the most common methods for quick note taking was called shorthand—a system made up of symbols that stood for certain words or meanings.

The field of mathematics also needed to develop a method that would make it easier to read and write small or large numbers. Consequently, a shorter, more concise method for writing such numbers was developed. Now mathematicians, scientists, and engineers use **scientific notation**, also called **exponential notation**, to express very small or large numbers. Scientific notation also makes computation of numbers significantly easier. Let's view the definition and format of scientific notation.

Scientific Notation	A method of writing large and small numbers in the form $a \times 10^n$ where $1 \leq a < 10$ and n is an integer.

In plain terms, $a \times 10^n$ is the product of a positive number that is greater than one but less than ten and a power of base 10. The power may be positive or negative.

To express numbers larger than 1 in scientific notation

Step 1: Locate the decimal in the number and move it to the left until only one non-zero digit remains to its left. The new number is the a part of the expression $a \times 10^n$.

Step 2: Count the number of places that the decimal was moved in the first step. This number of places will equal the n part of the expression.

Example 1 The distance from the earth to the sun is 93,000,000 miles. Write this number in scientific notation.

$93{,}000{,}000. = 9.3 \times 10^n$

Step 1: Recall that the decimal is actually to the right of the last zero in 93,000,000, even though it is not written. The decimal should be moved so that only one number is to its left.

$93,\ 000,\ 000. = 9.3 \times 10^7$

7 654 321

Step 2: Count the number of places the decimal was moved to the left; that number replaces the n in the notation.

To express numbers less than 1 in scientific notation

Follow the same steps for numbers less than one as for numbers that are greater than ten, except that the decimal must be moved to the *right* until only one non-zero number is to the left of the decimal. The number of positions that the decimal must be moved replaces the n in the notation. However, the exponent will be negative.

Example 2 Write 0.00237 in scientific notation.

$0.00237 = 2.37 \times 10^n$

Step 1: Move the decimal so that only one non-zero number is to the left.

$0.00237 = 2.37 \times 10^{-3}$

123

Step 2: Count the number of places you moved the decimal to the right. Notice that when you move decimals to the right, the exponent will be negative.

As seen in the second example, an exponent can be negative. However, you may sometimes want to work with a positive exponent rather than with a negative one. **Negative exponents** can be rewritten as positive exponents by inverting the number. Remember that 10^{-3} can also be written as $\frac{10^{-3}}{1}$. Use 10^0 to represent one in the denominator, then simplify:

$$\frac{10^{-3}}{1} = \frac{10^{-3}}{10^0} = \frac{10^{-3} \times 10^3}{10^0 \times 10^3} = \frac{1}{10^3} = \frac{1}{1000}$$

A quicker way to get the same result is to invert the fraction and make the exponent positive:

$$10^{-3} = \frac{10^{-3}}{1} = \frac{1}{10^3} = \frac{1}{1000}$$

Therefore, the following definition is derived:

Negative Exponent Rule	For all real numbers a, $a \neq 0$, and for all positive integers n: $a^{-n} = \frac{1}{a^n}$.

Consider the following example:

Example 3 Simplify: 10^{-3}.

$= \frac{1}{10^3}$ **Step 1:** Invert the negative exponent. The positive exponent is formed.

$= \frac{1}{1000}$ **Step 2:** Simplify the denominator.

Now that you can write numbers in scientific notation, you can also perform some computations. When you multiply numbers written in scientific notation, multiply the integers or decimal numbers first; then multiply the terms with base 10 by adding their exponents.

Example 4 Multiply: $(2.3 \times 10^3)(3 \times 10^4)$

$= (2.3 \times 3)(10^3 \times 10^4)$ **Step 1:** Regroup the numbers so that the integer or decimal numbers are in one set of parentheses and exponents are in the other.

$= (6.9)(10^3 \times 10^4)$ **Step 2:** Multiply the integer or decimal number.

$= (6.9)(10^7)$ **Step 3:** Multiply the exponential parts by adding the exponents.

Answer: Thus, the solution is 6.9×10^7.

Since multiplication and division are inverses of each other, it is necessary to discuss dividing numbers that are written in scientific notation. When you divide numbers written in scientific notation, you should divide the integers or decimal numbers first.

Next, divide the exponents. Remember that you subtract exponents when dividing.

Example 5 Divide: $\frac{(5.6 \times 10^9)}{(7 \times 10^4)}$

$\frac{5.6}{7} \times \frac{10^9}{10^4}$	**Step 1:** Regroup the numbers by placing the integers or decimal numbers in one fraction and the exponents in another fraction.
$= 0.8 \times \frac{10^9}{10^4}$	**Step 2:** Divide the decimal or integer number.
$= 0.8 \times 10^5$	**Step 3:** Subtract the exponential parts.
$= 0.8 \times 10^5$ $= \frac{8 \times 10^5}{1 \times 10^1}$ $= 8 \times 10^{-1} \times 10^5$	**Step 4:** Note that the number is not in scientific notation because 0.8 is not a number between 1 and 10. To fix this, move the decimal point one place to the right. Because you are moving to the right one place, it is the same as multiplying by 10^{-1}.
$= 8 \times 10^4$	**Step 5:** To simplify, add the exponents, which combines the powers of 10.

Answer: Thus, the solution is 8×10^4.

Problem Set

Are the following expressions written in scientific notation?

1. 0.8×10^5

2. 32.5×10^2

3. 7.5×10^3

4. 8.0×10^6

Express the following numbers in scientific notation:

5. 73,200

6. 684,000

7. 15,000,000

8. 82,500,000

9. 0.008

10. 0.00654

Evaluate each and express the answer in scientific notation:

11. $(4.0\times10^3)(2.1\times10^2)$

12. $(3.0\times10^3)(1.2\times10^{15})$

13. $(4.0 \times 10^6)(1.1 \times 10^6)$

14. $(3.8 \times 10^{11})(2.2 \times 10^4)$

15. $\frac{4.75 \times 10^2}{2.5 \times 10^5}$

16. $\frac{7.75 \times 10^{33}}{3.1 \times 10^3}$

17. $\frac{7.68 \times 10^{20}}{3.2 \times 10^2}$

18. $(4.72 \times 10^5)(5.5 \times 10^2)$

19. $\frac{9.6 \times 10^{15}}{12.0 \times 10^5}$

20. $\frac{2.5 \times 10^{14}}{5.0 \times 10^5}$

HA1-240: Identifying the Degree of Polynomials and Simplifying by Combining Like Terms

Recall that a monomial is an algebraic expression containing only one term. In this lesson, you will learn to identify the degree of polynomials and simplify by combining like terms.

Polynomial	An algebraic expression consisting of a monomial or the sum or difference of monomials

Binomials and trinomials are special types of polynomials. A binomial is an algebraic expression containing two terms separated by an addition or subtraction sign. A trinomial is an algebraic expression containing three terms that are each separated by an addition or subtraction sign. 1

Consider the following examples:

Monomial	Binomial	Trinomial	Polynomial
$3xy$	$2a + b$	$a + b + c$	$2x + y^3 - z + 5$
$-2a$	$4x - 5y^2$	$2x^2 - 3y + z$	$a - b + 3x - 4d^2 + h$
6	$6z - 3$	$5x + d - 2s$	$5f + 2h - h - 6i$

The **degree** of a monomial is found by adding the exponents of the variables. Note that coefficients do not have a degree; they are considered to be **constant**. A constant has a degree of 0. Recall that when a variable does not have an exponent, it is assumed to have an exponent of 1.

Degree of a Term	The sum of the exponents of its variables

Example 1 State the degree of $3x^2yz^3$.

Step 1: The exponents of the variables x^2, y^1, and z^3 are 2, 1, and 3. Add the exponents $2 + 1 + 3 = 6$.

Answer: 6

Example 2 State the degree of $54ab^2c^5$.

Step 1: Add the exponents of the variables a^1, b^2, and c^5, or $1 + 2 + 5 = 8$.

Answer: 8

Degree of a Polynomial	The highest degree of any of its terms

Example 3 State the degree of $3ab + 5c^2d^3$.

Step 1: The binomial's first term, $3ab$, has a degree of 2. The second term, $5c^2d^3$, has a degree of 5. The highest degree is 5, so the degree for the whole expression is 5.

Answer: 5

Example 4 State the degree of $4x + 5y^2 + 7z^3$.

Step 1: The first term, $4x$, has a degree of 1. The second term, $5y^2$, has a degree of 2. The third term, $7z^3$, has a degree of 3. Therefore, this trinomial has a degree of 3.

Answer: 3

Recall that in a previous lesson you learned to combine like terms. For example, in the expression $2x + 5y - 3x$, the only terms that can be combined are $2x$ and $-3x$. However, some expressions have terms with several variables containing different degrees. The process for combining terms in these types of expressions is the same as the process you used to simplify $2x + 5y - 3x$. Combine only those terms that share the same variable raised to the same degree. For example, in the expression, $4x^2y + 5xy + 2xy^2 + 3xy$, only the $5xy$ and $3xy$ can be combined because they are the only terms with the same variables raised to the same degree. The unlike terms remain as they are. Therefore, the simplified expression is $4x^2y + 8xy + 2xy^2$.

Now that you are working with complicated expressions with different variables, exponents, and terms, terms must be written in a certain order. Mathematicians put complex expressions either in *descending* or *ascending* order. To put an algebraic expression in order, choose one variable and look for its occurrence in all the other terms. Then, put all terms for the variable in ascending or descending order, according to the degree of that variable. This should be done for all variables, and will help you to not eliminate any terms from the expression. Even though the examples below put the simplified expressions in order based on ***x***, you can use any variable if you put all terms in ascending or descending order according to that variable.

Example 5 Simplify $6x^2 - 5x + 2x^2 + 3x^3$ and write in *descending* order with respect to *x*.

$= 8x^2 - 5x + 3x^3$ **Step 1:** Combine $6x^2 + 2x^2$.

$= 3x^3 + 8x^2 - 5x$ **Step 2:** Put in descending order based on x (x^3, x^2, x).

Answer: $3x^3 + 8x^2 - 5x$

Example 6 Simplify $-4xy^2z + 2x^3 + 3x^4y^3z^2 + 5x^2y$ and write in *ascending* order with respect to *x*.

Step 1: Combine like terms; there are none.

Step 2: Concentrate on the variable x only. In the term $-4xy^2z$, x has an exponent of 1; in $2x^3$, x has an exponent of 3; in $3x^4y^3z^2$, x has an exponent of 4; in $5x^2y$, x has an exponent of 2.

Step 3: Put the terms in ascending order according to *x*.

Answer: $-4xy^2z + 5x^2y + 2x^3 + 3x^4y^3z^2$

Example 7 Simplify $3x^4 - x^7 + 8x^7 - 3x + 7x$ and write in *ascending* order with respect to x.

$= 3x^4 + 7x^7 + 4x$ **Step 1:** Combine like terms.

$= 4x + 3x^4 + 7x^7$ **Step 2:** Arrange terms in ascending order according to x.

Answer: $4x + 3x^4 + 7x^7$

Problem Set

1. Find the degree of the following polynomial:

$$9m^5n^4p$$

2. Find the degree of the following polynomial:

$$17x^4y^2z$$

3. Find the degree of the following polynomial:

$$9m^5n^3p$$

4. Find the degree of the following polynomial:

$$17x^3y^2z$$

5. Is the following expression a monomial, binomial, or trinomial? $5xy$

6. Is the following expression a monomial, binomial, or trinomial? $6xy^2 + 3x$

7. Is the following expression a monomial, binomial, or trinomial? $3x^2 + 4x + 5$

8. Find the degree of the following polynomial:

$$3x^2y^5 - 11x^6y^8 + 12x^3 + 5y^5$$

9. Find the degree of the following polynomial:

$$6x^5y^4 + 5x^{10}y^8 + 3x^9y^7 + 2x^5y^2$$

10. Find the degree of the following polynomial:

$$12x^3y^3 + 3x^2y^2 + 2xy$$

11. Arrange the following polynomial in descending order with respect to x:

$$7mx - 8mx^3 + 10m^3x^2$$

12. Arrange the following polynomial in ascending order with respect to x:

$$-2x^2y + 3xy^3 + x^3$$

13. Arrange the following polynomial in descending order with respect to x:

$$2mx - 6mx^3 + 10mx^2$$

14. Arrange the following polynomial in ascending order with respect to x:

$$3ax - 5ax^3 + 12a^3x^2$$

15. Arrange the following polynomial in descending order with respect to x:

$$5x^4 + 6x^2 + 10 + 7x^3$$

16. Choose the polynomial that is written in descending order with respect to x:

$3x^4 - 5x^6 - 2x^4 + 6x^6$ $4x^3y^4 - 2x^2y - 4x$

$9x - 5 + 6x^3 - x^5$ $-2x + 4x^3 - 7x^2 + 8$

17. Choose the polynomial that is written in descending order with respect to x:

$4x^2 - 3 + 7x^5 + 2x^3$ $4x^3 + 6x^4 + 3x^2$

$4x^4 + 4x^3 + 8x$ $7 + 5x^2 + 3x^4 - 2x$

18. Simplify and write in descending order with respect to x:

$$-3x^3y^2 + 2x^2y^3 - 2x^3y^2 + 5x^2y^3$$

19. Simplify and write in descending order with respect to x:

$$-4x^3y^2 + 6x^2y^3 - 2x^3y^2 + 2x^2y^3$$

20. Simplify and write in descending order with respect to x:

$$-6x^3y^2 + 3x^2y^3 - 2x^3y^2 + x^2y^3$$

HA1-245: Adding and Subtracting Polynomials

Once you know how to combine terms, you are ready to add and subtract polynomials.

Say you want to add the trinomial $7x - 2y + 5z$ together with the binomial $-5x + 8y$. It would look like this: $(7x - 2y + 5z) + (-5x + 8y)$.

Since you are simply adding, or putting together, the two polynomials, you may drop the parentheses and combine like terms.

$$\begin{aligned} &= 7x - 2y + 5z - 5x + 8y \\ &= (7x - 5x) + (-2y + 8y) + 5z \\ &= 2x + 6y + 5z \end{aligned}$$

With the like terms grouped together, it is easier to see exactly which terms can be added or subtracted. Once the expression is simplified, you get the solution $2x + 6y + 5z$.

The following examples are more complex because they contain exponents. Recall that terms are considered like terms only if they have the same variable(s) raised to the same degree(s). Therefore, $2x^3$ and $5x^3$ are like terms; in contrast, $3x^3$ and $5y^3$ are not like terms. $4xy^2$ and $2xy$ are not like terms, but $4xy^2$ and $2xy^2$ are. If the expression contains variables with different exponents, you must rearrange the expression so that each variable is in ascending or descending order according to its exponents.

Example 1 Simplify: $(3x^2 - 6y^2 - 4z^2) + (-2y^2 + x^2 - z^2)$

$3x^2 - 6y^2 - 4z^2 - 2y^2 + x^2 - z^2$	**Step 1:**	Drop the parentheses since you are adding.
$= (3x^2 + x^2) + (-6y^2 - 2y^2) + (-4z^2 - z^2)$	**Step 2:**	Regroup the like terms and rewrite the expression.
$= 4x^2 - 8y^2 - 5z^2$	**Step 3:**	Simplify the expression by adding or subtracting the like terms.
	Answer:	$4x^2 - 8y^2 - 5z^2$

When you subtract polynomials, recall that subtracting means "to take the opposite of." In the next example, you will add the opposite of each term in the *second* polynomial. Remember that the opposite of –4 is 4, while the opposite of 3 is –3. Now, consider the following example.

Example 2 Simplify: $(4d^3 + 6f^2 - 9d^4) - (2f^2 - 4d^4 + 5d^3)$

$= (4d^3 + 6f^2 - 9d^4) + (-2f^2 + 4d^4 - 5d^3)$	**Step 1:**	Change this to an addition problem by making the subtraction sign an addition sign AND changing each term in the second polynomial to its opposite.
$= 4d^3 + 6f^2 - 9d^4 - 2f^2 + 4d^4 - 5d^3$	**Step 2:**	Drop the parentheses.
$= (4d^3 - 5d^3) + (6f^2 - 2f^2) + (-9d^4 + 4d^4)$	**Step 3:**	Group the like terms.
$= (-d^3 + 4f^2 - 5d^4)$	**Step 4:**	Add or subtract the grouped terms.

$= 4f^2 - d^3 - 5d^4$	**Step 5:**	Rewrite the terms in ascending (or descending) order according to one variable.
	Answer:	$4f^2 - d^3 - 5d^4$

Problem Set

Add as indicated:

1. $(7b^2 + 9b) + (5b^2 - 6b - 8)$
2. $(7x^2 - 5x + 6) + (-4x - 2x^2 + 8)$
3. $(-13m^2 + 8m - 27) + (6m^2 - 8)$
4. $(6b^2 + 10b) + (4b^2 - 5b - 6)$
5. $(8m^2 + 6m + 4) + (3m^2 - 8m + 1)$
6. $(6x^2 - 8x + 4) + (-3x - 2x^2 + 6)$
7. $(5m - 8x + 2z) + (7m - x - 4z)$
8. $(5x^2 + 12x + 2) + (3x^2 + 3x - 7)$

Subtract as indicated:

9. $(5n^2 - 4n + 5) - (3n^2 + 7n + 6)$
10. $(9x^2 - 5x - 4) - (7x^2 + 8x + 6)$
11. $(7n + 4m + a) - (-3a - 7n + m)$
12. $(3a - 2b + 3c) - (-a - 3b - 4c)$
13. $(7n^2 - 4n + 1) - (2n^2 + 3n - 5)$
14. $(8x^2 - x + 2) - (5x^2 + 4x - 5)$
15. $(-13m^2 + 8m - 27) - (6m^2 - 8)$
16. $(6b^2 + 10b) - (4b^2 - 5b - 6)$

Solve the following:

17. This weekend Maria earned $6x + 3y$ dollars babysitting and $15x - 7y$ dollars walking dogs. How much money did Maria earn?

18. When a ball is thrown, it takes $2x^2 + 4x + 3$ seconds to fall to the ground. Once it hits the ground, the ball rolls for $16x - 9$ seconds. How long was the ball in motion?

19. A rectangle has a base of $2x + 13xy + 4y$ inches and a height of $4x - 2xy - 7y$ inches. What is the perimeter of the rectangle in inches?

20. Carlos owned $7x^2 + 6x - 4$ baseball cards. He lost $2x^2 + 3$ cards and sold $4x - 13$ cards. How many baseball cards does Carlos still have?

HA1-255: Multiplying Two Binomials Using the FOIL Method

When you multiply two binomials, you will use the distributive property in a slightly different way. The **FOIL** method will help you do this.

FOIL Method	A method used to multiply two binomials, in the following order:		
	F	First	Multiply the first terms of each binomial.
	O	Outer	Multiply the outer terms: the first term of the first binomial and the second term of the second binomial.
	I	Inner	Multiply the inner terms: the second term of the first binomial and the first term of the second binomial.
	L	Last	Multiply the second terms of each binomial.

When you multiply binomials using FOIL, the solution is almost always a trinomial, a polynomial with three terms. In some cases you may get another binomial; this occurs when the middle terms of the trinomial cancel each other or equal zero. Let's use FOIL to solve the following problems:

Example 1 Multiply: $(2x+y)(4x-3y)$

F	$(2x+y)(4x-3y) = 8x^2$	**Step 1:**	Multiply the first terms of each binomial, in this case $2x$ and $4x$.
O	$(2x+y)(4x-3y) = -6xy$	**Step 2:**	Multiply the outer terms of each binomial, in this case $2x$ and $-3y$.
I	$(2x+y)(4x-3y) = 4xy$	**Step 3:**	Multiply the inner terms of each binomial, in this case y and $4x$.
L	$(2x+y)(4x-3y) = -3y^2$	**Step 4:**	Multiply the last terms of each binomial, in this case y and $-3y$.
	$-6xy + 4xy = -2xy$	**Step 5:**	Combine the outer and inner terms to get the middle term of the trinomial.

Answer: $8x^2 - 2xy - 3y^2$

Note: Remember that the outer and inner terms must be combined using addition or subtraction. The resulting middle term becomes part of the solution.

Example 2 Multiply: $(2x+2)(x+5)$

F $2x \cdot x = 2x^2$

O $2x \cdot 5 = 10x$

I $2 \cdot x = 2x$

L $2 \cdot 5 = 10$

$10x + 2x \Rightarrow 12x$

Answer: $2x^2 + 12x + 10$

Example 3 Multiply: $(2x-4)(x-5)$

F	$2x \cdot x = 2x^2$	
O	$2x \cdot (-5) = -10x$	$-14x$
I	$(-4) \cdot x = -4x$	
L	$(-4) \cdot (-5) = 20$	

Answer: $2x^2 - 14x + 20$

Example 4 Multiply: $(x-5)(x+5)$

F	$x \cdot x = x^2$	
O	$x \cdot 5 = 5x$	0, so the terms cancel
I	$x \cdot (-5) = -5x$	
L	$-5 \cdot 5 = -25$	

Answer: $x^2 - 25$

Note: The example above shows a binomial that is simplified using the FOIL method, resulting in another binomial rather than a trinomial. Note that this occurs when the last terms of both binomials are opposites.

Problem Set

Multiply the following binomials using the FOIL method:

1. $(x+5)(x+6)$
2. $(y+9)(y+7)$
3. $(y+2)(y+11)$
4. $(c-2)(c-3)$
5. $(d-6)(d-3)$
6. $(k-5)(k-2)$
7. $(k-1)(k-6)$
8. $(x+1)(x-3)$
9. $(s-8)(s+5)$
10. $(s+5)(s-9)$
11. $(n-7)(n+7)$
12. $(s+8)(s-8)$
13. $(2x-3)(x-3)$
14. $(x+3)(3x+4)$
15. $(2x-9)(x+3)$
16. $(x+4)(3x-1)$
17. $(y-4)(5y+7)$
18. $(2y+4)(2y+9)$
19. $(x+3y)(3x-4y)$
20. $(2x-9y)(x-3y)$

Solve the following:

21. The length of a rectangular swimming pool is $(x+3)$ feet and the width is $(x-5)$ feet. How many square feet are contained in the area of the swimming pool?

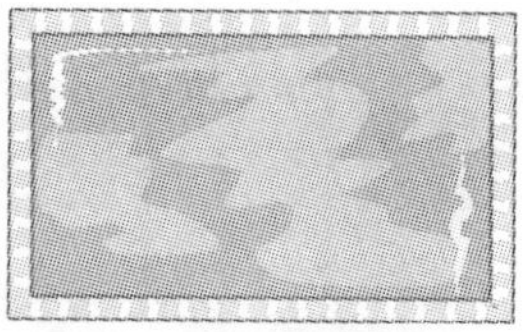

$(x-5)$

$(x+3)$

22. Chad runs $(2g+15)$ miles each morning. If he runs $g-2$ minutes per mile, how many minutes does Chad run each morning?

HA1-260: Squaring a Binomial and Finding the Product of a Sum and Difference

Recall that when you see a base with an exponent such as x^2, you know that x squared means x times x (or x multiplied by x). The exponent 2 means to use the base as a factor two times, so that $x^2 = x \cdot x$. By substituting the number 6 for the variable x, you get $6^2 = 6 \cdot 6$, which equals 36. Exponents can also be squared through the application of the Power of a Power rule, as in $(y^3)^2 = y^3 \cdot y^3 = y^6$.

Binomials (polynomials that have two terms separated by either a plus or minus sign) can also be squared. Suppose you are asked to solve the expression $(2x + 3y)^2$. This means that $(2x + 3y)$ will be used as a factor two times. You will find it easier to solve the binomial $(2x + 3y)^2$ when it is in the expanded form, $(2x + 3y)(2x + 3y)$. Once a squared binomial is written in its expanded form, you can solve it using the **FOIL** method. For the first example, let's try solving $(2x + 3y)^2$. Do you remember what each letter in the acronym FOIL means? **F** = first terms, **O** = outer terms, **I** = inner terms, and **L** = last terms.

Example 1 Simplify: $(2x + 3y)^2$

$= (2x + 3y)(2x + 3y)$ — **Step 1:** Write the expression in expanded form.

F $2x \cdot 2x = 4x^2$ — **Step 2:** Multiply using FOIL.

O $2x \cdot 3y = 6xy$

I $3y \cdot 2x = 6xy$

L $3y \cdot 3y = 9y^2$

$= 4x^2 + 6xy + 6xy + 9y^2$ — **Step 3:** Combine like terms.

$= 4x^2 + 12xy + 9y^2$

Answer: Therefore, the solution is $(2x + 3y)^2 = 4x^2 + 12xy + 9y^2$.

A formula can also be used to simplify the square of a binomial.

Square of a Binomial	$(a - b)^2 = a^2 - 2ab + b^2$ $(a + b)^2 = a^2 + 2ab + b^2$

Look at the following examples.

Example 2 Simplify: $(x + 7)^2$

$(x + 7)^2$ — **Step 1:** Identify, as a special binomial, the square of a binomial.

$a = x$ — **Step 2:** Identify a and b.

$b = 7$

$= x^2 + 2 \cdot x \cdot 7 + 7^2$ $= x^2 + 14x + 49$	**Step 3:** Use the formula to simplify: $(a+b)^2 = a^2 + 2ab + b^2$

Answer: $x^2 + 14x + 49$

Example 3 Simplify: $(3x-4)^2$

$(3x-4)^2$	**Step 1:** Identify, as a special binomial, the square of a binomial.
$a = 3x$ $b = 4$	**Step 2:** Identify a and b.
$= (3x)^2 - 2(3x)(4) + 4^2$ $= 9x^2 - 24x + 16$	**Step 3:** Use the formula to simplify: $(a-b)^2 = a^2 - 2ab + b^2$

Answer: $9x^2 - 24x + 16$

Next, look at the product $(x+y)(x-y)$. Notice that the factors are very similar: the first factor is the sum of x and y and the second is the difference of x and y. The solution to this binomial is referred to as the **difference of two squares.**

Example 4 Simplify: $(x+y)(x-y)$

F $(x)(x) = x^2$ **O** $(x)(-y) = -xy$ **I** $(y)(x) = xy$ ($-xy + xy \Rightarrow 0$) **L** $(y)(-y) = -y^2$	**Step 1:** Multiply using FOIL.

Answer: $x^2 - y^2$

The sign between the two terms will always be a negative.

A formula can be used to simplify the product of a sum and difference.

Product of a Sum and Difference	$(a-b)(a+b) = a^2 - b^2$

Look at the following examples.

Example 5 Simplify: $(2a+3)(2a-3)$

$(2a+3)(2a-3)$	**Step 1:** Identify as a special binomial (the product of a sum and difference). Therefore, the middle terms will cancel.
$2a \cdot 2a = 4a^2$	**Step 2:** Multiply the first terms.
$3 \cdot -3 = -9$	**Step 3:** Multiply the last terms.
	Answer: $4a^2 - 9$

Example 6 Simplify: $(x+4y)(x-4y)$

$(x+4y)(x-4y)$	**Step 1:** Identify as a special binomial (the product of a sum and difference). Therefore, the middle terms will cancel.
$x \cdot x = x^2$	**Step 2:** Multiply the first terms.
$4y \cdot -4y = -16y^2$	**Step 3:** Multiply the last terms.
	Answer: $x^2 - 16y^2$

Problem Set

1. Simplify: $(a+1)(a-1)$

2. Simplify: $(2-n)(2+n)$

3. Simplify: $(x+3)(x-3)$

4. Simplify: $(n+5)(n-5)$

5. Simplify: $(d-12)(d+12)$

6. Simplify: $(a+1)^2$

7. Simplify: $(b-1)^2$

8. Simplify: $(r+2)^2$

9. Simplify: $(d-7)^2$

10. Simplify: $(k+9)^2$

11. Simplify: $(5y-7)^2$

12. Simplify: $(4d+11)^2$

13. Simplify: $(3m-8)^2$

14. Simplify: $(x-y)^2$

15. Simplify: $(3+k)^2$

16. Simplify: $(5k+8)^2$

17. Simplify: $(6m-4)^2$

18. Simplify: $(4x-3y)(4x+3y)$

19. Simplify: $(3x-4y)^2$

20. The formula for the area of a square is giver by $Area = (side)^2$. Which expression represents the area of the square?

$2s+6$

$4s^2 + 24s + 36$ $\quad$ $4s^2 + 36$

$4s^2 + 12s + 36$ $\quad$ $4s^2 - 36$

HA1-265: Writing a Number in Prime Factorization and Finding the Greatest Common Factor

Suppose a bag contains 24 cookies. How many ways could the cookies be shared so that none are left over? You could have one group of 24 cookies, 24 groups of 1 cookie, 2 groups of 12, 12 groups of 2, 3 groups of 8, 8 groups of 3, 4 groups of 6, or 6 groups of 4 cookies. Many possibilities! In this example, 1, 2, 3, 4, 6, 8, 12 and 24 are called **factors** of 24 because when each of these is multiplied with another, the product is 24.

Factor	A number that is being multiplied

Consider the number 50. The numbers 2 and 25 are factors of 50 because 2 x 25 = 50, and 50 is divisible by both 2 and 25 with a remainder of zero. Are there other factors of 50? Note that $1 \cdot 50 = 50$ and $5 \cdot 10 = 50$; therefore, the numbers 1, 2, 5, 10, and 25 are all factors of 50.

Depending on the number of factors, a given number can be classified as **prime** or **composite**. Every whole number greater than one is either a prime or composite number.

Prime Number	An integer greater than one that has only two factors, one and the number itself

For example, the number 5 is a prime number because it has only two factors, 5 and 1. The first few prime numbers are 2, 3, 5, 7, 11, 13, 17, 19. Euclid, a Greek mathematician, proved that there are an infinite number of primes.

A composite number can always be expressed as the product of two or more prime numbers.

Composite Number	An integer greater than one that has more than two factors

The number 8 is a composite number since it has four factors: 1, 2, 4, and 8. Examples of other composite numbers are 4, 6, 8, 9, 10, 12, 15, 16, 18. From this information, you can deduce:

- the numbers 0 and 1 are neither prime nor composite;
- zero has an endless number of factors (every number is a factor of zero, since any number multiplied by zero is zero);
- the number 1 has only one factor, itself.

Several methods of **prime factorization** can be used to express a product of factors that are prime.

Prime Factorization	A composite number written as the product of prime numbers.

One method is called the *factor tree*. This is a diagram resembling a tree that begins with two easily divisible factors of a number. If possible, these factors are factored again and the process continues until all factors are prime.

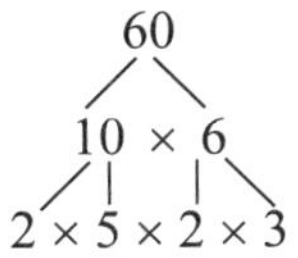

Example 1 Find the prime factors of 180 using a factor tree.

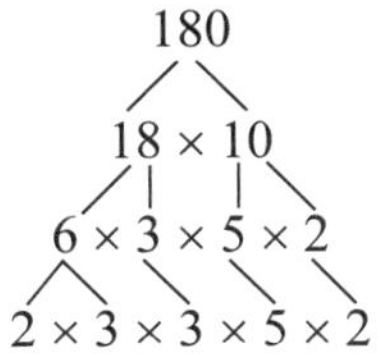

Step 1: Express the number as a product of two factors. In this example, 18 and 10 are used but two different factors could have been used. If one or both factors are composite, then that number must be factored again.

$= 2 \times 2 \times 3 \times 3 \times 5$

Step 2: Rewrite the prime numbers in ascending order.

Sometimes you may need to factor a negative integer. Write any negative integer as the product of –1 and a whole number. This example will be solved using a different method, involving division by prime numbers.

Example 2 Find the prime factors of –120.

$= -1 \times 120$

Step 1: Express –120 as a product of two factors.

$= -1 \times 2 \times 60$

$= -1 \times 2 \times 2 \times 30$

$= -1 \times 2 \times 2 \times 2 \times 15$

Step 2: Factor each composite number as a product of two factors.

$= -1 \times 2 \times 2 \times 2 \times 3 \times 5$

Step 3: Ensure the factors are written in ascending or descending order.

Once you have found the prime factors of a number, an exponent may be used to express the power of a factor if any factor appears more than once. So, if you reconsider the solution to –120 in the second example, the final answer can be rewritten as $-1 \times 2^3 \times 3 \times 5$. This is known as the exponential form.

Suppose you wanted to know the common factor of two or more numbers.

Common Factor	The same factor for two or more integers

For example, because 3 is a factor of both 6 and 9, it is a common factor for those numbers.

The **Greatest Common Factor (GCF)** is the greatest integer that is a factor of all integers in the list.

Greatest Common Factor (GCF)	The largest integer that will divide into two or more numbers evenly with no remainder

Example 3 Find the greatest common factor of 48 and 60.

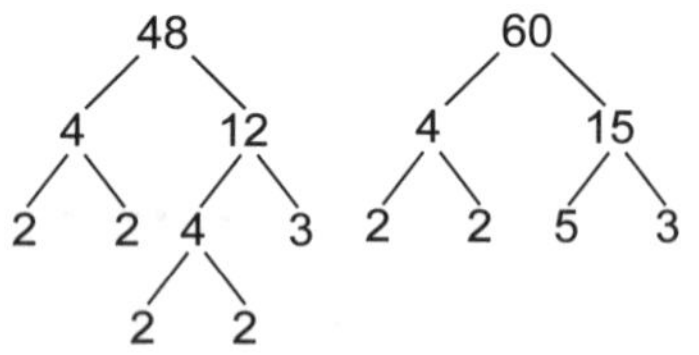

Step 1: Use factor trees or repeated division by prime numbers to find the prime factors of each number.

60 = 2, 2, 3, 5
48 = 2, 2, 2, 2, 3

Step 2: Write the list of prime factors for each number.

60 = 2, 2, 3, 5
48 = 2, 2, 2, 2, 3

Step 3: Underline the common factors.

$2 \cdot 2 \cdot 3 = 12$

Step 4: Multiply these common factors to calculate the greatest common factor.

Answer: 12

Problem Set

1. Find the prime factorization of 36.

$2^2 \cdot 3^2$ $2^4 \cdot 3$ $2^3 \cdot 3^2$ 3^2

2. Find the prime factorization of 98.

$2^2 \cdot 3^3$ $2^2 \cdot 7^2$ $2^3 \cdot 3^2$ $2 \cdot 7^2$

3. Find the prime factorization of 84.

$2^2 \cdot 3 \cdot 7$ $2 \cdot 3 \cdot 7$ $3^2 \cdot 7$ $2 \cdot 3^2 \cdot 7$

4. Is 7 prime, composite, or neither?

prime neither composite

5. Is 15 prime, composite, or neither?

prime composite neither

6. Is 17 prime, composite, or neither?

neither prime composite

7. Is 13 prime, composite, or neither?

composite prime neither

8. Find the prime factorization of 125.

5^3 3^5 $25 \cdot 5$ $25 \cdot 25$

9. Find the prime factorization of 132.

$6 \cdot 22$ $2^2 \cdot 3 \cdot 11$ $4 \cdot 33$ $2 \cdot 6 \cdot 11$

10. Find the prime factorization of 480.

$2^5 \cdot 3 \cdot 5$ $5^2 \cdot 3 \cdot 5$ $2 \cdot 3 \cdot 5$ $32 \cdot 15$

Find the GCF of the following:

11. 28 and 42

12. 63 and 144

13. 88 and 154

14. 32 and 48

15. 81 and 243

16. 54 and 81

17. 99 and 243

18. 24, 72, and 96

19. 42, 63, and 105

20. 60, 75, and 120

HA1-270: Factoring the Greatest Common Monomial Factor from a Polynomial

When factoring polynomial expressions, it is always important to find the greatest common factor of each of the monomial terms of the polynomial. This is usually the first step one takes when factoring polynomials. The **greatest common factor of a set of monomials** is the largest factor that can be divided evenly into all monomials in the set.

Greatest Common Factor of a Set of Monomials (GCF)	The largest factor that can be divided evenly into all monomials in the set

For example, the GCF of $6x$ and $8xy$ is $2x$. This can be determined by writing out all of the factors of $6x$ and of $8xy$ and then multiplying together the factors that are in common. Since $6x$ has factors 2, 3, and x, and $8xy$ has factors 2, 2, 2, x, and y, we can see that the only common factors are 2 and x. Therefore, the GCF is equal to $2x$.

Now, take for example the polynomial $3x^2 + 12xy + 24xy^2$. We want to factor out the GCF from the polynomial. First, determine what the greatest common factor of the three monomial terms is: $3x^2$, $12xy$, and $24xy^2$. Since each of the three terms contains the factors 3 and x and no other common factor, the GCF of $3x^2$, $12xy$, and $24xy^2$ is $3x$. Next we factor out $3x$ from the polynomial and get $3x^2 + 12xy + 24xy^2 = 3x(x + 4y + 8y^2)$.

Example 1 Find the GCF of $15x^2y$, $10x^3y^2$, and $25xy^3$.

$15x^2y = 3 \cdot 5 \cdot x \cdot x \cdot y$	**Step 1:**	Factor $15x^2y$.
$10x^3y^2 = 2 \cdot 5 \cdot x \cdot x \cdot x \cdot y \cdot y$	**Step 2:**	Factor $10x^3y^2$.
$25xy^3 = 5 \cdot 5 \cdot x \cdot y \cdot y \cdot y$	**Step 3:**	Factor $25xy^3$.
$\text{GCF} = 5 \cdot x \cdot y$ $= 5xy$	**Step 4:**	Since all terms have 5, x, and y in common, the GCF is $5xy$.

Answer: GCF of $15x^2y$, $10x^3y^2$, and $25xy^3 = 5xy$

Example 2 Factor out the GCF of $5x - 15$.

$5x$ and 15	**Step 1:**	Find the monomial terms in the binomial.
$5x = 5 \cdot x$	**Step 2:**	Factor $5x$.
$15 = 3 \cdot 5$	**Step 3:**	Factor 15.
$\text{GCF} = 5$	**Step 4:**	Since both terms have 5 in common, the GCF is 5.
$5x - 15 = 5(x - 3)$	**Step 5:**	Factor 5 from each term.

Answer: $5x - 15 = 5(x - 3)$

Example 3 Factor out the GCF of $-2z^4 - 10z$.

$2z^4$ and $10z$	**Step 1:** Find the two monomial terms in the binomial.
$2z^4 = 2 \cdot z \cdot z \cdot z \cdot z$	**Step 2:** Factor $2z^4$.
$10z = 2 \cdot 5 \cdot z$	**Step 3:** Factor $10z$.
$\text{GCF} = 2 \cdot z$ $= 2z$	**Step 4:** Since both terms have 2 and z in common, the GCF is $2z$.
$-2z^4 - 10z = -2z(z^3 + 5)$	**Step 5:** Since both terms are negative, we will factor out $-2z$ from each term.

Answer: $-2z^4 - 10z = -2z(z^3 + 5)$

Example 4 Factor out the GCF of $6xy^2 + 39xy$.

$6xy^2$ and $39xy$	**Step 1:** Find the monomial factors in the binomial.
$6xy^2 = 2 \cdot 3 \cdot x \cdot y \cdot y$	**Step 2:** Factor $6xy^2$.
$39xy = 3 \cdot 13 \cdot x \cdot y$	**Step 3:** Factor $39xy$.
$\text{GCF} = 3 \cdot x \cdot y$ $= 3xy$	**Step 4:** Since both terms have 3, x, and y in common, the GCF is $3xy$.
$6xy^2 + 39xy = 3xy(2y + 13)$	**Step 5:** Factor $3xy$ from each term.

Answer: $6xy^2 + 39xy = 3xy(2y + 13)$

Example 5 Factor out the GCF of $15x^3y - 45x^2y + 30xy$.

$15x^3y$, $45x^2y$, and $30xy$	**Step 1:** Find the monomial terms in the trinomial.
$15x^3y = 3 \cdot 5 \cdot x \cdot x \cdot x \cdot y$	**Step 2:** Factor $15x^3y$.
$45x^2y = 3 \cdot 3 \cdot 5 \cdot x \cdot x \cdot y$	**Step 3:** Factor $45x^2y$.
$30xy = 2 \cdot 3 \cdot 5 \cdot x \cdot y$	**Step 4:** Factor $30xy$.
$\text{GCF} = 3 \cdot 5 \cdot x \cdot y$ $= 15xy$	**Step 5:** Since all terms have 3, 5, x, and y in common, the GCF is $15xy$.
$15x^3y - 45x^2y + 30xy = 15xy(x^2 - 3x + 2)$	**Step 6:** Factor $15xy$ from each term.

Answer: $15x^3y - 45x^2y + 30xy = 15xy(x^2 - 3x + 2)$

Problem Set

1. Find the GCF of $10y^2$ and $15y^3$.
2. Factor out the GCF of $9a - 18$.
3. Find the GCF of $11p^5$ and $5p^2$.
4. Factor out the GCF of $4b + 8$.
5. Find the GCF of $6w^5$ and $4w^4$.
6. Factor out the GCF of $16 - 4w$.
7. Find the GCF of $14a^2b^3$, $21a^3b^2$, and $7a^4b^4$.
8. Factor out the GCF of $35p + 14$.
9. Find the GCF of $10w^5y^3$, $8w^4y^5$, and $4w^6y$.
10. Factor out the GCF of $12m + 24$.
11. Find the GCF of $2a^2$, $14b^5$, and $10c^3$.
12. Factor out the GCF of $2n - 10$.
13. Find the GCF of $20x^2y$, $8x^3y^2$, and $12xy^3$.
14. Factor out the GCF of $6x + 15$.
15. Find the GCF of $9x^2z^5$, $6x^3z^4$, and $15x^4z^3$.
16. Factor out the GCF of $20x^3 - 5x^4$.
17. Factor out the GCF of $7b^4 + 14b^{10}$.
18. Factor out the GCF of $4x^3y - 2x^4y^2$.
19. Factor out the GCF of $-5x^3 - 20x^5$.
20. Factor out the GCF of $3x^5 + 7x^4$.

Factor out the GCF of the following:

21. $8x^3y^4 + 16x^4y^5$ = 4(2x
22. $3x^3y^4 - 6x^4y^3$
23. $5b^2c^3 - 5bc$
24. $12y^9 - 8y^7$
25. $8x^4 - 2x^2$
26. $3a^2b + 9ab^2$
27. $x^4 - 2x^6$
28. $8x^2y^3 + 8xy^4$
29. $4d^8 + 16d^{10}$
30. $9pq + 18p$
31. $5y^2z^3 - 2y^3z^4 + y^2$
32. $8x^2y - 4xy - 4xy^2$
33. $16x^4 + 8x^3 + x^2$
34. $5d^4 - 25d^5 + 10d^6$
35. $18p^3q^3 + 9p^2q - 9p^4q^2$
36. $2w^6 + 6w^5 + w^4$
37. $-2x^4y^3 - 2x^3 - 16x^4y$
38. $3a^3 + 6a^2 - 3a$
39. $7a^3b^3 - 14a^2b^2 - 7a^2b$
40. $-12x^4 - 6x^2 - 2x$

HA1-271: Factoring Trinomials and Differences of Squares Using Algebra Tiles

Algebra tiles can be used to factor trinomials and the difference of squares. Various expressions in algebra are represented by tiles of particular colors and shapes. A large blue square represents the expression x^2. A green rectangle represents the variable x. A red rectangle represents the opposite of the variable x, or $-x$. A yellow square represents 1, and a red square represents -1.

Expression	Tile
x^2	
x	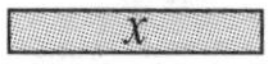
$-x$	
1	
-1	

The first step in factoring with algebra tiles is to select the tiles that represent the given trinomial based on the tile values stated above, then place the tiles in a design that represents the factored trinomial. After obtaining the design for the factored trinomial, the factors are found by reading the values of the tile edges along the left of the design and across the top of the design. The only tiles that will appear along the left and across the top are tiles that represent variable expressions. The values of the edges of these tiles are described below.

Values of Tile Edges	Tile
The value of each edge of the blue square, or x^2 tile, is x.	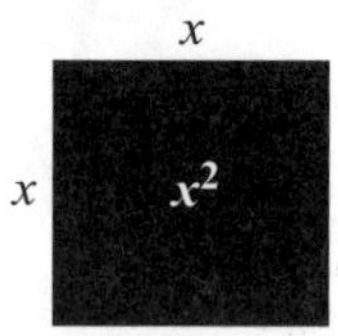
The values of the edges of the green rectangle, or x tile, are 1 and x. The short side has a value of 1, and the long side has a value of x.	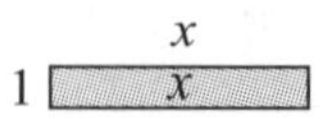
The values of the edges of the red rectangle, or x tile, are -1 and x. The short side has a value of -1, and the long side has a value of x.	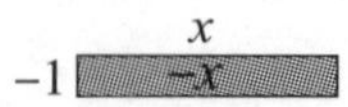

When factoring with algebra tiles, sometimes it is necessary to add a zero pair of rectangles or squares. Recall that a zero pair of rectangles consists of a green rectangle and a red rectangle and a zero pair of squares consists of a green square and a red square. A green rectangle and a red rectangle represent $x + (-x)$, which is equal to zero; a green square and a red square represent $1 + (-1)$, which is equal to zero. Zero can be added to an expression without changing its value.

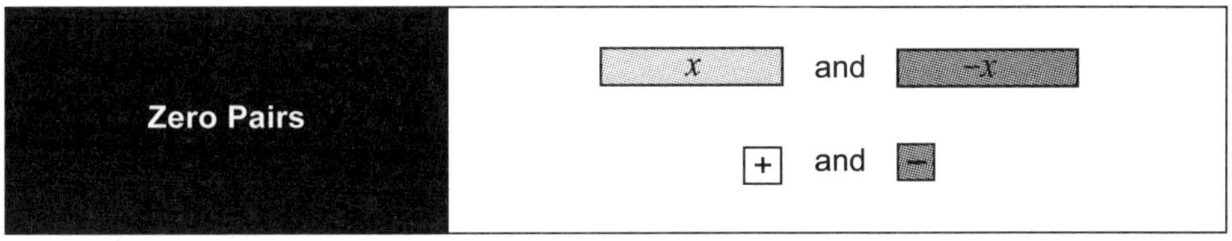

Example 1 The given tile design represents the factorization of a trinomial. What is the trinomial and what are its factors?

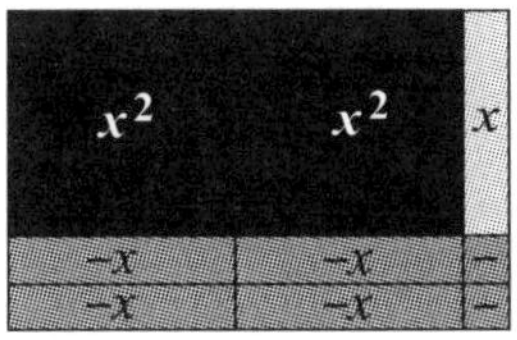

$2x^2 + x - 4x - 2 = 2x^2 - 3x - 2$

Step 1: Determine the expression modeled, find the sum, and simplify.

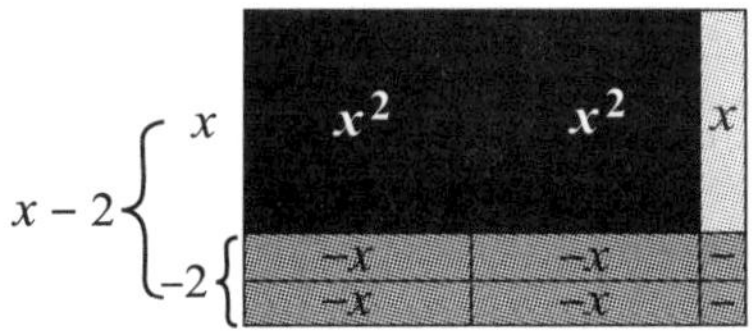

$(x - 2)$ is a factor of the trinomial.

Step 2: Read the value of the tile edges on the left of the design to find one of the factors.

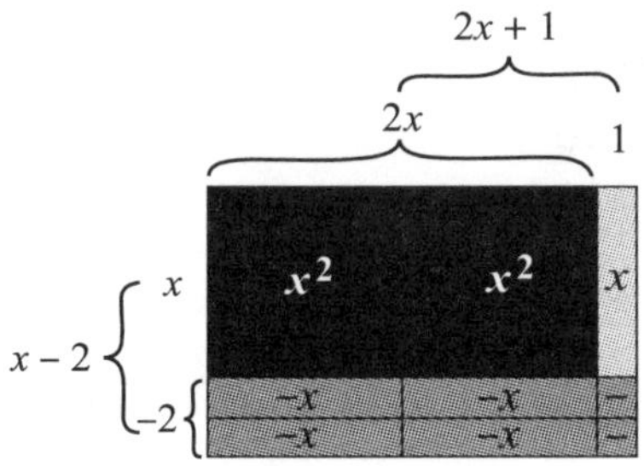

$(2x + 1)$ is a factor of the trinomial.

Step 3: Read the value of the tile edges on the top of the design to find the other factor.

Answer: $2x^2 - 3x - 2 = (x - 2)(2x + 1)$

Example 2 Create the tile design for the factored trinomial $x^2 - 5x + 6$, then use the tile design to find its factored form.

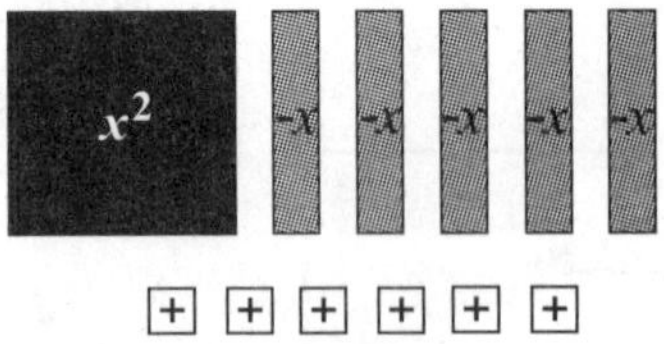

Step 1: Model the trinomial using algebra tiles.

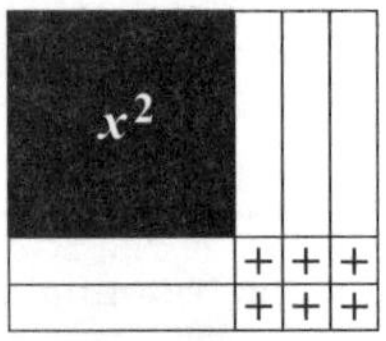

Step 2: Create the tile design for the factored trinomial by placing the blue square in the upper left corner and the yellow squares in the lower right corner. Use trial and error when placing the yellow squares.

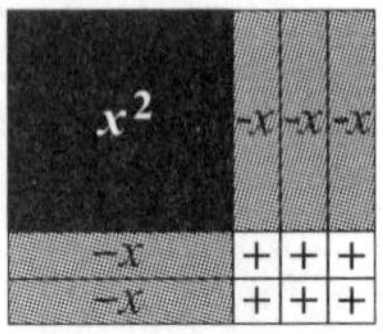

Step 3: Place the rectangular tiles in the design. This is the tile design for the factored trinomial.

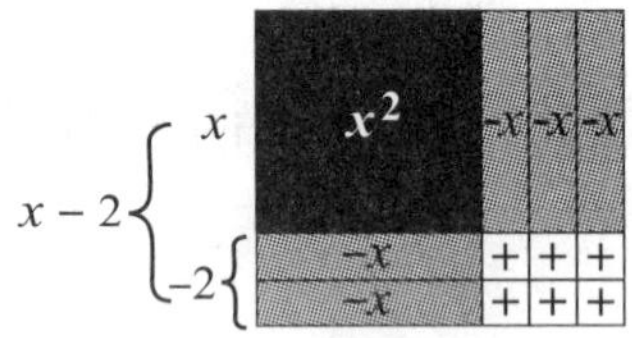

$(x - 2)$ is a factor of the trinomial

Step 4: Read the value of the tile edges on the left of the design to find one of the factors.

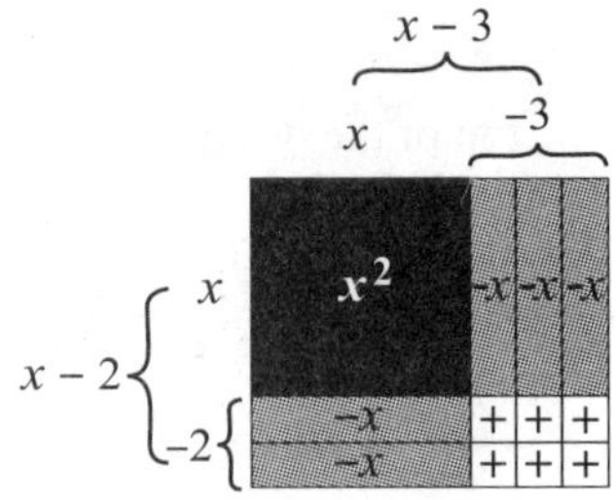

$(x - 3)$ is a factor of the trinomial

Step 5: Read the value of the tile edges across the top of the design to find the other factor.

Answer:

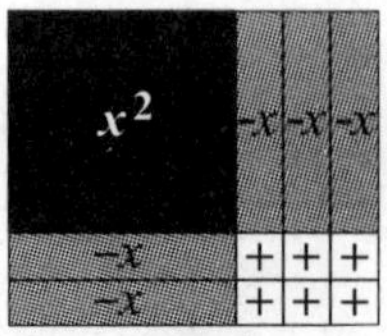

$(x - 2)(x - 3)$

Example 3 Create the tile design for the factored trinomial $x^2 + x - 6$, then use the tile design to find its factored form.

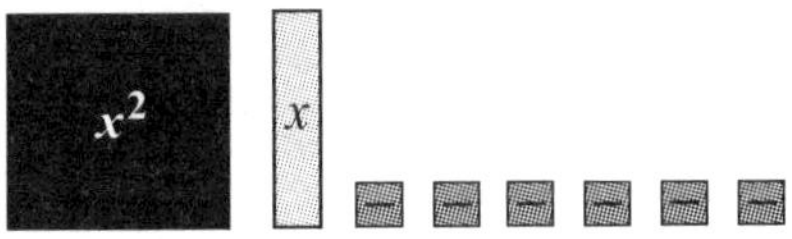

Step 1: Model the trinomial using algebra tiles.

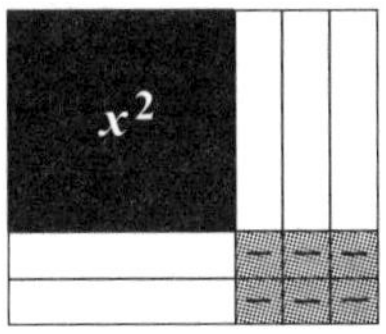

Step 2: Create the tile design for the factored trinomial by placing the blue square in the upper left corner and the red squares in the lower right corner. Use trial and error when placing the red squares.

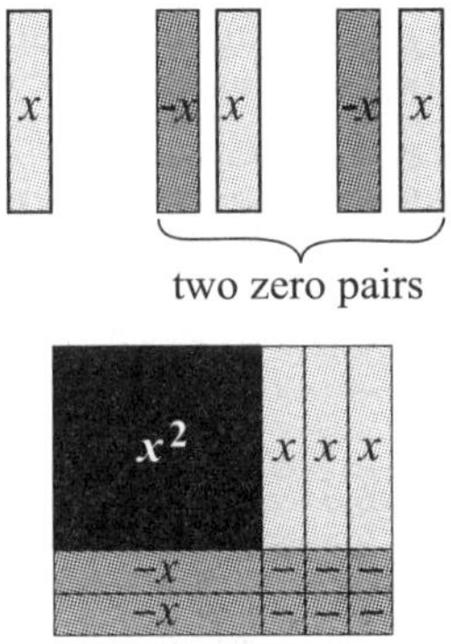

Step 3: Place the rectangular tiles in the design.

Note: Since there is only one rectangle, zero pairs will be added to fill the empty spaces. This is the tile design for the factored trinomial.

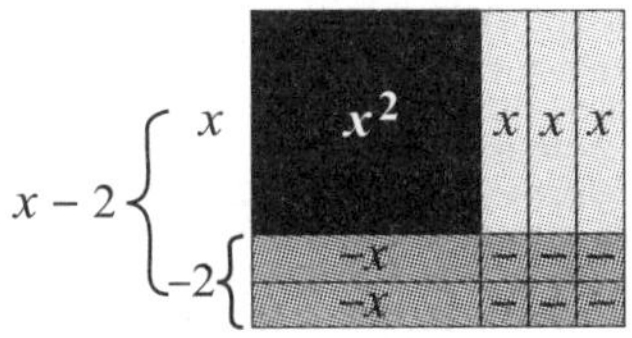

$(x - 2)$ is a factor of the trinomial.

Step 4: Read the value of the tile edges on the left of the design to find one of the factors.

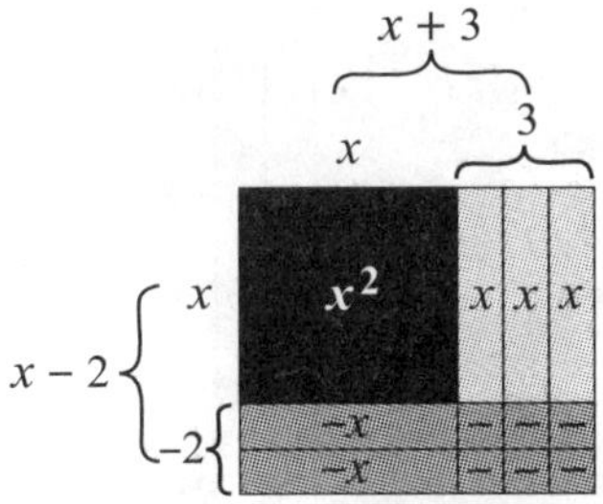

$(x + 3)$ is a factor of the trinomial.

Step 5: Read the value of the tile edges across the top of the design to find the other factor.

Answer:

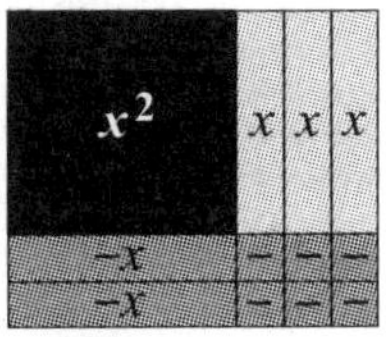

$(x - 2)(x + 3)$

Example 4 Create the tile design for the factored expression $9x^2 - 16$, then use the tile design to find its factored form.

Step 1: Model the expression using algebra tiles.

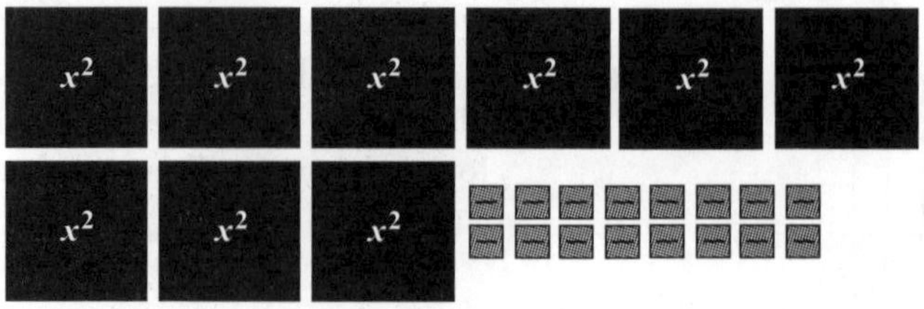

Step 2: Place the blue tiles in the upper left corner of the design in a square array. This signifies that the number of rows and columns is equal. Place the red squares in the lower right corner of the design in a square array.

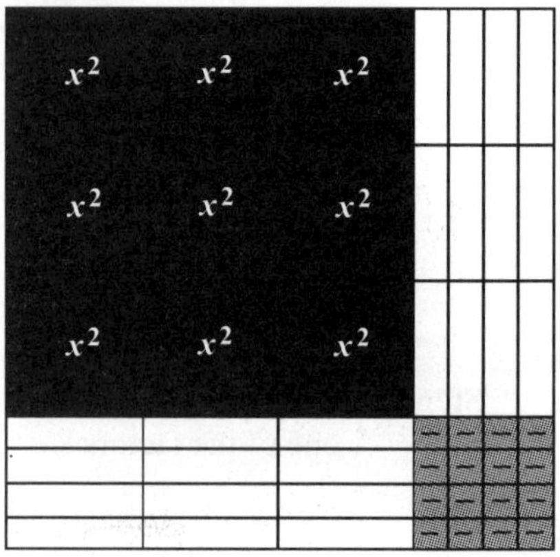

Step 3: Fill in the empty spaces with rectangles. Since there are no rectangles in the model of the expression, use zero pairs to fill the spaces. This is the tile design for the factored expression.

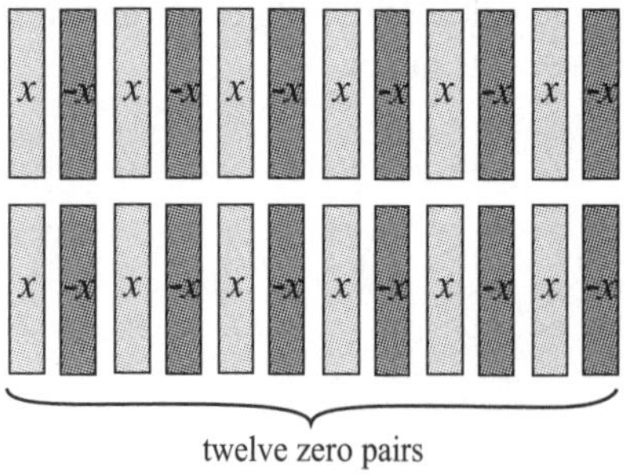

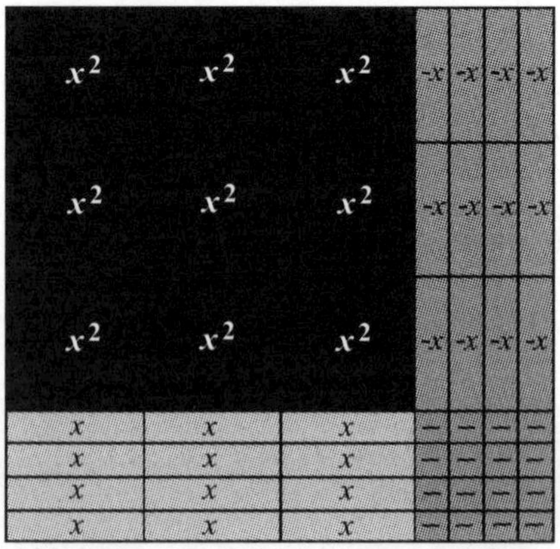

Step 4: Read the value of the tile edges on the left of the design to find one of the factors.

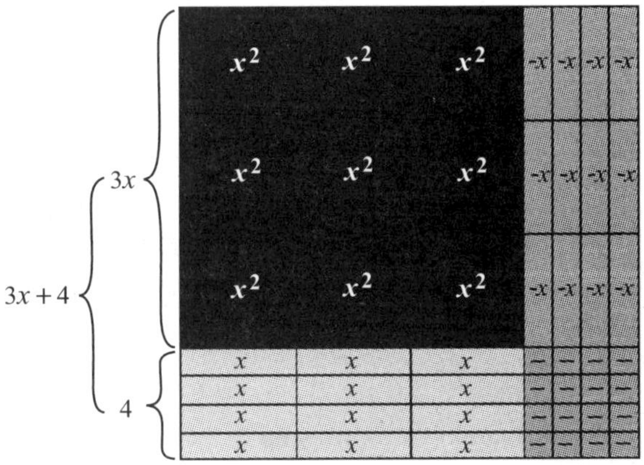

$(3x+4)$ is a factor of the expression.

Step 5: Read the value of the tile edges across the top of the design to find the other factor.

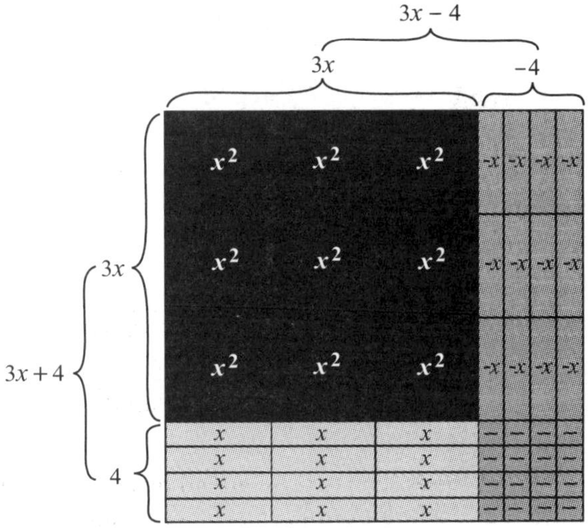

$(3x+4)$ is a factor of the expression.

Answer:

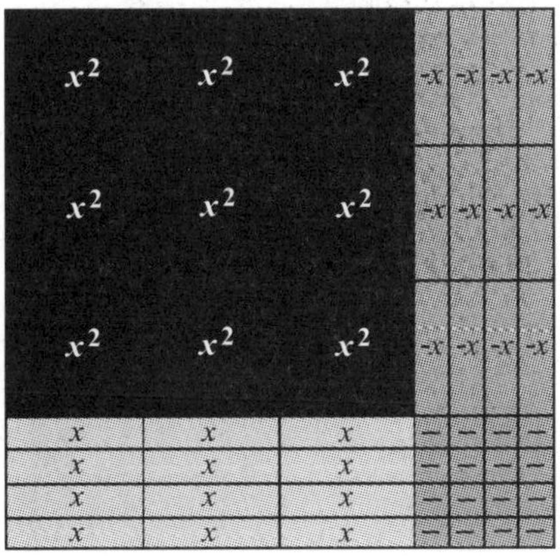

$(3x+4)(3x-4)$

Example 5

Create the tile design for the factored trinomial $3x^2 - 8x - 3$, then use the tile design to find its factored form.

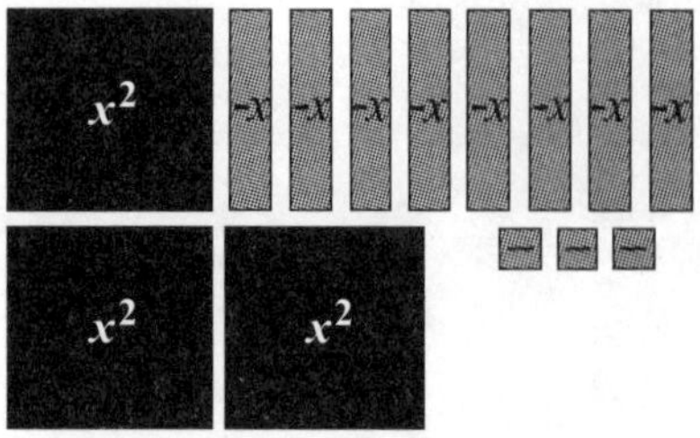

Step 1: Model the trinomial using algebra tiles.

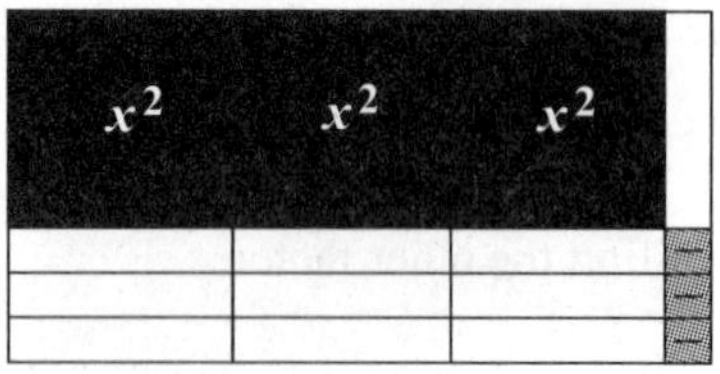

Step 2: Create the tile design for the factored trinomial by placing the blue square in the upper left corner and the red squares in the lower right corner. Use trial and error when placing the red squares.

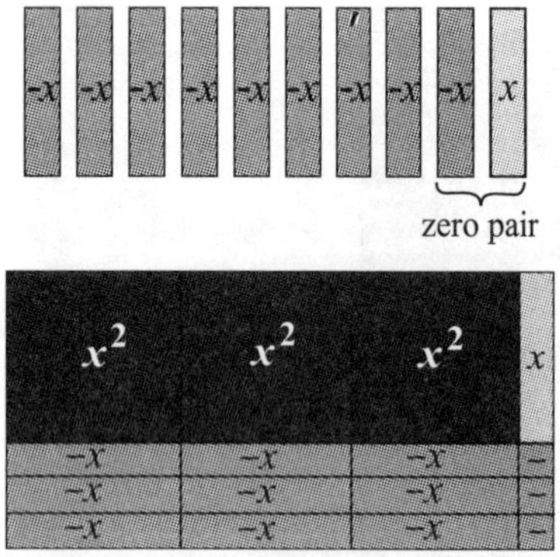

Step 3: Now place the rectangular tiles in the tile design. One zero pair is needed to fill the spaces for rectangles. This is the tile design for the factored trinomial.

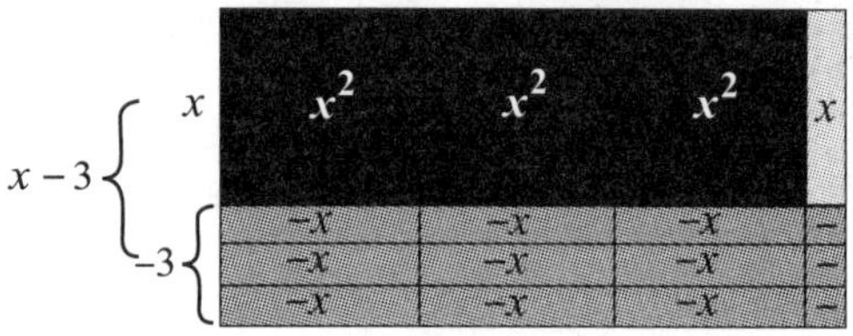

$(x - 3)$ is a factor of the trinomial.

Step 4: Read the value of the tile edges on the left of the design to find one of the factors.

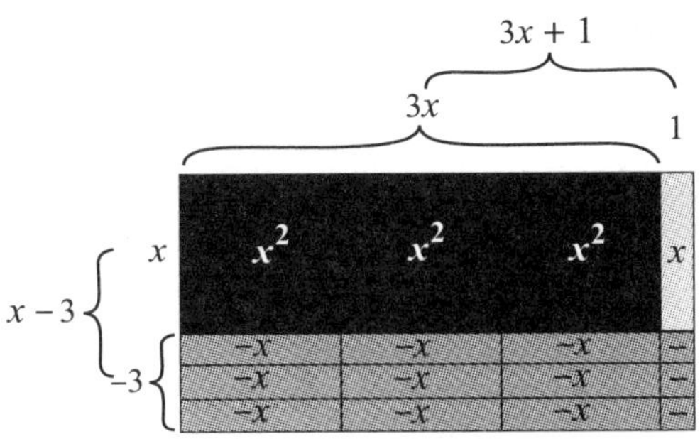

(3*x* + 1) is a factor of the trinomial.

Step 5: Read the value of the tile edges across the top of the design to find the other factor.

Answer:

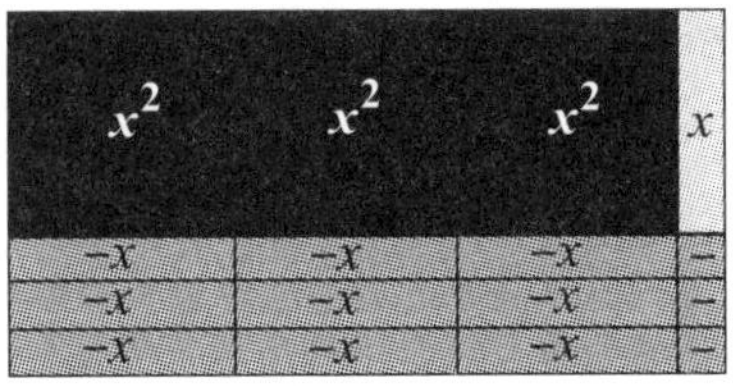

$(x-3)(3x+1)$

Problem Set

Each given tile design represents the factorization of a trinomial. What is the trinomial?

1.

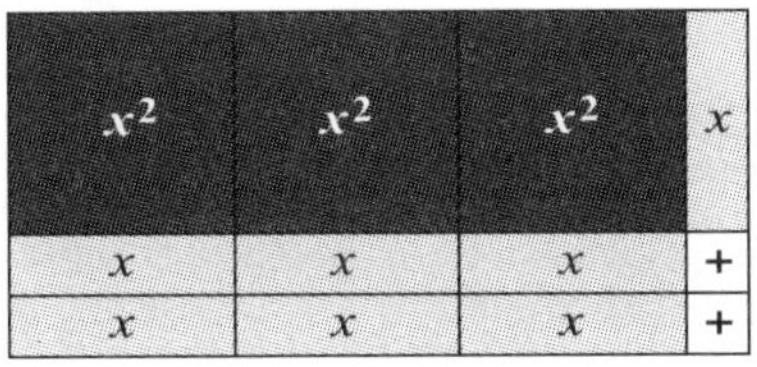

2.

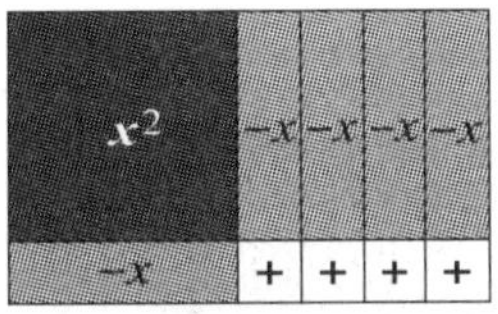

3.

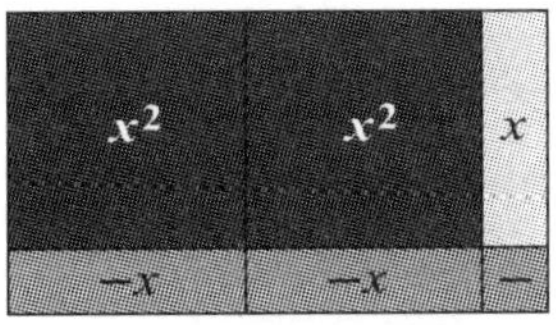

4.

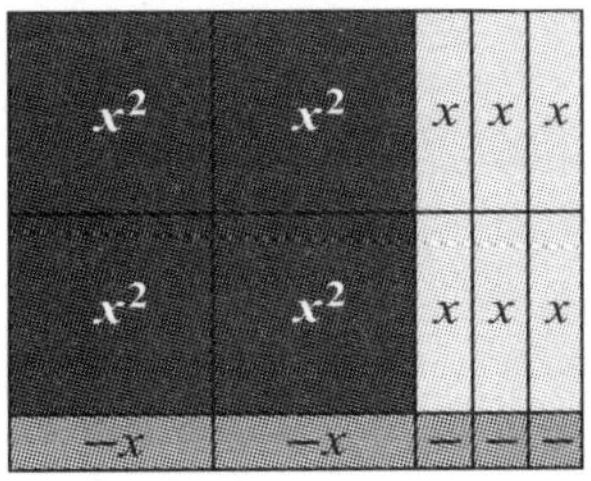

5.

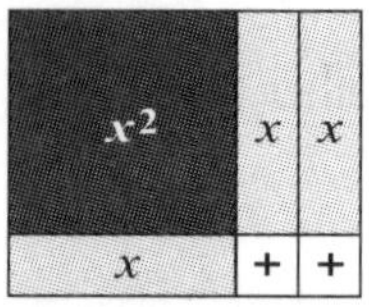

6.

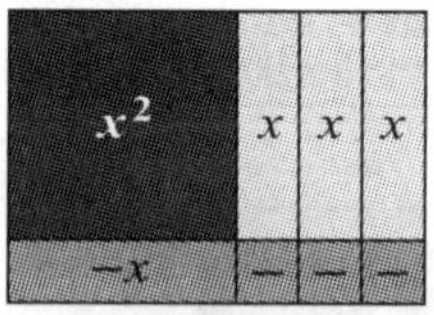

7.

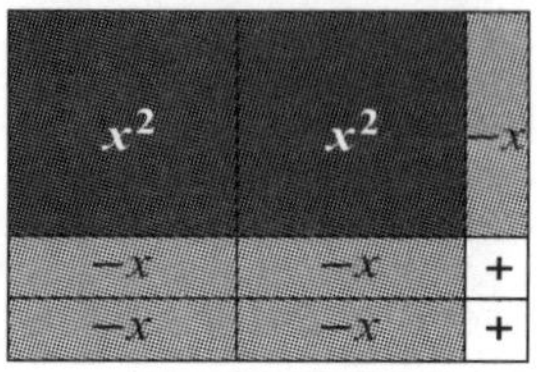

8.

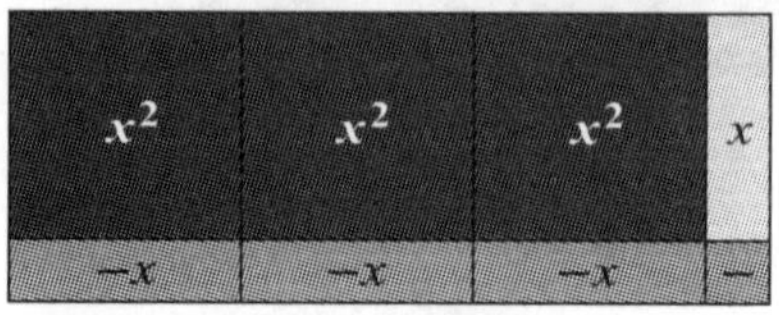

Create a tile design and find the factors to represent the factorization for each of the following expressions:

9. $x^2 + 4x + 3'$

10. $x^2 - 3x + 2$

11. $x^2 + 4x + 4$

12. $x^2 - 2x + 1$

13. $x^2 + 7x + 12$

14. $x^2 - 5x + 6$

15. $x^2 - 6x + 8$

16. $x^2 - x - 2$

17. $x^2 + x - 6$

18. $x^2 - 2x - 3$

19. $x^2 + 3x - 4$

20. $x^2 - 2x - 8$

21. $x^2 + x - 12$

22. $x^2 - 4x - 5$

23. $x^2 - 4x - 12$

24. $x^2 - 4$

25. $x^2 - 25$

26. $4x^2 - 9$

27. $9x^2 - 25$

28. $25x^2 - 4$

29. $x^2 - 16$

30. $9x^2 - 16$

31. $2x^2 + 5x + 2$

32. $2x^2 - 3x + 1$

33. $3x^2 + 5x - 2$

34. $2x^2 + 5x - 3$

35. $3x^2 - 7x + 2$

36. $3x^2 + x - 2$

37. $3x^2 + 8x + 4$

38. $3x^2 - 7x - 6$

39. $2x^2 - 7x - 4$

40. $2x^2 - x - 3$

HA1-275: Factoring the Difference Between Two Squares and Perfect Square Trinomials

Previously, you learned how to multiply two binomials to get the difference of two squares and perfect square trinomials.

Recall the formulas:

Product of a Sum and Difference: $(a-b)(a+b) = a^2 - b^2$ (Difference of two squares)

Square of a Binomial: $(a-b)^2 = a^2 - 2ab + b^2$ (Perfect Square Trinomials)

$(a+b)^2 = a^2 + 2ab + b^2$

Example 1 Multiply: $(x+5)^2$

$(x+5)^2 = (x+5)(x+5)$	**Step 1:** Rewrite the binomial in expanded form.
$= x^2 + 5x + 5x + 25$	**Step 2:** Use FOIL to multiply the two binomials.
$= x^2 + 10x + 25$	**Step 3:** Combine like terms.
	Answer: $x^2 + 10x + 25$

Let's see if $x^2 + 10x + 25$ is a perfect square trinomial.

- Is the first term a perfect square? Yes; $x^2 = x \cdot x$
- Is the second term twice the product of the square roots of the first and third terms? Yes; $10x = 2 \cdot 5 \cdot x$
- Is the third term a perfect square? Yes; $25 = 5^2$

Since the answer to all three questions is "yes", the trinomial is a perfect square.

Example 2 Determine whether or not $9y^2 + 42y + 49$ is a perfect square trinomial.

$9y^2 = 3y \cdot 3y$	**Step 1:** Is the first term a perfect square? Yes.
$42y = 2 \cdot 3y \cdot 7$	**Step 2:** Is the second term twice the product of the square roots of the first and third terms? Yes.
$49 = 7^2$	**Step 3:** Is the third term a perfect square? Yes.
	Answer: The trinomial $9y^2 + 42y + 49$ is a perfect square, which when factored gives $(3y+7)^2$.

Now that you can recognize if a trinomial is a perfect square we need to factor the trinomial into two binomial factors or one binomial factor squared.

Example 3 Factor: $a^2 - 8a + 16$

$a^2 - 8a + 16 = a^2 - 2(4a) + 4^2$

Step 1: Determine if this is a perfect square trinomial. Since the first term and the last term are perfect squares and the middle term is two times the product of the first and last terms, this is a perfect square trinomial.

$= (a - 4)^2$

Step 2: The first term of the binomial is a and the last term is 4. Subtraction is used because the middle term in the original trinomial had subtraction.

Answer: $(a - 4)^2$

Example 4 Factor: $x^2 - 20xy + 100y^2$

$x^2 - 20xy + 100y^2 = x^2 - 2(10xy) + (10y)^2$

Step 1: Determine if this is a perfect square trinomial. Since the first term and the last term are perfect squares and the middle term is two times the product of the first and last terms, this is a perfect square trinomial.

$= (x - 10y)^2$

Step 2: The first term of the binomial is x and the last term is $10y$. Subtraction is used because the middle term in the original trinomial had subtraction.

Answer: $(x - 10y)^2$

Recall that in some cases the multiplication of two binomials will not result in a trinomial. When this happens, the middle term cancels or equals zero; therefore, the middle term is not written. The result is a difference of two squares. A solution in the form $(a - b)(a + b)$ or $(a + b)(a - b)$ will always be the difference of two squares. Let's try to recognize binomials that are the difference of two squares.

Example 5 Determine if $9y^2 - 25$ is the difference of two squares.

$9y^2 = 3y \cdot 3y = (3y)^2$

Step 1: Is the first term a perfect square? Yes.

$25 = 5^2$

Step 2: Is the second term a perfect square? Yes.

Answer: The binomial $9y^2 - 25$ is the difference of two squares, which when factored gives $(3y - 5)(3y + 5)$.

Example 6 Factor: $49a^2 - 16b^2$

$49a^2 - 16b^2 = (7a)^2 - (4b)^2$	**Step 1:**	Determine if this binomial is the difference between two squares. Yes, since the first and last terms are perfect squares.
$= (7a - 4b)(7a + 4b)$	**Step 2:**	The first term of the two binomial factors is $7a$ and the last term is $4b$.

Answer: $(7a - 4b)(7a + 4b)$

Problem Set

1. Is the binomial the difference of two perfect squares? Answer "yes" or "no".
$$g^2 - 25$$

2. Is the binomial the difference of two perfect squares? Answer "yes" or "no".
$$n^2 + 36$$

3. Is the binomial the difference of two perfect squares? Answer "yes" or "no".
$$4x^2 - 16$$

4. Is the binomial the difference of two perfect squares? Answer "yes" or "no".
$$m^2 + 9$$

5. Is the binomial the difference of two perfect squares? Answer "yes" or "no".
$$9 - 49r^2$$

6. Factor: $n^2 - 121$

7. Factor: $x^2 - 9$

8. Factor: $x^2 - 196$

9. Factor: $m^2 - 144$

10. Factor: $b^2 - 81$

11. Factor: $4r^2 - 49$

12. Factor: $9b^2 - 16$

13. Factor: $25c^2 - 81$

14. Factor: $9x^2 - 49y^2$

15. Is the polynomial a perfect square trinomial? Answer "yes" or "no".
$$w^2 - 20w + 100$$

16. Is the polynomial a perfect square trinomial? Answer "yes" or "no".
$$u^2 + 9u + 144$$

17. Is the polynomial a perfect square trinomial? Answer "yes" or "no".
$$z^2 - 6z + 9$$

18. Factor: $d^2 - 10d + 25$

19. Factor: $d^2 + 4d + 4$

20. Factor: $4r^2 - 28rs + 49s^2$

HA1-276: Factoring Sums and Differences of Cubes

Previously, we have learned how to factor the difference of two squares $(a^2 - b^2)$.

Difference of Squares	$a^2 - b^2 = (a+b)(a-b)$

Now we will learn two more factorization formulas: the **sum of cubes** $(a^3 + b^3)$ and the **difference of cubes** $(a^3 - b^3)$.

Sum of Cubes	$a^3 + b^3 = (a+b)(a^2 - ab + b^2)$

Difference of Cubes	$a^3 - b^3 = (a-b)(a^2 + ab + b^2)$

Example 1 Factor the binomial $x^3 + y^3$.

$$a^3 + b^3 = (a+b)(a^2 - ab + b^2)$$
$$= (x+y)(x^2 - xy + y^2)$$

Step 1: Use the formula for the sum of cubes with $a = x$ and $b = y$.

Answer: $x^3 + y^3 = (x+y)(x^2 - xy + y^2)$

Example 2 Factor the binomial $c^3 - 7^3$.

$$a^3 - b^3 = (a-b)(a^2 + ab + b^2)$$
$$= (c-7)(c^2 + c \cdot 7 + 7^2)$$

Step 1: Use the formula for the difference of cubes with $a = c$ and $b = 7$.

$$= (c-7)(c^2 + 7c + 49)$$

Step 2: Simplify.

Answer: $c^3 - 7^3 = (c-7)(c^2 + 7c + 49)$

Example 3 Factor the binomial $x^3 + 27$.

$x^3 + 27 = x^3 + 3^3$	**Step 1:** Notice that $27 = 3^3$.
$a^3 + b^3 = (a+b)(a^2 - ab + b^2)$ $= (x+3)(x^2 - x \cdot 3 + 3^2)$	**Step 2:** Use the formula for the sum of cubes with $a = x$ and $b = 3$.
$= (x+3)(x^2 - 3x + 9)$	**Step 3:** Simplify.

Answer: $x^3 + 27 = (x+3)(x^2 - 3x + 9)$

Example 4 Factor the binomial $64 - 125z^3$.

$64 - 125z^3 = 4^3 - 5^3z^3$	**Step 1:** Notice that $64 = 4^3$ and $125 = 5^3$.
$4^3 - (5z)^3 =$	**Step 2:** Group the 5 and the z terms, since they have the same power.
$a^3 - b^3 = (a-b)(a^2 + ab + b^2)$ $= (4-5z)(4^2 + 4(5z) + (5z)^2)$	**Step 3:** Use the formula for the difference of cubes with $a = 4$ and $b = 5z$.
$= (4-5z)(16 + 20z + 25z^2)$	**Step 4:** Simplify.

Answer: $64 - 125z^3 = (4-5z)(16 + 20z + 25z^2)$

Example 5 Factor the binomial $x^6 - (4y)^6$.

Step 1: Notice that $x^6 = x^3 \cdot x^3$ and $(4y)^6 = (4y)^3 \cdot (4y)^3$.

$$x^6 - (4y)^6 = x^3 \cdot x^3 - (4y)^3 \cdot (4y)^3$$

Step 2: Combining terms we get $x^6 = x^3 \cdot x^3 = (x^3)^2$ and $(4y)^6 = (4y)^3 \cdot (4y)^3 = [(4y)^3]^2$. Now we can use the sum of cubes and difference of squares with $a = x^3$ and $b = (4y)^3$.

$$(x^3)^2 - [(4y)^3]^2 = (x^3 + (4y)^3)(x^3 - (4y)^3)$$

Step 3: Apply the sum of cubes formula for $(x^3 + (4y)^3)$ and the difference of cubes formula for $(x^3 - (4y)^3)$ with $a = x$ and $b = 4y$.

$$x^3 + (4y)^3 = (x + 4y)(x^2 - x(4y) + (4y)^2)$$
$$x^3 - (4y)^3 = (x - 4y)(x^2 + x(4y) + (4y)^2)$$

Step 4: Multiply both expressions together and simplify.

$$\begin{aligned}(x^3 + (4y)^3)(x^3 - (4y)^3) &= (x + 4y)(x^2 - x(4y) + (4y)^2)(x - 4y)(x^2 + x(4y) + (4y)^2)\\ &= (x + 4y)(x^2 - 4xy + 16y^2)(x - 4y)(x^2 + 4xy + 16y^2)\end{aligned}$$

Step 5: Regroup the factors so the binomial factors are written first.

$$(x + 4y)(x - 4y)(x^2 - 4xy + 16y^2)(x^2 + 4xy + 16y^2)$$

Answer: $x^6 - (4y)^6 = (x + 4y)(x - 4y)(x^2 - 4xy + 16y^2)(x^2 + 4xy + 16y^2)$

Problem Set

Factor completely:

1. $y^3 - z^3$
2. $d^3 - e^3$
3. $m^3 + p^3$
4. $g^3 + h^3$
5. $k^3 - m^3$
6. $q^3 - r^3$
7. $s^3 + v^3$
8. $w^3 + z^3$
9. $c^3 + 11^3$
10. $f^3 + 4^3$
11. $5^3 + n^3$
12. $7^3 + s^3$
13. $w^3 - 1^3$
14. $6^3 - k^3$
15. $r^3 - 13^3$
16. $c^3 + 216$
17. $e^3 + 343$
18. $f^3 + 27$
19. $h^3 + 512$
20. $m^3 - 8$
21. $n^3 - 1,000$
22. $x^3 - 729$
23. $s^3 - 1,728$
24. $216y^3 + 1,000$
25. $729 - 8x^3$
26. $343c^3 + 27$
27. $1,331x^3 + 512$
28. $729x^3 + 343$
29. $27f^3 - 1,000$
30. $64c^3 - 3,375$
31. $d^6 - 64e^6$
32. $y^6 - 729$
33. $b^6 - (4m)^6$
34. $c^6 - (5g)^6$
35. $d^6 - (7h)^6$
36. $(5k)^6 - (3n)^6$
37. $(4y)^6 - (9x)^6$
38. $(12a)^6 - (5b)^6$
39. $(10s)^6 - (8r)^6$
40. $(11m)^6 - (3f)^6$

HA1-280: Factoring $x^2 + bx + c$ When c is Greater Than Zero

In this lesson, you will continue your factoring quest. Before learning how to factor a trinomial, let's review a concept that you will find useful when you begin to factor—multiplying binomials. Recall that the acronym FOIL is used to multiply binomials.

Example 1 Multiply: $(x + 3)(x + 4)$

$x \cdot x = x^2$	F	Multiply first terms.
$x \cdot 4 = 4x$	O	Multiply outer terms.
$3 \cdot x = 3x$	I	Multiply inner terms.
$3 \cdot 4 = 12$	L	Multiply last terms.

Answer: $x^2 + 4x + 3x + 12 = x^2 + 7x + 12$

Example 2 Multiply: $(x - 2)(x - 6)$

$x \cdot x = x^2$	F	Multiply first terms.
$x \cdot (-6) = -6x$	O	Multiply outer terms.
$(-2) \cdot x = -2x$	I	Multiply inner terms.
$(-2) \cdot (-6) = 12$	L	Multiply last terms.

Answer: $x^2 - 6x - 2x + 12 = x^2 - 8x + 12$

Recall that algebraic expressions such as $x^2 + 9x + 18$ and $x^2 - 5xy + 6y^2$ are called "trinomials", because these expressions contain three terms. Given the trinomial, your quest is to find the binomial factors that equal the trinomial. However, the second terms of the binomials have dual functions—when multiplied, the product equals the last term of the trinomial, and when added, the sum equals the middle term of the trinomial.

Let's look more closely at the third terms, also referred to as the c terms, in the trinomials $x^2 + 9x + 18$ and $x^2 - 5xy + 6y^2$. The numbers $+18$ and $+6y^2$ are greater than zero. When the c term is greater than zero, this indicates that both second terms in the binomials are either positive or negative. Based on the middle term, we can apply the correct signs to the solution. Use the following hints to remember this concept.

> If the middle term of the trinomial is **positive**, the second terms of both binomials are *positive*.
> If the middle term of the trinomial is **negative**, the second terms of both binomials are *negative*.

After using the hints above to determine the signs correctly, make sure that the product of the second terms of the binomials equals the third term of the trinomial.In Examples 1 and 2, you were given the factors, which you then multiplied by using the FOIL method. Now, you need to learn to reverse the process. When you are given a trinomial to factor, you must determine which two binomial factors, when multiplied, yield that trinomial.

Example 3	Factor the trinomial $x^2 + 10x + 24$.
$x \cdot x$	**Step 1:** Determine which two terms are multiplied to get x^2.
	Step 2: Determine which two terms are multiplied to get 24. Remember that the terms you select must not only equal 24 when multiplied, but their sum also must equal the coefficient of the middle term, 10. Use the following procedure to accomplish this.
$1 \cdot 24$ $2 \cdot 12$ $3 \cdot 8$ $4 \cdot 6$	**Step 3:** Make a systematic list of all of the two-combination factors that equal 24. Since the middle term is positive, you only need to list the positive factors of 24.
$4 + 6 = 10$	**Step 4:** From the list, determine which set of factors, when added, equal the coefficient of the middle term of the trinomial, 10.
$(x + \_\_)(x + \_\_)$	**Step 5:** Write the factors that you found for the first term as the first terms of each binomial. You can also insert the signs at this point. You know that the signs will be positive, since the middle term is positive.
$(x + 4)(x + 6)$	**Step 6:** Write the factors you found to equal the third term as the second terms of each binomial.
	Answer: $(x + 4)(x + 6) = x^2 + 10x + 24$

To practice factoring trinomials, let's solve more examples in which the third term is greater than zero.

Example 4	Factor: $x^2 + 12x + 36$
$x \cdot x$	**Step 1:** Determine which two factors are multiplied to get x^2.
$1 \cdot 36$ $2 \cdot 18$ $3 \cdot 12$ $4 \cdot 9$ $6 \cdot 6$ √ yes	**Step 2:** Determine which two numbers equal 36 when multiplied, and equal 12 when added. Since the middle term is positive, you only need to list the positive factors of 36.
	Answer: $x^2 + 12x + 36 = (x + 6)(x + 6)$

Example 5	Factor: $x^2 – 13x + 30$
$x \cdot x$	**Step 1:** Determine which two factors are multiplied to get x^2.
$-1 \cdot -30$ $-2 \cdot -15$ $-3 \cdot -10$ √ yes $-5 \cdot -6$	**Step 2:** Determine which two numbers equal 30 when multiplied, and equal –13 when added. Note that, in this example, the middle term is negative and the last term is positive; therefore, you only need to test the negative factors of 30.
	Answer: $x^2 - 13x + 30 = (x - 3)(x - 10)$

Example 6 Factor: $x^2 - 9x + 18$

$x \cdot x$	**Step 1:**	Determine which two factors are multiplied to get x^2.
$-1 \cdot -18$ $-2 \cdot -9$ $-3 \cdot -6$ √ yes	**Step 2:**	Determine which two numbers equal 18 when multiplied, and equal -9 when added. Since the middle term is negative, only test the negative factors of 18.

Answer: $x^2 - 9x + 18 = (x-3)(x-6)$

Problem Set

1. Which of the following is a factor of the trinomial?

$z^2 - 10z + 25$

$z-2$ $\quad$ $z-5$

$z-10$ $\quad$ $z+5$

2. Which of the following is a factor of the trinomial?

$x^2 - 9x + 20$

$x-4$ $\quad$ $x+4$

$x-9$ $\quad$ $x+5$

3. Which of the following is a factor of the trinomial?

$x^2 + 12x + 20$

$x+10$ $\quad$ $x+4$

$x-2$ $\quad$ $x-10$

4. Which of the following is a factor of the trinomial?

$x^2 + 6x + 9$

$x-3$ $\quad$ $x+2$

$x+3$ $\quad$ $x-2$

5. Which of the following is a factor of the trinomial?

$x^2 - 12xy + 35y^2$

$x+7y$ $\quad$ $x-7$

$x+5y$ $\quad$ $x-5y$

6. Which of the following is a factor of the trinomial?

$x^2 + 15x + 50$

$x+5$ $\quad$ $x+25$

$x-5$ $\quad$ $x-10$

7. Which of the following is a factor of the trinomial?

$x^2 + 8x + 16$

$x+8$ $\quad$ $x-4$

$x+2$ $\quad$ $x+4$

8. Which of the following is a factor of the trinomial?

$x^2 - 15xy + 56y^2$

$x+8y$ $\quad$ $x-7$

$x-8$ $\quad$ $x-8y$

9. Which of the following is a factor of the trinomial?

$x^2 + 7x + 12$

$x+2$ $\quad$ $x+4$

$x+12$ $\quad$ $x+6$

10. Which of the following is a factor of the trinomial?

$x^2 + 17x + 16$

$x+2$ $\quad$ $x+1$

$x+4$ $\quad$ $x+8$

11. Factor: $x^2 + 11x + 28$

12. Factor: $c^2 + 11c + 24$

13. Factor: $x^2 + 12x + 35$

14. Factor: $k^2 + 4k + 3$

15. Factor: $x^2 + 11x + 30$

16. Factor: $c^2 + 24c + 144$

17. Factor: $x^2 + 10x + 21$

18. Factor: $x^2 - 20xy + 96y^2$

19. Factor: $x^2 - 8x + 15$

20. Given that the area of a rectangle is $x^2 + 14x + 48$, find the dimensions of the rectangle.

HA1-285: Factoring $x^2 + bx + c$ When c is Less Than Zero

You have been working mainly with trinomials that have a positive third term. When you had to find the two binomials that made up the trinomial, you knew the factors were either both positive or both negative. In this lesson, the third term of the trinomial will be less than zero, or negative. Think for a moment about how a negative product is derived when you multiply. Remember that one of the two factors must be negative. Recall that the two factors that you multiply to find the third term are the same factors that you must then add in order to obtain the middle term of the trinomial. With this in mind, your one additional step will be to determine which factor is negative. Let's follow the process by working the first example together.

Example 1 Factor: $x^2 - 5x - 24$

$x \cdot x$

Step 1: Determine which factors are multiplied to get x^2.

$1 \cdot -24$ $\quad -1 \cdot 24$
$2 \cdot -12$ $\quad -2 \cdot 12$
$3 \cdot -8$ √ yes $\quad -3 \cdot 8$
$4 \cdot -6$ $\quad -4 \cdot 6$

Step 2: Determine which two numbers will have a product of –24, and a difference of –5. If you cannot derive the factors, list all of the positive and negative pairs of factors that equal –24.

$(x + 3)(x - 8)$

Step 3: Show the factors in binomial form.

F $\quad x \cdot x = x^2$
O $\quad x \cdot (-8) = -8x$
I $\quad 3 \cdot x = 3x$
L $\quad 3 \cdot (-8) = -24$

Step 4: Use the FOIL method to check.

Answer: $x^2 - 5x - 24 = (x + 3)(x - 8)$

Let's work through some examples.

Example 2 Factor: $x^2 - 4x - 32$

$x \cdot x$

Step 1: Determine which factors are multiplied to get x^2.

$1 \cdot -32$
$2 \cdot -16$
$4 \cdot -8$ √ yes
$-1 \cdot 32$
$-2 \cdot 16$
$-4 \cdot 8$

Step 2: Determine which two numbers will have a product of –32, and a difference of –4.

$(x + 4)(x - 8)$

Step 3: Show the factors in binomial form.

F $x \cdot x = x^2$ O $x \cdot (-8) = -8x$ I $4 \cdot x = 4x$ L $(4)(-8) = -32$	**Step 4:**	Use the FOIL method to check. $x^2 - 8x + 4x - 32 = x^2 - 4x - 32$

Answer: $x^2 - 4x - 32 = (x + 4)(x - 8)$

Example 3 Factor: $x^2 + 5x - 14$

$x \cdot x$	**Step 1:**	Determine which factors are multiplied to get x^2.
$1 \cdot -14$ $7 \cdot -2$ √ yes $-1 \cdot 14$ $-2 \cdot 7$	**Step 2:**	Determine which two numbers will have a product of –14 and a difference of 5.
$(x + 7)(x - 2)$	**Step 3:**	Show the factors in binomial form.
F $x \cdot x = x^2$ O $x \cdot (-2) = -2x$ I $7 \cdot x = 7x$ L $(7)(-2) = -14$	**Step 4:**	Use the FOIL method to check. $x^2 - 2x + 7x - 14 = x^2 + 5x - 14$

Answer: $x^2 + 5x - 14 = (x + 7)(x - 2)$

Example 4 Factor: $x^2 + xy - 20y^2$

$x \cdot x$	**Step 1:**	Determine which factors are multiplied to get x^2.
$1y \cdot -20y$ $2y \cdot -10y$ $4y \cdot -5y$ $-1y \cdot 20y$ $-2y \cdot 10y$ $-4y \cdot 5y$ √ yes	**Step 2:**	Determine which two numbers will have a product of $-20y^2$ and a difference of 1.
$(x - 4y)(x + 5y)$	**Step 3:**	Show the factors in binomial form.

F	$x \cdot x = x^2$	**Step 4:**	Use the FOIL method to check.
O	$x \cdot (5y) = 5xy$		$x^2 + 5xy - 4xy - 20y^2 = x^2 + xy - 20y^2$
I	$x \cdot (-4y) = -4xy$		
L	$(-4y)(5y) = -20y^2$		

Answer: $x^2 + xy - 20y^2 = (x - 4y)(x + 5y)$

Problem Set

1. Which of the following is a factor of the trinomial?

$$x^2 - 3x - 4$$

$x - 2$ $\quad$ $x + 1$

$x - 3$ $\quad$ $x + 4$

2. Which of the following is a factor of the trinomial?

$$x^2 - 4x - 5$$

$x - 1$ $\quad$ $x - 5$

$x - 3$ $\quad$ $x - 4$

3. Which of the following is a factor of the trinomial?

$$x^2 + 2x - 8$$

$x + 2$ $\quad$ $x - 8$

$x + 4$ $\quad$ $x - 4$

4. Which of the following is a factor of the trinomial?

$$x^2 + x - 12$$

$x - 4$ $\quad$ $x - 3$

$x + 3$ $\quad$ $x - 12$

5. Which of the following is a factor of the trinomial?

$$x^2 + 6x - 40$$

$x + 40$ $\quad$ $x + 4$

$x - 4$ $\quad$ $x - 10$

6. Which of the following is a factor of the trinomial?

$$x^2 - 9x + 20$$

$x + 4$ $\quad$ $x - 5$

$x + 5$ $\quad$ $x - 10$

7. Which of the following is a factor of the trinomial?

$$y^2 + 2y - 80$$

$y + 10$ $\quad$ $y + 8$

$y - 10$ $\quad$ $y + 2$

8. Which of the following is a factor of the trinomial?

$$y^2 - 5y - 50$$

$y + 2$ $\quad$ $y - 10$

$y + 10$ $\quad$ $y - 5$

9. Which of the following is a factor of the trinomial?

$$x^2 + 6x - 72$$

$x - 6$ $\quad$ $x - 12$

$x - 8$ $\quad$ $x + 6$

10. Which of the following is a factor of the trinomial?

$$x^2 - 17x - 60$$

$x + 3$ $\quad$ $x - 12$

$x + 5$ $\quad$ $x + 20$

Factor each of the following terms:

11. $w^2 + 9w - 36$

12. $w^2 - 9w - 36$

13. $w^2 - 18w - 40$

14. $w^2 - 6w - 40$

15. $w^2 + 34w - 72$

16. $w^2 - 14w - 72$

17. $x^2 + 4x - 45$

18. $w^2 - 21wy - 100y^2$

19. $w^2 - 46wy - 96y^2$

20. $w^2 - 29wy - 96y^2$

HA1-290: Factoring $ax^2 + bx + c$

In previous lessons, you factored polynomials in the form $x^2 + bx + c$, with the c term being greater than or less than zero. Essentially, you had to find the two factors that, when multiplied, equal the value of the c term, and when added, equal the coefficient of the b term. This lesson will add steps to the process you used to determine the binomial factors of trinomials.

You will notice that a trinomial has an additional variable in the term—an a. The variable a indicates that a coefficient is part of the first term. This means that you must give more attention to finding the first terms of the binomial factors. Then, follow the same steps you used in previous lessons to find the b and c terms of the trinomial. After factoring, use the FOIL method to check the two resulting binomials to ensure that they equal the original trinomial. Let's begin by factoring a trinomial that has a coefficient in the first term.

Example 1 Determine the binomial factors of the trinomial $3x^2 - 20x - 7$.

$1x \cdot 3x$

Step 1: Find the factors of the first term. Next, think about two factors that equal the coefficient 3. Since 3 is a prime number, it has only two possible factors: 1 and 3. Thus, you have the first factors of the binomials.

$-1 \cdot 7$

$-7 \cdot 1$

Step 2: Find the factors of the last term.

Step 3: Try a combination of factors of the first and last terms of the binomials until you find the middle term of the trinomial. Since the first terms of the binomials now have coefficients, the order in which you write the second terms is very important.

Note: Remember that the factors will probably be binomials, so try to put the answer in binomial form.

$(3x - 1)(x + 7)$

Step 4: Let's try using –1 as the second term in the first binomial, and 7 as the second term in the second binomial.

$$\left.\begin{aligned} 3x \cdot 7 &= 21x \\ -1 \cdot (x) &= -x \end{aligned}\right\} \rightarrow \begin{array}{r} 21x \\ -1x \\ \hline 20x \end{array}$$

Step 5: Using the second and third steps of FOIL, multiply to see if you get the same middle term as the trinomial.

$(3x + 1)(x - 7) = -21x$

$1 \cdot (x) = x$

$$\left.\begin{aligned} 3x \cdot (-7) &= -21x \\ 1 \cdot (x) &= x \end{aligned}\right\} \rightarrow \begin{array}{r} -21x \\ +\ 1x \\ \hline -20x \end{array}$$

Step 6: The middle term you need is $-20x$ rather than $20x$. Since the binomial factor pair you tested did not equal the middle term of the trinomial, test another set of factors. In this case, the only other factor pair is 1 and –7. Since 20 is the correct coefficient, do not change the order of the second term; simply change the sign.

Answer: $3x^2 - 20x - 7 = (3x + 1)(x - 7)$

Example 2 Find the binomial factors of the trinomial $9x^2 + 20x + 4$.

$9x \cdot x$ $3x \cdot 3x$	**Step 1:** Find the factors of the first term.
$1 \cdot 4$ $-1 \cdot -4$ $2 \cdot 2$ $-2 \cdot -2$	**Step 2:** Find the factors of the last term.
$(9x + 2)(x + 2)$	**Step 3:** Try a combination of the factors of the first and last terms of the binomials until you find the middle term of the trinomial.
	Answer: $9x^2 + 20x + 4 = (9x + 2)(x + 2)$

Example 3 Find the binomial factors of the trinomial $4x^2 - 8xy - 21y^2$.

$4x \cdot x$ $2x \cdot 2x$	**Step 1:** Find the factors of the first term.
$-1y \cdot 21y$ $1y \cdot -21y$ $3y \cdot -7y$ $-3y \cdot 7y$	**Step 2:** Find the factors of the last term.
$(2x - 7y)(2x + 3y)$	**Step 3:** Try a combination of the factors of the first and last terms of the binomials until you find the middle term of the trinomial.
	Answer: $4x^2 - 8xy - 21y^2 = (2x - 7y)(2x + 3y)$

Problem Set

For each problem, determine which of the choices is a factor of the given trinomial:

1. $12x^2 + 7x + 1$

$3x + 4$ $\quad 4x + 4$

$3x + 1$ $\quad 4x + 3$

2. $2x^2 + 13x + 15$

$x + 3$ $\quad 2x + 5$

$x + 5$ $\quad 2x - 3$

3. $8x^2 - 14xy + 3y^2$

$4x + y$ $\quad 4x - 3y$

$2x - 3y$ $\quad 2x + 3y$

4. $6y^2 - 41y - 7$

$y + 1$ $\quad$ $y + 7$

$y - 7$ $\quad$ $6y + 7$

5. $3x^2 - 11x + 6$

$x - 2$ $\quad$ $3x - 3$

$2x - 3$ $\quad$ $3x - 2$

6. $12y^2 + 31y + 20$

$5y + 2$ $\quad$ $3y + 4$

$4y - 3$ $\quad$ $4y + 3$

7. $8y^2 + 14y + 3$

$4y + 1$ $\quad$ $2y + 1$

$2y - 3$ $\quad$ $4y + 3$

8. $15x^2 - xy - 2y^2$

$5x + 2y$ $\quad$ $x + 2y$

$3x - y$ $\quad$ $3x + y$

9. $2x^2 + 7x + 5$

$2x + 5$ $\quad$ $x + 7$

$2x + 1$ $\quad$ $x + 5$

10. $7x^2 + 9x + 2$

$7x + 1$ $\quad$ $x + 1$

$7x + 9$ $\quad$ $x + 2$

Factor each of the following trinomials:

11. $9x^2 + 3x - 2$

12. $15x^2 - x - 2$

13. $2x^2 - 9x + 4$

14. $21x^2 + 5x - 6$

15. $6x^2 + 5x - 4$

16. $9x^2 + 6x - 8$

17. $8x^2 - 2x - 1$

18. $6x^2 + 25xy + 14y^2$

19. $15x^2 + 34xy + 15y^2$

20. $15x^2 - 13xy + 2y^2$

HA1-291: Factoring Quadratic Equations Using the Graphing Calculator

Trinomial expressions like $x^2 - 5x - 14$ can often be factored into the product of two binomials. In this case we have the factorization: $x^2 - 5x - 14 = (x + 2)(x - 7)$. In this lesson, you will learn how to use the TI-83 Plus™ graphing calculator to graphically factor trinomial expressions.

To use the calculator, we utilize the fact that when a trinomial expression like $x^2 - 5x - 14$ has factors of the form $x + 2$ and $x - 7$, we then know that $x = -2$ and $x = 7$ are where the graph crosses the x-axis and the points $(-2, 0)$ and $(7, 0)$ are the x-intercepts.

We will use the graphing power of the calculator to help us find those intercepts, and thereby find the factors of the trinomial.

The process has five steps:

- Step 0: (Settings) Initialize the calculator to a standard set of values so that we have a common starting point.
- Step 1: (Enter the expression) Enter the symbolic form of the expression to be analyzed into the calculator.
- Step 2: (Graph) Obtain a graph of the form y = expression.
- Step 3: (Check) Use other features of the graphing calculator to verify the x-intercepts of the graph.
- Step 4: (Factor) Use information about the x-intercepts to find the factorization.

Note: Before doing any of these examples you should check your calculator's settings to be sure that it will function as described above.

Press the MODE key and be sure that your settings appear as shown below.

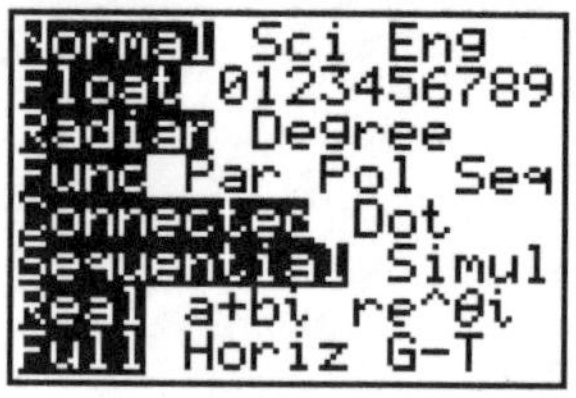

Next press the 2nd ZOOM keys and be sure that your settings appear as shown below.

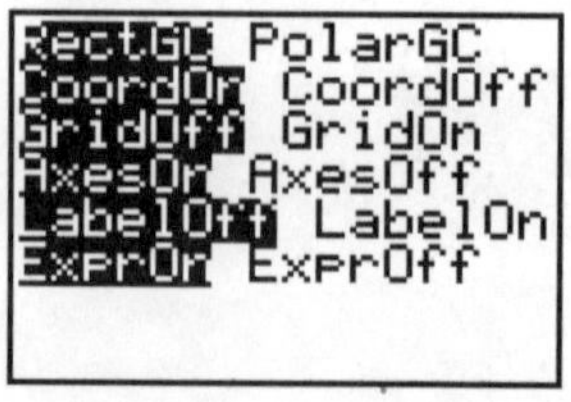

Next press the ZOOM key then the 6 key to select the standard window setting.

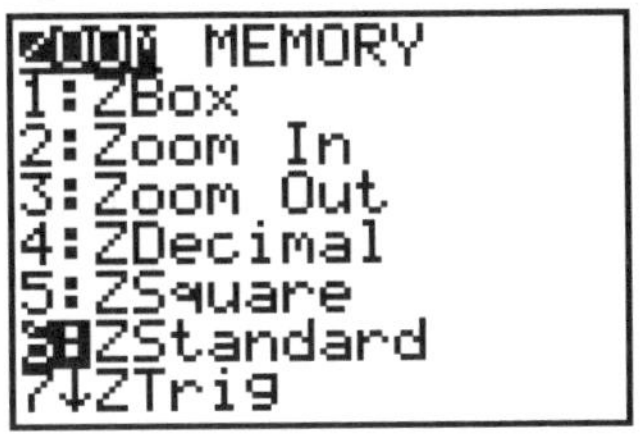

Press the WINDOW key to get the following screen.

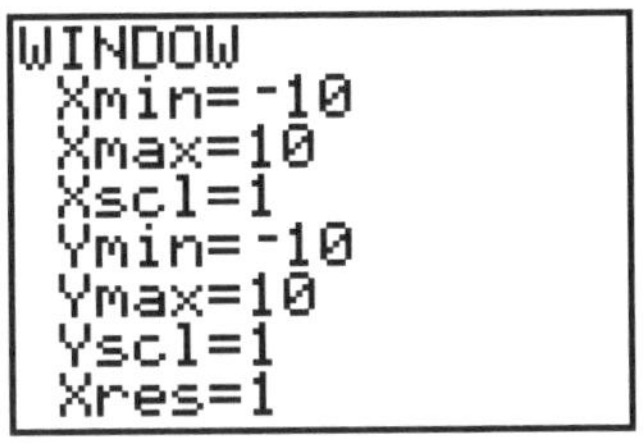

Example 1 Use the graphing calculator to factor $x^2 - 6x - 16$.

Step 0: (Settings) Make sure you have the default settings as directed above.

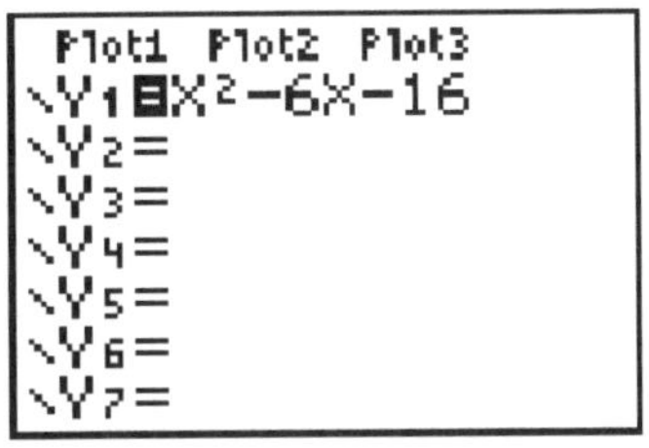

Step 1: (Enter the expression) Press the Y= key and enter the expression $x^2 - 6x - 16$ into the Y_1 slot.

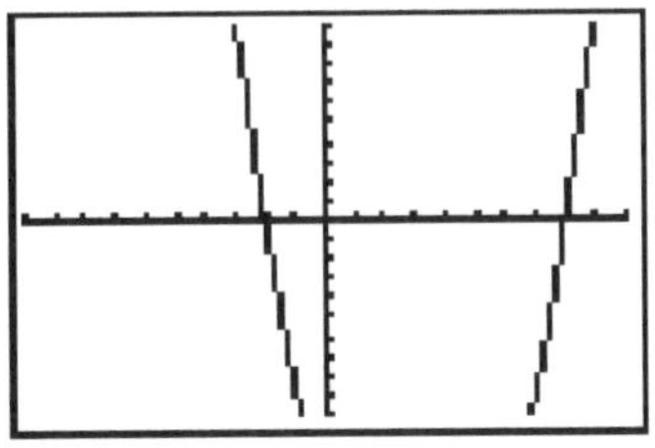

Step 2: (Graph) Press the GRAPH key. We see that the x-intercepts are $(-2, 0)$ and $(8, 0)$.

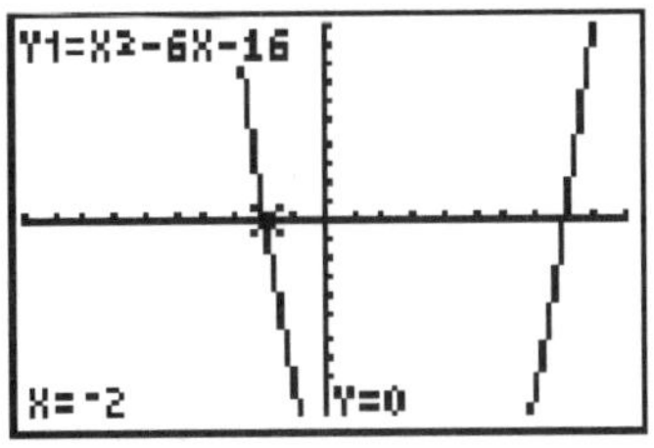

Step 3: (Check) You can check to see that $(-2, 0)$ and $(8, 0)$ are the intercepts. To do this, press TRACE and then − 2 followed by ENTER and you will get the screen to the left. Notice that at the bottom of the screen when $x = -2$, $y = 0$; this verifies the first intercept.

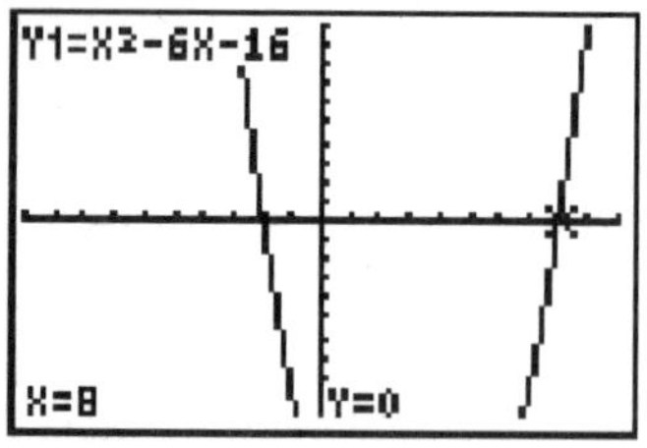

Now type in [8] followed by [ENTER], and at the bottom of the screen we see that when $x = -2$ that $y = 0$, verifying the second intercept.

Step 4: (Factor) Since the intercepts are (–2, 0) and (8, 0), we can factor the expression as:

$x^2 - 6x - 16 = (x - (-2))(x - 8) = (x + 2)(x - 8)$.

Answer: The expression $x^2 - 6x - 16$ factored is $(x + 2)(x - 8)$.

Example 2 Use the graphing calculator to factor $4x^2 + 7x - 15$.

Step 0: (Settings) Make sure you have the default settings as described in the beginning of the lesson.

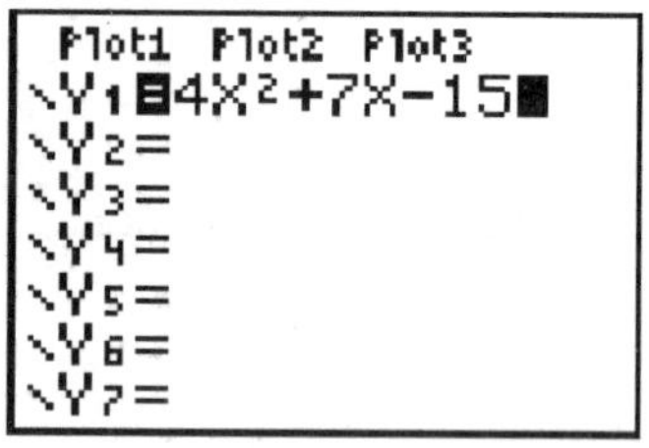

Step 1: (Enter the expression) Press the [Y=] key and enter the expression $4x^2 + 7x - 15$ into the Y_1 slot.

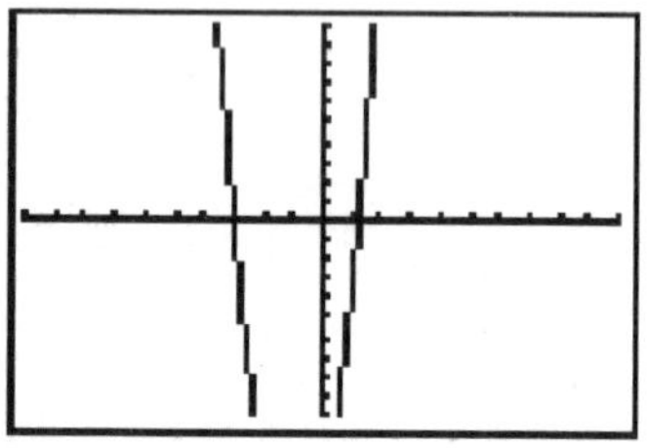

Step 2: (Graph) Press the [GRAPH] key. We see that one x-intercepts falls between $x = 1$ and $x = 2$, and one falls near $x = -3$.

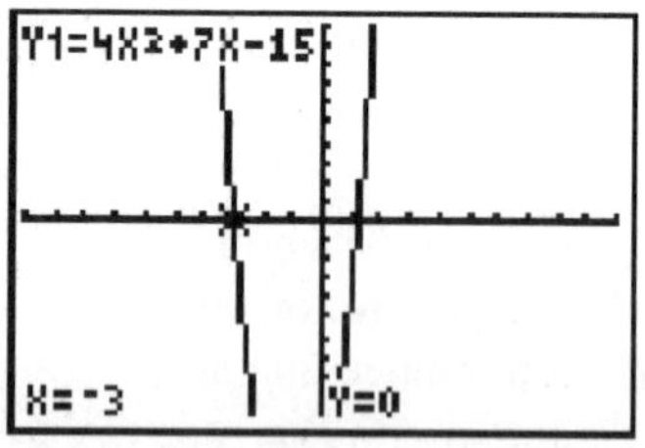

Step 3: (Check) We can verify the intercept at (–3, 0) as we did in the previous example. To do this, press [TRACE] and then [–][3] followed by [ENTER] and you will get the screen to the left. This verifies that (–3, 0) is one of the intercepts.

The second intercept isn't as easily guessed, so we will use an alternative method of finding the intercept.

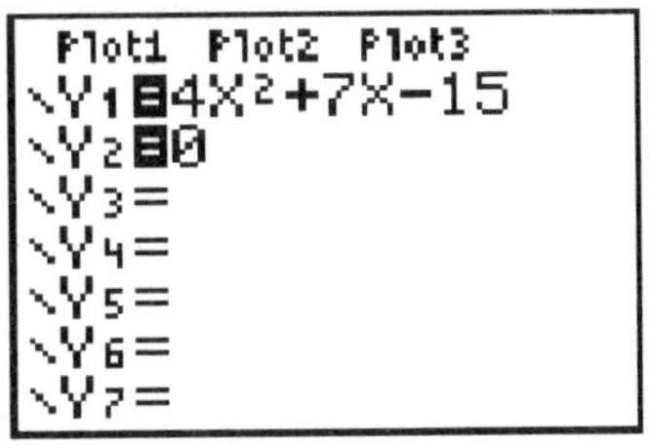

Recall that the x-axis is the graph of the line $y = 0$, so we will enter the equation $y = 0$ into the calculator by pressing the [Y=] key and enter the expression 0 into the Y_2 slot as shown to the left.

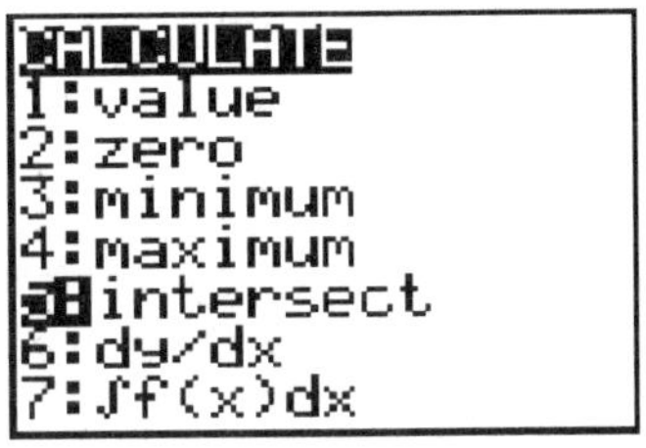

Next, we will find the place where the graph of Y_1 and Y_2 intersect (this will be the second x-intercept).
This is done by pressing [2nd][TRACE] followed by [5].

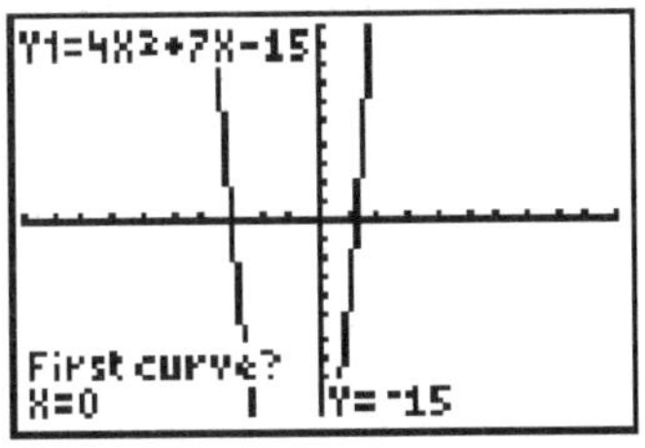

You will be prompted for the first curve with a screen like the one to the left. Press [ENTER].

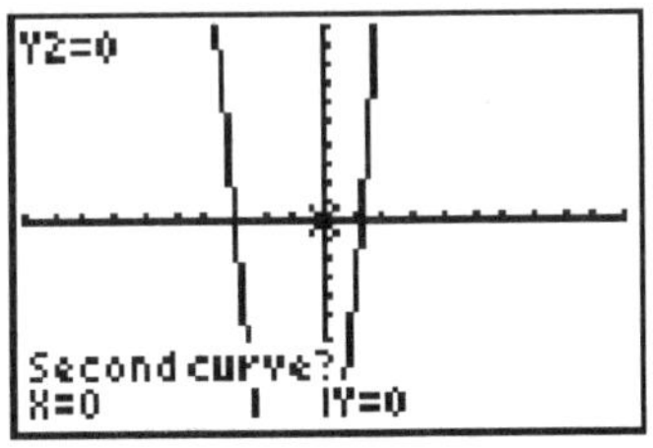

You will be prompted for the second curve with a screen like the one to the left. Press [ENTER] again.

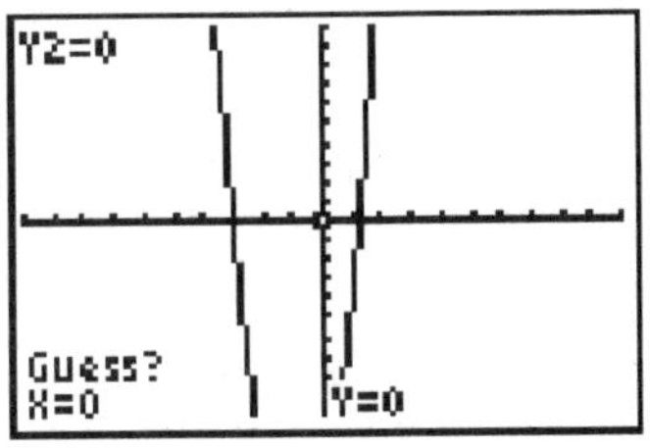

You will be prompted for a guess with a screen like the one to the left. Use the [◄] and [►] keys to move the highlighted point as close as possible to the intercept you are trying to verify and press [ENTER].

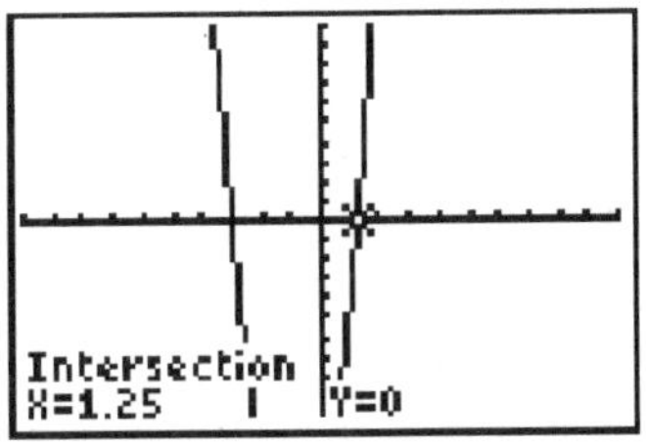

The calculator should display a screen like the one to the left, verifying that the second intercept is at $x = 1.25 = \frac{5}{4}$.

Step 4: (Factor) Now, since the intercepts are $\left(\frac{5}{4}, 0\right)$ and $(-3, 0)$, we can factor the expression as:

$$4x^2 + 7x - 15 = 4\left(x - \left(\frac{5}{4}\right)\right)(x - (-3)) = 4\left(x - \frac{5}{4}\right)(x + 3)$$
$$= (4x - 5)(x + 3).$$

Answer: The expression $4x^2 + 7x - 15$ can be factored as $(4x - 5)(x + 3)$.

Example 3 Use the graphing calculator to factor $-2x^2 + 5x + 12$.

Step 0: (Settings) Make sure you have the default settings as described in the beginning of the lesson.

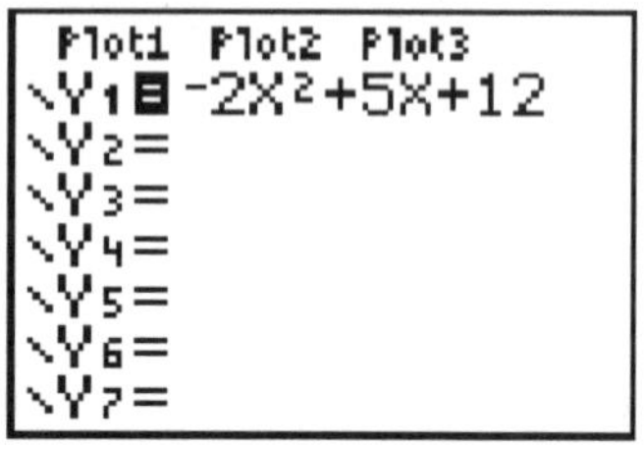

Step 1: (Enter the expression) Press the [Y=] key and enter the expression $-2x^2 + 5x + 12$ into the Y_1 slot.

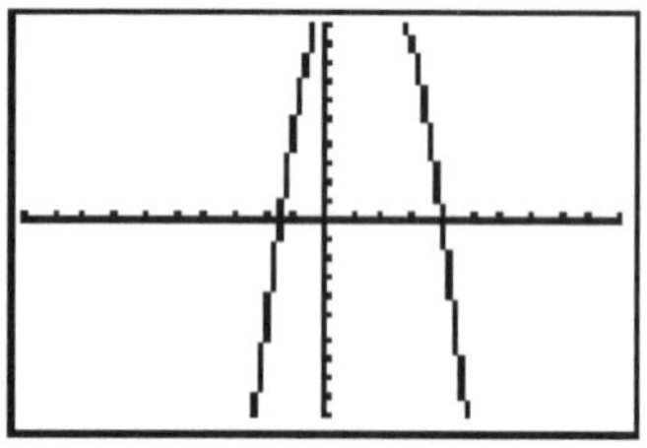

Step 2: (Graph) Press the [GRAPH] key. We see that one x-intercepts falls between $x = -2$ and $x = -1$, and one falls near $x = 4$.

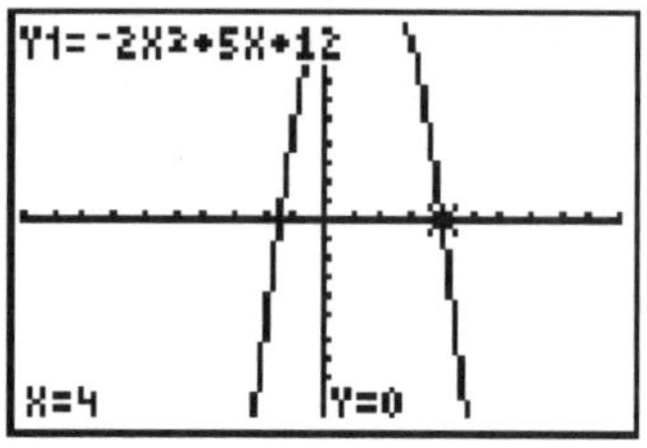

Step 3: (Check) We can verify the intercept at (4, 0) as we did in the previous example. To do this, press [TRACE] and then [4] followed by [ENTER] to get the screen shown to the left. This verifies that (4, 0) is one of the intercepts. The second intercept isn't as easily guessed, so we will use an alternative method of finding the intercept.

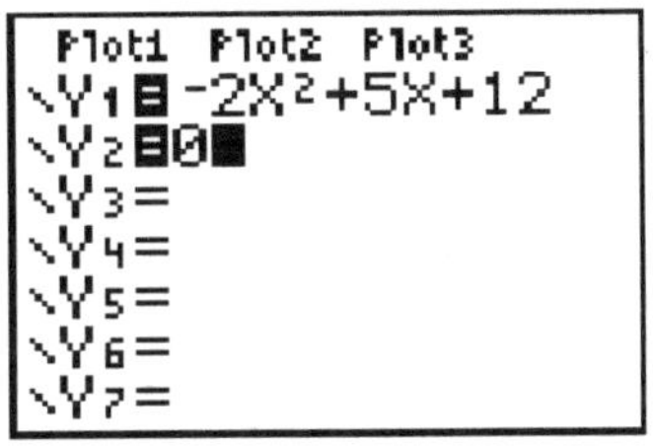

Recall that the x-axis is the graph of the line $y = 0$, so we will enter the equation $y = 0$ into the calculator by pressing the [Y=] key and entering the expression 0 into the Y_2 slot as shown to the left.

Next, we will find the place where the graph of Y_1 and Y_2 intersect (this will be the second x-intercept). This is done by pressing [2nd][TRACE] followed by [5].

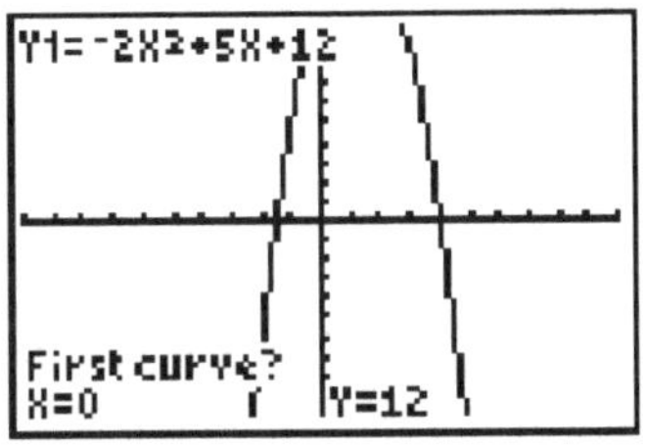

You will be prompted for the first curve with a screen like the one to the left. Press [ENTER].

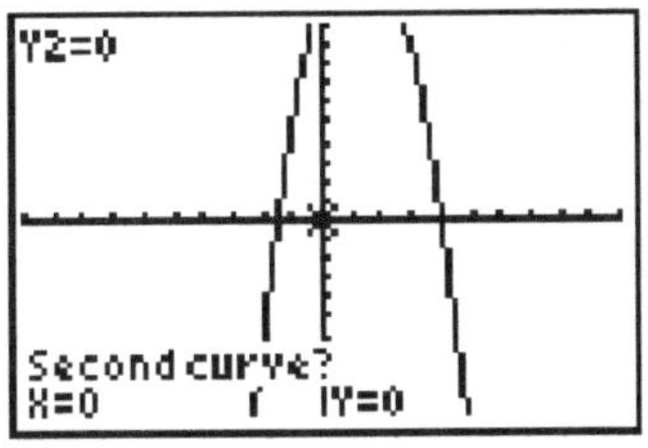

You will be prompted for the second curve with a screen like the one to the left. Press [ENTER] again.

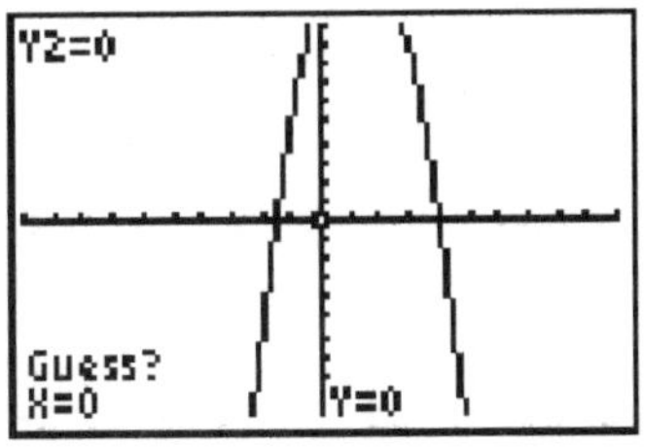

You will be prompted for a guess with a screen like the one to the left. Use the [◄] and [►] keys to move the highlighted point as close as possible to the intercept you are trying to verify and press [ENTER].

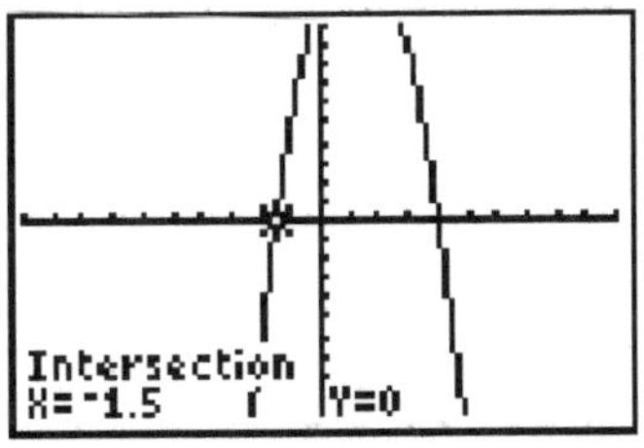

The calculator should display a screen like the one to the left, verifying that the second intercept is at $x = -1.5 = -\frac{3}{2}$.

Step 4: (Factor) Since the intercepts are $\left(-\frac{3}{2}, 0\right)$ and $(4, 0)$, we can factor the expression as:

$$-2x^2 + 5x + 12 = -2\left(x - \left(-\frac{3}{2}\right)\right)(x-4) = -2\left(x + \frac{3}{2}\right)(x-4)$$
$$= -(2x+3)(x-4).$$

Answer: The expression $-2x^2 + 5x + 12$ can be factored as $-(2x+3)(x-4)$.

Problem Set

Evaluate.

1. Use the graphing calculator to factor $x^2 - 11x + 18$.

2. Use the graphing calculator to factor $t^2 - 3t - 18$.

3. Use the graphing calculator to factor $-x^2 - 5x + 14$.

4. Use the graphing calculator to factor $z^2 - 37z + 252$.

5. Use the graphing calculator to factor $y^2 + 21y + 68$.

6. Use the graphing calculator to factor $2x^2 - 9x - 5$.

7. Use the graphing calculator to factor $-4x^2 + 19x - 12$

8. Use the graphing calculator to factor $10b^2 - 47b + 42$.

9. Use the graphing calculator to factor $-6x^2 + 13x + 5$.

10. Use the graphing calculator to factor $3x^2 + 3x + 2$.

HA1-295: Factoring by Removing a Common Factor and Grouping

Think of properties and concepts that you have used in algebra to help you solve equations or simplify expressions. One method that you have used is the Distributive Property. Recall that this property converts one term into two or more terms. Consider the expression $7(a + 2)$. Using the Distributive Property, $7(a + 2)$ becomes $7a + 14$. Although the number of terms increases, the Distributive Property simplifies the expression, making it easier to manipulate.

Now, consider grouping terms. Remember that if you are given an expression such as $2x + 4y - x + 5y$, it can be grouped by like terms to form a new expression, $x + 9y$. Note that the original expression has four different terms, but once it is simplified, it has only two terms.

As you continue to study the factoring process, you will be able to simplify expressions even faster once you remove common factors and group like terms. You already have learned these concepts in previous lessons; you are now going to combine these steps to ensure that each expression is in its simplest terms.

Look carefully at the expression $2x(5x + 9) + 7(5x + 9)$. The binomial $(5x + 9)$ appears in two places: once being multiplied by $2x$, and again being multiplied by 7. Because $(5x + 9)$ is a factor of both terms, it is referred to as the **common factor**. To simplify, you *remove* the common binomial factor $(5x + 9)$ by writing it as a stand-alone term inside a set of parentheses. Then, you regroup the other terms, making sure you follow the rules regarding placement of positive and negative signs. Therefore, $2x(5x + 9) + 7(5x + 9) = (5x + 9)(2x + 7)$.

Try to select the common binomial factor in the examples below.

Example 1 Factor: $3x(5x + 9) - 13(5x + 9)$

$5x + 9$	**Step 1:**	Find the common binomial factor.
$(5x + 9)(3x - 13)$	**Step 2:**	Group the remaining terms and put in parentheses.
	Answer:	$(5x + 9)(3x - 13)$

Example 2 Factor: $3(a + 7) + c(7 + a) - f(a + 7)$

$(a + 7)$	**Step 1:**	Find the common binomial factor.
$(a + 7)(3 + c - f)$	**Step 2:**	Group the remaining terms and put in parentheses. *Note: Although $(7 + a)$ in the second term is not written in the same order as $(a + 7)$, they are still common terms according to the Commutative Property.*
	Answer:	$(a + 7)(3 + c - f)$

Now consider four-term polynomials in which the common factor is not obvious.

Example 3 Factor: $25xy + 30x + 15y + 18$

$\mathbf{25xy} + \mathbf{30x} + 15y + 18$	**Step 1:**	Determine whether or not the first two terms contain a common factor.
$5x(5y + 6) + \mathbf{15y} + \mathbf{18}$	**Step 2:**	Check the last two terms to see if there is another common factor.

$5x(\mathbf{5y+6})+3(\mathbf{5y+6})$	**Step 3:**	Place the common factors in one set of parentheses.
$(5y+6)(5x+3)$	**Step 4:**	Place the remaining terms in separate parentheses.
	Answer:	$(5y+6)(5x+3)$

Example 4 Factor: $63xy+21x+45y+15$

$63xy+21x+45y+15$	**Step 1:**	Determine whether or not the first two terms contain a common factor.
$21x(3y+1)+45y+15$	**Step 2:**	Check the last two terms to see if there is another common factor.
$21x(3y+1)+15(3y+1)$	**Step 3:**	Place the common factors in one set of parentheses.
$(21x+15)(3y+1)$	**Step 4:**	Place the remaining terms in separate parentheses.
	Answer:	$(21x+15)(3y+1)$

Problem Set

1. Which of the following is a factor of the polynomial below?
$2x(y+2)+4(y+2)$

$4x+8$ $4y+8$
$4y+2$ $2x+4$

2. Which of the following is a factor of the polynomial below?
$3a(b-1)+2(b-1)$

$3a+1$ $2b-2$
$3a+2$ $3a-1$

3. Which of the following is a factor of the polynomial below?
$4y(x+2)-3(x+2)$

$3x+2$ $4y-3$
$4y+2$ $4y+3$

4. Which of the following is a factor of the polynomial below?
$x(3y+2)+5(3y+2)$

$x+2$ $5x+2$
$3x+2$ $x+5$

5. Which of the following is a factor of the polynomial below?
$5a(b+5)-(b+5)$

$5a+5$ $5a$
$5a-1$ $5b+5$

6. Which of the following is a factor of the polynomial below?
$2x^2+6x+xy+3y$

$2x+y$ $2x+3$
$y+3$ $3x+y$

7. Which of the following is a factor of the polynomial below?
$3x^2-15x+2xy-10y$

$x+2y$ $2y-5$
$3x+2$ $x-5$

8. Which of the following is a factor of the polynomial below?
$4xy-y+12x-3$

$4x-1$ $x-3$
$y-1$ $y-3$

9. Which of the following is a factor of the polynomial below?
$3ay-15bx+5ax-9by$

$3y+5$ $a-3b$
$a+3b$ $3y-5y$

10. Which of the following is a factor of the polynomial below?
$4a^2+3a+4ab+3b$

$4a+b$ $a+3$
$b+3$ $a+b$

HA1-300: Factoring a Polynomial Completely

In the past, you mastered many tasks that seemed difficult at the time—tying your shoestrings, riding a bike, driving a car. Each of these tasks required learning a series of steps, one by one, and finally combining all of them to perform the tasks competently. In algebra, you must also master many skills and then combine those skills to solve complex problems.

Factoring is an important skill in algebra. You have learned the techniques for factoring polynomials that have been presented and developed in previous lessons. Now, it is important to use all that you have learned about factoring in order to factor polynomials completely. "Completely" is the key word here. A polynomial is *factored completely* when it is written as a prime polynomial—one that initially cannot be simplified or factored, or one that is written in its simplest form.

When factoring polynomials, you have learned:

Factoring Process	Example
Factor out a common factor.	$4x + 6 = 2(2x + 3)$
Factor the difference of squares.	$a^2 - b^2 = (a + b)(a - b)$
Factor a trinomial using FOIL.	$x^2 + 7x + 12 = (x + 3)(x + 4)$
Factor polynomials by grouping.	$3x^2 + 3xy + 2x + 2y = 3x(x + y) + 2(x + y)$ $= (x + y)(3x + 2)$

Let's factor some polynomials completely.

Example 1 Factor: $6y^2 - 96$

$= 6(y^2 - 16)$ **Step 1:** Factor out the common factor 6.

$= 6(y + 4)(y - 4)$ **Step 2:** Factor $y^2 - 16$, the difference of squares. The polynomial cannot be simplified further.

Answer: $6y^2 - 96 = 6\,(y + 4)\,(y - 4)$

Example 2 Factor: $8x^4 - 128$

$= 8(x^4 - 16)$ **Step 1:** Factor out the common factor.

$= 8(x^2 + 4)(x^2 - 4)$ **Step 2:** Factor the difference of squares.

$= 8(x^2 + 4)(x + 2)(x - 2)$ **Step 3:** Factor the difference of squares. The polynomial cannot be simplified further.

Answer: $8x^4 - 128 = 8\,(x^2 + 4)\,(x + 2)\,(x - 2)$

Example 3 Factor: $3y^2 + 30y - 72$

$= 3(y^2 + 10y - 24)$	**Step 1:** Factor out the common factor.
$= 3(y-2)(y+12)$	**Step 2:** Factor using FOIL. The polynomial cannot be simplified further.
	Answer: $3y^2 + 30y - 72 = 3(y-2)(y+12)$

Example 4 Factor: $3xy^2 - 12x + ay^2 - 4a$

$= 3x(y^2 - 4) + ay^2 - 4a$	**Step 1:** Factor out the common factor in the first two terms.
$= 3x(y^2 - 4) + a(y^2 - 4)$	**Step 2:** Factor out the common factor from the last terms.
$= 3x(y-2)(y+2) + a(y-2)(y+2)$	**Step 3:** Factor the difference of squares.
$= (3x + a)(y-2)(y+2)$	**Step 4:** Group like terms. The terms outside the parentheses can be grouped together first. Next, write the terms $(y-2)$ and $(y+2)$ only once because when the $3x$ and a are distributed, you get the expanded factored form again.
	Answer: $3xy^2 - 12x + ay^2 - 4a = (3x + a)(y-2)(y+2)$

Problem Set

Factor completely:

1. $5x^2 - 10x + 5$
2. $3x^2 - 30x + 75$
3. $a + 6a^2b - ab$
4. $32x^2 + 24b^2 - 40z^2$
5. $5x^2 + 20x - 60$
6. $2m^2 + 4m + 2$
7. $9x^2 + 45x$
8. $2x^2 - 98$
9. $8ab - 64a$
10. $x^3 - 4x$
11. $b^4 - 9b^2$
12. $-x^3 + 4x^2 + 12x$
13. $-4x^3 - 32x^2 - 64x$
14. $-3x^5 - 12x^4 - 9x^3$
15. $7ax^2 - 28a + ax^2 - 4a$
16. $x^3 - xy^2$
17. $x^3 - x^2 - 5x + 5$
18. $10x^4 - 34x^3 + 12x^2$
19. $27x^3 - 18x^2 - 45x$
20. $-8x^2 - 28x - 12$

HA1-305: Solving Polynomial Equations by Factoring

An algebraic equation is a mathematical sentence that uses an equal sign (=) to relate one expression to another. Mastering algebraic equations will help you solve polynomial equations. There are two types of polynomial equations: **linear equations** and **quadratic equations.**

Linear Equation	An equation in the form $ax + b = 0$
Quadratic Equation	An equation in the form $ax^2 + bx + c = 0,\ a \neq 0$

Though you must still solve for an unknown variable when solving polynomial equations, you will find that factoring makes solving the equation easier. You also can use any of the properties discussed in earlier lessons to solve polynomial equations.

A polynomial equation can usually be solved by factoring, and then applying the **zero-product property**.

Zero-Product Property	If a and b are real numbers and $ab = 0$, then $a = 0$ or $b = 0$ or both a and $b = 0$.

This property states that if the product of two factors equals zero, then one or both factors must equal zero. This means that you must test each factor of a polynomial equation to see which factor equals zero.

Let's use the zero-product property to solve a quadratic equation.

Example 1 Solve: $x^2 - 16 = 0$

$(x+4)(x-4) = 0$

Step 1: Recall that you should try factoring the polynomial equation first. You should recognize immediately that the equation is the difference of two squares. The factored binomials are at left.

$$x + 4 = 0 \qquad x - 4 = 0$$
$$x + 4 - 4 = 0 - 4 \qquad x - 4 + 4 = 0 + 4$$
$$x = -4 \qquad x = 4$$

Step 2: Now, apply the zero-product property. Do this by setting each factor equal to 0, one at a time. Keep in mind that the answers you find are not the actual solutions to the equation: they are part of the replacement set. The **replacement set** is the set of numbers that may be substituted for a variable. The replacement set for the equation above is $\{-4, 4\}$.

Check: $x = -4$

$$x^2 - 16 = 0$$
$$(-4)^2 - 16 = 0$$
$$16 - 16 = 0 \text{ true}$$

Check: $x = 4$

$$x^2 - 16 = 0$$
$$(4)^2 - 16 = 0$$
$$16 - 16 = 0 \text{ true}$$

Step 3: Next, you always should check each member of the replacement set by substituting it into the equation in the form $ax^2 + bx + c = 0$. If the equation is true, then the substituted number is a member of the solution set. The **solution set** is the set of all numbers from the replacement set that make the equation true. Let's test the replacement set $\{4, -4\}$.

Answer: Substituting either 4 or –4 makes the equation true; therefore, the solution set is $\{-4, 4\}$.

If an equation is not in the $ax^2 + bx + c = 0$ form, you must still simplify the equation before you can factor, as in the following example. Before you can factor, you must use the additive inverse to get all terms on one side; whether you add or subtract the number depends on its sign.

Example 2 Solve: $x^2 - 5x = -6$

$x^2 - 5x + 6 = -6 + 6$

Step 1: Use the additive inverse to move the –6 to the left member of the equation.

$x^2 - 5x + 6 = 0$

Step 2: Set the equation equal to 0.

$(x - 2)(x - 3) = 0$

Step 3: Factor the trinomial.

$x - 2 = 0$ or $x - 3 = 0$

$x = 2$ $\quad x = 3$

Step 4: Set each factor equal to 0 and solve for x. The replacement set is $\{2, 3\}$.

Check: $x = 2$

$x^2 - 5x + 6 = 0$

$(2)^2 - 5(2) + 6 = 0$

$4 - 10 + 6 = 0$

$-6 + 6 = 0$

$0 = 0$ true

Check: $x = 3$

$x^2 - 5x + 6 = 0$

$(3)^2 - 5(3) + 6 = 0$

$9 - 15 + 6 = 0$

$-6 + 6 = 0$

$0 = 0$ true

Step 5: Substitute members of the replacement set into the equation to determine whether or not they are actual solutions to the equation. If the equation is true after the substitution has been made, then that solution is a member of the solution set.

Answer: The substitution of 2 or 3 makes the equation true; therefore, the solution set is $\{2, 3\}$.

Let's use factoring to solve several different polynomials. Note that solutions are not always whole numbers: they may also be fractions or decimals.

Example 3 Solve: $x^2 + 9x = 0$

$x(x + 9) = 0$

Step 1: Factor out the common factor, which is x.

$x + 9 = 0$ or $x = 0$

$x = -9$

Step 2: Set each factor equal to 0 and solve for x. The replacement set is $\{-9, 0\}$.

Check: $x = -9$

$x^2 + 9x = 0$

$(-9)^2 + 9(-9) = 0$

$81 - 81 = 0$

$0 = 0$ true

Check: $x = 0$

$x^2 + 9x = 0$

$0^2 + 9(0) = 0$

$0 + 0 = 0$

$0 = 0$ true

Step 3: Substitute members of the replacement set into the equation to determine whether or not they are actual solutions to the equation.

Answer: The substitution of –9 or 0 makes the equation true; therefore, the solution set is $\{-9, 0\}$.

Example 4 Solve: $8x^2 - 4x = 24$

$$8x^2 - 4x - 24 = 24 - 24$$

Step 1: Subtract 24 from both sides of the equation.

$$8x^2 - 4x - 24 = 0$$

Step 2: Set the equation equal to 0.

$$(2x + 3)(4x - 8) = 0$$

Step 3: Factor the trinomial.

$$2x + 3 = 0 \qquad 4x - 8 = 0$$
$$2x + 3 - 3 = 0 - 3 \qquad 4x - 8 + 8 = 0 + 8$$
$$2x = -3 \qquad 4x = 8$$
$$\frac{2x}{2} = -\frac{3}{2} \qquad \frac{4x}{4} = \frac{8}{4}$$
$$x = -\frac{3}{2} \qquad x = 2$$

Step 4: Set each factor equal to 0 and solve for x. The replacement set is $\left\{-\frac{3}{2}, 2\right\}$.

Check: $x = -\frac{3}{2}$

$$8\left(-\frac{3}{2}\right)^2 - 4\left(-\frac{3}{2}\right) = 24$$
$$8\left(\frac{9}{4}\right) - 4\left(-\frac{3}{2}\right) = 24$$
$$18 - (-6) = 24$$
$$24 = 24 \text{ true}$$

Check: $x = 2$

$$8(2)^2 - 4(2) = 24$$
$$8(4) - 8 = 24$$
$$32 - 8 = 24$$
$$24 = 24 \text{ true}$$

Step 5: Substitute members of the replacement set into the equation to determine whether or not they are actual solutions of the equation.

Answer: The substitution of $-\frac{3}{2}$ or 2 makes the equation true; thus, the solution set is $\left\{-\frac{3}{2}, 2\right\}$.

Problem Set

Solve:

1. $w^2 - 9 = 0$

2. $x^2 + 5x + 6 = 0$

3. $x^2 - 169 = 0$

4. $x^2 + 7x + 12 = 0$

5. $x^2 + 12x + 27 = 0$

6. $x^2 - 2x - 35 = 0$

7. $x^2 + 3x - 54 = 0$

8. $x^2 - 5x - 6 = 0$

9. $5x(x + 2) = 0$

10. $3x(x - 6) = 0$

11. $a^2 + 64 = -16a$

12. $x^2 + x = 72$

13. $x^2 = 24 - 10x$

14. $2x^2 = 14x$

15. $2x^2 = 72$

16. $4x^2 - 24x + 20 = 0$

17. $20 + 6x = 2x^2$

18. $x^3 + x^2 - (4x + 4) = 0$

19. $x^3 + x^2 + (x + 1) = 0$

20. $30x^2 - x = 20$

HA1-310: The Practical Use of Polynomial Equations

Some students find word problems—sometimes called story problems—difficult to solve. You must approach these problems with logic, as well as skill. You need to use common sense, knowing that if you logically break down the problem into workable parts, you can find the solution. Often, key words in the problem let you know whether you must add, subtract, multiply, or divide. Finally, add the math skills you have already learned to help solve these types of problems. Let's consider how to use polynomial equations to solve word problems.

Example 1 Three consecutive integers have the sum of 33. What are the integers?

x = 1st integer
$(x + 1)$ = 2nd integer
$(x + 2)$ = 3rd integer

Step 1: Choose a variable to represent the unknown quantity.

$$x + (x + 1) + (x + 2) = 33$$

Step 2: Set up the equation.

$$3x + 3 = 33$$

Step 3: Combine like terms.

$$3x + 3 - 3 = 33 - 3$$
$$3x = 30$$
$$x = 10$$

Step 4: Solve for x.

$$x + 1 = 11$$

Step 5: Solve for the second integer.

$$x + 2 = 12$$

Step 6: Solve for the third integer.

Answer: The three consecutive integers are 10, 11, and 12.

Example 2 A rectangle has an area of 135 square inches. What is the rectangle's length and width if one side is 6 inches longer than the other side?

x = width
$(x + 6)$ = length

Step 1: Choose a variable to represent the unknown quantity. Use the formula $A = lw$.

$$\ell w = A$$

Step 2: Set up the equation.

$$(x + 6)x = 135$$
$$x^2 + 6x = 135$$

Step 3: Distribute x to terms in the parentheses.

$$x^2 + 6x - 135 = 0$$

Step 4: Put equation into $ax^2 + bx + c = 0$ format.

$$(x + 15)(x - 9) = 0$$

Step 5: Factor the trinomial.

$x + 15 = 0$ or $x - 9 = 0$
$x = -15$ $\quad$ $x = 9$

Step 6: Set each factor equal to 0 then solve for x.

width $= x = 9$ inches length $= x + 6 = 9 + 6$ $= 15$ inches	**Step 7:**	Substitute the answers into the given measures. However, you do not need to substitute –15, because measurements cannot be negative. Thus, substituting 9 for x in both measurements, you get a width of 9 inches and length of 15 inches. You can check your answers by substituting both numbers into the original equation. When this is done, the equation is true; therefore, the answer is correct.
	Answer:	The length is 15 inches and the width is 9 inches.

Example 3 A square has an area of 81 square inches. What is the square's perimeter?

$A = s \cdot s$ or $A = s^2$ $s = \pm\sqrt{81} = \pm 9$ Therefore, $s = 9$	**Step 1:**	Before you can determine the perimeter of the square, you must find the length of one side. Use the formula for the area of a square to determine the length of one of the sides. Since the area is 81 in.^{2}, take the square root of 81 to find the length of one side. We know that 81 has two square roots: 9 and –9. However, in practical problems like this, we can see that 9 is the only viable solution because a square cannot have a side whose length is less than or equal to zero.
$P = 4s$ $P = 4(9)$ $P = 36$ in.	**Step 2:**	Substitute the value of s into the perimeter formula to determine the perimeter of the square, which is 36 in. *Note: The units for perimeter are not squared.*

When solving word problems, remember to be very precise when you write down information. Next, set up your equation and solve for x. Always go back and reread the question. Then use the first information that you wrote and the solution for x to completely answer the word problem.

Example 4 Four consecutive multiples of 4 have a sum of 88. What are the four integers?

x = first number $(x + 4)$ = second number $(x + 8)$ = third number $(x + 12)$ = fourth number	**Step 1:**	Choose a variable to represent the unknown quantity. Recall that the numbers will increase by four each time.
$x + (x + 4) + (x + 8) + (x + 12) = 88$	**Step 2:**	Set up the equation.
$4x + 24 = 88$	**Step 3:**	Simplify by combining like terms.
$4x + 24 - 24 = 88 - 24$ $4x = 64$	**Step 4:**	Isolate the variable term by subtracting 24 from both sides.
$\frac{4x}{4} = \frac{64}{4}$ $x = 16$	**Step 5:**	To solve for x, divide both sides by 4.

$x = 16$ $x + 4 = 16 + 4 = 20$ $x + 8 = 16 + 8 = 24$ $x + 12 = 16 + 12 = 28$	**Step 6:**	Substitute x with 16 to find the first, second, third and fourth numbers.
$16 + 20 + 24 + 28 = 88$	**Step 7:**	Verify that the sum of the integers is 88.
	Answer:	The four multiples are 16, 20, 24, and 28.

Problem Set

1. Find two consecutive positive integers whose product is 6. Separate the integers with a comma.

2. Find two consecutive negative integers whose product is 182. Separate the integers with a comma.

3. Find two consecutive positive integers whose product is 132. Separate the integers witha comma.

4. Find two consecutive negative integers whose product is 306. Separate the integers with a comma.

5. Find two consecutive positive even integers whose product is 288. Separate the integers with a comma.

6. Find two consecutive negative odd integers whose product is 399. Separate the integers with a comma.

7. Find two consecutive positive even integers whose product is 168. Separate the integers with a comma.

8. Find two consecutive negative odd integers whose product is 323. Separate the integers with a comma.

9. The formula $D = \frac{n(n-3)}{2}$ is used to find the number of diagonals, D, of a polygon with n sides. Find the number of sides of a polygon with 2 diagonals.

10. The formula $D = \frac{n(n-3)}{2}$ is used to find the number of diagonals, D, of a polygon with n sides. Find the number of sides of a polygon with 14 diagonals.

11. The area of a rectangular garden is 63 square feet. If the length is 2 feet more than the width, find the width of the garden.

12. The area of a rectangular garden is 45 square feet. If the length is 4 feet more than the width, find the length of the garden.

13. The area of a rectangular garden is 36 square feet. If the length is 5 feet more than the width, find the width of the garden.

14. The area of a rectangular garden is 120 square feet. If the length is 2 feet more than the width, find the length of the garden.

15. The area of a rectangular garden is 54 square feet. If the length is 3 feet more than the width, find the width of the garden.

16. The area of a rectangular garden is 96 square feet. If the length is 4 feet more than the width, find the length of the garden.

17. The area of a rectangular garden is 44 square feet. If the length is 7 feet more than the width, find the width of the garden.

18. The altitude of a triangle is two meters less than the base. If the area of the triangle is 40 square meters, find the length of the altitude.

19. The sum of the squares of two positive consecutive integers is 61. Find the integers. Separate the integers with a comma.

20. The sum of the squares of two positive consecutive integers is 313. Find the integers. Separate the integers with a comma.

HA1-355: Dividing Polynomials

Dividing polynomials, or mathematical expressions with more than one term, is similar to procedures that you have already learned, like addition of fractions and long division. Dividing polynomials by monomials, or mathematical expressions with only one term, is like adding fractions.

Dividing a Polynomial by a Monomial	To divide a polynomial by a monomial, divide each term in the polynomial by a monomial. $\frac{a+b}{c} = \frac{a}{c} + \frac{b}{c}$ when $c \neq 0$

These examples show how these division processes work.

Example 1 Divide: $(4a^3 + 24a) \div (8a)$

$= \frac{4a^3 + 24a}{8a}$

$= \frac{4a^3}{8a} + \frac{24a}{8a}$

Step 1: In order to solve this problem write the quotient in fraction form and divide each term of the polynomial $4a^3 + 24a$ by the monomial $8a$.

$= \frac{4a^3}{8z} + \frac{24a}{8a}$

$= \frac{{}^{1}\cancel{4}a^{\cancel{3}\,2}}{{}_{2}\cancel{8}\,{}_{1}\cancel{a}} + \frac{{}^{3}\cancel{24}\cancel{a}}{{}_{1}\cancel{8}\cancel{a}}$

$= \frac{a^2}{2} + 3$

Step 2: Now we must simplify the equation and solve.

Answer: $\frac{a^2}{2} + 3$

Example 2 Divide: $(15x^5 + 5x^3 + 20x^2 + 5) \div 5x$

$= \frac{15x^5 + 5x^3 + 20x^2 + 5}{5x}$

$= \frac{15x^5}{5x} + \frac{5x^3}{5x} + \frac{20x^2}{5x} + \frac{5}{5x}$

Step 1: In order to solve this problem write the quotient in fraction form and divide each term of the polynomial $15x^5 + 5x^3 + 20x^2 + 5$ by the monomial $5x$.

$$= \frac{15x^5}{5x} + \frac{5x^3}{5x} + \frac{20x^2}{5x} + \frac{5}{5x}$$

Step 2: Now we must simplify the equation and solve.

$$= \frac{{}^{3}\cancel{15}x^{\cancel{5}\,4}}{{}_{1}\cancel{5}\cancel{x}} + \frac{{}^{1}\cancel{5}x^{\cancel{3}\,2}}{{}_{1}\cancel{5}\cancel{x}} + \frac{{}^{4}\cancel{20}\cancel{x^2}}{{}_{1}\cancel{5}\cancel{x}} + \frac{{}^{1}\cancel{5}}{{}_{1}\cancel{5}x}$$

$$= 3x^4 + x^2 + 4x + \frac{1}{x}$$

Answer: $3x^4 + x^2 + 4x + \frac{1}{x}$

Now let's try dividing by a binomial. One way to do this is by using long division, which is similar to long division with whole numbers; you must be careful to include all variable terms in the dividend. Make sure all terms are written in descending exponential order. Use zero as the coefficient of any missing terms. If there is a remainder, you can add it to the quotient in the form of remainder over divisor.

Example 3 Divide: $(x^2 - 5x + 4) \div (x - 1)$

$$x - 1\overline{)\,x^2 - 5x + 4}$$

Step 1: Rewrite the problem as a long division problem.

$$\begin{array}{r} x \phantom{{}-5x+4} \\ x - 1\overline{)\,x^2 - 5x + 4} \end{array}$$

Step 2: Find the term that will equal x^2 when multiplied by x. Place that term above $5x$.

$$\begin{array}{r} x \phantom{{}-5x+4} \\ x - 1\overline{)\,x^2 - 5x + 4} \\ x^2 - 1x \phantom{{}+4} \end{array}$$

Step 3: Multiply x by each term of the divisor, aligning terms with the same exponents.

$$\begin{array}{r} x \phantom{{}-5x+4} \\ x - 1\overline{)\,x^2 - 5x + 4} \\ \underline{-(x^2 - 1x)} \phantom{{}+4} \\ -4x + 4 \end{array}$$

Step 4: Subtract the partial product from the dividend and bring down the next term.

$$\begin{array}{r} x - 4 \\ x - 1\overline{)\,x^2 - 5x + 4} \\ \underline{-(x^2 - 1x)} \phantom{{}+4} \\ -4x + 4 \\ \underline{-(-4x + 4)} \\ 0 \end{array}$$

Step 5: Now find the term that will equal $-4x$ when multiplied by x. Subtract the partial product.

Answer: $x - 4$

Example 4 Divide: $\dfrac{4x^2 - 9x + 15}{x - 6}$

$x - 6\overline{)\,4x^2 - 9x + 15}$

Step 1: In order to solve this problem we must perform long division.

$$
\begin{array}{r}
4x + 15 \\
x - 6\overline{)\,4x^2 - 9x + 15} \\
-(4x^2 - 24x) \\
\hline
15x + 15 \\
-(15x - 90) \\
\hline
105
\end{array}
$$

Step 2: Perform the long division and simplify.

Answer: $4x + 15 + \dfrac{105}{x - 6}$

Problem Set

Divide:

1. $(12m^2 + 8m) \div 4m$
2. $(18n + 6n^2) \div 3n$
3. $(24y^3 - 6y) \div 6y$
4. $(12b^2 - 15) \div 3$
5. $\dfrac{5c^3 + 10c^2 - 4}{5}$
6. $(36y^4 - 12y^2) \div 6y$
7. $\dfrac{6b^3 + 12b^2 - 24b}{12b}$
8. $(18b^3 + 27b) \div 9$
9. $\dfrac{100m^6 - 50m^5 - 5}{10}$
10. $\dfrac{24x^6 - 12x^2 + x + 6}{6x}$
11. $(y^2 - 3y + 8) \div (y - 5)$
12. $(4x^3 + 4) \div (x - 1)$
13. $(c^2 - 2c + 3) \div (c - 7)$
14. $(8k^2 - 30) \div (k - 2)$
15. $(x^2 + 10x + 21) \div (x + 3)$
16. $(a^2 - 8a - 16) \div (a + 4)$
17. $(y^2 - 8y + 16) \div (y - 4)$
18. $(20x^3 - 27x^2 + x + 6) \div (4x - 3)$
19. $(6x^3 + x^2 - 2x + 8) \div (2x + 3)$
20. $(20w^2 + 30w + 18) \div (5w + 5)$

HA1-362: Solving Work Problems

Work problems involve two or more people or objects working together to complete a job. Information such as how long it takes each person to do the job alone or how long it takes the people to do the job together is given. Additional information about the completion of the job is then determined.

Suppose Fred can paint a house in 5 days and Jeff can paint the same house in 6 days. If Fred paints for 1 day, then he completes $\frac{1}{5}$ of the job. Fred's rate of work is $\frac{1}{5}$. If Jeff paints for 1 day, then he completes $\frac{1}{6}$ of the job. Jeff's rate of work is $\frac{1}{6}$.

Rate of Work	The amount of work done in one unit of time

If Fred paints for 2 days, then he completes $\frac{1}{5}(2)$ or $\frac{2}{5}$ of the entire job. Therefore, the amount of work completed by Fred is $\frac{2}{5}$. If Jeff paints for 2 days, then he completes $\frac{1}{6}(2)$ or $\frac{1}{3}$ of the entire job. Therefore, the amount of work completed by Jeff is $\frac{1}{3}$. Generally, if r represents the rate of work, t represents the time spent on the project, and w represents the amount of work completed, the following formula can be used.

Amount of Work Completed	Amount of work completed (w) = rate of work (r) • number of units of time (t) or $w = r \cdot t$

A work problem often asks a question about completing an entire job with two or more people or objects working together; each person or object completes a fractional part. When the entire job is finished, the sum of the fractional parts is equal to one.

In the previous example, after Fred and Jeff paint the house together for 2 days, they have completed $\frac{2}{5} + \frac{1}{3}$ or $\frac{11}{15}$ of the job. Since the sum of the parts completed by each painter is less than one, they have not yet completed the entire job. In order to determine the time required to complete the entire job, let t denote the time they paint together and set the sum of the parts completed by each equal to one, which represents the entire job. Therefore, the fractional part that Fred completes is $\frac{1}{5}(t)$ or $\frac{t}{5}$ and the fractional part that Jeff completes is $\frac{1}{6}(t)$ or $\frac{t}{5}$.

Part completed by Fred	plus	part completed by Jeff	equals	the entire job.
$\frac{t}{6}$	$+$	$\frac{t}{5}$	$=$	1

Solve the equation to determine the amount of time it takes Fred and Jeff to paint the entire house together.

$$\begin{aligned} \frac{t}{6}+\frac{t}{5} &= 1 \\ 30\left(\frac{t}{6}+\frac{t}{5}\right) &= 30(1) \\ 5t+6t &= 30 \\ 11t &= 30 \\ \frac{11t}{11} &= \frac{30}{11} \\ t &= \frac{30}{11} \end{aligned}$$

Fred and Jeff can paint the entire house in $\frac{30}{11}$ or $2\frac{8}{11}$ days if they work together.

Let's look at some examples that involve finding the rate of work, time, or the amount of work completed.

Example 1 Finney can clean the house in 3 hours. At this rate, how much of the house can she clean in 2 hours?

Step 1: Find Finney's rate of work.

Since Finney can clean the house in 3 hours, she can clean $\frac{1}{3}$ of the house in 1 hour. Therefore, her rate of work is $\frac{1}{3}$.

Step 2: The problem states that the time she spends cleaning the house is 2 hours. Organize the rate of work and the time in a table.

	Rate of Work	Time	Amount of Work Completed
Finney	$\frac{1}{3}$	2	

Step 3: Use the formula $r \cdot t = w$ to find the amount of work that can be completed in 2 hours.

$$\begin{aligned} r \cdot t &= w \\ \frac{1}{3} \cdot 2 &= w \\ \frac{2}{3} &= w \end{aligned}$$

Step 4: Fill the table using the information obtained in Step 3.

	Rate of Work	Time	Amount of Work Completed
Finney	$\frac{1}{3}$	2	$\frac{1}{3}(2) = \frac{2}{3}$

Answer: Finney can clean $\frac{2}{3}$ of the house in 2 hours.

Example 2 Doug is installing a fence around his back yard. In 5 hours, he has completed $\frac{1}{6}$ of the fence. Determine Doug's rate of work.

Let r = Doug's rate of work.

	Rate of Work	Time	Amount of Work Completed
Doug	r	5	

Step 1: Let r denote Doug's rate of work. The problem states that he has been working for 5 hours. Organize the given information in a table.

	Rate of Work	Time	Amount of Work Completed
Doug	r	5	$5r$

Step 2: Use the formula $r \cdot t = w$ to find an expression for the amount of work that can be completed in 5 hours. Place that expression in the table.

$$5r = \frac{1}{6}$$

$$\frac{1}{5}(5r) = \frac{1}{5}\left(\frac{1}{6}\right)$$

$$r = \frac{1}{30}$$

Step 3: The problem states that the amount of work completed by Doug is $\frac{1}{6}$. The table shows that the amount of work completed by Doug is $5r$. Set these expressions equal to each other and solve for r.

Answer: Doug completes $\frac{1}{30}$ of the fence per hour.

Example 3 Mr. Brown can shuck all of the ears of corn in a bushel in 4 hours. His son can do the same job in 6 hours. How long will it take Mr. Brown and his son to shuck all of the ears of corn if they work together?

Since Mr. Brown can shuck all of the corn in 4 hours, he can shuck $\frac{1}{4}$ of the corn in 1 hour. Therefore, his rate of work is $\frac{1}{4}$.

Since Mr. Brown's son can shuck all of the corn in 6 hours, he can shuck $\frac{1}{6}$ of the corn in 1 hour. Therefore, his rate of work is $\frac{1}{6}$.

Step 1: Determine the rates of work for Mr. Brown and his son.

	Rate of Work	Time	Amount of Work Completed
Mr. Brown	$\frac{1}{4}$	t	
son	$\frac{1}{6}$	t	

Step 2: Let t represent the time they work. Organize the information about their rates of work and times they work in a table.

	Rate of Work	Time	Amount of Work Completed
Mr. Brown	$\frac{1}{4}$	t	$\frac{1}{4}(t) = \frac{t}{4}$
son	$\frac{1}{6}$	t	$\frac{1}{6}(t) = \frac{t}{6}$

Step 3: Use the formula $r \cdot t = w$ to find expressions for the amount of work completed in t hours by each person. Place the expressions in the table.

$$\frac{t}{4} + \frac{t}{6} = 1$$

Step 4: The sum of the work completed is equal to one. Use this to write an equation.

$$\begin{aligned} 12\left(\frac{t}{4} + \frac{t}{6}\right) &= 12(1) \\ 3t + 2t &= 12 \\ 5t &= 12 \\ \frac{5t}{5} &= \frac{12}{5} \\ t &= 2.4 \end{aligned}$$

Step 5: Solve the equation for t.

Answer: If they work together, they can shuck all of the ears of corn in 2.4 hours.

Example 4 Martha can clean the kitchen in 30 minutes. If her brother Mark helps her, the job can be completed in 20 minutes. How long would it take Martha's brother to clean the kitchen alone?

Since Martha can clean the kitchen in 30 minutes, she can clean $\frac{1}{30}$ of the kitchen in 1 minute. Therefore, her rate of work is $\frac{1}{30}$. The amount of time it takes Mark to clean the kitchen is unknown. Therefore, the variable r will be used to denote his rate of work.

Step 1: Determine the rates of work for Martha and her brother.

	Rate of Work	Time	Amount of Work Completed
Martha	$\frac{1}{30}$	20	
brother	r	20	

Step 2: The problem states that they both work for 20 minutes. Organize the information about their rates of work and the times they work in a table.

	Rate of Work	Time	Amount of Work Completed
Martha	$\frac{1}{30}$	20	$\frac{1}{30}(20) = \frac{2}{3}$
brother	r	20	$r(20) = 20r$

Step 3: Use the formula $r \cdot t = w$ to find an expression for the amount of work completed by each person in 20 minutes. Place the expressions in the table.

$$\frac{2}{3} + 20r = 1$$

Step 4: The sum of the work completed is equal to one. Use this to write an equation.

$$3\left(\frac{2}{3} + 20r\right) = 3(1)$$
$$2 + 60r = 3$$
$$2 - 2 + 60r = 3 - 2$$
$$60r = 1$$
$$r = \frac{1}{60}$$

Step 5: Solve the equation for r.

$$r \cdot t = w$$
$$\frac{1}{60} \cdot t = 1$$
$$60\left(\frac{1}{60} \cdot t\right) = 60(1)$$
$$t = 60$$

Step 6: Use the formula $r \cdot t = w$ to determine the time it takes Martha's brother to do the whole job alone. Use 1 to represent the whole job.

Answer: It takes Martha's brother 60 minutes to clean the kitchen alone.

Example 5 A large hose can fill a swimming pool in 10 hours. A small hose can fill the same pool in 20 hours. The large hose begins filling the pool alone at 5 a.m., then the small hose is turned on at 9 a.m. to help fill the pool. At what time will the swimming pool be full?

Since the large hose can fill the pool in 10 hours, it can fill $\frac{1}{10}$ of the pool in 1 hour. Therefore, the rate of work for the large hose is $\frac{1}{10}$. Since the small hose can fill the pool in 20 hours, it can fill $\frac{1}{20}$ of the pool in 1 hour. Therefore, its rate of work is $\frac{1}{20}$.

Step 1: Determine the rates of work for the large hose and for the small hose.

Step 2: Let t represent the amount of time the large hose is turned on. Since the small hose is turned on 4 hours later, the amount of time the small hose runs is $t - 4$ hours. Organize the information about the times and the rates of work in a table.

	Rate of Work	Time	Amount of Work Completed
Large hose	$\frac{1}{10}$	t	
Small hose	$\frac{1}{20}$	$t-4$	

Step 3: Use the formula $r \cdot t = w$ to find expressions for the amount of work completed by each hose. Place the expressions in the table.

	Rate of Work	Time	Amount of Work Completed
Large hose	$\frac{1}{10}$	t	$\frac{1}{10}(t)$
Small hose	$\frac{1}{20}$	$t-4$	$\frac{1}{20}(t-4)$

Step 4: The sum of the work completed is equal to one. Use this to write an equation.

$$\frac{1}{10}(t) + \frac{1}{20}(t-4) = 1$$

Step 5: Solve the equation.

$$\begin{aligned}
20\left(\frac{1}{10}(t) + \frac{1}{20}(t-4)\right) &= 20(1) \\
2t + t - 4 &= 20 \\
3t - 4 &= 20 \\
3t - 4 + 4 &= 20 + 4 \\
3t &= 24 \\
\frac{3t}{3} &= \frac{24}{3} \\
t &= 8
\end{aligned}$$

Step 6: Determine what the time will be 8 hours after 5 a.m.

Eight hours from 5 a.m. is 1 p.m.

Answer: The pool will be full at 1 p.m.

Problem Set

1. Melissa can paint her room in 7 hours. At this rate, how much of the room can she paint in 4 hours?

2. A carpenter can build shelves for a kitchen in 6 days. At this rate, how much of the shelving project can he complete in 4 days?

3. Mr. Williams can fertilize his lawn in 45 minutes. At this rate, how much of the lawn can he fertilize in 15 minutes?

4. A bricklayer can put up the walls of a house in 15 days. At this rate, how much of the walls can he build in 12 days?

5. Mary Jane is making a skirt. In 2 hours, she has completed $\frac{3}{4}$ of the job. Determine Mary Jane's rate of work.

6. Kirk is building a model car. In 3 hours, he has completed $\frac{1}{2}$ of the job. Determine Kirk's rate of work.

7. Juan is building shelves in his garage. In 5 days, he has completed $\frac{3}{4}$ of the job. Determine Juan's rate of work.

8. Ms. Ott is cleaning her windows. In 10 minutes, she has completed $\frac{1}{4}$ of the job. Determine Ms. Ott's rate of work.

9. Joey can prepare the food to cook for dinner in 45 minutes. His sister can prepare the food for the same dinner in 90 minutes. How many minutes will it take them to prepare the food to cook for dinner if they work together?

10. A carpet layer can carpet a house in 10 hours. His assistant can carpet the same house in 15 hours. How many hours will it take them to carpet the house if they work together?

11. Mr. Clark can clean his pool in 30 minutes. His son can clean the pool in 50 minutes. How many minutes will it take them to clean the pool if they work together?

12. A brick layer can put pavers on a pool deck in 12 hours. His assistant can put pavers on the same pool deck in 15 hours. How many hours will it take them to put down the pavers if they work together?

13. It takes an experienced worker 4 hours to lay a sidewalk. If his assistant helps, the job can be completed in 3 hours. How many hours would it take the assistant to do the job alone?

14. It takes Manny 40 minutes to wash and clean his car. If his brother helps, they can wash and clean the car in 24 minutes. How many minutes would it take his brother to wash and clean the car alone?

15. It takes Martha's mother 10 minutes to fold and put away the laundry. If Martha helps, they can fold and put away the laundry in 8 minutes. How many minutes would it take Martha to fold and put away the laundry alone?

16. A skilled carpenter can build cabinets for a kitchen in 3 days. If his assistant helps, they can complete the same job in 2 days. How many days would it take the assistant to build the cabinets alone?

17. A large pipe can fill a reservoir in 20 hours. A small pipe can do the same job in 30 hours. If the small pipe begins filling the reservoir at 4 a.m. and the large pipe is turned on at 2 p.m., then at what time will the reservoir be filled?

18. Jenny can plant her flower garden in 10 hours. Her mom can do the same job in 20 hours. If Jenny begins planting the garden at 10 a.m. and her mom begins helping her at 2 p.m., then at what time will the job be completed?

19. An expert artist can decorate a set of ceramic vases in 4 hours. Her daughter can decorate the same set of vases in 12 hours. If the daughter begins decorating the vases at 10 a.m. and the artist begins helping her at 12 p.m., then at what time will the job be completed?

20. A secretary can type her memos in 45 minutes. Her clerk can do the same job in 60 minutes. If the secretary begins typing the memos at 9 a.m. and the clerk begins helping her at 9:10 a.m., then at what time will the typing job be completed?

HA1-370: Graphing Ordered Pairs on a Coordinate Plane

Graphs are common in many areas of our lives. In mathematics, we represent a number of different things with graphs, like equations, solutions to equations, or solutions to systems of equations. It is important to know how to read them. In this lesson, you will learn some basic concepts that will help you read graphs. A graph begins with something called a **coordinate system**. A coordinate system is made up of two number lines that intersect at zero, and are arranged so that they make a right angle. To understand how the coordinate system is set up, let's look at a regular, horizontal number line.

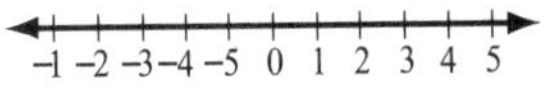

Notice how the zero is between the negative numbers and the positive numbers. The zero is called the **origin**, and the negative numbers go to the left, while the positive numbers go to the right. Now, let's look at a vertical number line.

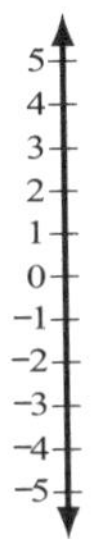

The zero is still between the positive and negative numbers, but this time the positive numbers go up, and the negative numbers go down. If you connect these two number lines at their origins, you get a coordinate system, like the one to the right.

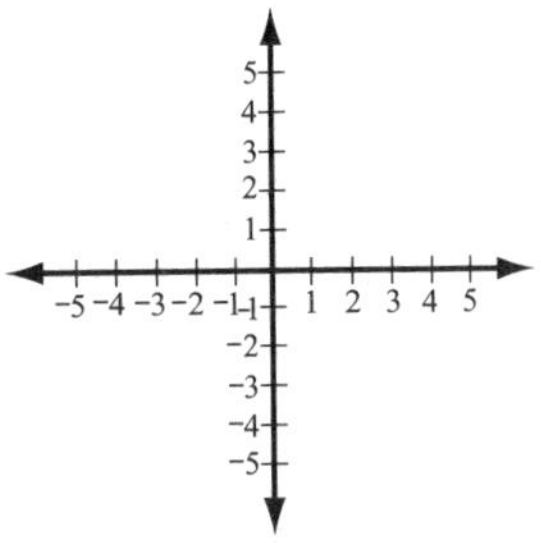

The point where the two number lines meet is called the **origin** of the coordinate system, and it is where both number lines are equal to zero. The horizontal number line is called the ***x*-axis**, and the vertical number line is called the ***y*-axis**.

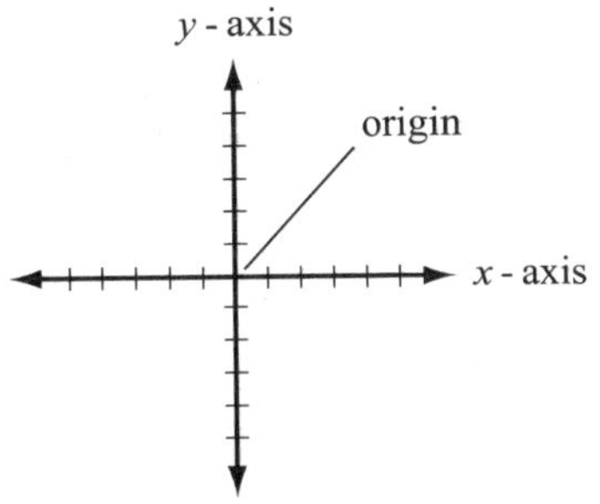

The coordinate system can be divided into four sections, called quadrants. The quadrants are named: Quadrant I (or QI for short), Quadrant II (QII), Quadrant III (QIII), and Quadrant IV (QIV). The top right quadrant is QI, and the others are numbered in a counter-clockwise fashion.

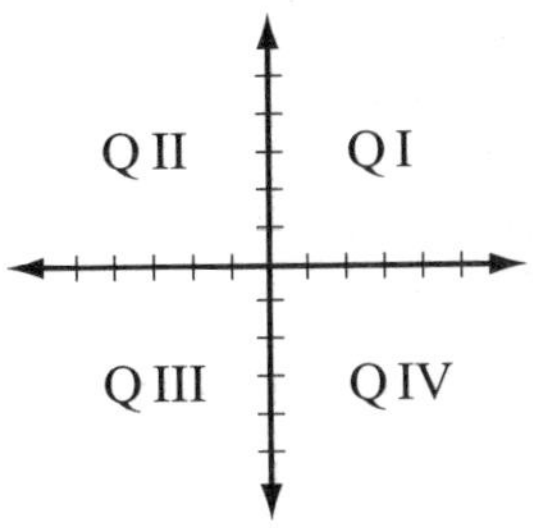

Every point in a coordinate system has a unique location, and a unique name based on that location. The name of a point is called an **ordered pair**, and it is written as (x, y). The x is the number that corresponds to the location of the point relative to the x-axis, and it is called the x-coordinate of the point. The y is the number that corresponds to the location of the point relative to the y-axis, and it is called the y-coordinate of the point. It is important to always write the x-coordinate first, followed by the y-coordinate.

Example 1 Graph the ordered pair (3,–2) and identify the quadrant where it is located.

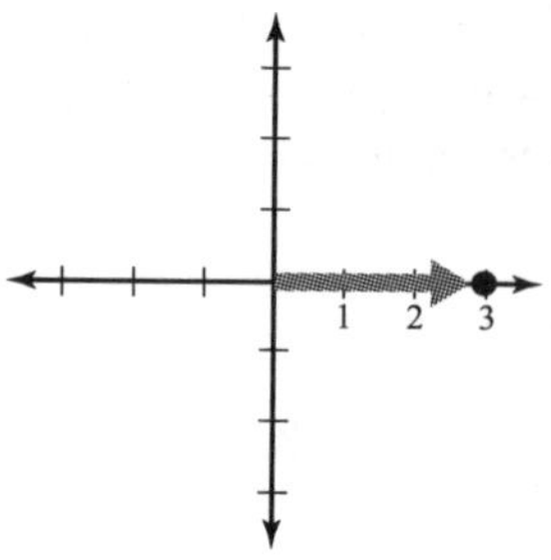

Step 1: Starting at the origin, move three units to the right.

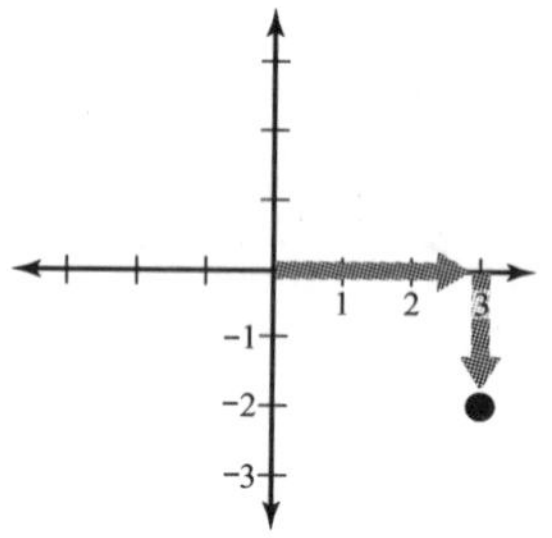

Step 2: Move down two units and place your point.

Answer: The ordered pair (3,–2) is in the 4th quadrant.

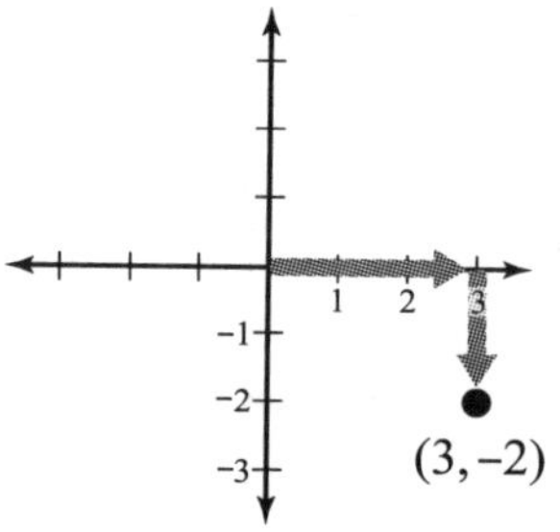

Example 2 Graph the ordered pair (–2, 3).

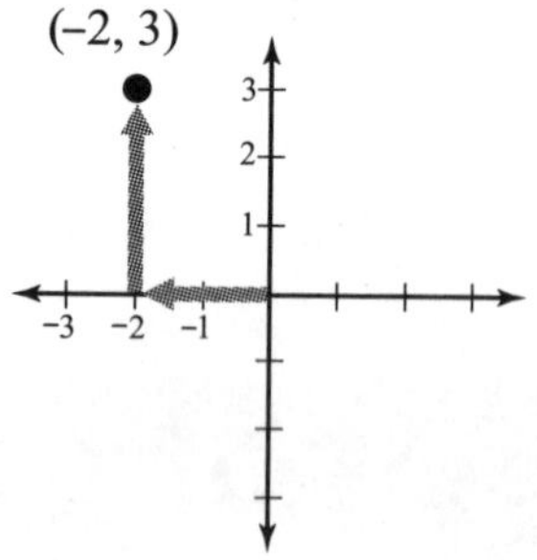

Step 1: Start at the origin; move two units to the left.
Step 2: Move three units up and place the point.

Answer: The ordered pair (–2, 3) is in Quadrant II.

There are two more important things to note. First, if you choose a point on the x-axis or on the y-axis, then that point is not considered to be in any of the four quadrants; it is on an axis. Second, the coordinates of the origin itself are (0, 0).

Example 3 Find the coordinates of the ordered pairs for each letter on the graph below.

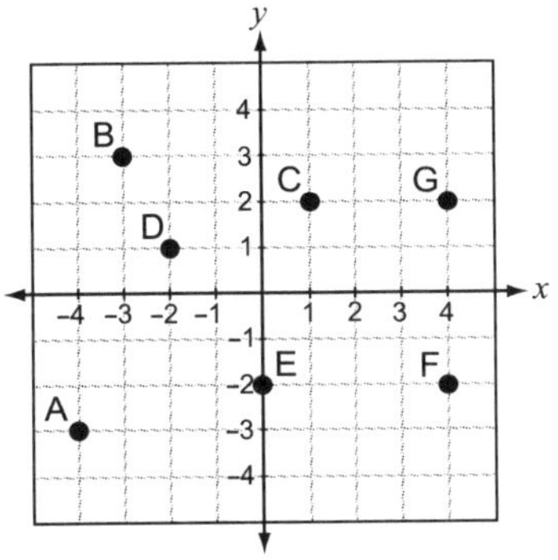

Answer: See the following table for answers.

For extra practice, look at the ordered pairs in the table below. Cover up the last column to see if you can determine the quadrant for each ordered pair.

Point	Coordinate	Quadrant
A	(–4,–3)	QIII
B	(–3, 3)	QII
C	(1, 2)	QI
D	(–2, 1)	QII
E	(0,–2)	Quadrantal Point (a point that lies on an axis)
F	(4,–2)	QIV
G	(4, 2)	QI

Problem Set

Solve:

1. What is the ordered pair for point L?

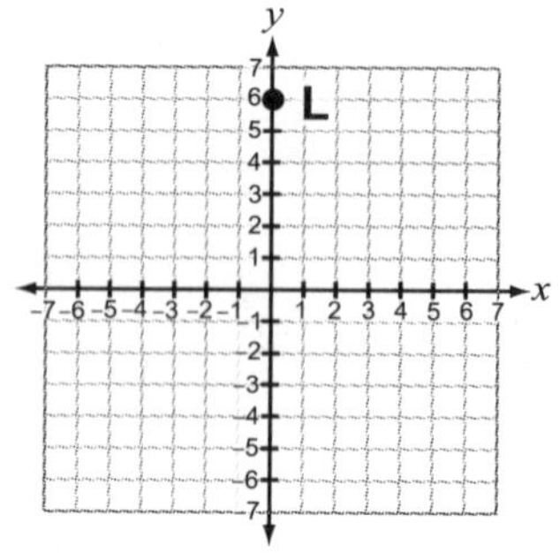

2. What is the ordered pair for point V?

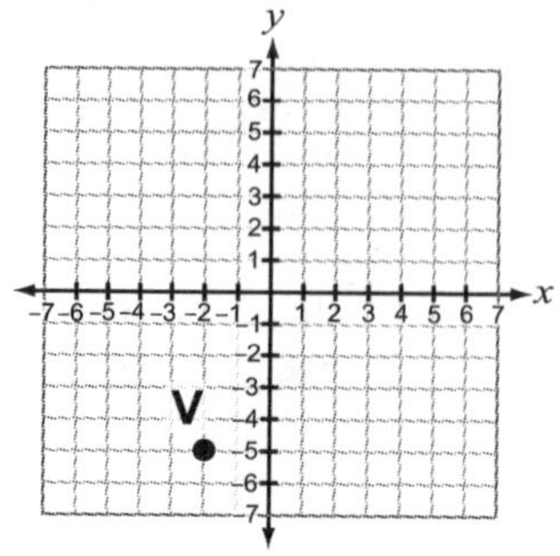

3. What is the ordered pair for point T?

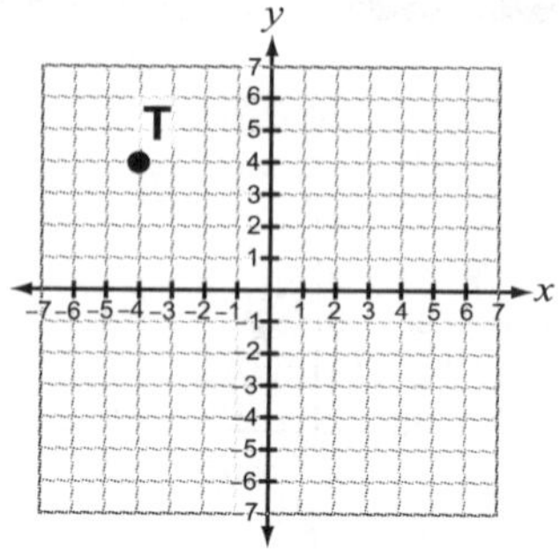

4. What is the ordered pair for point Z?

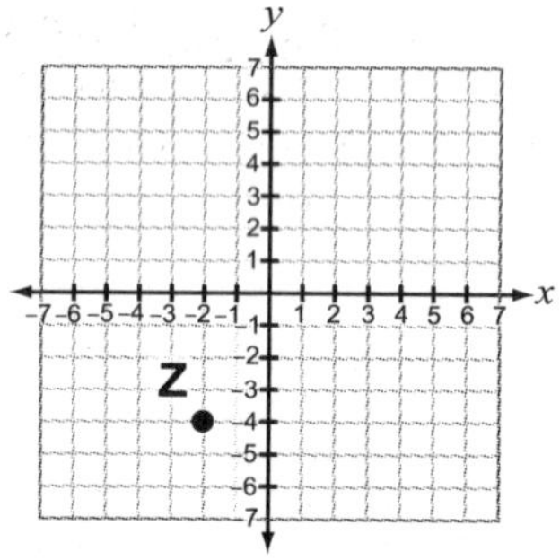

5. What is the ordered pair for point H?

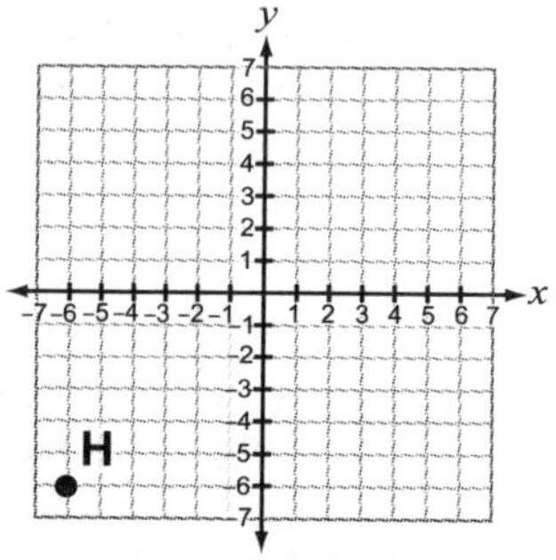

6. What is the ordered pair for point B?

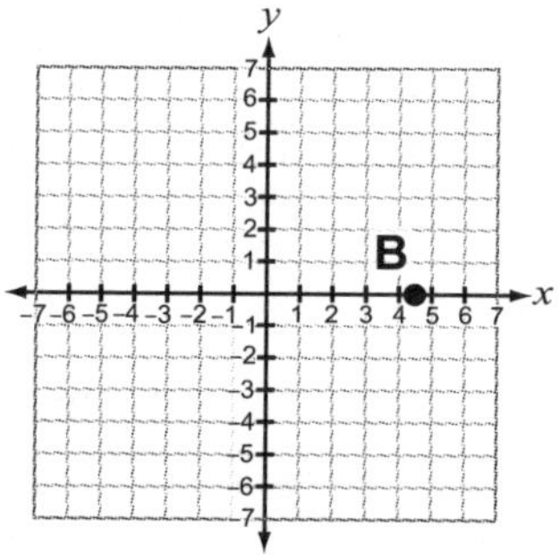

7. What is the ordered pair for point D?

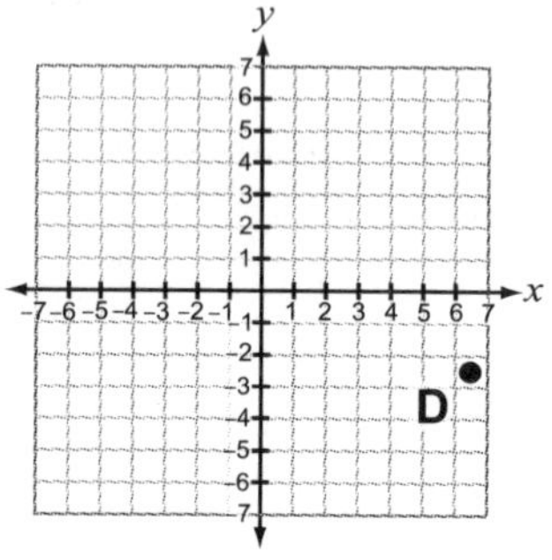

8. What is the ordered pair for point E?

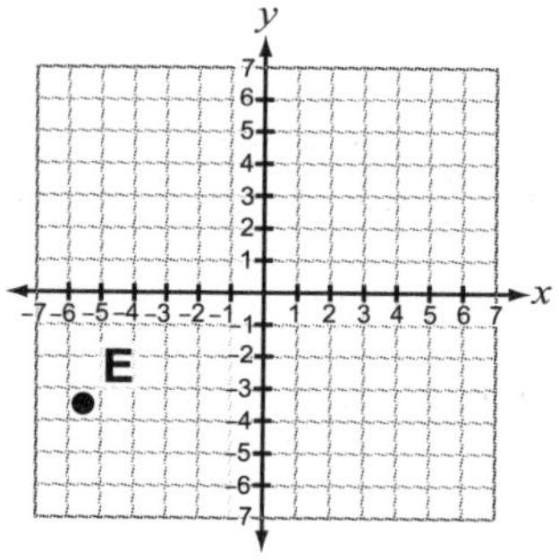

9. Which quadrant or axis contains the ordered pair (–2, 0)?

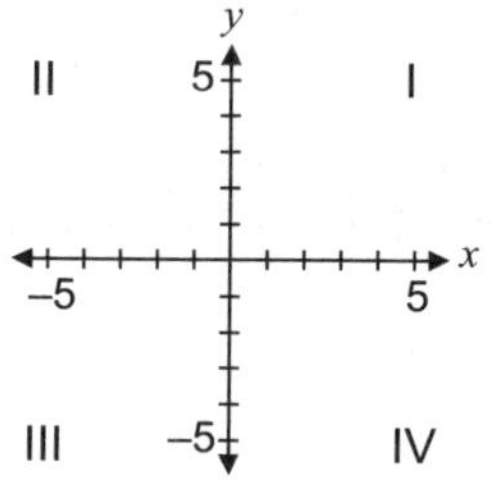

10. Which quadrant contains the ordered pair (–1,–4)?

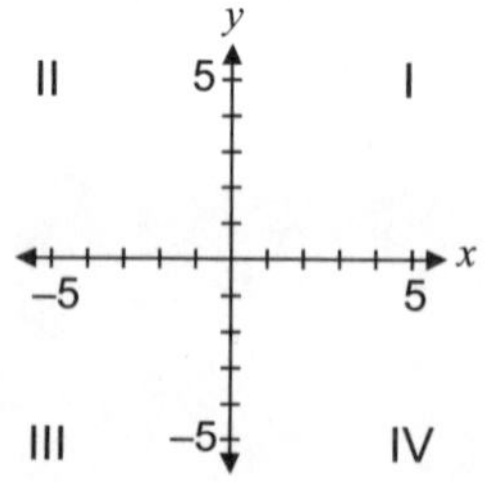

HA1-375: Identifying Solutions of Equations in Two Variables

Recall that an equation with only one variable has only one solution. For example, let's look at the equation $x + 5 = 8$. If you solve this equation, you get:

$$\begin{aligned} x + 5 &= 8 \\ x + 5 - 5 &= 8 - 5 \\ x &= 3 \end{aligned}$$

and so you see that the solution to the equation $x + 5 = 8$ is 3. You don't stop there, though. You need to check this answer to make sure that it is correct. To check our answer, you go back to the original equation, and substitute our answer for the variable. In this case, you want to substitute 3 for x, which gives us:

$$\begin{aligned} 3 + 5 &= 8 \\ 8 &= 8 \end{aligned}$$

Since it is true that 8 equals 8, we have verified that the answer is correct. In this lesson, you will move to the next step–finding solutions of equations with two variables. Here is an example of an equation with two variables, both to the first power: $x + y = 8$. The two variables are x and y. We say that they are both "to the first power" because they have no exponents. An equation like this one has many solutions. One possible solution is $x = 3$ and $y = 5$, because if you substitute 3 for x and 5 for y, you get $3 + 5 = 8$, which is true. When you have a pair of numbers for x and y like this, where both numbers must be used at the same time, you can write the numbers as an ordered pair. For example, you would write the pair $x = 3$ and $y = 5$ as the ordered pair (3, 5). Therefore, (3, 5) is a solution to the equation $x + y = 8$.

Remember that in ordered pairs, the x-value always comes first, and the y-value always comes second. Let's consider the ordered pair (1, 7) in relation to the equation $x + y = 8$. You want to determine if this ordered pair is a solution for this equation. To do this, you substitute 1 for x and 7 for y. This gives us $1 + 7 = 8$, which is true. Therefore, (1, 7) is also a solution for this equation. At this point, you should begin to see other possible solutions for this equation. As a matter of fact, the solutions for $x + y = 8$ are any pairs of numbers whose sum is 8. Can you think of other solutions?

Let's try some other examples. Remember that a solution to an equation with two variables is any ordered pair (x, y) that will make the equation true when the x-value is substituted for x in the equation, and the y-value is substituted for y.

Example 1 Determine whether or not the ordered pair (–3, 5) is a solution for $3x + y = -4$.

$3(-3) + (5) = -4$ — **Step 1:** Substitute the ordered pair (–3, 5) for x and y.

$$\begin{aligned} -9 + 5 &= -4 \\ -4 &= -4 \text{ true} \end{aligned}$$

Step 2: Simplify the equation.

Answer: The solution makes the equation true. Therefore, (–3, 5) is a solution.

Example 2 Determine whether or not the ordered pair (4, 0) is a solution for the equation $2x - 4y = 7$.

$2(4) - 4(0) = 7$ — **Step 1:** Substitute the ordered pair (4, 0) for x and y.

$$\begin{aligned} 8 - 0 &= 7 \\ 8 &= 7 \text{ false} \end{aligned}$$

Step 2: Simplify the equation.

Answer: Note that 8 does not equal 7. The statement is false. Therefore, (4, 0) is not a solution to this equation.

Example 3 Determine whether or not the ordered pair (–5, 2) is a solution for the equation $8 - y = 6$.

$8 - (2) = 6$	**Step 1:**	Substitute the ordered pair (–5, 2) for x and y. Because there is no x in the equation, you can only substitute the y value.
$6 = 6$	**Step 2:**	Simplify the equation.

Answer: The equation is true; therefore, $y = 2$ is a solution. This is an equation with only one variable, so there is only one solution for the variable. Any ordered pair that has 2 for a y-value will be a solution to this equation.

Example 4 Determine whether the ordered pairs below are solutions for the equation $-x + 2y = 7$: (–7, 0); (5, 6); (0, –7).

$-(-7) + 2(0) = 7$ $7 + 0 = 7$ $7 = 7$ true	**Step 1:**	Substitute the ordered pair (–7, 0), which gives $7 = 7$. Therefore, (–7, 0) is a solution, since the equation is true.
$-(5) + 2(6) = 7$ $-5 + 12 = 7$ $7 = 7$ true	**Step 2:**	Use the ordered pair (5, 6), which gives $7 = 7$. The equation is true. This ordered pair is also a solution.
$-(0) + 2(-7) = 7$ $0 - 14 = 7$ $-14 = 7$ false	**Step 3:**	Use the ordered pair (0, –7). This gives $-14 = 7$, which is not true. This is not a solution to the equation.

Answer: The ordered pairs (–7, 0) and (5, 6) are solutions.
Note: This equation has two variables, so there are an infinite number of solutions.

Problem Set

1. Determine whether the ordered pair $(2, -3)$ is a solution to the given equation:
$$x + 3y = 11$$
Not a solution Solution

2. Determine whether the ordered pair $(6, 1)$ is a solution to the given equation:
$$2x - 3y = 9$$
Not a solution Solution

3. Determine whether the ordered pair $(1, -2)$ is a solution to the given equation:
$$x + 5y = -9$$
Solution Not a solution

4. Determine whether the ordered pair $(2, 4)$ is a solution to the given equation:
$$x - 4y = 0$$
Not a solution Solution

5. Determine whether the ordered pair $(2, -3)$ is a solution to the given equation:

$$2x + 3y = 13$$

Solution Not a solution

6. Determine whether the ordered pair $(-2, -8)$ is a solution to the given equation:

$$x + y = -10$$

Not a solution Solution

7. Determine whether the ordered pair $(5, -1)$ is a solution to the given equation:

$$5x + 4y = 21$$

Not a solution Solution

8. Determine whether the ordered pair $(2, 0)$ is a solution to the given equation:

$$6x + y = 12$$

Solution Not a solution

9. Determine whether the ordered pair $(-1, -3)$ is a solution to the given equation:

$$x - 5y = -14$$

Solution Not a solution

10. Determine whether the ordered pair $(-2, 3)$ is a solution to the given equation:

$$-2y - x = -1$$

Solution Not a solution

11. Which ordered pair is a solution for the equation?

$$3x + 2y = 2$$

$(-4, 7)$ $(2, 2)$ $(7, -4)$

12. Which ordered pair is a solution for the equation?

$$2x - 3y = 6$$

$(6, 2)$ $(-2, 0)$ $(0, 3)$

13. Which ordered pair is a solution for the equation?

$$5x - y = 17$$

$(3, -2)$ $(0, -22)$ $(2, 2)$

14. Which ordered pair is a solution for the equation?

$$2x - 4y = -4$$

$(-1, -4)$ $(1, 2)$ $(-4, -1)$

15. Which ordered pair is a solution for the equation?

$$x + 3y = 1$$

$(0, 1)$ $(1, 0)$ $(2, 2)$

16. Which ordered pair is a solution for the equation?

$$2x - y = -14$$

$(10, -2)$ $(-4, -6)$ $(-2, 10)$

17. Which ordered pair is a solution for the equation?

$$2x + 5y = 8$$

$(1, 2)$ $(-1, 2)$ $(-6, 0)$

18. An airplane averaged 225 miles per hour on a trip. The equation $d = 225t$ can be used to calculate the distance, d, that the plane flies in t hours. On a trip to Mississippi, the plane flew for two hours and traveled 450 miles. Does this information satisfy the equation? Answer yes or no.

19. Harold averages 17 points per game in basketball. The equation $p = 17g$ can be used to calculate the total points, p, that he scores in g games. After 8 games, he scored a total of 135 points. Does this information satisfy the equation? Answer yes or no.

20. Heather averages 88 points per test in History. The equation $p = 88t$ can be used to calculate the points, p, that she totals in t tests. After 4 tests, she scored 352 points. Does this information satisfy the equation? Answer yes or no.

HA1-380: Graphing Linear Equations

There are many ways to combine the various areas of mathematics, and the different combinations allow us to do different things. One useful combination is algebra and geometry. By pairing up these two areas of mathematics, you can display solutions to equations using the coordinate system. Graphing solutions of linear equations using this combination is the topic of this lesson.There are two ideas you should review before proceeding. First, remember how to verify solutions to two-variable equations. If you have an equation $5x + 2y = 9$, and you want to know if the ordered pair (1, 2) is a solution, substitute the x-value, 1, for x in the equation, and the y-value, 2, for y. Then simplify the resulting equation. Doing this would give you the following:

$$\begin{aligned} 5(1) + 2(2) &= 9 \\ 5 + 4 &= 9 \\ 9 &= 9 \end{aligned}$$

Since it is true that $9 = 9$, the ordered pair (1, 2) is indeed a solution to the equation $5x + 2y = 9$. Take this process one step further. Suppose that you want an x-value of 3 in your solution to this equation. To find the complete solution, you need to determine the y-value. To do so, substitute the x-value, 3, for x in the equation. This gives you $5(3) + 2y = 9$. This is an equation with one variable, which you can then solve:

$$\begin{aligned} 5(3) + 2y &= 9 \\ 15 + 2y &= 9 \quad \text{simplify} \\ 2y &= -6 \quad \text{subtract 15 from both sides} \\ y &= -3 \quad \text{divide both sides by 2} \end{aligned}$$

Therefore, if you want x to be 3, then y must be –3. You have found a new solution to the equation, the ordered pair (3, –3). You can also reverse this process: if you know what the y-value is, you can determine the x-value by substituting the value of y in the equation and solving for x.

The second idea to review is the concept of a coordinate system. Remember that a coordinate system is two number lines that intersect at their origins. The horizontal number line is the x-axis, and the vertical number line is the y-axis. Remember that the positive numbers on the x-axis go the right, while the negative numbers go the left. For the y-axis, the positive numbers go up, and the negative numbers go down. The point where the two number lines meet is called the origin. Here is a picture of a coordinate system.

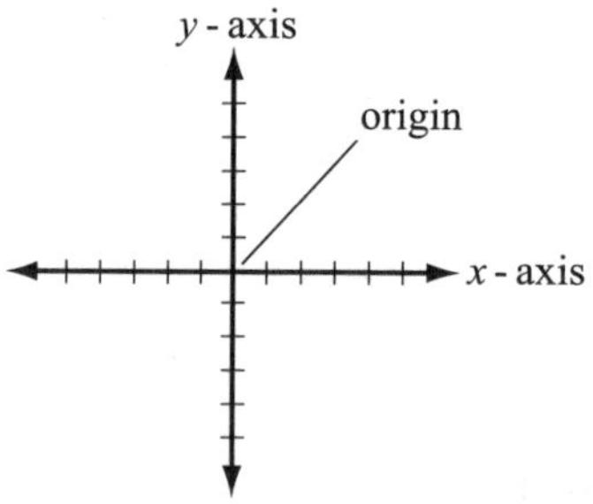

Also, remember that any point anywhere in a coordinate system has a name. The name of a point is called the coordinates of the point, and any pair of coordinates can be plotted. In this lesson we will focus on linear equations.

Linear Equation	A linear equation is an equation that can be written in the form $Ax + By = C$, where A, B, and C are real numbers and A and B are not both zero.

An example of a linear equation is the equation $5x + 2y = 9$. To understand the definition of a linear equation, think about the characteristics of a linear equation. One characteristic is that you must have at least one variable; either x or y, or both. This is why the definition says that A and B cannot both be zero; if both are zero, there are no variables. The other characteristic is that there are no visible exponents on the variables. This classifies the equation as linear.

You know that the solutions to linear equations are ordered pairs, and you also know that any linear equation has infinitely many ordered pairs as solutions. Since any given linear equation has an infinite number of solutions, there is no way to write down all the ordered pairs that are solutions. The graph of a linear equation is a way to represent all the ordered pairs that are solutions of the given linear equation, without actually listing every single ordered pair.

Example 1 Graph the line that contains the points (0, 3), (1, 0), (–1, 6), (2, –3).

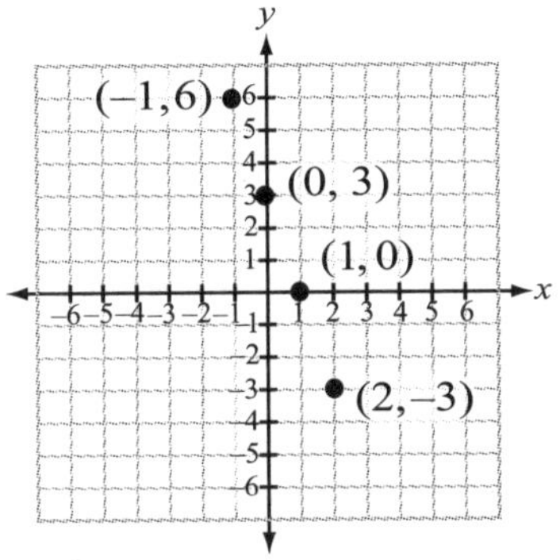

Step 1: Plot and label each ordered pair on the graph.

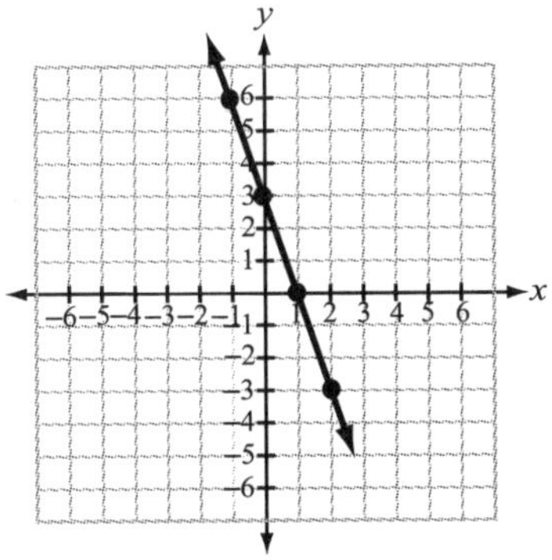

Step 2: Draw the line that connects the points.

In this example, you were given a set of points, you plotted them and connected them with a line. Putting arrows on both ends of the line indicates that the line continues forever in both directions, meaning that the line is infinite. Any linear equation can be graphed in the same way. The graph of any linear equation is a line (notice the similarity between the words "linear" and "line"). To graph a linear equation, you need to draw a line. As you just saw in Example 1, to draw a line you need a collection of points. Then you can connect them and put arrows at the ends.

Recall the previous linear equation: $5x + 2y = 9$. You already know that the ordered pairs (1, 2) and (3,–3) are solutions for this equation, so you have two points on the line that will be the graph of this equation. To draw the graph, simply plot the two points and connect them with a line.

The result is the graph that you see here.

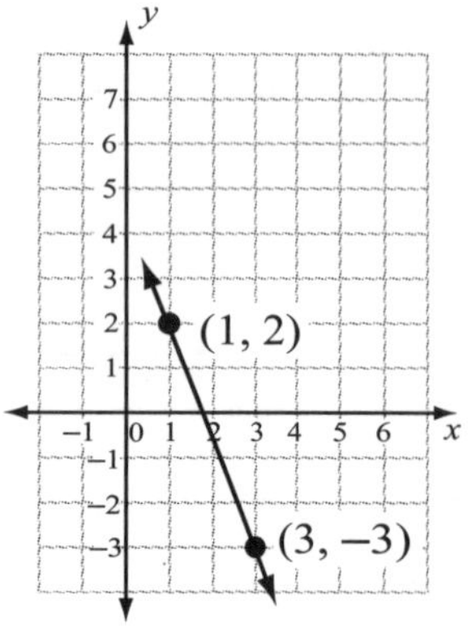

Once you have the graph of a linear equation, you can choose any point on that line, and the coordinates of that point will be a solution to the equation that is represented by the graph. For example, look at the new point shown on this graph, the point (–1, 7).

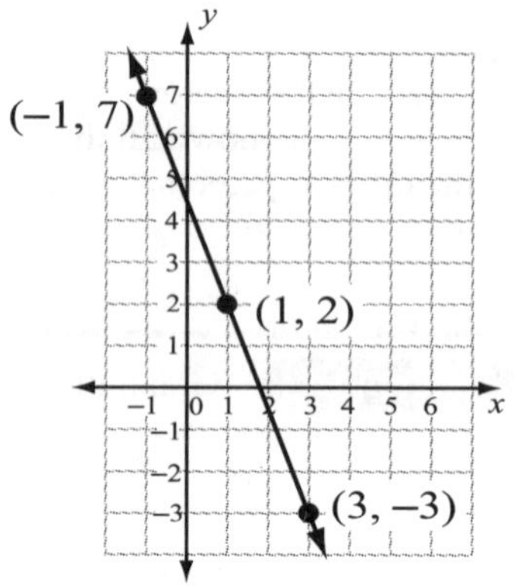

This coordinate pair should be a solution to the equation. To verify that it is a solution, substitute –1 for x and 7 for y, as follows.

$$\begin{aligned} 5(-1) + 2(7) &= 9 \\ -5 + 14 &= 9 \\ 9 &= 9 \end{aligned}$$

Because $9 = 9$ is true, you have verified that the point (–1, 7) is a solution to the equation that you graphed. This will hold true for any point that you choose on the line.

Look at another example.

Example 2 Graph the equation $2x + 3y = 6$.

Step 1: You need to find solutions to this equation so that you will have ordered pairs that you can plot. To do this, create a table. In the left column, select some numbers to use as x-values. This example uses –3, 0, and 3, although any numbers will do. Now find the corresponding y-values.

x	y
–3	
0	
3	

Step 2: To find the corresponding y-values for the table, take each x-value individually, substitute it for x in the equation, and solve for y. You will find that the y-values are 4, 2, and 0, respectively.

$$\begin{aligned} 2(-3) + 3y &= 6 \\ -6 + 3y &= 6 \\ -6 + 6 + 3y &= 6 + 6 \\ 3y &= 12 \\ \frac{3y}{3} &= \frac{12}{3} \\ y &= 4 \end{aligned}$$

$$\begin{aligned} 2(0) + 3y &= 6 \\ 0 + 3y &= 6 \\ 3y &= 6 \\ \frac{3y}{3} &= \frac{6}{3} \\ y &= 2 \end{aligned}$$

$$\begin{aligned} 2(3) + 3y &= 6 \\ 6 + 3y &= 6 \\ 6 - 6 + 3y &= 6 - 6 \\ 3y &= 0 \\ y &= 0 \end{aligned}$$

Step 3: Fill in the y-values in the table, and write down the resulting ordered pairs.

x	y	Ordered Pair
-3	4	$(-3, 4)$
0	2	$(0, 2)$

Step 4: Plot the ordered pairs and connect them with a line, putting arrows on the ends of the line.

Answer:

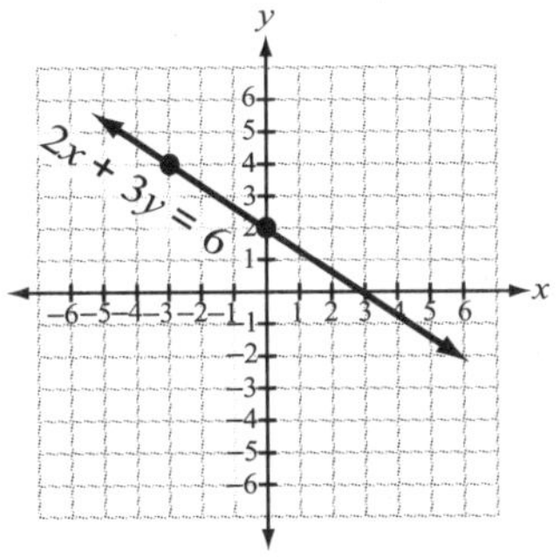

The x-intercept and the y-intercept are important points on any graph. They are the points, if any, at which the graph of an equation crosses, or intercepts, the x- and y-axes.

x-Intercept	The x-intercept is the point where the graph of an equation crosses the x-axis.

y-Intercept	The y-intercept is the point where the graph of an equation crosses the y-axis.

Look at the line in Example 2. The line crosses the x-axis at the point (3, 0), so the x-intercept is 3. The graph crosses the y-axis at the point (0, 2), so the y-intercept is 2. If you think about x-intercepts in general, you know that they will always be points that lie on the x-axis. Because they lie on the x-axis, their y-coordinates will always be zero. Likewise, y-intercepts will always be points that lie on the y-axis, and will therefore always have x-coordinates of zero. Look at the intercepts that you just found, (3,0) and (0,2), and notice where the zeroes are in these points.

One of the reasons that x- and y-intercepts are important is that they can be used to easily graph linear equations, as you will see in the next example.

Example 3 Graph the equation $x + 2y = 4$ by finding the x-intercept and the y-intercept.

Step 1: You are being asked to graph the equation by finding the x- and y-intercepts. So, instead of just filling in some random x-values in the T table, fill in 0 for x, which will give you the y-intercept, and fill in 0 for y, which will give you the x-intercept.

x	y
0	
	0

Step 2: Find the missing values for the table. First substitute 0 for x and find the missing y-value, then substitute 0 for y to find the missing x-value.

$$\begin{aligned}(0)+2y &= 4 \\ 2y &= 4 \\ y &= 2\end{aligned} \qquad \begin{aligned}x+2(0) &= 4 \\ x+0 &= 4 \\ x &= 4\end{aligned}$$

Step 3: Fill in the table and find the ordered pairs. You will find that the y-intercept is (0, 2), and the x-intercept is (4, 0).

x	y	Ordered Pair
0	2	(0, 2)
4	0	(4, 0)

Step 4: Plot the points and draw a line.

Answer:

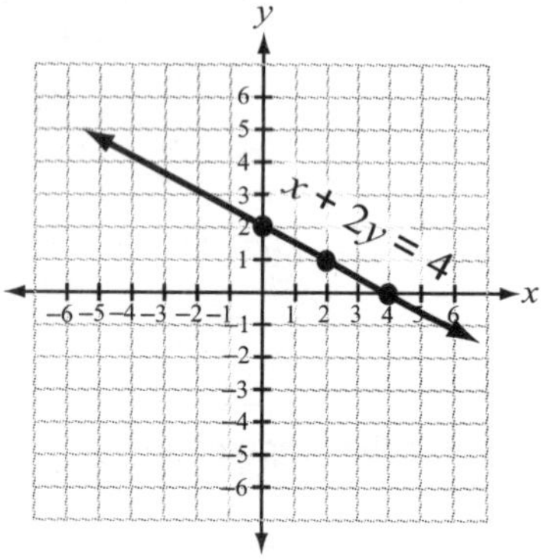

It is recommended that you always find three ordered pair solutions. If you make an error during calculations, the points will not line up. You must then check your work.

Let $x = 2$, then

$$\begin{aligned}2+2y &= 4 \\ 2y &= 2 \\ y &= 1\end{aligned}$$

The third point is (2, 1). Notice it is on our graphed solution.

The next example shows a special type of linear equation that has no y-intercept.

Example 4 Graph the equation $x = -2$.

Step 1: Create a table. As you look at this equation, though, you will see that the only x-value that is ever valid is –2. You cannot substitute any other value for x in this equation and have the equation hold true; therefore, you can only put –2 in the x column of the table.

x	y
–2	
–2	
–2	

Step 2: Fill in the y-values. Since there is no y in the equation at all, substituting the x-values for x won't tell you anything. Since there is no y in the equation, the y-values can be anything. You can just fill in some random numbers for y-values, and determine the resulting ordered pairs.

x	y	Ordered Pair
–2	–3	(–2, –3)
–2	0	(–2, 0)

Step 3: Plot the points and draw the line.

Answer:

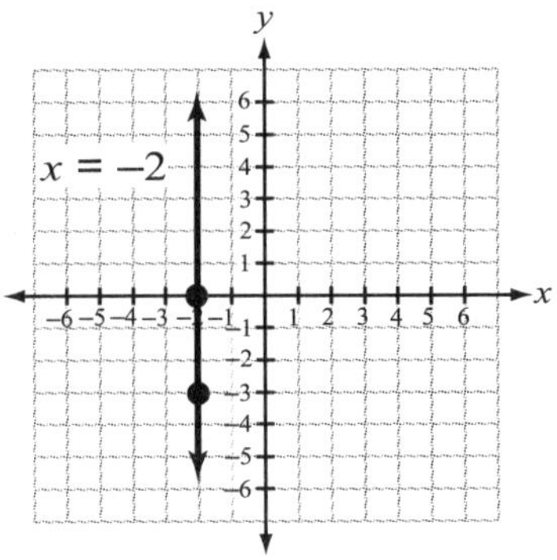

Notice in the above equation $(x = -2)$ there was no y-variable, which means no y-intercept. No y-intercept means the graphed solution does not cross the y-axis, making the graph a vertical line. Similarly, a horizontal line would not cross the x-axis, so its equation would not have an x-variable.

Problem Set

Graph the following equations:

1. $x + y = 7$
2. $x - y = -2$
3. $x - y = 3$
4. $2x - 6 = 2$
5. $3y + 18 = 0$
6. $2x + y = -4$
7. $4x - y = 4$
8. $2x + y = -8$
9. $2x + 8y = -16$
10. $4x - 3y = 12$

HA1-382: Solving Linear Equations Using the Graphing Calculator

Linear equations can be solved using the TI-83 Plus™ graphing calculator. To do this, we separate the equations and find the x-coordinate of the point where the graphs intersect. The calculator allows us to find this intersection point easily.

Our process has five steps:

- Step 0: (Settings) Initialize the calculator to a standard set of values so that we have a common starting point.
- Step 1: (Enter the expressions) Enter the symbolic form of the two sides of the equation to be analyzed.
- Step 2: (Graph) Obtain the graphs of the form y = left-hand side and y = right-hand side.
- Step 3: (Identify Intersection Point) Use other Intersect feature of the graphing calculator to verify the x-coordinate of the point where the graphs intersect.
- Step 4: (Solve) Use information about the x-coordinate of the intersection point to solve the original equation.

Note: Before doing any of these examples you should check your calculator's settings to be sure that it will function as described above.

Press the [MODE] key and be sure that your settings appear as shown below.

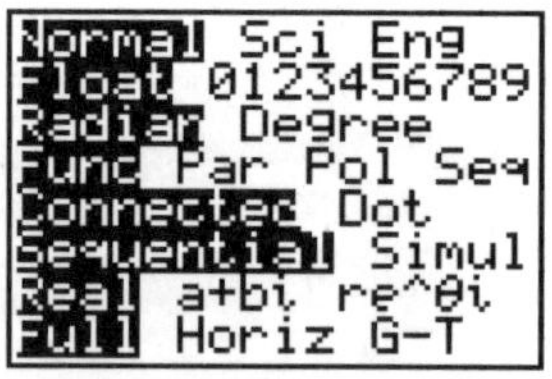

Next press the [2nd][ZOOM] keys and be sure that your settings appear as shown below.

Next press the [ZOOM] key then the [6] key to select the standard window setting.

Press the [WINDOW] key to get the following screen.

Example 1 Use the graphing calculator to solve the equation $5x + 3 = 7$.

Step 0: (Settings) Make sure you have the default settings described above.

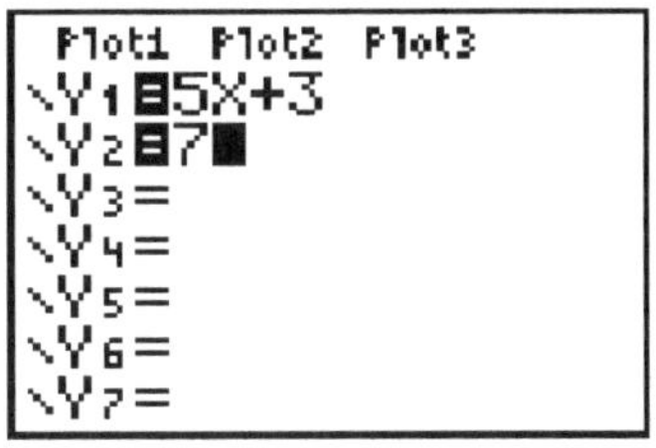

Step 1: (Enter the expression) Press the [Y=] key and enter the expression $5x + 3 = 7$ into the Y_1 slot and 7 into the Y_2 slot.

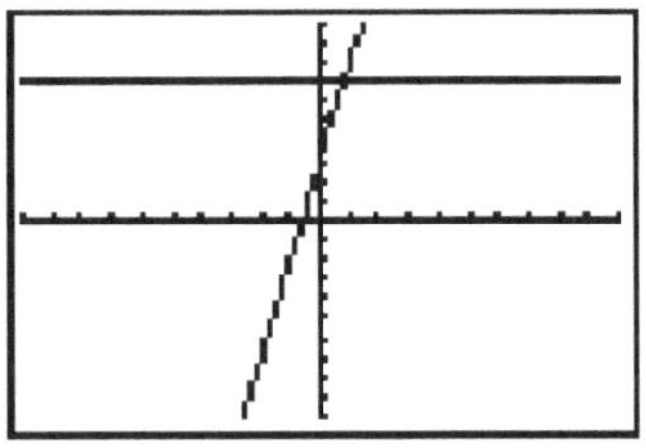

Step 2: (Graph) Press the [GRAPH] key. From the graph, we see that the two graphs intersect at a point in the first quadrant.

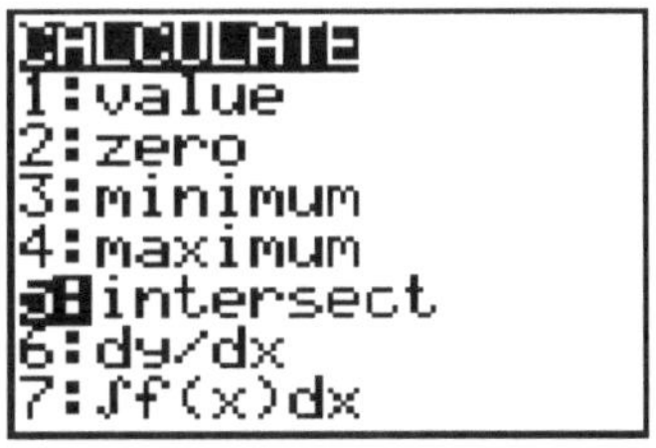

Step 3: (Identify Intersection Point) Next, we will find where the graphs of Y_1 and Y_2 intersect (the x-coordinate of this point is the solution to our original equation). This is done by pressing [2nd][TRACE] followed by [5].

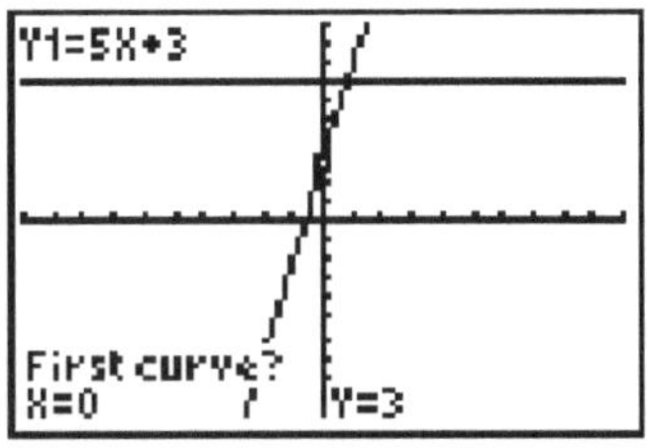

You will be prompted for the first curve with a screen like the one to the left. Press [ENTER].

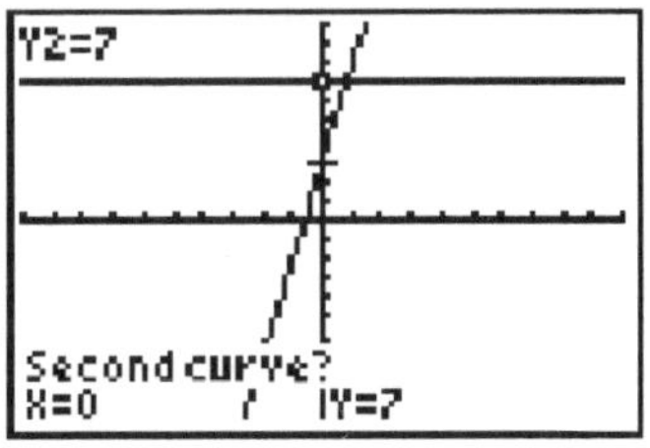

Now you will be prompted for the second curve with a screen like the one to the left. Press [ENTER] again.

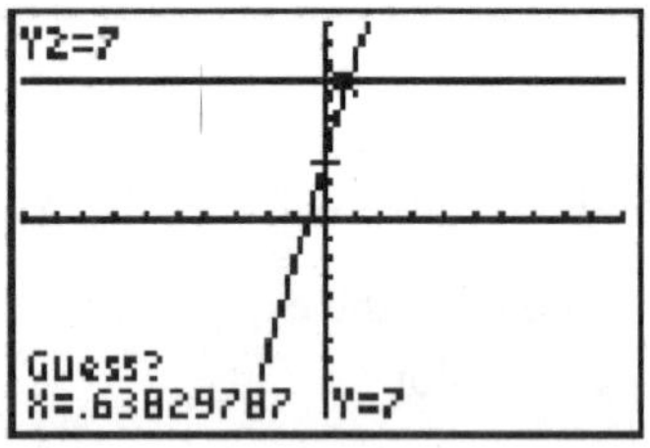

You will be prompted for a guess with a screen like the one to the left. Use the ◄ and ► keys to move the highlighted point as close as possible to the intercept you are trying to verify and press ENTER.

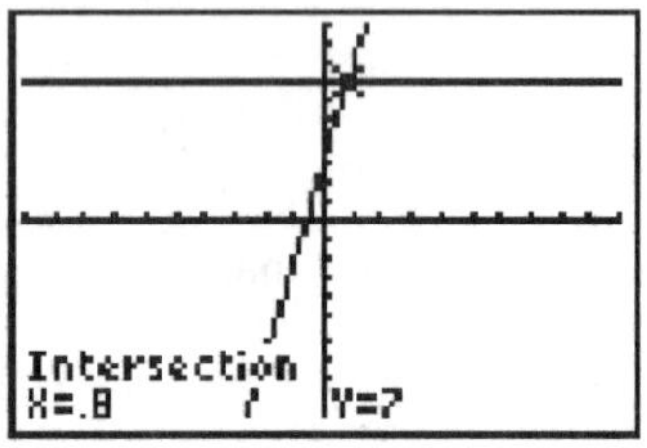

You should get a screen like the one to the left.

This gives an intersection point whose x-coordinate is 0.8. Convert this to fraction form by pressing the keys 2nd MODE to return to the home screen. Press the keys X,T,Θ,n MATH 1 to get the screen to the left.

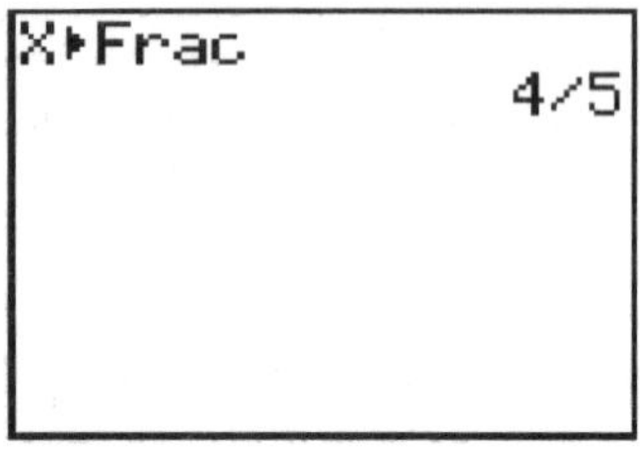

Finally, press ENTER and you will see a screen like the one to the left.

So, our solution as a fraction is $x = \frac{4}{5}$.

Step 4: (Solve) The equation $5x + 3 = 7$ is satisfied by the value $x = 0.8 = \frac{4}{5}$.

Answer: The solution of $5x + 3 = 7$ is $x = 0.8 = \frac{4}{5}$.

Example 2 Use the graphing calculator to solve the equation $0.5(x - 2.4) = 4(0.2 - 0.5x) + 0.3$.

Step 0: (Settings) Make sure you have the default settings as described in the beginning of the lesson.

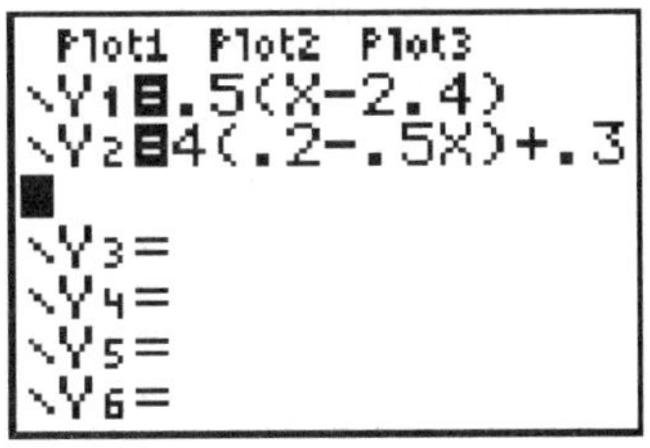

Step 1: (Enter the expression) Press the [Y=] key and enter the expression $0.5(x-2.4)$ into the Y_1 slot and $4(0.2-0.5x)+0.3$ into the Y_2 slot.

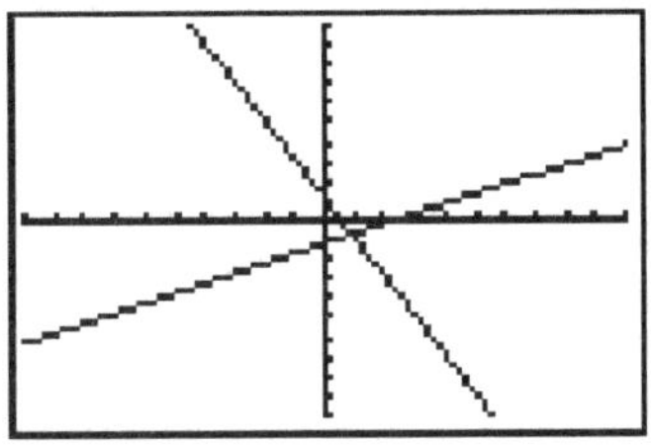

Step 2: (Graph) Press the [GRAPH] key. We see that the two graphs intersect at a point in the fourth quadrant.

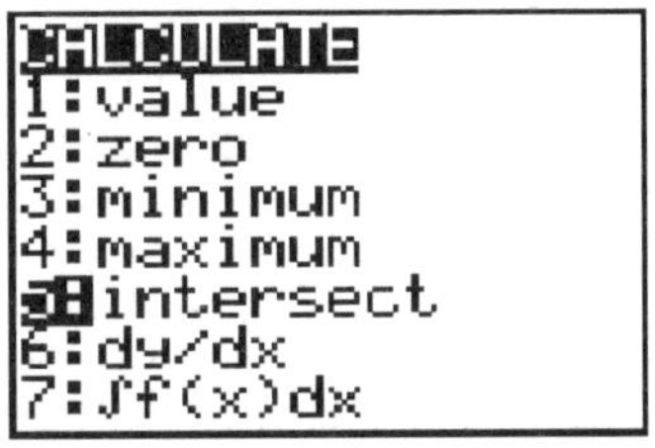

Step 3: (Identify Intersection Point) Next, find the place where the graph of Y_1 and Y_2 intersect (the x-coordinate of this point is the solution to our original equation). This is done by pressing [2nd][TRACE] followed by [5].

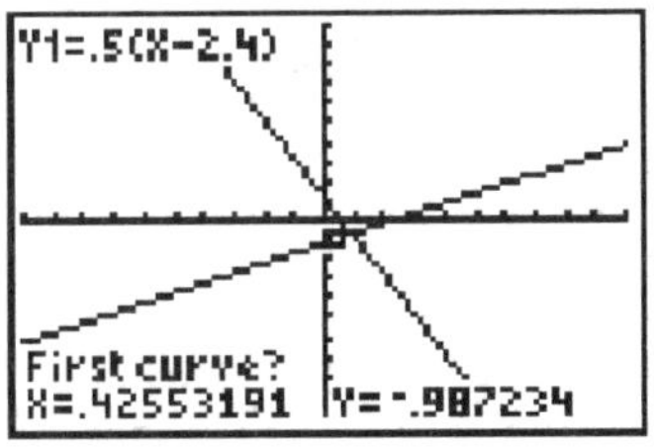

You will be prompted for the first curve with a screen like the one to the left. Press [ENTER].

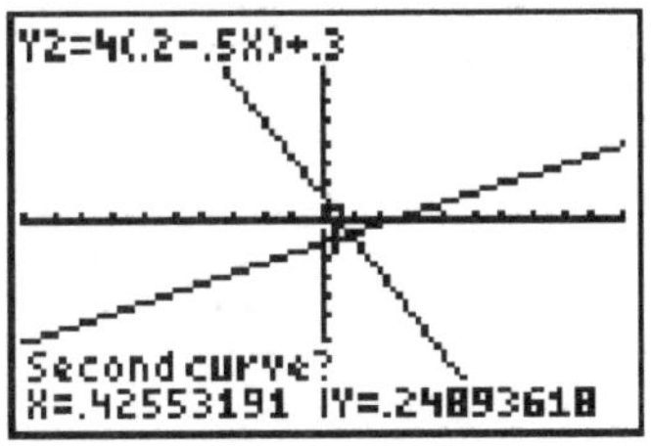

Now you will be prompted for the second curve with a screen like the one to the left. Press [ENTER] again.

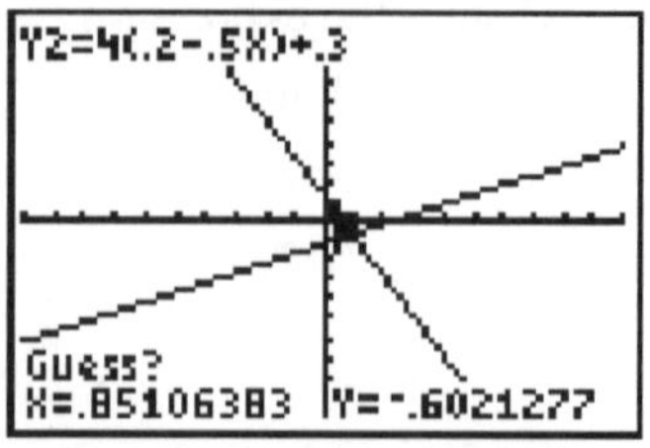

You will be prompted for a guess with a screen like the one to the left. Use the ◄ and ► keys to move the highlighted point as close as possible to the intercept you are trying to verify and press ENTER.

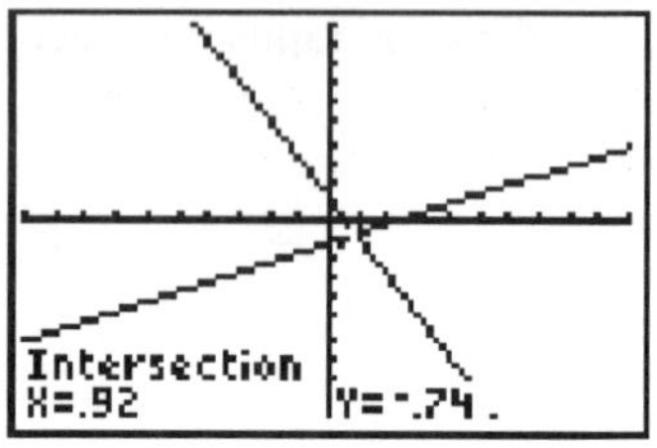

The calculator should display a screen like the one to the left.

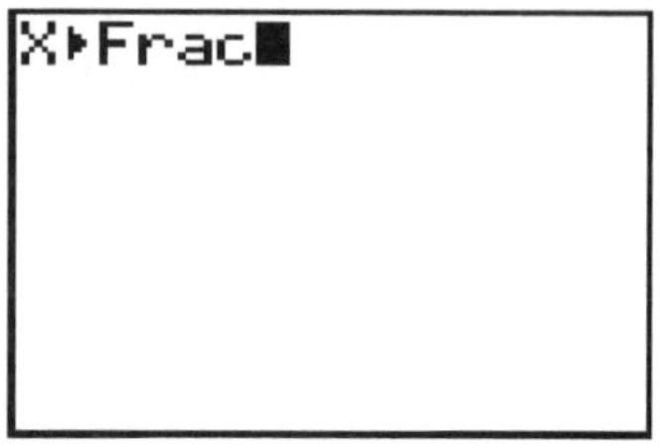

This gives us an intersection point with an x-coordinate of 0.92. Convert this to fraction form by pressing the keys 2nd MODE to return to the home screen. Next, press the keys X,T,Θ,n MATH 1 to get a screen like the one shown to the left.

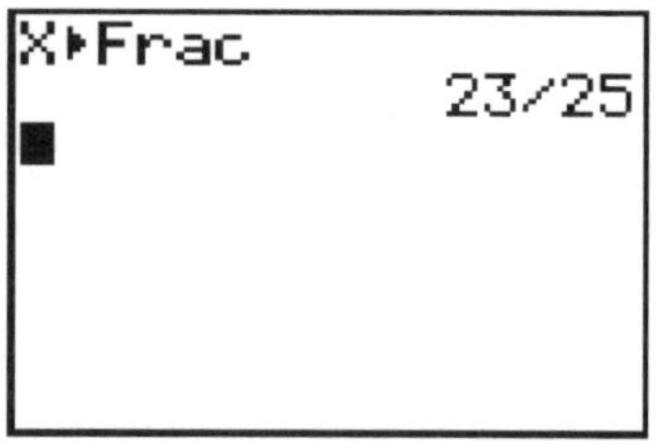

Finally, press ENTER and you will get a screen like the one to the left.

So, our solution expressed as a fraction is $x = 0.92 = \frac{23}{25}$.

Step 4: (Solve) The equation $0.5(x - 2.4) = 4(0.2 - 0.5x) + 0.3$ is satisfied by the value $x = 0.92 = \frac{23}{25}$.

Answer: The solution of $0.5(x - 2.4) = 4(0.2 - 0.5x) + 0.3$ is $x = 0.92 = \frac{23}{25}$.

Example 3 Use the graphing calculator to solve the equation $5x - 23 = 40 - 2x$.

Step 0: (Settings) Make sure you have the default settings as described in the beginning of the lesson.

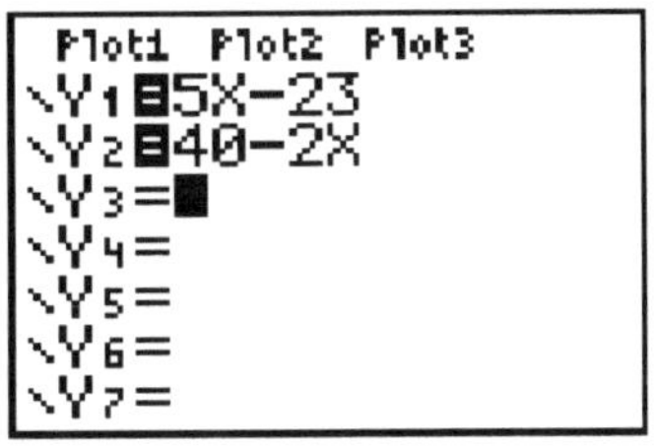

Step 1: (Enter the expression) Press the [Y=] key and enter the expression $5x - 23$ into the Y_1 slot and $40 - 2x$ into the Y_2 slot.

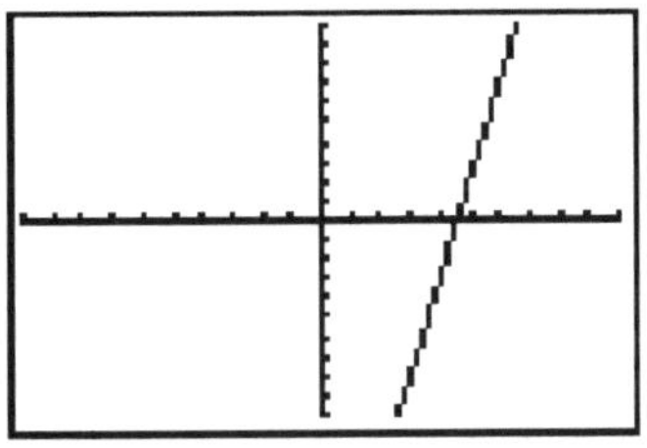

Step 2: (Graph) Press the [GRAPH] key. Since we can only see one graph, the other graph (and the intersection point) must not be visible in this window.

Resize the window using the [ZOOM][3] keys and then pressing [ENTER] again.

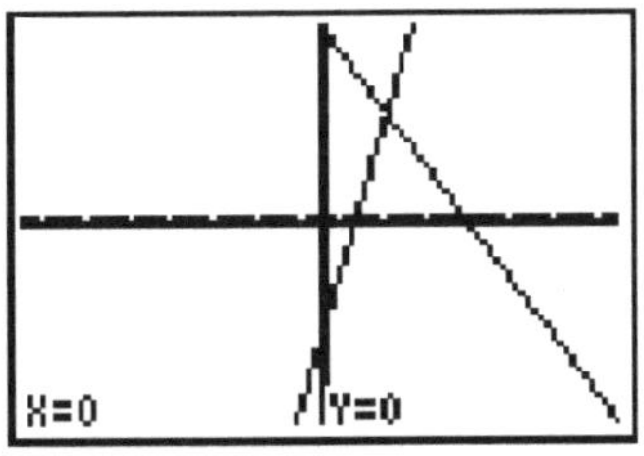

You should see the other graph appear as in the screen shown to the left.

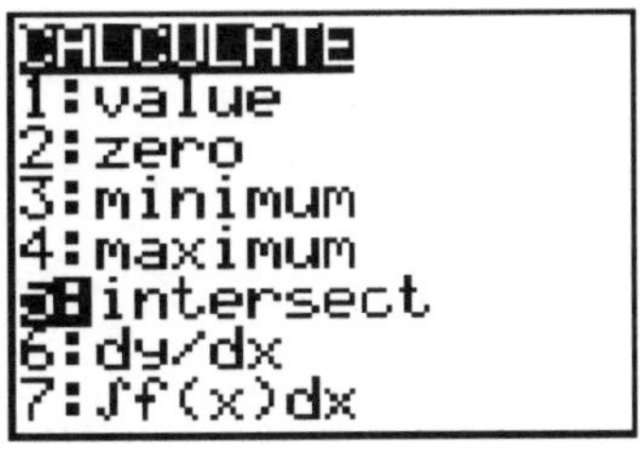

Step 3: (Identify Intersection Point) Next, find the place where the graphs of Y_1 and Y_2 intersect (the x-coordinate of this point is the solution to our original equation). This is done by pressing [2nd][TRACE] followed by [5].

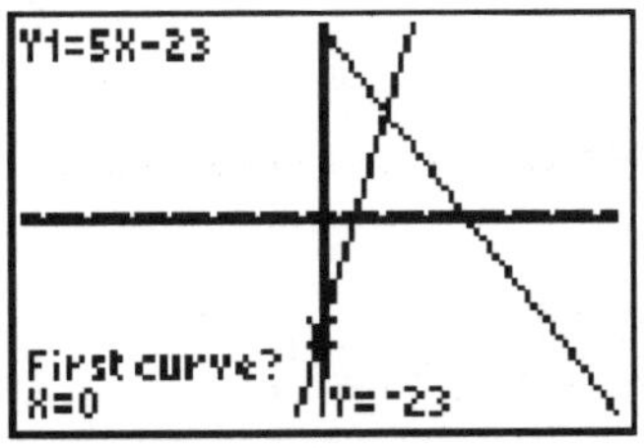

You will be prompted for the first curve with a screen like the one to the left. Press [ENTER].

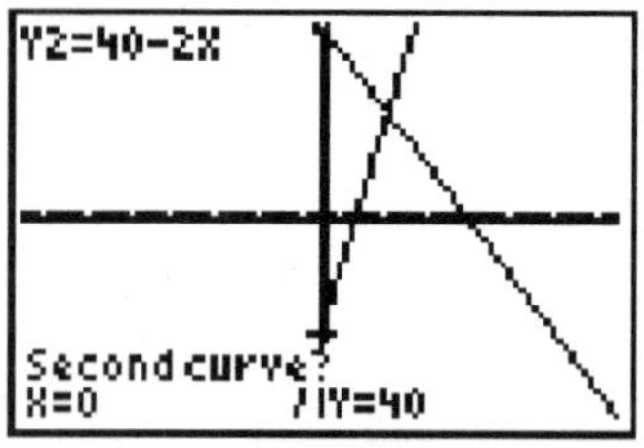

Now you will be prompted for the second curve with a screen like the one to the left. Press [ENTER] again.

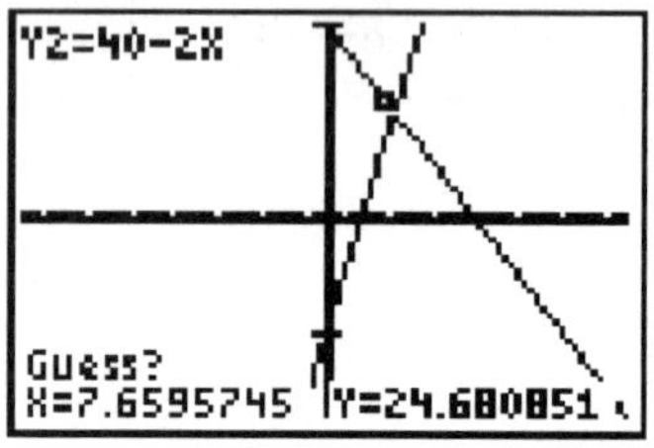

You will be prompted for a guess with a screen like the one to the left. Use the [◄] and [►] keys to move the highlighted point as close as possible to the intercept you are trying to verify and press [ENTER].

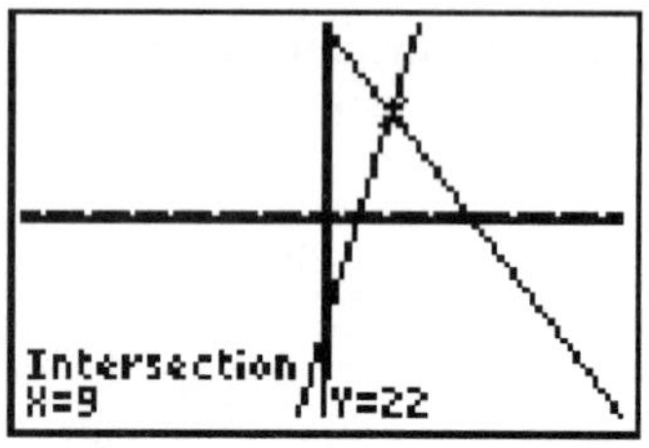

The calculator should display a screen like the one to the left.

This gives us an intersection point whose x-coordinate is 9. Therefore, our solution is $x = 9$.

Step 4: (Solve) The equation $5x - 23 = 40 - 2x$ is satisfied by the value $x = 9$.

Answer: The solution of $5x - 23 = 40 - 2x$ is $x = 9$.

Problem Set

1. Solve using the graphing calculator:
$$5x = 2x + 6$$

2. Solve using the graphing calculator:
$$27 = 6x - 3$$

3. Solve using the graphing calculator:
$$9(x - 8) - 4(x - 13) = 3x + 2$$

4. Solve using the graphing calculator:
$$-4x - 8 + x = 3 - 5x - 5 + 4x$$

5. Solve using the graphing calculator:
$$\frac{1}{3}x - 14 = 24$$

6. Solve using the graphing calculator:
$$\frac{x+6}{4} - \frac{3x-4}{2} = \frac{-1-2x}{6}$$

7. Solve using the graphing calculator:
$$5b + \frac{4}{3} = \frac{2}{9}$$

8. Solve using the graphing calculator:
$$2.4x - 11.01 = 7.23$$

9. Solve using the graphing calculator:
$$\frac{x+1}{6} + \frac{3x+1}{5} = 3 - \frac{7x+1}{2}$$

10. Solve using the graphing calculator:
$$2.43x + 3.1 = 1.73(2x - 16) - 0.03(5x - 146)$$

HA1-385: Finding the Slope of a Line from Its Graph or from the Coordinates of Two Points

Think about the word "slope." What comes to mind? There are many images that can be associated with this word. One of them might be a tall, pointy roof on a house, or a snow-covered ski slope, or a tall, steep bridge or road. All these things have a common characteristic: a slanted surface. In each case, it is possible to talk about how steep that slanted surface is. This lesson discusses the idea of slope mathematically, that is, the measure of how steep a line is.

In previous lessons, you learned about different types of lines. Some lines slanted upward from left to right, while others slanted downward from left to right. Some lines were vertical, while others were horizontal. The slant of a line is referred to as its **slope**.

Slope of a Line	Slope of a line is the ratio of the change in the y-coordinates to the corresponding change in the x-coordinates. For any two points (x_1, y_1) and (x_2, y_2) on a line, the slope is found as $m = \frac{(y_2 - y_1)}{(x_2 - x_1)}$

Example 1 Find the slope of the line that contains the points (1, 0) and (0, 2).

$$m = \frac{y_2 - y_1}{x_2 - x_1}$$

$$m = \frac{2-0}{0-1}$$

Step 1: Substitute the given coordinates into the formula.

$$m = \frac{2}{-1}$$

$$m = -2$$

Step 2: Simplify the expression.

Answer: The slope of the line with the ordered pairs (1, 0) and (0, 2) is –2.

Now that you have seen a concrete example of how to calculate slope, let's go back to the general idea. Another common way to think about slope is $\frac{\text{rise}}{\text{run}}$. The "rise" refers to the vertical change, which is the change in the y-coordinates. It is called "rise" because it indicates how many spaces up (or down) you have to count to get from one point on the line to another point. The "run" refers to the horizontal change, or the change in the x-coordinates. It is called "run" because it indicates how many spaces to the right (or left) you have to count to get from one point to another point.

Consider the line that you have drawn here, through the points $(-2, -3)$ and $(1, 4)$.

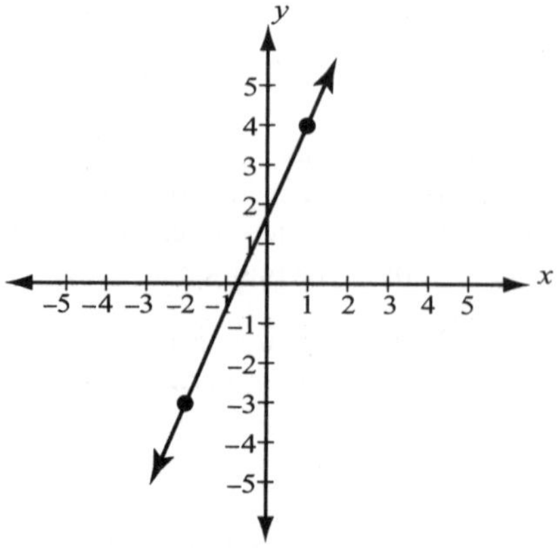

Using these points in the slope formula you see that the slope of this line is $\frac{7}{3}$. Suppose that you use these two points, and the portion of the line between these points, to draw a right triangle. Beginning with the point that is further to the left, (–2, –3), you draw one leg of the triangle going straight up, until you are even with the other point. Then you draw the other leg of the triangle going to the right to touch the second point, (1, 4). The result is what you see in the next graph.

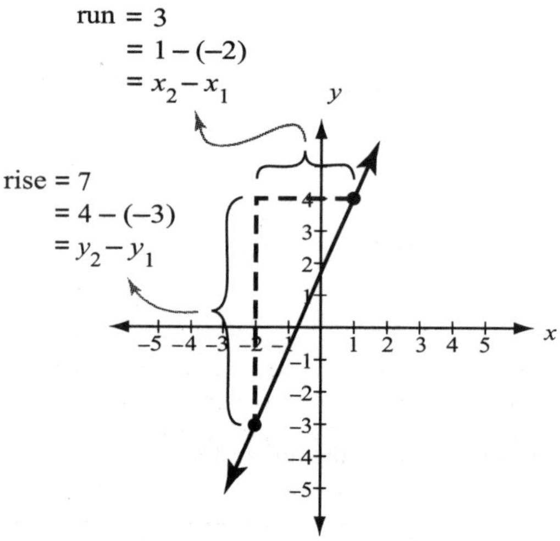

The vertical leg of the triangle is the rise. Looking at the graph, you can see that the length of this leg is 7 spaces. That is because you have to count up 7 spaces to get from the point (–2, –3) to the point (1, 4). Notice that this length is equal to the difference in the y-coordinates of the two points: $4 - (-3) = 7$. The horizontal leg of the triangle is the run. Looking at the graph, you see that the length of this leg is 3 spaces, because you have to count 3 spaces to the right to get from the vertical leg to the point (1, 4). Notice how this length is equal to the difference in the x-coordinates: $1 - (-2) = 3$. You can use this $\frac{\text{rise}}{\text{run}}$ to find the slope of a line that has already been graphed.

The next example illustrates how to use this rise/run method.

Example 2 Find the slope of the line graphed below.

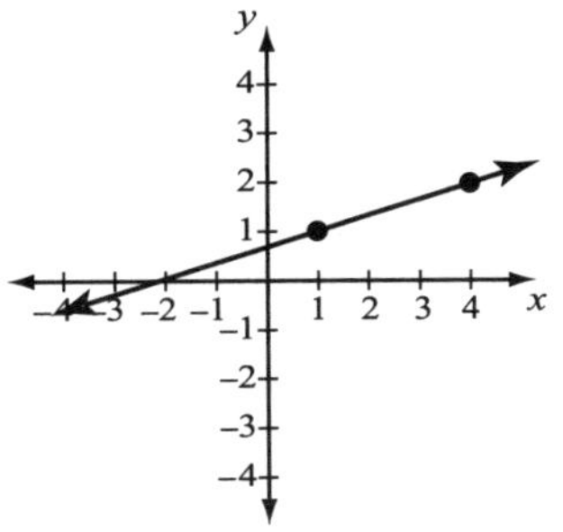

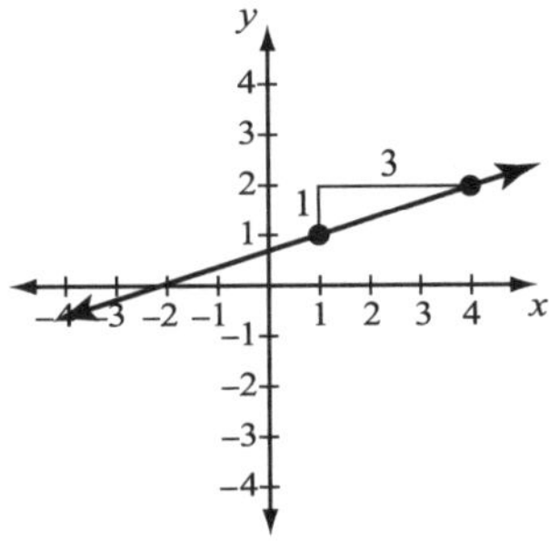

Step 1: Make the triangle as discussed above by drawing a vertical line up from the first point until it is even with the second point. Then draw a horizontal line from there to the second point.

$\frac{\text{rise}}{\text{run}} = \frac{1}{3}$

Step 2: Write the slope by putting the number of rise units, 1, over the number of run units, 3, to form the fraction $\frac{1}{3}$.

Answer: The slope is $\frac{1}{3}$.

Note that in Example 2 that, although the ordered pairs are not explicitly given, you could use the slope formula to find the slope of the line. In Example 2, instead of being given two points, you have a graph of a line. Once you choose two points on the line and determine their coordinates, you can use the coordinates in the slope formula. This is what you see in Example 3.

Example 3 Find the slope of the line graphed below.

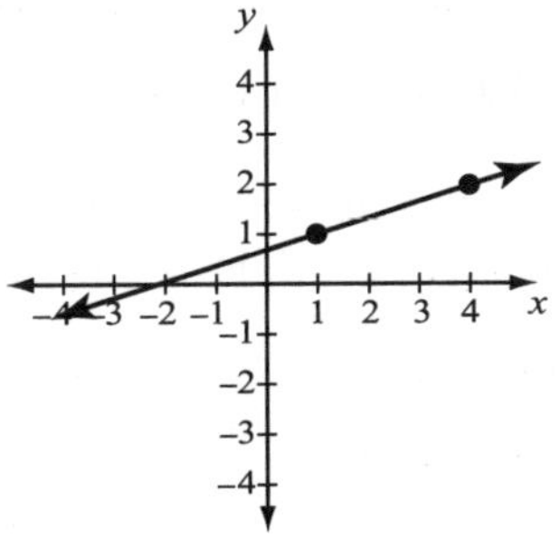

(1, 1) and (4, 2)

Step 1: Determine the coordinates of the indicated points.

$$m = \frac{2-1}{4-1} = \frac{1}{3}$$

Step 2: Use these coordinates in the slope formula.

Answer: The slope is $\frac{1}{3}$.

One thing to remember about the rise/run method is that the direction in which you count is very important. In Example 2, you counted up 1, and so you rose 1. If you had counted down instead, you would still have called it a rise, but you would have made it negative. For example, if you had to count down 3, you would say that the rise is –3, and you would put –3 in the numerator when you form the fraction in Step 2. Likewise, in Example 2, you counted to the right 3, and so you ran 3. If you had counted to the left instead, you would have had a negative run. For example, if you had to count to the left 5, you would say that the run is –5, and you would put –5 in the denominator when you form the fraction.

Why is the slope of a line important? The numerical value of the slope of the line tells you about the slant, or tilt, of the line, and how steep the line is. Notice that in Example 1, the slope that you found was negative, and, in Example 2, the slope was positive. Compare the graphs of the two lines from those examples. Looking at the two lines, you see that the first line has a negative slope, and the line is drawn down from left to right. The second graph has a positive slope, and the line is drawn up from left to right. Also, looking at the actual numerical values of the slopes, you see that $|-2|$ is greater than $\left|\frac{1}{3}\right|$, and the line with slope –2 is steeper. These observations follow the general rules. A line with negative slope goes down from left to right. A graph with positive slope goes up from left to right. The larger the absolute value of the slope of a line, the steeper the line. Let's consider two special cases: a vertical line and a horizontal line. Here is the graph of a vertical line, with the points (–2, 3) and (–2, 1) labeled.

Example 4 Find the slope of the line graphed below.

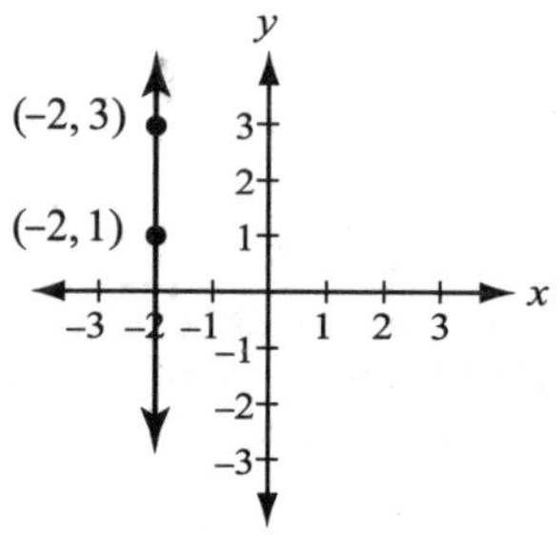

$$m = \frac{1-3}{-2-(-2)} = \frac{-2}{0}$$

Step 1: Calculate the slope using the slope formula.

Step 2: By calculating the slope of this line, you see that a zero results in the denominator of the fraction, which makes the fraction undefined. This will always happen with vertical lines, since the x-coordinates of all points on any vertical line are always the same. Therefore, vertical lines have undefined slope.

Answer: The slope is undefined.

Example 5 Find the slope of the line graphed below.

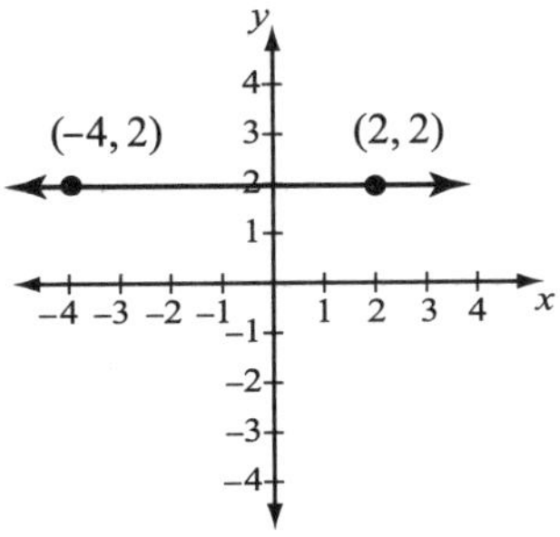

$$m = \frac{2-2}{2-(-4)} = \frac{0}{6} = 0$$

Step 1: Calculate the slope using the slope formula.

Step 2: By calculating the slope of this line, you see that a zero results in the numerator of the fraction, which makes the fraction itself equal to 0. This will always happen with horizontal lines, since the *y*-coordinates of all points on any horizontal line are always the same. Therefore, horizontal lines have a slope of 0.

Answer: The slope is 0.

Previously, you have chosen which point to use as (x_1, y_1) and which point to use as (x_2, y_2). It actually doesn't matter which set of coordinates you use as (x_1, y_1) and (x_2, y_2), as long as you don't change your mind halfway through the problem.

Example 6 Find the slope of the line graphed below, using the rise/run method:

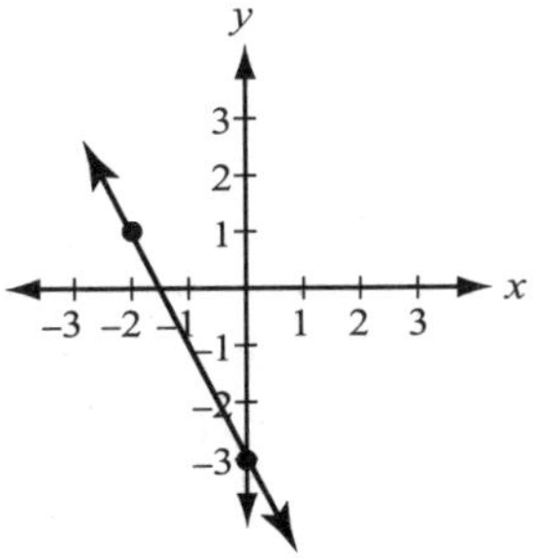

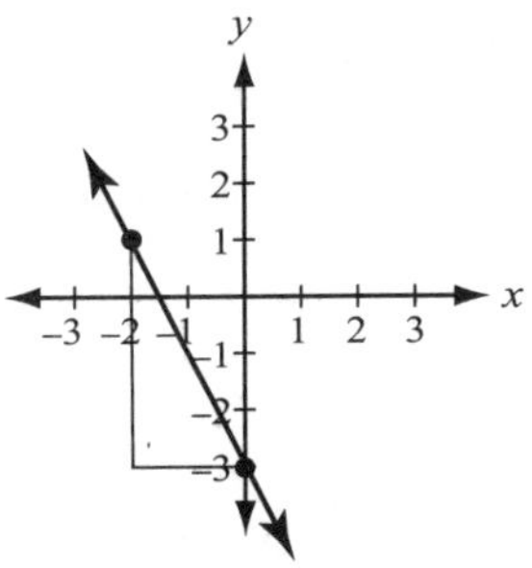

Step 1: Draw a vertical line down from the first point until it is even with the second point. Then draw a horizontal line from there to the second point.

Step 2: Find the rise by determining the number of units it takes to move downward from the first point to be even with the second point. The rise will be negative, because you are moving downward.

Step 3: Find the run by determining the number of units it takes to move right to the second point. The run will be positive, because you are moving to the right.

$m = \frac{\text{rise}}{\text{run}} = \frac{-4}{2} = -2$	**Step 4:**	Put the number for the rise over the number for the run to form the fraction $\frac{-4}{2}$. Simplify the fraction. The slope is –2.
$m = \frac{y_2 - y_1}{x_2 - x_1}$ $m = \frac{-3-1}{0-(-2)}$ $m = \frac{-4}{2}$ $m = -2$	**Step 5:**	Using the coordinates: Substitute the coordinates of the points plotted on the graph into the slope formula.The coordinates are (–2, 1) and (0, –3).

Answer: The slope is –2.

Problem Set

1. Find the slope of the line containing the given points: (0, 0) and (2, 2)

2. Find the slope of the line containing the given points: (2, 3) and (3, 6)

3. Find the slope of the line containing the given points: (2, 1) and (6, 9)

4. Find the slope of the line containing the given points: (3, 9) and (4, 16)

5. Find the slope of the line containing the given points: (1, 6) and (3, 16)

6. Find the slope of the line containing the given points: (1, 2) and (3, 4)

7. Find the slope of the line containing the given points: (0, 3) and (5, 3)

8. Find the slope of the line containing the given points: (3, 2) and (5, 8)

9. Find the slope of the line containing the given points: (3, 0) and (4, 6)

10. Find the slope of the line containing the given points: (4, 4) and (6, 6)

11. Find the slope of the line containing the given points: (–10, 0) and (8,–6)

12. Find the slope of the line containing the given points: (–4, 2) and (–4,3)

13. Find the slope of the line containing the given points: (–6, 3) and (4, 1)

14. Find the slope of the line containing the given points: (1, 0) and (0,–4)

15. Find the slope of the line containing the given points: (0, 1) and (3,–2)

16. Find the slope of the line containing the given points: (–6,–7) and (–2,–8)

17. Find the slope of the line containing the given points: (–4,–2) and (–1, 0)

18. Find the slope of the line from its graph:

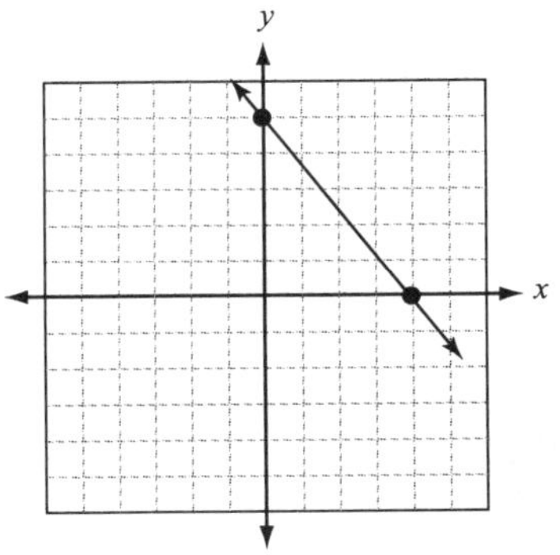

19. Find the slope of the line from its graph:

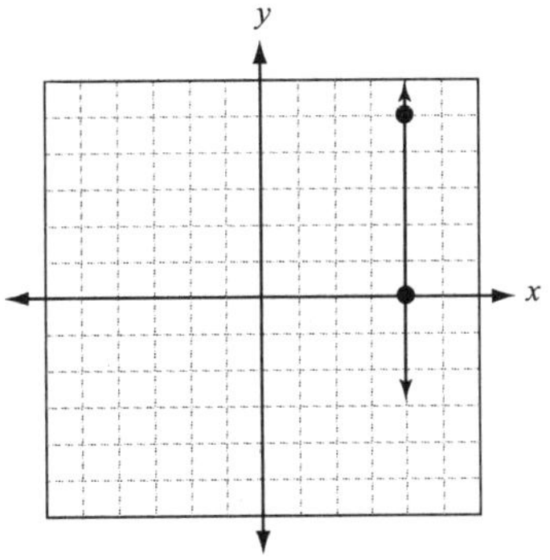

20. Find the slope of the line from its graph:

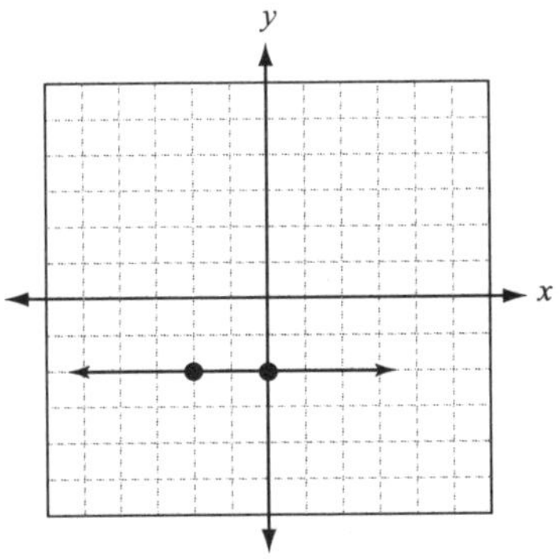

HA1-394: Interchanging Linear Equations Between Standard Form and Slope-Intercept Form

Linear equations, like $3x + y = 5$, have graphs in the coordinate plane that are represented by lines. If the linear equation is written in a different form, we can determine the slope and the y-intercept of the line. For example, the linear equation $3x + y = 5$ can be rewritten as $y = -3x + 5$. In this form, we can determine that the slope of the line is –3 (falls 3 units for each unit moved to the right) and the y-coordinate of the y-intercept is 5 [intersects the y-axis at the point (0,5)].

Notice that this information can be read directly from the form of the equation when it is written in the specific form $y = mx + b$, where $m = 3$ and $b = 5$. This form of an equation of a line is called the **Slope-Intercept Form**. The original form of the equation of the line $3x + y = 5$, is called the **Standard Form** ($Ax + By = C$, where $A = 3$, $B = 1$, and $C = 5$).

Note: The procedure for changing a line from standard form to slope-intercept form is to solve the equation for y and then to write the other part of the equation in the form $mx + b$.

Standard Form of a Linear Equation	$Ax + By = C$, where A, B, and C are real numbers (written in integer form when possible). $A \geq 0$, A and B are not both zero.

Slope-Intercept Form of a Linear Equation	$y = mx + b$, where m is the slope and b is the y-coordinate of the y-intercept $(0, b)$.

Example 1 Change the equation $4x + 5y = 10$ into slope-intercept form.

$$4x - 4x + 5y = 10 - 4x$$
$$5y = 10 - 4x$$

Step 1: Subtract $4x$ from both sides of the equation.

$$\frac{5y}{5} = \frac{10}{5} - \frac{4x}{5}$$
$$y = 2 - \frac{4}{5}x$$

Step 2: Divide both sides of the equation by 5.

$$y = -\frac{4}{5}x + 2$$

Step 3: Write in the form $y = mx + b$.

Answer: $y = -\frac{4}{5}x + 2$

Note: In general, when changing a line given in slope-intercept form into standard form, we move all the terms with the variables x and y to one side of the equation, and the constant to the other side of the equation.

Example 2 Write the equation $y = 4x - 3$ in standard form.

$-4x + y = 4x - 4x - 3$ | **Step 1:** Subtract $4x$ from both sides of the equation.

$-4x + y = -3$

$-1(-4x + y) = (-1)(-3)$ | **Step 2:** Multiply both sides of the equation by -1 to make the coefficient of x positive.

$4x - y = 3$

Answer: $4x - y = 3$

Example 3 Write the equation of the line $3x - 5y = 4$ in slope-intercept form. Then use this form to find the slope and y–intercept of the line.

$3x - 3x - 5y = -3x + 4$ | **Step 1:** Subtract $3x$ from both sides of the equation.

$-5y = -3x + 4$

$\frac{-5y}{-5} = \frac{-3x}{-5} + \frac{4}{-5}$ | **Step 2:** Divide both sides of the equation by -5.

$y = \frac{3}{5}x - \frac{4}{5}$

$m = \frac{3}{5}$ | **Step 3:** Determine the slope and y-intercept using $y = mx + b$.

y-intercept: $\left(0, -\frac{4}{5}\right)$

Answer: The line $3x - 5y = 4$ written in slope-intercept form is $y = \frac{3}{5}x - \frac{4}{5}$.

The line has slope $m = \frac{3}{5}$ and the y-intercept is $\left(0, -\frac{4}{5}\right)$.

Example 4 Write the equation $y = -\frac{1}{2}x - 1$ in standard form.

$\frac{1}{2}x + y = -\frac{1}{2}x + \frac{1}{2}x - 1$ | **Step 1:** Add $\frac{1}{2}x$ to both sides of the equation.

$\frac{1}{2}x + y = -1$

$2\left(\frac{1}{2}x + y\right) = 2(-1)$ | **Step 2:** Multiply both sides by 2 since A, the coefficient of x, must be written in integer form when possible.

$x + 2y = -2$

Answer: $x + 2y = -2$

Example 5 Write the equation $y = \frac{1}{4}x + \frac{5}{6}$ in standard form.

$$-\frac{1}{4}x + y = \frac{1}{4}x - \frac{1}{4}x + \frac{5}{6}$$ **Step 1:** Subtract $\frac{1}{4}x$ from both sides of the equation.

$$-\frac{1}{4}x + y = \frac{5}{6}$$

$$12\left(-\frac{1}{4}x + y\right) = 12\left(\frac{5}{6}\right)$$

$$-3x + 12y = 10$$

Step 2: Since the coefficients of x and y (A and B) must be written in integer form when possible, multiply both sides by 12, the LCD of $\frac{1}{4}$ and $\frac{5}{6}$.

$$-1(-3x + 12y) = (-1)(10)$$ **Step 3:** Multiply both sides by -1 since A, the coefficient of x, must be positive.

Answer: $3x - 12y = -10$

Problem Set

Write the following equations in slope-intercept form:

1. $6x + 3y = 7$ **2.** $14x - 7y = -4$ **3.** $12x + 3y = 24$ **4.** $16x + 4y = -20$

5. $30x + 6y = 42$ **6.** $18x + 3y = 5$ **7.** $15x - 5y = 40$

Write the following equations in standard form:

8. $y = -8x + 4$ **9.** $y = -12x + 7$ **10.** $y = -11x - 32$ **11.** $y = -7x - 8$

12. $y = 2x + 5$ **13.** $y = 10x - 8$ **14.** $y = 6x + 12$ **15.** $y = 9x - 1$

Write the following equations in slope-intercept form:

16. $2x - 4y = 8$ **17.** $7x - 3y = 12$ **18.** $4x - 7y = 21$ **19.** $10x + 8y = 32$

20. $11x - 9y = 27$ **21.** $3x + 4y = 16$ **22.** $12x + 5y = 25$

Write the following equations in standard form:

23. $y = -\frac{3}{5}x - 8$ **24.** $y = -\frac{3}{4}x + 9$ **25.** $y = -\frac{5}{6}x - 4$ **26.** $y = -\frac{7}{9}x + 8$

27. $y = -\frac{8}{3}x + 16$ **28.** $y = -\frac{5}{8}x - 7$ **29.** $y = -\frac{4}{7}x + 9$ **30.** $y = -\frac{9}{2}x - 13$

31. $y = \frac{1}{3}x + \frac{1}{5}$ **32.** $y = \frac{3}{4}x - \frac{2}{5}$ **33.** $y = \frac{2}{9}x + \frac{1}{3}$ **34.** $y = \frac{5}{6}x - \frac{3}{4}$

35. $y = \frac{3}{8}x + \frac{1}{6}$ **36.** $y = \frac{3}{7}x - \frac{2}{3}$ **37.** $y = \frac{8}{5}x + \frac{4}{9}$ **38.** $y = \frac{21}{4}x - \frac{1}{2}$

39. $y = \frac{2}{3}x + \frac{1}{6}$ **40.** $y = \frac{3}{4}x + \frac{1}{7}$

HA1-395: Finding the Equation of a Line Parallel or Perpendicular to a Given Line

In previous lessons, you learned how to find the slope of a line given two points or a line on a graph. You also learned the different forms for the equation of a line. First, let's review the different forms for the equation of a line.

Standard Form for the Equation of a Line	$Ax + By = C$, where A, B, and C are real numbers, $A \geq 0$, and A and B are not both zero.

Slope-intercept Form for the Equation of a Line	$y = mx + b$, where m is the slope of the line and $(0, b)$ is the y-intercept.

Recall that the importance of writing an equation in slope-intercept form is that you can determine the slope and y-intercept from the equation of the line.

This is useful in determining whether two lines are parallel or perpendicular. Equations of lines that are parallel or perpendicular to a given line can be written using the relationship between their slopes.

Parallel Lines	Lines that lie in the same plane that do not intersect

For example, lines j and k are parallel lines.

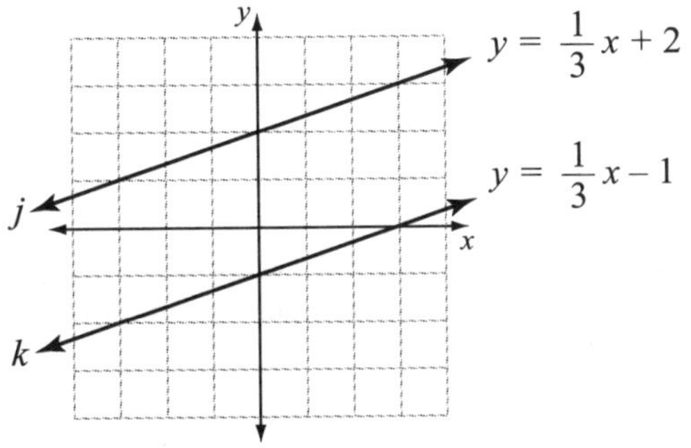

The equation of line j is $y = \frac{1}{3}x + 2$, and the equation of line k is $y = \frac{1}{3}x - 1$. Both of these equations are written in the form $y = mx + b$, where m is the slope and $(0, b)$ is the y-intercept. Notice that both lines have a slope of $\frac{1}{3}$ and their y-intercepts are different. Parallel lines have the same slope and different y-intercepts. **If m_1 and m_2 are slopes of parallel lines, then $m_1 = m_2$.**

Equations of lines that are perpendicular to a given line can be written using the relationship between their slopes.

Perpendicular Lines	Lines that intersect to form right angles, or 90-degree angles

For example, lines l and n are perpendicular lines.

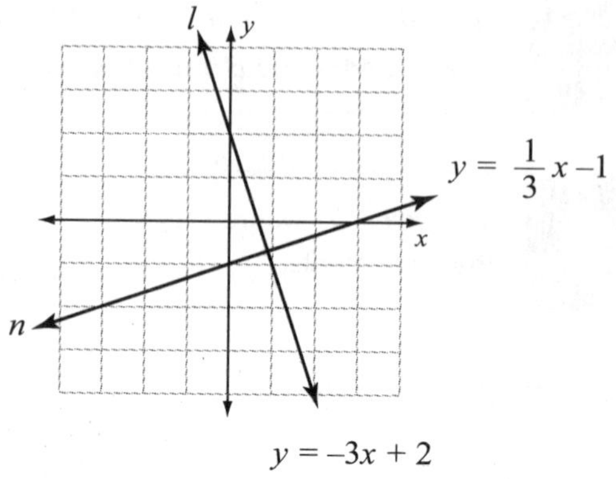

The equation of line l is $y = -3x + 2$, and the equation of line n is $y = \frac{1}{3}x - 1$. Both of these equations are written in the form $y = mx + b$, where m is the slope and $(0, b)$ is the y-intercept. The slope of line l is -3 and the slope of line n is $\frac{1}{3}$. The numbers -3 and $\frac{1}{3}$ are opposite reciprocals of each other. This means that their signs are opposites and their absolute values are reciprocals. Perpendicular lines have slopes that are opposite reciprocals of each other.

If m_1 and m_2 are slopes of perpendicular lines, then m_2 is the opposite reciprocal of m_1.

Example 1 Write an equation of a line that is parallel to $y = 2x - 4$.

$y = mx + b$ $y = 2x - 4$ $m_1 = 2$	**Step 1:** Find the slope of the given line using the slope-intercept form of the equation of a line.
$m_2 = 2$	**Step 2:** Find the slope of the parallel line using the fact that parallel lines have the same slope.
$y = mx + b$ $y = 2x + b$	**Step 3:** Substitute into the slope-intercept form of the equation of a line and use any value for b, except for the value in the given equation.
Note that b can be any value except -4.	**Answer:** $y = 2x + 1$

Example 2 Find the slope of a line that is perpendicular to the line whose graph is shown.

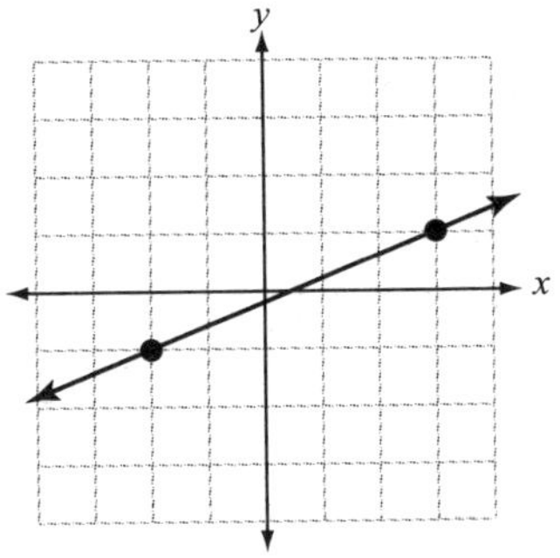

Step 1: Find the slope of the given line using $m = \frac{\text{rise}}{\text{run}}$.

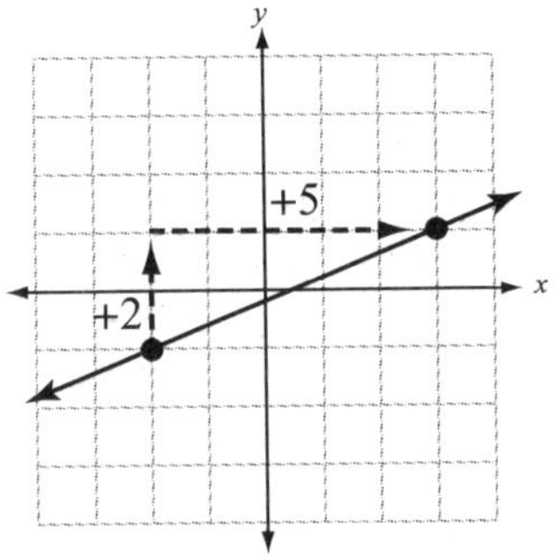

The rise is 2 and the run is 5.

$$m_1 = \frac{2}{5}$$

Step 2: Find the opposite reciprocal of m_1.

The opposite reciprocal of $\frac{2}{5}$ is $-\frac{5}{2}$.

Step 3: m_2 is the opposite reciprocal of m_1.

$$m_2 = -\frac{5}{2}$$

Answer: $-\frac{5}{2}$

Example 3 Determine if the lines of the given equations are parallel, perpendicular, or neither.

$$x - 2y = 6$$
$$y = 2x + 3$$

$$\begin{aligned} x - 2y &= 6 \\ -x + x - 2y &= -x + 6 \\ -2y &= -x + 6 \\ \frac{-2y}{-2} &= \frac{-x}{-2} + \frac{6}{-2} \\ y &= \frac{1}{2}x - 3 \\ y &= mx + b \\ m_1 &= \frac{1}{2} \end{aligned}$$

Step 1: Find the slope of the first line by putting the equation in slope-intercept form.

$$\begin{aligned} y &= 2x + 3 \\ y &= mx + b \\ m_2 &= 2 \end{aligned}$$

Step 2: Find the slope of the second line.

Note: The equation is in slope-intercept form

The lines are not parallel because the slopes are not equal. The opposite reciprocal of 2 is $-\frac{1}{2}$. Therefore, the lines are not perpendicular because the opposite reciprocal of m_2 is not equal to m_1.

Step 3: Compare the slopes.

Answer: Neither

Example 4 Write an equation in slope-intercept form for a line that is parallel to $5x + 2y = -8$ and passes through the point $(-4, 3)$.

$$\begin{aligned} 5x + 2y &= -8 \\ -5x + 5x + 2y &= -5x - 8 \\ 2y &= -5x - 8 \\ \frac{2y}{2} &= \frac{-5x}{2} - \frac{8}{2} \\ y &= -\frac{5}{2}x - 4 \\ y &= mx + b \\ m_1 &= -\frac{5}{2} \end{aligned}$$

Step 1: Find the slope of the given line by putting the equation in slope-intercept form.

$$m_2 = -\frac{5}{2}$$

Step 2: If m_1 and m_2 are slopes of parallel lines, then $m_1 = m_2$.

$y = mx + b$

$3 = -\frac{5}{2}(-4) + b$

$3 = 10 + b$

$3 - 10 = 10 - 10 + b$

$-7 = b$

Step 3: Since the parallel line passes through the point (–4, 3), substitute $x = -4$, $y = 3$, and $m = -\frac{5}{2}$ into the slope-intercept form and solve for b.

$y = mx + b$

$y = -\frac{5}{2}x - 7$

Step 4: Now substitute the values for m and b into the slope-intercept form.

Answer: $y = -\frac{5}{2}x - 7$

Example 5 A city planner is drawing the roads for a development on a coordinate plane. One of the roads passes through the points (1, 9) and (5, 2). Find the equation in slope-intercept form of a second road that is perpendicular to the given road and passes through the point (7, 4).

$m_1 = \frac{y_2 - y_1}{x_2 - x_1}$

$m_1 = \frac{2 - 9}{5 - 1}$

$m_1 = -\frac{7}{4}$

Step 1: Find the slope of the line for the first road using the formula for finding the slope of a line given two points on the line.

The opposite reciprocal of $-\frac{7}{4}$ is $\frac{4}{7}$.

Step 2: Find the opposite reciprocal of m_1.

$m_2 = \frac{4}{7}$

Step 3: m_2 is the opposite reciprocal of m_1.

$y = mx + b$

$4 = \frac{4}{7}(7) + b$

$4 = 4 + b$

$4 - 4 = 4 - 4 + b$

$0 = b$

Step 4: Since the perpendicular line passes through the point (7, 4), substitute $x = 7$, $y = 4$, and $m = \frac{4}{7}$ into the slope-intercept form and solve for b.

$y = mx + b$

$y = \frac{4}{7}x + 0$

$y = \frac{4}{7}x$

Step 5: Now substitute the values for m and b into the slope-intercept form.

Answer: The equation of the second road is $y = \frac{4}{7}x$.

Problem Set

Solve:

1. Which equation of a line is parallel to the graph of $y = 8x - 3$?

$y = -8x + 5$ $y = 8x + 5$

$y = -\frac{1}{8}x + 5$ $y = \frac{1}{8}x + 5$

2. Which equation of a line is perpendicular to the graph of $y = 5x + 1$?

$y = \frac{1}{5}x + 2$ $y = -\frac{1}{5}x + 2$

$y = -5x + 2$ $y = 5x + 2$

3. Which equation of a line is parallel to the graph of $y = \frac{1}{2}x - 12$?

$y = \frac{1}{2}x - 10$ $y = -\frac{1}{2}x - 10$

$y = 2x - 10$ $y = -2x - 10$

4. Which equation of a line is parallel to the graph of $y = -4x + 2$?

$y = 4x + 9$ $y = -4x + 9$

$y = \frac{1}{4}x + 9$ $y = -\frac{1}{4}x + 9$

5. Which equation of a line is parallel to the graph of $y = -\frac{3}{7}x$?

$y = -\frac{7}{3}x - 9$ $y = \frac{3}{7}x - 9$

$y = -\frac{3}{7}x - 9$ $y = \frac{7}{3}x - 9$

6. Which equation of a line is perpendicular to the graph of $y = -3x + 6$?

$y = -3x - 2$ $y = 3x - 2$

$y = \frac{1}{3}x - 2$ $y = -\frac{1}{3}x - 2$

7. Which equation of a line is perpendicular to the graph of $y = \frac{2}{9}x$?

$y = \frac{2}{9}x + 2$ $y = -\frac{9}{2}x + 2$

$y = \frac{9}{2}x + 2$ $y = -\frac{2}{9}x + 2$

8. Which equation of a line is perpendicular to the graph of $y = -\frac{1}{2}x - 7$?

$y = -2x - 4$ $y = 2x - 4$

$y = -\frac{1}{2}x - 4$ $y = \frac{1}{2}x - 4$

9. Find the slope of a line that is parallel to the line whose graph is shown.

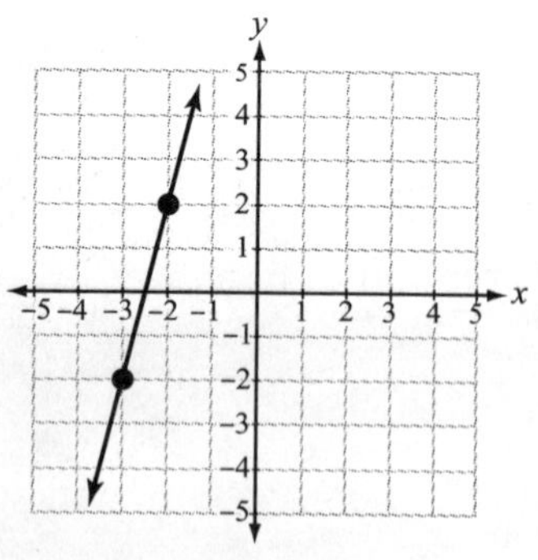

10. Find the slope of a line that is parallel to the line whose graph is shown.

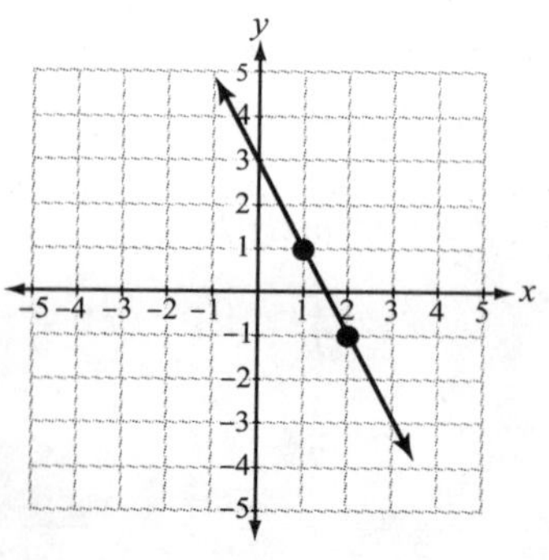

11. Find the slope of a line that is parallel to the line whose graph is shown.

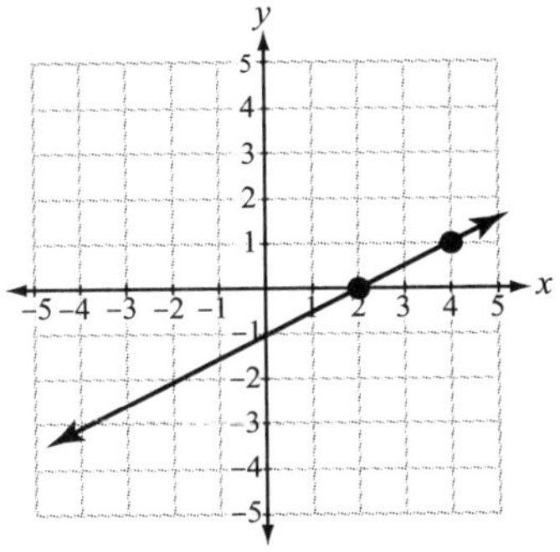

12. Find the slope of a line that is perpendicular to the line whose graph is shown.

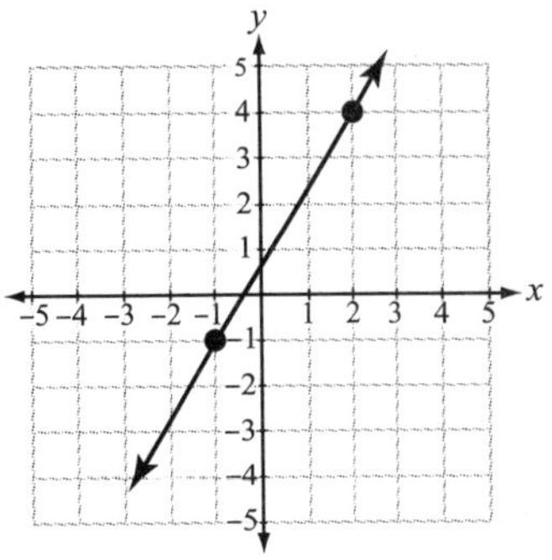

13. Find the slope of a line that is perpendicular to the line whose graph is shown.

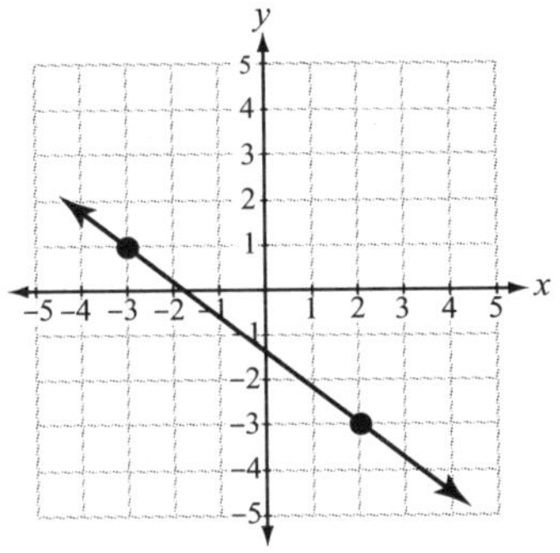

14. Find the slope of a line that is perpendicular to the line whose graph is shown.

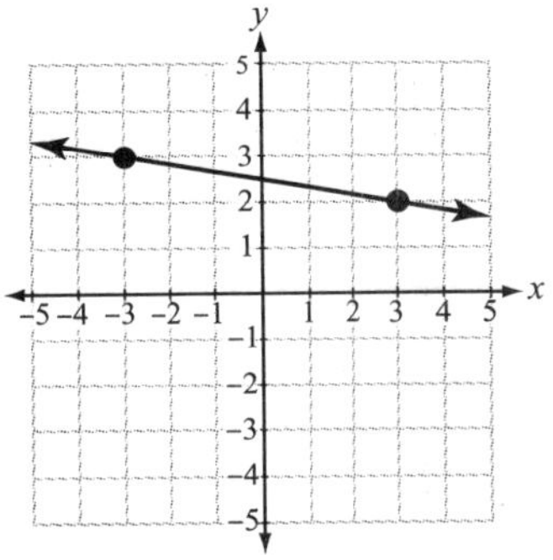

15. Find the slope of a line that is parallel to the line whose graph is shown.

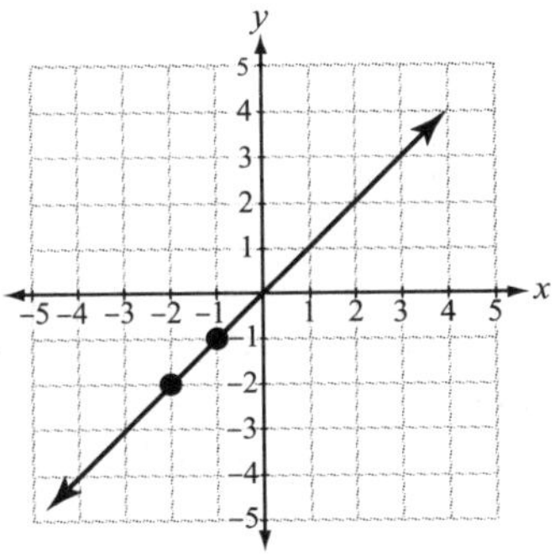

16. Determine if the lines of the given equations are parallel, perpendicular, or neither.

$$y = \frac{1}{2}x - 4$$
$$2y = x - 12$$

17. Determine if the lines of the given equations are parallel, perpendicular, or neither.

$$y = 3x + 2$$
$$2x - 6y = 9$$

18. Determine if the lines of the given equations are parallel, perpendicular, or neither.

$$5x + y = -1$$
$$x - 5y = 10$$

19. Determine if the lines of the given equations are parallel, perpendicular, or neither.

$$-3x + 2y = -8$$
$$y = \frac{3}{2}x$$

20. Determine if the lines of the given equations are parallel, perpendicular, or neither.

$$10x + 6y = 25$$
$$5x - 3y = 10$$

21. Determine if the lines of the given equations are parallel, perpendicular, or neither.

$$-3x - 6y = 19$$
$$y = 2x + 1$$

22. Determine if the lines of the given equations are parallel, perpendicular, or neither.

$$y = \frac{9}{2}x - 14$$
$$2y = 20x + 9$$

23. Determine if the lines of the given equations are parallel, perpendicular, or neither.

$$7x - 4y = -2$$
$$-14x + 8y = 5$$

24. Write an equation in slope-intercept form for a line that is parallel to $y = -3x - 1$ and passes through the point $(1, 4)$.

25. Write an equation in slope-intercept form for a line that is parallel to $y = \frac{3}{4}x + 9$ and passes through the point $(-8, 1)$.

26. Write an equation in slope-intercept form for a line that is parallel to $x + y = -3$ and passes through the point $(-7, 0)$.

27. Write an equation in slope-intercept form for a line that is perpendicular to $y = 4x + 5$ and passes through the point $(-4, -2)$.

28. Write an equation in slope-intercept form for a line that is perpendicular to $y = x - 7$ and passes through the point $(4, 1)$.

29. Write an equation in slope-intercept form for a line that is perpendicular to $x + 2y = 4$ and passes through the point $(-6, 8)$.

30. Write an equation in slope-intercept form for a line that is perpendicular to $-5x - 2y = 10$ and passes through the point $(5, 4)$.

31. The sides of a railroad track lie in parallel lines. When placed on a coordinate plane, one of the parallel sides lies on the line $x + 3y = 12$ and the other parallel side passes through the point $(3, 6)$. Find an equation of the other parallel side.

32. Two airplanes are observed on a screen that is on a coordinate plane. The first airplane is traveling at a constant speed so that its path lies on the line $6x + 5y = 10$. The second airplane is flying on a linear path that contains the point $(6, 12)$ and is perpendicular to the path of the first plane. Find an equation of the path of the second airplane.

33. The streets of a neighborhood are drawn on a coordinate plane. A man walked in the neighborhood along a street containing the points (5, 2) and (9, –2). A woman walked along a parallel street in the same neighborhood that contains the point (0, 3). Find an equation of the line containing the street on which the woman walked.

34. A piece of tile with perpendicular sides is placed on a coordinate plane as shown. Point A is at (–3, –1) and point B is at (1, –4). Find an equation of the line that contains points A and D.

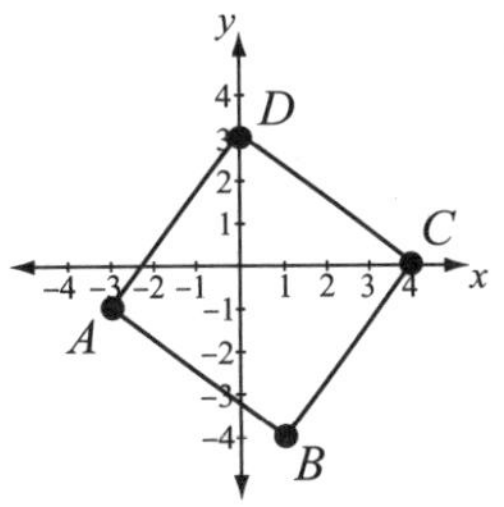

35. The equation of the line that models the population growth of deer in forest A is $y = 5x + 20$. The line that models the population growth of deer in forest B contains the point (2, 50) and is parallel to the line modeling population growth of deer in forest A. Find an equation of the line that models the population growth of deer in forest B.

36. Supply and demand equations are studied in economics classes. The supply equation for a certain product is $y = \frac{1}{5}x + 10$. The demand equation for this product describes a line perpendicular to the graph of the supply equation and it contains the point (5, 799). Find the demand equation for this product.

37. A plumber is laying two pipes that are perpendicular to each other. Pipe P contains the points (1, 1) and (4, 3). Pipe Q lies on a line perpendicular to pipe P and it contains the point (4, 8). Find the equation of the line containing pipe Q.

38. The lines on a highway lie in parallel lines. When placed on a coordinate plane, the equation describing the line on the left side is $x + 3y = 9$. The line on the right contains the point (3, 8) and is parallel to the line on the left. Find an equation of the line on the right.

39. Two trains are traveling on parallel railroad tracks. When placed on a coordinate plane, the first train passes through the points (–1, 2) and (3, 4). The second train passes through the point (6, –4). Find an equation of the line containing the path of the second train.

40. Mariah used a coordinate plane to draw perpendicular lines for an art project. The equation of the first line she drew was $x + y = 8$. The second line contained the point (4, 7) and was perpendicular to the first line. Find an equation of the second line.

HA1-398: Graphing Linear Equations Using Slope and y-intercept or Slope and a Point

In order to sketch the graph of a line, $y = mx + b$, we need to find two distinct points that lie on the line.

If we are given one point and the slope of the line, we can find a second point on the line using the slope.

- First, locate and plot the point that is given.
- Next, use the slope $m = \frac{\text{rise}}{\text{run}}$ to find the second point.
- Finally, draw a line through the two points to sketch the graph.

Using these steps, you can find a second point on a graph if you know one point.

Example 1 Sketch the graph of the line that passes through the point (–2, 4) with slope, $m = -3$.

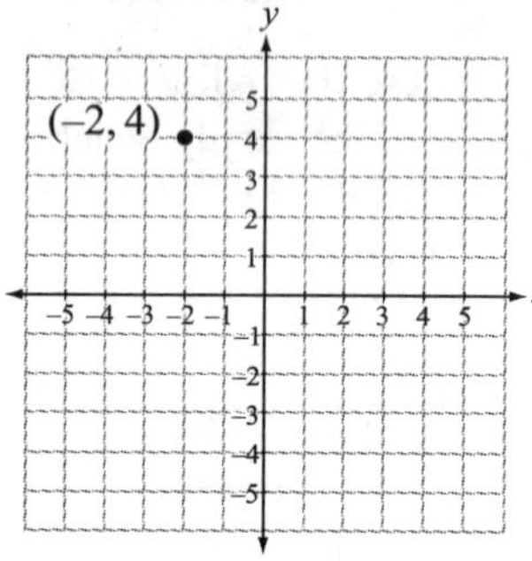

Step 1: Plot the given point, $(-2, 4)$.

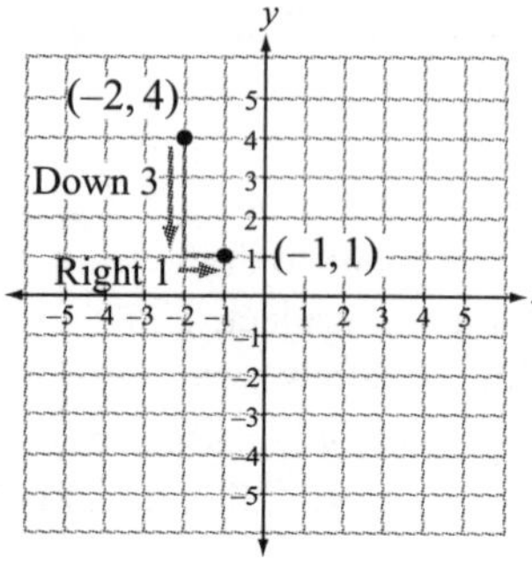

Step 2: Use the slope to find the second point. Since the slope, m, is equal to –3 or $-\frac{3}{1}$, move from the point $(-2, 4)$ down 3 units and to the right 1 unit to locate the second point, $(-1, 1)$.

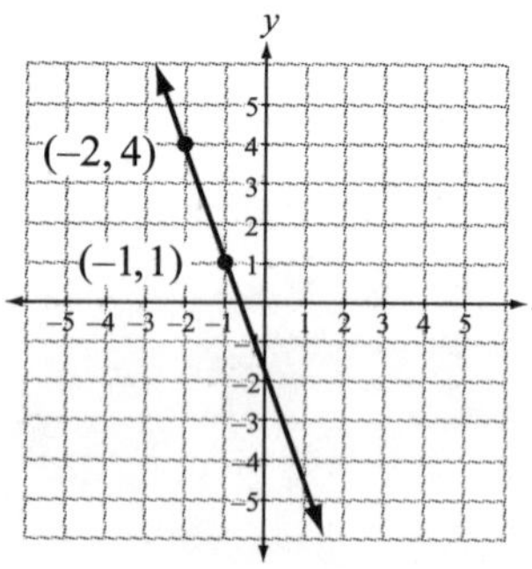

Step 3: Draw a line through the two points.

Answer: See graph in Step 3.

Example 2 Sketch the graph of the line $y = \frac{3}{2}x - 5$ using a point and the slope.

$m = \frac{3}{2}$

Step 1: Determine the slope of the line. Since the equation of the line is in slope-intercept form $y = mx + b$, we can see that the slope is $\frac{3}{2}$.

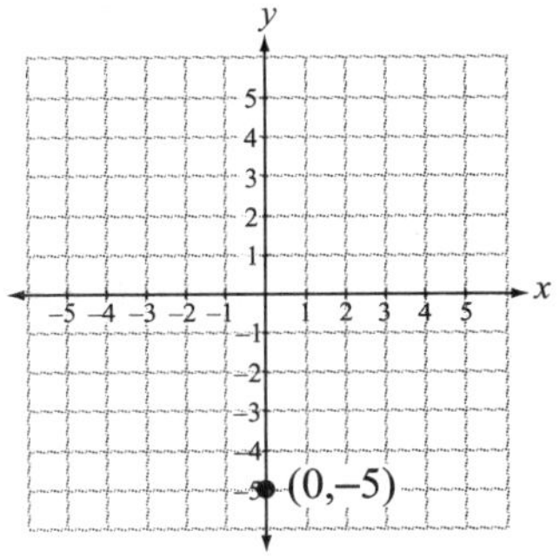

Step 2: Find a point on the line. Since the line is in slope-intercept form $y = mx + b$, we can see that the y-coordinate of the y-intercept is –5. Therefore, the point on the line is the y-intercept $(0, -5)$

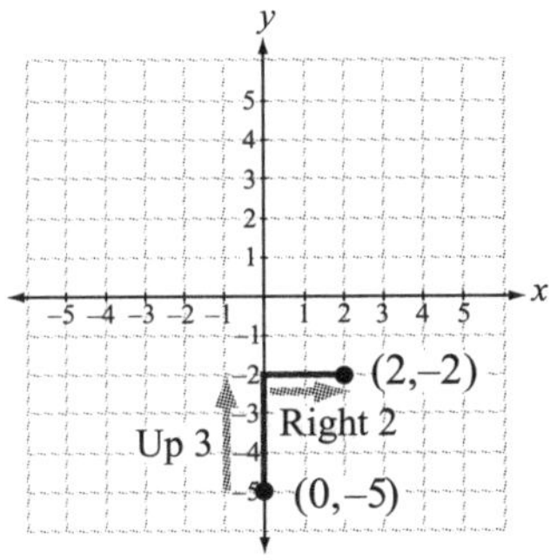

Step 3: Use the slope to find the second point. Since the slope, m, is equal to $\frac{3}{2}$, we move from the point $(0, -5)$ up 3 units and to the right 2 units to locate the second point, $(2, -2)$.

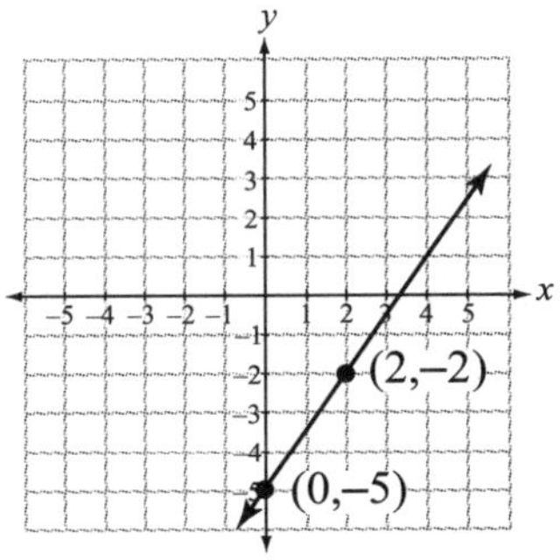

Step 4: Draw a line through the two points.

Answer: See graph in Step 4.

Example 3 Sketch the graph of the line that passes through $(-3, -5)$ with slope $\frac{7}{4}$.

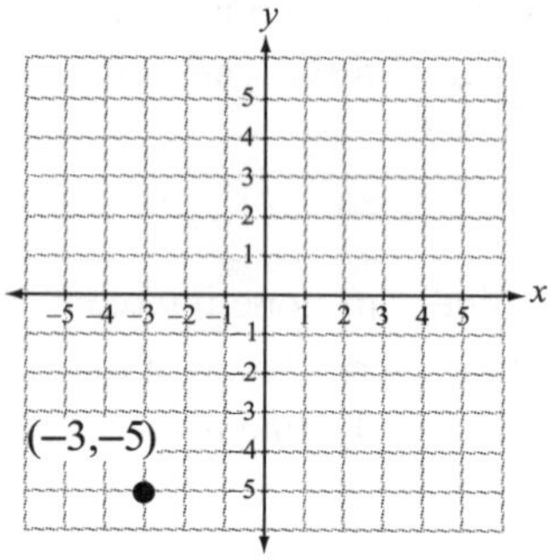

Step 1: Plot the given point, $(-3, -5)$.

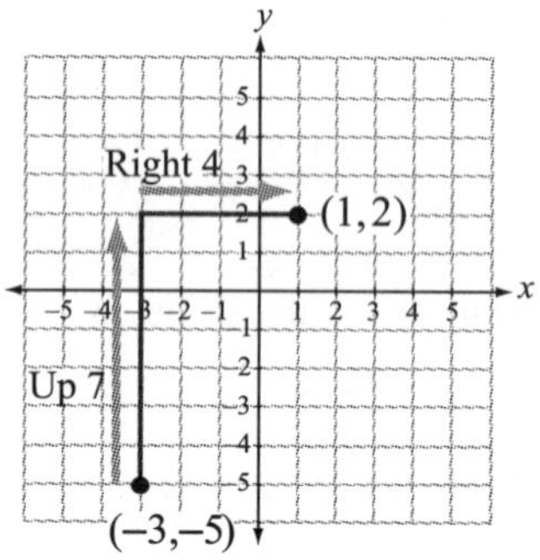

Step 2: Use the slope to find the second point. Since the slope, m, is $\frac{7}{4}$, we move from the point $(-3, -5)$ up 7 units and to the right 4 units to locate the second point, (1, 2).

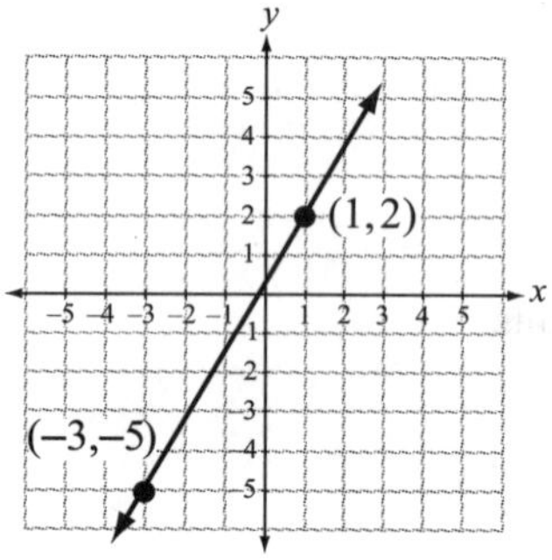

Step 3: Draw a line through the two points.

Answer: See graph in Step 3.

Example 4 Sketch the graph of the line that passes through $(-3, 1)$ with slope $-\frac{4}{5}$.

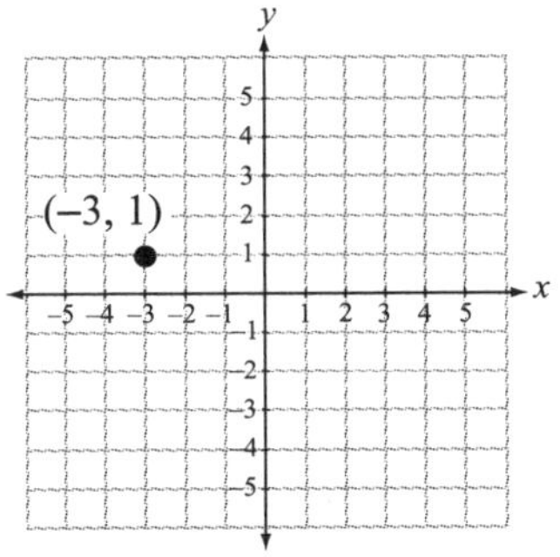

Step 1: Plot the given point $(-3, 1)$.

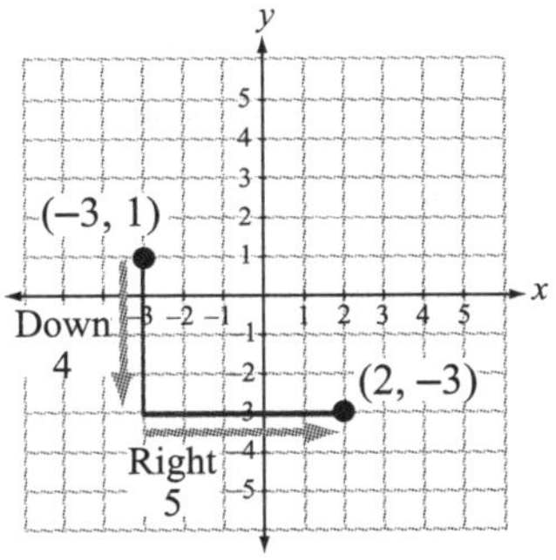

Step 2: Use the slope to find the second point. Since the slope, m, is equal to $-\frac{4}{5}$, we move from the point $(-3, 1)$ down 4 units and to the right 5 units to locate the second point, $(2, -3)$.

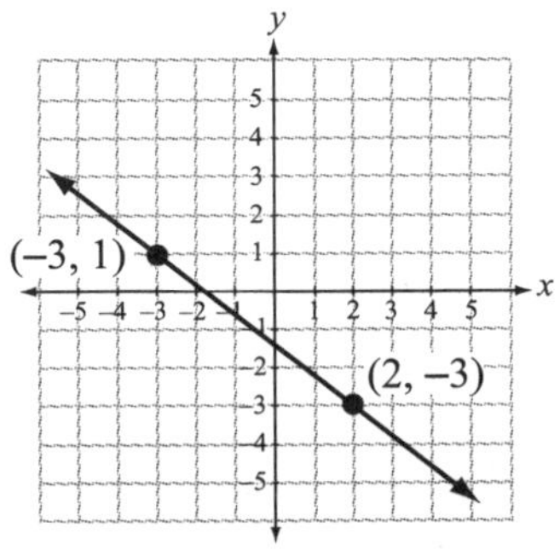

Step 3: Draw a line through the two points.

Answer: See graph in Step 3.

Example 5 Angela bought a condominium in 2003 for $120,000. The condominium increased in value by $5,000 every two years for the first seven years. Find the graph that represents the value of the condominium from the year 2003 to the year 2008.

$$m = \frac{5,000}{2}$$

Step 1: Determine the slope from the information given. The problem states that the condominium increases in value by $5,000 every 2 years. Therefore, the slope, m, is $\frac{5,000}{2}$.

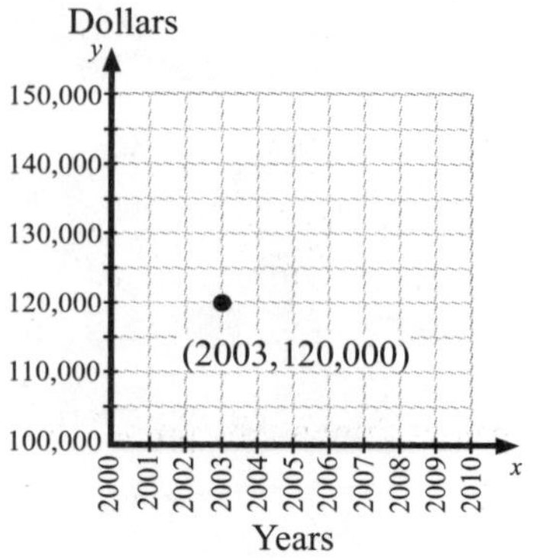

Step 2: Establish a point from the information given. Since the condominium was bought in the year 2003 for \$120,000, the point on the line is (2003, 120,000).

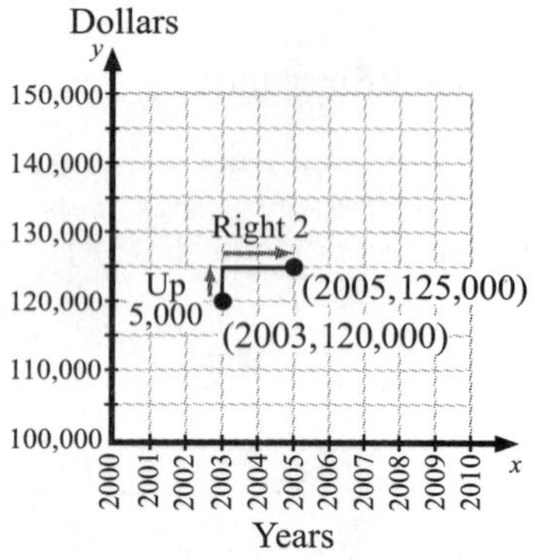

Step 3: Use the slope and point to find the second point. Since the slope is $\frac{5,000}{2}$, move up 5,000 units and to the right 2 units to locate the second point, (2005, 125,000).

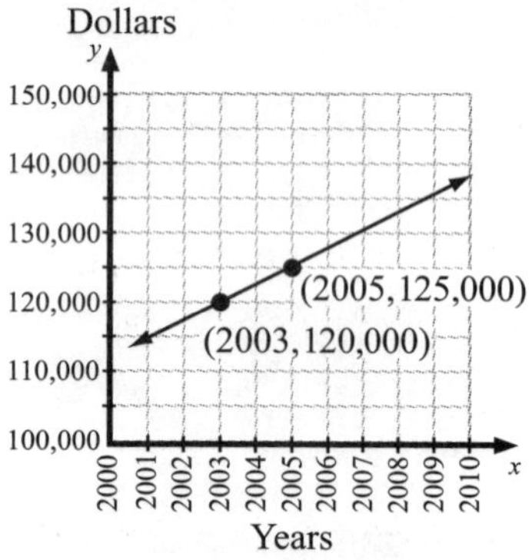

Step 4: Draw a line through the 2 points.

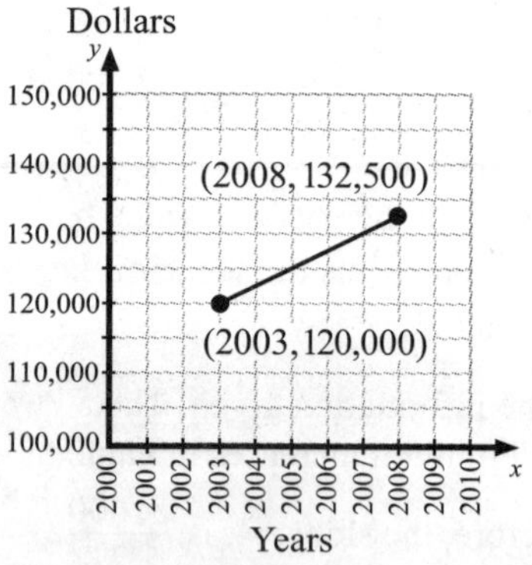

Step 5: Delete the sections to the left of 2003 and to the right of 2008.

Answer: See graph in Step 5.

Problem Set

1. Graph the line that passes through the point (3, 1) and has a slope of 2.

2. Graph the line that passes through the point (–1, 4) and has a slope of 2.

3. Graph the line that passes through the point (–3, –2) and has a slope of 3.

4. Graph the line that passes through the point (5, –4) and has a slope of 1.

5. Graph the line that passes through the point (–4, 0) and has a slope of 1.

6. Graph the line that passes through the point (–5, –6) and has a slope of 2.

7. Graph the line that passes through the point (–4,–7) and has a slope of 8.

8. Graph the line that passes through the point (10, –3) and has a slope of 4.

9. Graph the equation using the slope and *y*-intercept: $y = 2x - 4$.

10. Graph the equation using the slope and *y*-intercept: $y = 3x + 1$.

11. Graph the equation using the slope and *y*-intercept: $y = 4x - 5$.

12. Graph the equation using the slope and *y*-intercept: $y = 2x + 5$.

13. Graph the equation using the slope and *y*-intercept: $y = 6x - 1$.

14. Graph the equation using the slope and *y*-intercept: $y = 5x - 5$.

15. Graph the equation using the slope and *y*-intercept: $y = 7x - 3$.

16. Graph the line that passes through the point (3, –2) and has a slope of $\frac{4}{3}$.

17. Graph the line that passes through the point (–4,–6) and has a slope of $\frac{1}{3}$.

18. Graph the line that passes through the point (–7, 2) and has a slope of $\frac{3}{4}$.

19. Graph the line that passes through the point (–10, –1) and has a slope of $\frac{5}{6}$.

20. Graph the line that passes through the point (–2, 4) and has a slope of $\frac{2}{5}$.

21. Graph the line that passes through the point (3, 0) and has a slope of $\frac{2}{3}$.

22. Graph the line that passes through the point (2, 5) and has a slope of $\frac{1}{2}$.

23. Graph the line that passes through the point (4, –3) and has a slope of $\frac{4}{5}$.

24. Graph the equation using the slope and *y*-intercept: $y = -\frac{2}{3}x + 1$.

25. Graph the equation using the slope and *y*-intercept: $y = -\frac{1}{2}x - 4$.

26. Graph the equation using the slope and *y*-intercept: $y = -\frac{3}{5}x + 4$.

27. Graph the equation using the slope and *y*-intercept: $y = -\frac{1}{4}x - 5$.

28. Graph the equation using the slope and *y*-intercept: $y = -\frac{1}{3}x + 2$.

29. Graph the equation using the slope and y-intercept: $y = -\frac{1}{6}x - 3$.

30. Graph the equation using the slope and y-intercept: $y = -\frac{2}{5}x + 3$.

31. Sean bought a new motorcycle in 2000 for \$11,500. The value of the motorcycle depreciates \$2,000 per year for the first three years. Find the graph that represents the depreciation of the motorcycle from the year 2000 to the year 2003.

32. Carolina bought a new car in 2000 for \$18,500. The value of the car depreciates \$2,500 per year for the first three years. Find the graph that represents the depreciation of value of the car from the beginning of 2000 to the beginning of 2003.

33. Bryant bought a new sports car in 2002 for \$45,000. The value of the car depreciates \$2,500 per year for the first three years. Find the graph that represents the depreciation of the value of the sports car from the beginning of 2002 to the beginning of 2005.

34. The initial temperature of a cup of coffee is 200° F. The cup of coffee is left on a table to cool down. The temperature of the coffee cools down by 40° per hour for the first three hours. Find the graph that represents the temperature of the cup of coffee over the first three hours.

35. A particular stock is said to increase \$50 each year, on average. Your initial investment in this stock is \$500. Find the graph that represents the value of your stock over the next ten years.

36. A particular stock is said to increase \$50 each year, on average. Your initial investment in this stock is \$600. Find the graph that represents the value of your stock over the next six years.

37. Sharon is saving her babysitting money for college. She currently has \$50 saved at home in her piggy bank. She plans to babysit this summer, charging \$5 per hour. Find the graph that represents the amount she has earned after babysitting 10 hours during the summer.

38. Marissa is saving her babysitting money for college. She currently has \$40 saved at home in her piggy bank. She plans to babysit this summer, charging \$5 per hour. Find the graph that represents the amount she has after babysitting 10 hours during the summer.

39. An organization is raising money throughout the school year. Their beginning balance is \$350. They anticipate earnings of \$125 for each fundraiser. Find the graph that represents the balance amount that they anticipate having after three fundraisers.

40. Over the past 5 years the total amount of rainfall in a certain region has been decreasing at the rate of $\frac{1}{3}$ inch per year. At the beginning of the 5-year period, the amount of rainfall per year was 6 inches. Find the graph of the rainfall amounts in this region over the 5-year period.

HA1-401: How Variations of "m" and "b" Affect the Graph of $y = mx + b$

Linear equations written in slope-intercept form, ($y = mx + b$), have graphs with slope m and y-intercept (0, b). The slope is written in the form $m = \frac{\text{rise}}{\text{run}}$. The "rise" refers to the vertical change in the y-coordinates. The "run" refers to the horizontal change in the x-coordinates. For example, the line $y = 3x + 2$ has slope $m = 3$ and y-intercept (0, 2).

The slope determines the direction of the line. In other words, the slope measures how fast the line rises or falls as it moves from left to right on the graph. Lines with positive slopes rise from left to right. Lines with negative slopes fall from left to right.

The slope also determines the inclination of the line. If the slope, m, is equal to 1, the line forms a $45°$ angle with respect to the x-axis. If the slope is greater than 1, the inclination of the line increases and the line rises toward the y-axis. If the slope is less than 1, the inclination of the line decreases and the line falls toward the x-axis. The slope of a horizontal line is neither positive nor negative and it neither falls nor rises. The slope of the vertical line is always undefined.

The y-coordinate, b, of the linear equation $y = mx + b$ determines the y-intercept of the line. If b is positive, the line is shifted b units up on the y-axis. If b is negative, the line is shifted b units down on the y-axis.

Example 1 Determine the direction of the equation $y = -5x$.

$m = -5$

Step 1: Determine the slope of the line.

Step 2: Since the slope is negative, the line falls downward from left to right.

Answer: The line falls downward from left to right.

Example 2 Sketch the graphs of the three lines $y = 3x + 2$, $y = x + 2$, and $y = -4x + 2$, and observe the effect of changing the slope of the line.

y-intercepts: (0, 2)

Step 1: Notice that these linear equations are all in slope-intercept form and have the same y-intercept.

$m_1 = 3$

$m_2 = 1$

$m_3 = -4$

Step 2: Notice the slope of each line. The first line, $y = 3x + 2$, has slope 3. The second line, $y = x + 2$, has slope 1 and the last line, $y = -4x + 2$, has slope -4.

Line 1: $y = 3x + 2$
First point: (0, 2)
Second point: (1, 5)

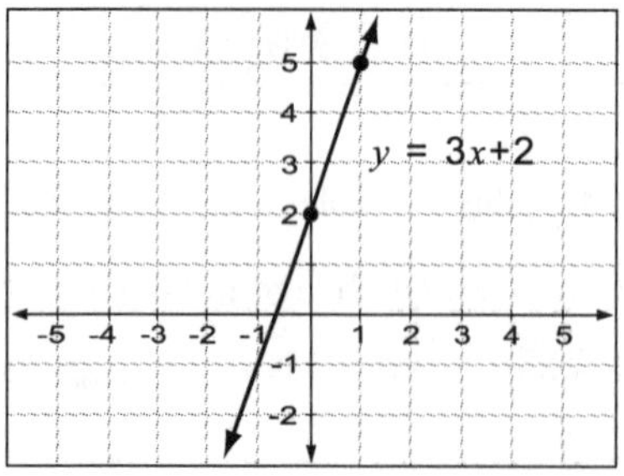

Step 3: To graph the first line, $y = 3x + 2$, start at the y-intercept, (0, 2), and move up 3 units and to the right 1 unit (since the slope is $m = 3 = \frac{3}{1}$) to locate the second point, (1, 5). Then, draw a line through the points.

Line 2: $y = x + 2$
First point: (0, 2)
Second point: (1, 3)

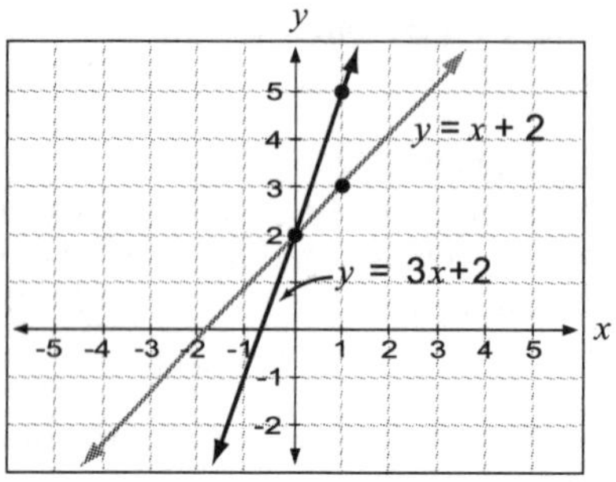

Step 4: To graph the next line, $y = x + 2$, start at the y-intercept (0, 2) and move up 1 unit and to the right 1 unit (since the slope of the second line is $m = 1 = \frac{1}{1}$) to locate the second point (1, 3). Then, draw a line through the points.

Line 3: $y = -4x + 2$
First point: (0, 2)
Second point: $(1, -2)$

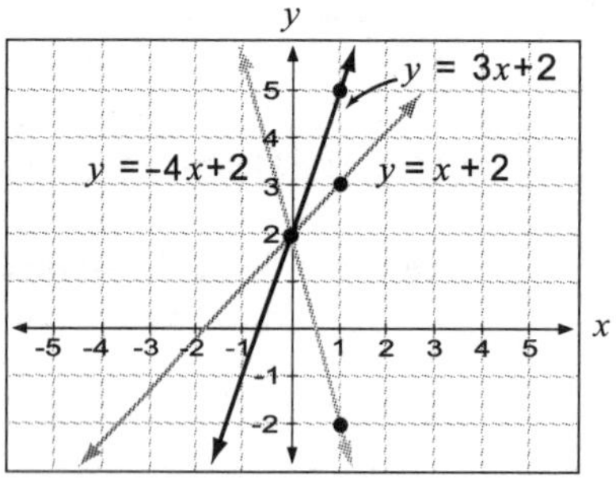

Step 5: To graph the last line, $y = -4x + 2$, start at the point (0, 2) and move down 4 units and to the right 1 unit (since the slope of the second line is $m = -4 = -\frac{4}{1}$) to locate the second point $(1, -2)$. Then, draw a line through the points.

Answer: As the absolute value of the slope gets larger, the line is closer to becoming a vertical line.

If the slope is held constant, then variations in b correspond to parallel shifts of the line that pass through their new y-intercept. Consider the following example.

Example 3 Sketch the lines $y = 3x + 2$, $y = 3x$, and $y = 3x - 4$ on the same coordinate system and observe the effect of changing the y-intercept.

Slope = $m = 3 = \frac{3}{1}$

Step 1: Notice that each of these lines is given in slope-intercept form and that each line has the same slope.

$b_1 = 2$
$b_2 = 0$
$b_3 = -4$

Step 2: Determine the y-coordinates of the y-intercepts for the three lines.

y_1 – intercept (0, 2)

y_2 – intercept (0, 0)

y_3 – intercept (0, -4)

Step 3: Determine the y-intercepts of the three lines. The first line has the y-intercept (0, 2) since $b_1 = 2$. The second line has the y-intercept (0, 0) since $b_2 = 0$. The last line has the y-intercept (0, –4) since $b_3 = -4$.

Line 1: $y = 3x + 2$
First point: (0, 2)
Second point: (1, 5)

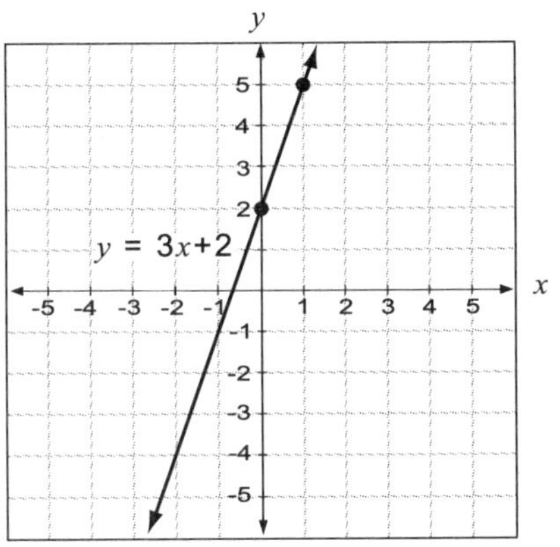

Step 4: To graph the first line, $y = 3x + 2$, start by plotting the point (0, 2) and move up 3 units and to the right 1 unit to the point (1, 5).

Line 2: $y = 3x$
First point: (0, 0)
Second point: (1, 3)

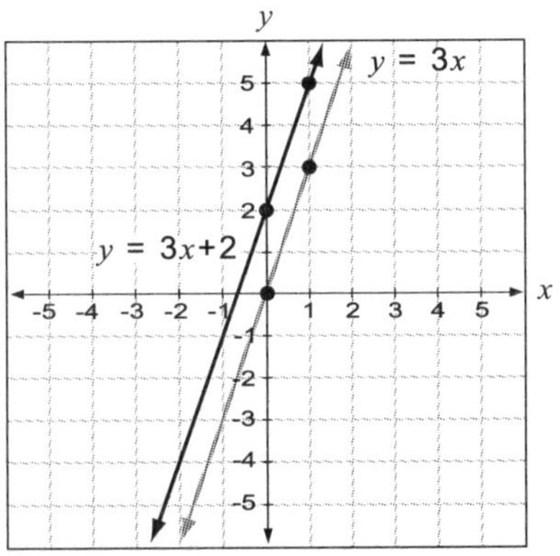

Step 5: To graph the second line, $y = 3x$, start by plotting the point (0, 0) and move up 3 units and to the right 1 unit to the point (1, 3).

Line 3: $y = 3x - 4$
First point: $(0, -4)$
Second point: $(1, -1)$

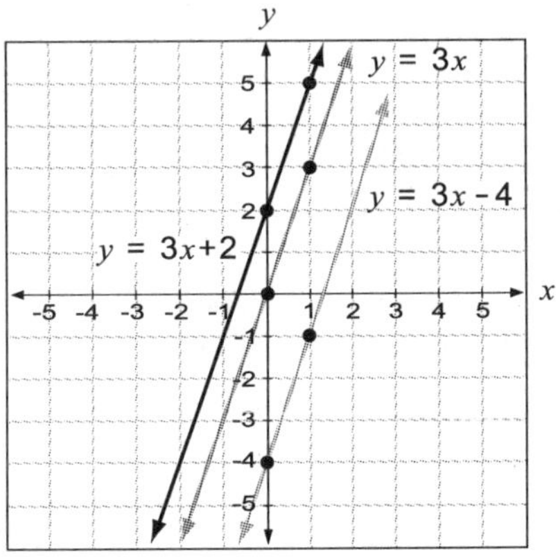

Step 6: To graph the last line, $y = 3x - 4$, start by plotting the point (0, –4) and move up 3 units and to the right 1 unit to the point (1, –1).

Answer: Note that the lines are parallel. The effect of changing the y-intercept causes the graphs to shift up or down.

Example 4 For the given equation of a line, $y = -\frac{2}{7}x + 5$, use the slope to estimate the measure of the angle formed with respect to the x-axis.

$m = -\frac{2}{7}$

Step 1: Determine the slope of the line.

$\left|-\frac{2}{7}\right| < 1$

Step 2: Take the absolute value of the slope and see if it is less than, greater than, or equal to 1.

Step 3: Since the absolute value of the slope is less than one, the angle is less than 45°.

Answer: The measure of the angle formed with respect to the x-axis is less than 45°.

Example 5 Compare the slopes and y-intercepts of the lines shown in the given graph.

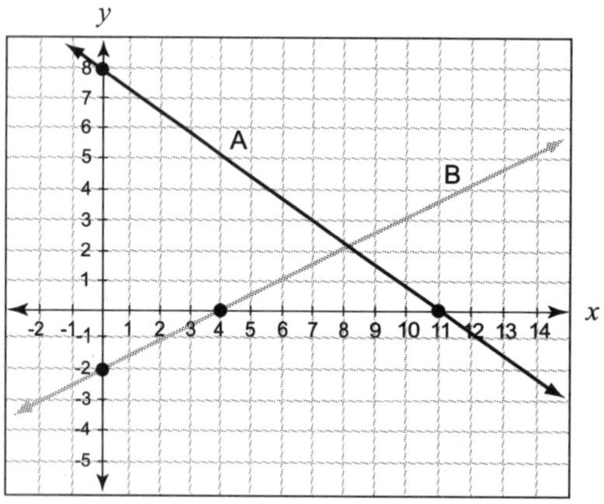

$m_A < 0$

$m_B > 0$

$m_A < 0 < m_B$

Step 1: The slope of line A is negative since it falls from left to right; the slope of line B is positive since it rises from left to right. Combining the statements, we see that the slope of line B is greater than zero, which is greater than the slope of line A.

$b_B < b_A$

Step 2: The value of b is the y-coordinate of the y-intercept. The y-intercept for line A is (0, 8) and the y-intercept for line B is (0, –2). Therefore, $b_A = 8$ and $b_B = -2$. This means that b_B is smaller than b_A.

$m_A < 0$

$m_B > 0$

$m_A < 0 < m_B$

$b_B < b_A$

Answer: The slope of line B is greater than the slope of line A, and the y-coordinate of line B is smaller than the y-coordinate of line A.

Problem Set

Evaluate.

1. Which of the following equations describes a line that rises upward from left to right and has positive direction?

$y = 0x$ $\quad$ $y = -10x$

$y = 10$ $\quad$ $y = 10x$

2. Which equation matches the line graphed below?

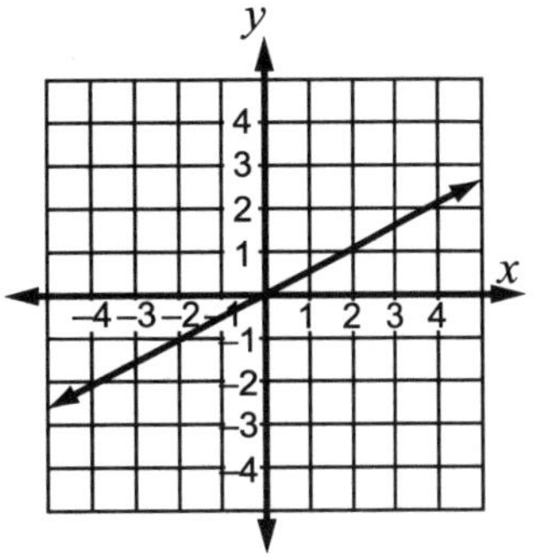

$y = -2x$ $\quad$ $y = \frac{1}{2}x$

$y = 0$ $\quad$ $y = -\frac{1}{2}x$

3. Which of the following equations describes a line that falls downward from left to right and has negative direction?

$y = -8$ $\quad$ $y = 8x$,

$y = \frac{3}{8}x$ $\quad$ $y = -8x$

4. Which equation matches the line graphed below?

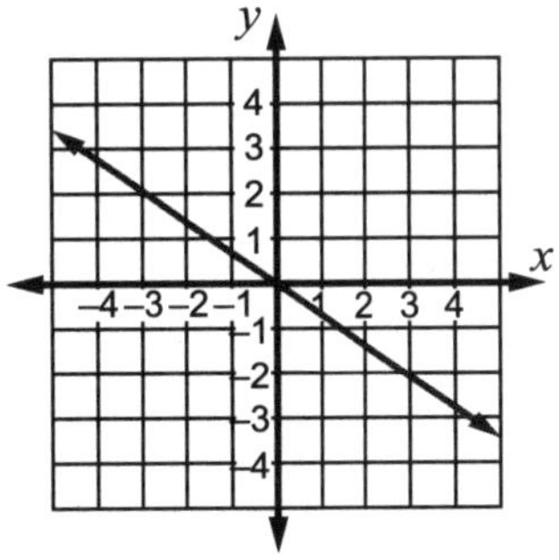

$y = 2x$ $\quad$ $y = -\frac{2}{3}x$

$y = -2$ $\quad$ $y = \frac{3}{2}x$

5. Which of the following equations describes a line that rises upward from left to right and has positive direction?

$y = -\frac{2}{7}x$ $\quad$ $y = 0$

$y = -\frac{1}{7}x$ $\quad$ $y = \frac{3}{7}x$

6. Which equation matches the line graphed below?

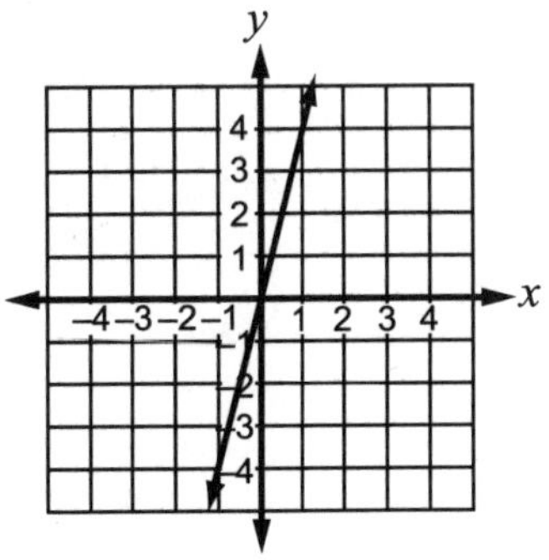

$y = -4x$ $\quad$ $y = 4x$

$y = -\frac{1}{4}x$ $\quad$ $y = 0$

7. Which of the following equations describes a line that falls downward from left to right and has negative direction?

$y = \frac{1}{7}x$ $\qquad$ $y = -7x$

$y = 7x$ $\qquad$ $y = -7$

8. Which equation matches the line graphed below?

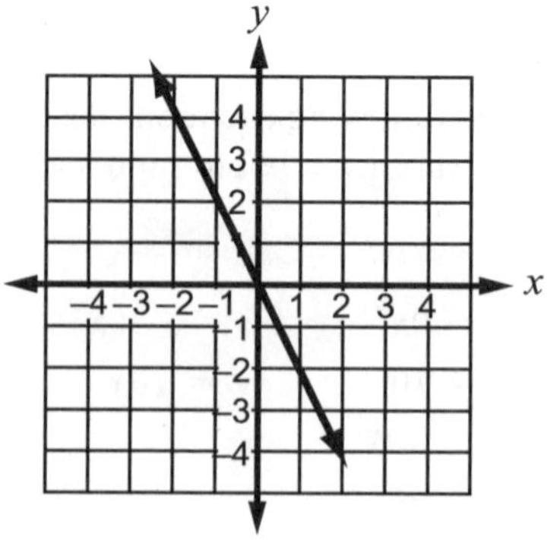

$y = 2x$ $\qquad$ $y = -2$

$y = -2x$ $\qquad$ $y = \frac{1}{2}x$

9. Which of the following equations describes a line that rises upward from left to right and has positive direction?

$y = -x$ $\qquad$ $y = x$

$y = -2x$ $\qquad$ $y = 1$

10. Which equation matches the line graphed below?

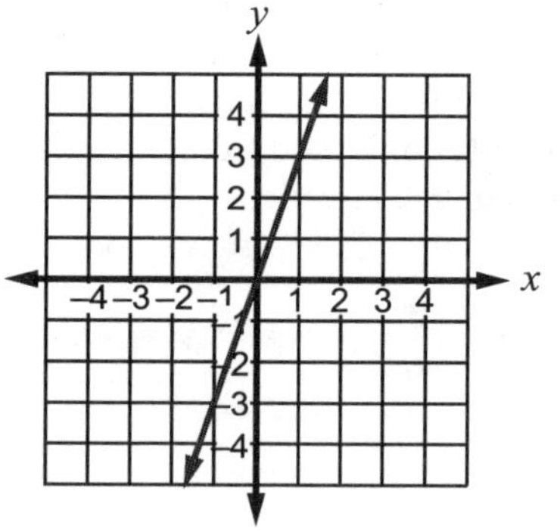

$y = \frac{8}{3}x$ $\qquad$ $y = -2x$

$y = -\frac{5}{2}x$ $\qquad$ $y = 0$

11. Which of the following equations describes a line that falls downward from left to right and has negative direction?

$y = -\frac{2}{7}$ $\qquad$ $y = -\frac{2}{7}x$

$y = \frac{7}{2}x$ $\qquad$ $y = \frac{2}{7}x$

12. Which equation matches the line graphed below?

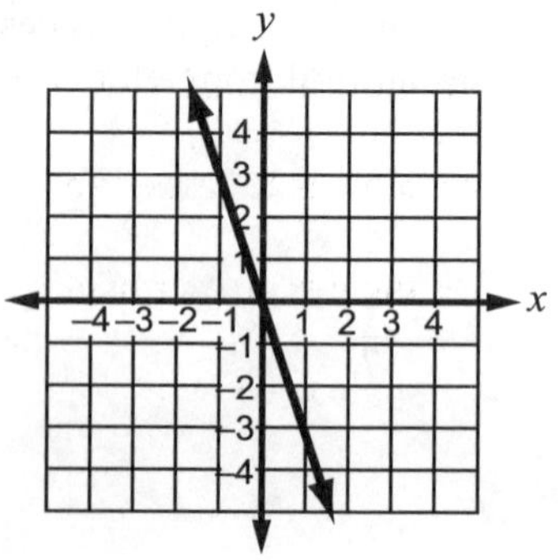

$y = -\frac{10}{3}$ $\qquad$ $y = -\frac{10}{3}x$

$y = \frac{10}{3}x$ $\qquad$ $y = 3x$

13. Which of the following equations describes a line that rises upward from left to right and has positive direction?

$y = -\frac{1}{9}x$ $\qquad$ $y = -9x$

$y = 9x$ $\qquad$ $y = 9$

14. Which equation matches the line graphed below?

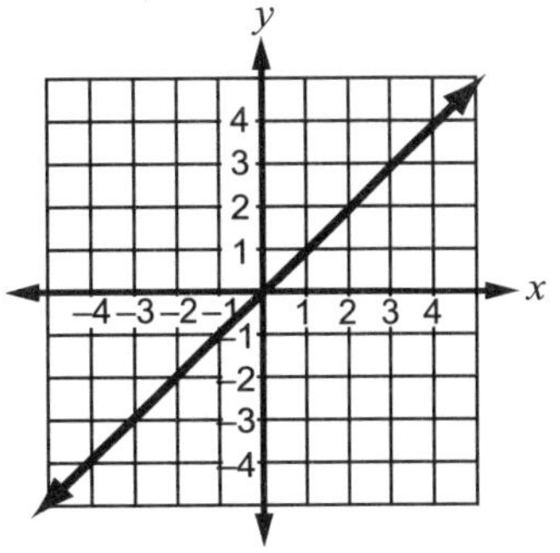

$y = x$ $\qquad$ $y = -x$

$y = 0x$ $\qquad$ $y = 1$

15. Which of the following equations describes a line that neither falls nor rises from left to right?

$y = 10x$ $\qquad$ $x = 0$

$y = 0$ $\qquad$ $y = -\frac{1}{10}x$

16. Use the given graph of the line $y = -\frac{1}{4}x$ as a point of reference to describe how we may obtain the graph of the line $y = -\frac{1}{4}x - 5$.

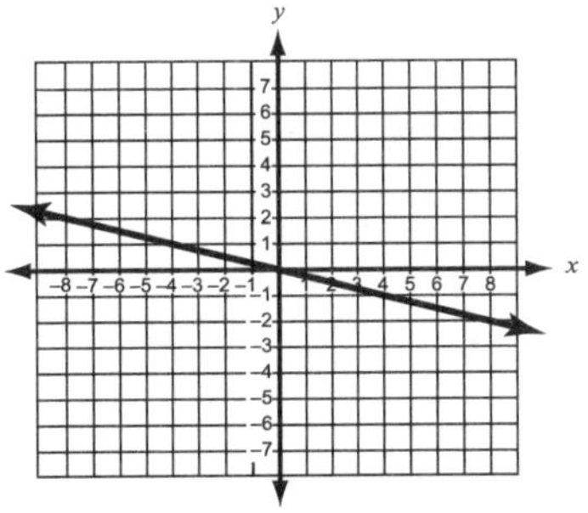

17. For the given equation of a line, $y = -\frac{1}{4}x$, use the slope to estimate the measure of the angle formed with respect to the x-axis.

The angle is equal to $0°$.

The angle is less than $45°$.

The angle is greater than $45°$.

The angle is equal to $45°$.

18. Use the given graph of the line $y = 2x$ as a point of reference to describe how we may obtain the graph of the line $y = 2x + 6$.

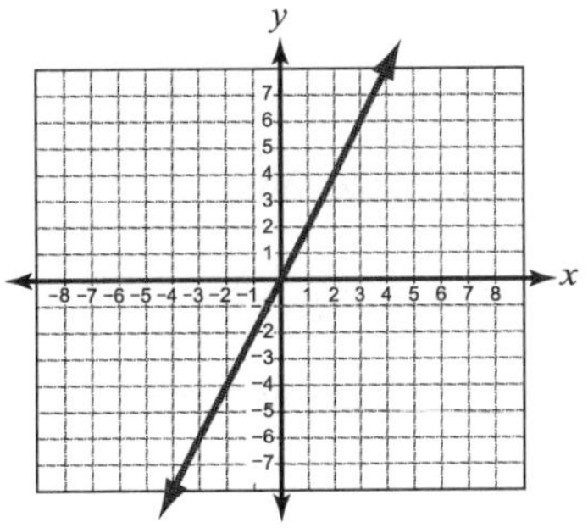

19. For the given equation of a line, $y = x - 10$, use the slope to estimate the measure of the angle formed with respect to the x-axis.

The angle is less than $45°$.

The angle is equal to $0°$.

The angle is greater than $45°$.

The angle is equal to $45°$.

20. Use the given graph of the line $y = -\frac{1}{2}x$ as a point of reference to describe how we may obtain the graph of the line $y = -\frac{1}{2}x - 4$.

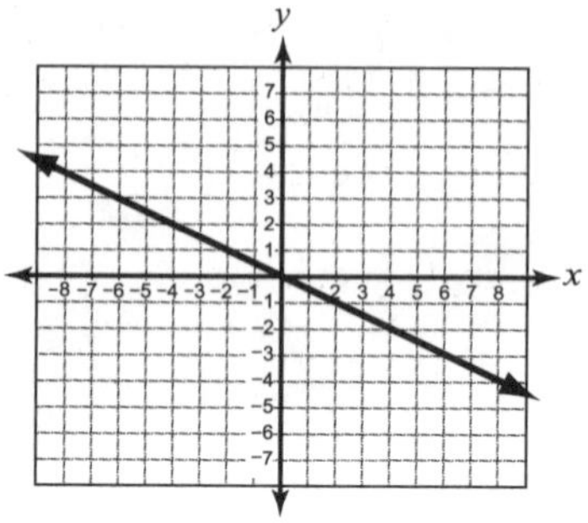

21. For the given equation of a line, $y = -2x + 3$, use the slope to estimate the measure of the angle formed with respect to the x-axis.

The angle is equal to $45°$.

The angle is equal to $0°$.

The angle is greater than $45°$.

The angle is less than $45°$.

22. Use the given graph of the line $y = -x$ as a point of reference to describe how we may obtain the graph of the line $y = -x + 3$.

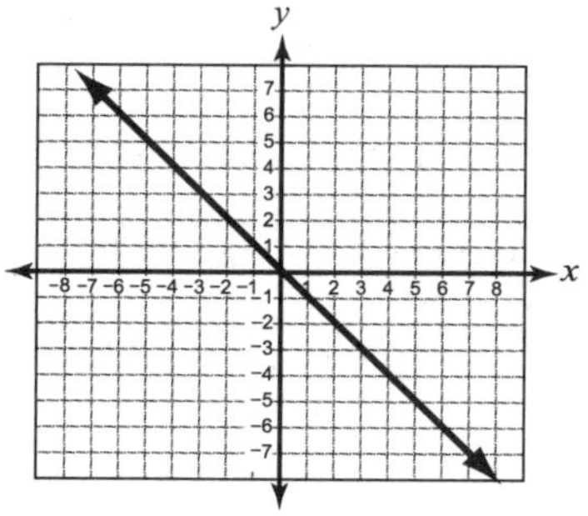

23. For the given equation of a line, $y = \frac{7}{5}x$, use the slope to estimate the measure of the angle formed with respect to the x-axis.

The angle is greater than $45°$.

The angle is equal to $45°$.

The angle is less than $45°$.

The angle is equal to $0°$.

24. Use the given graph of the line $y = x$ as a point of reference to describe how we may obtain the graph of the line $y = x - 6$.

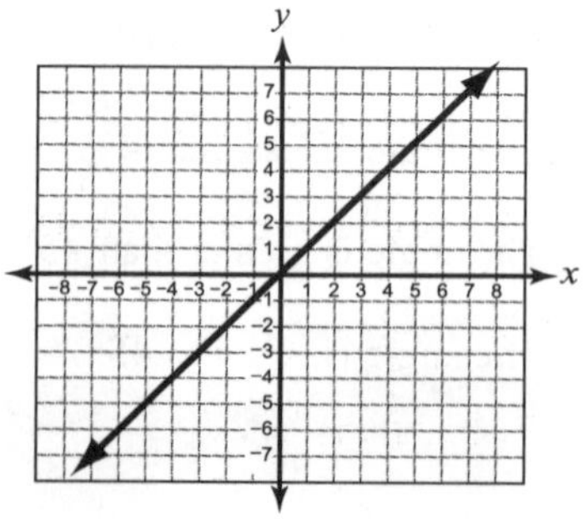

25. For the given equation of a line, $y = -x - 4$, use the slope to estimate the measure of the angle formed with respect to the x-axis.

The angle is less than 45°.

The angle is greater than 45°.

The angle is equal to 45°.

The angle is equal to 0°.

26. Use the given graph of the line $y = \frac{1}{2}x$ as a point of reference to describe how we may obtain the graph of the line $y = \frac{1}{2}x + \frac{8}{3}$.

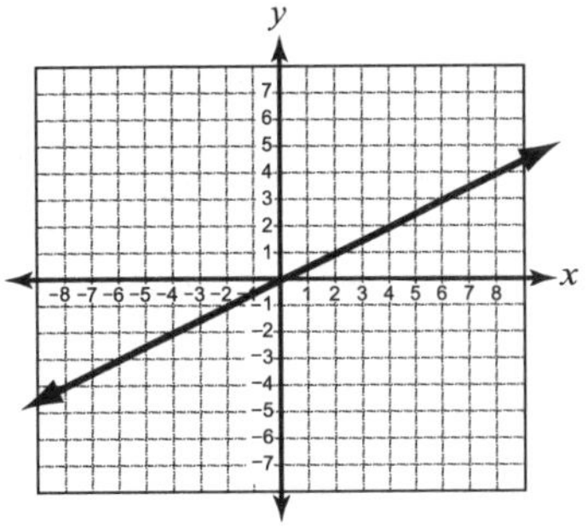

27. For the given equation of a line, $y = \frac{1}{8}x$ use the slope to estimate the measure of the angle formed with respect to the x-axis.

The angle is equal to 45°.

The angle is greater than 45°.

The angle is equal to 0°.

The angle is less than 45°.

28. Use the given graph of the line, $y = -\frac{1}{5}x$, as a point of reference to describe how we may obtain the graph of the line $y = -\frac{1}{5}x - \frac{10}{3}$.

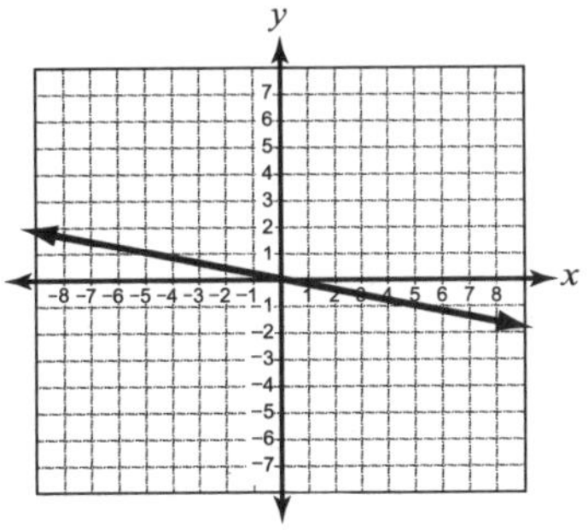

29. For the given equation of a line, $y = -3x - 12$, use the slope to estimate the measure of the angle formed with respect to the x-axis.

The angle is equal to 45°.

The angle is greater than 45°.

The angle is equal to 0°.

The angle is less than 45°.

30. Use the given graph of the line $y = 3x$ as a point of reference to describe how we may obtain the graph of the line $y = 3x + 3$.

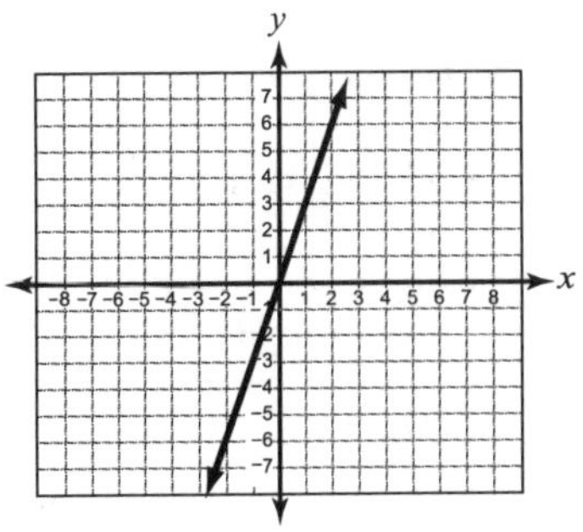

For problems 31-40, choose from the four answer sets given.

31. Compare the slopes and y-intercepts of the given graph.

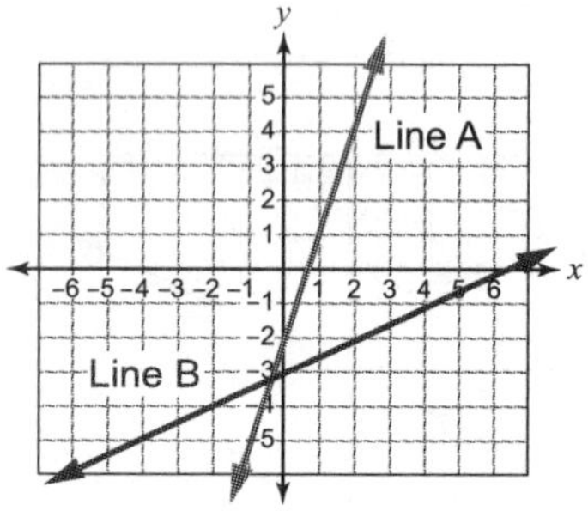

$m_A > m_B > 0$

$b_A > b_B$

$|m_A| > |m_B|$ $\quad b_A > b_B$

$m_A > 0 > m_B$

$m_B > m_A > 0$

$b_B < b_A$

$|m_A| < |m_B|$ $\quad 0 > m_B$

$0 > m_A$ $\quad b_A > b_B$

---

32. Compare the slopes and y-intercepts of the given graph.

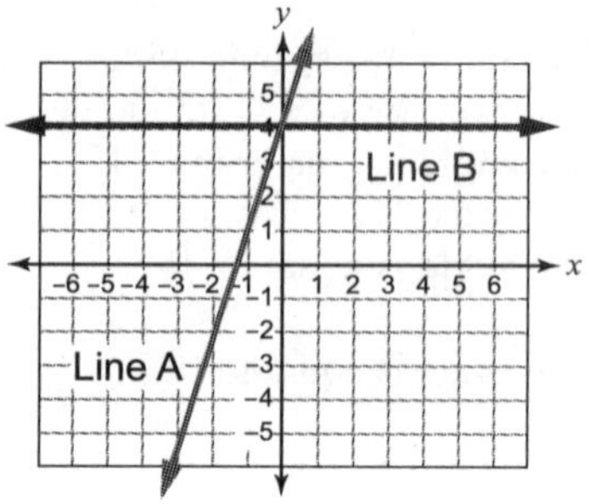

$|m_A| < |m_B|$ $\quad 0 = m_B$

$0 < m_A$ $\quad b_A = b_B = 4$

$m_B > m_A$

$b_B = b_A = 4$

$m_A > 0$ $\quad |m_A| > |m_B|$

$m_B = 0$ $\quad b_A > b_B$

$m_A > 0$ $\quad m_A > m_B$

$m_B = 0$ $\quad b_A = b_B = 4$

---

33. Compare the slopes and y-intercepts of the given graph.

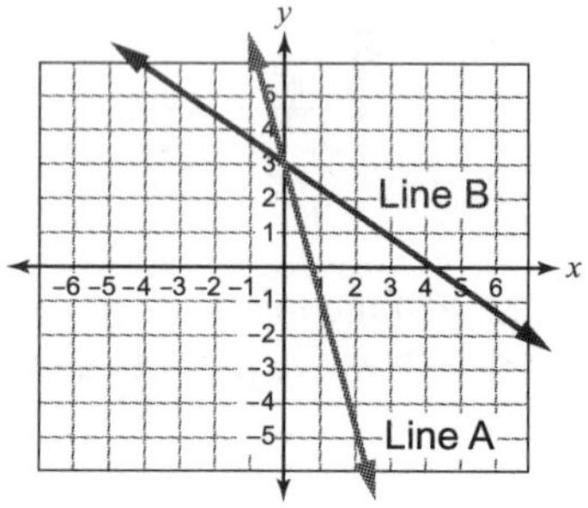

$m_A > m_B > 0$

$b_A = b_B = 3$

$|m_A| > |m_B|$ $\quad m_B < 0$

$m_A < 0$ $\quad b_A = b_B = 3$

$m_B > m_A > 0$

$b_B > b_A$

$m_A < 0$ $\quad |m_B| > |m_A|$

$m_B < 0$ $\quad b_A = b_B = 3$

---

34. Compare the slopes and y-intercepts of the given graph.

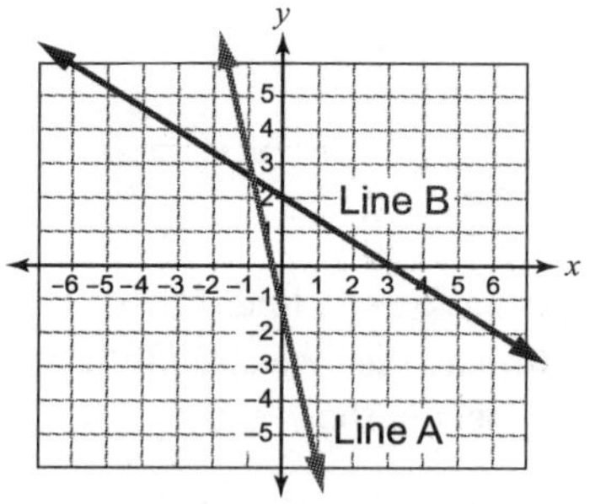

$|m_A| > |m_B|$ $\quad m_B < 0$

$m_A < 0$ $\quad b_B > b_A$

$m_B > m_A > 0$

$b_B > b_A$

$m_A > m_B > 0$

$b_A < b_B$

$|m_A| > |m_B|$ $\quad b_A < b_B$

$m_A > 0 > m_B$

---

35. Compare the slopes and y-intercepts of the given graph.

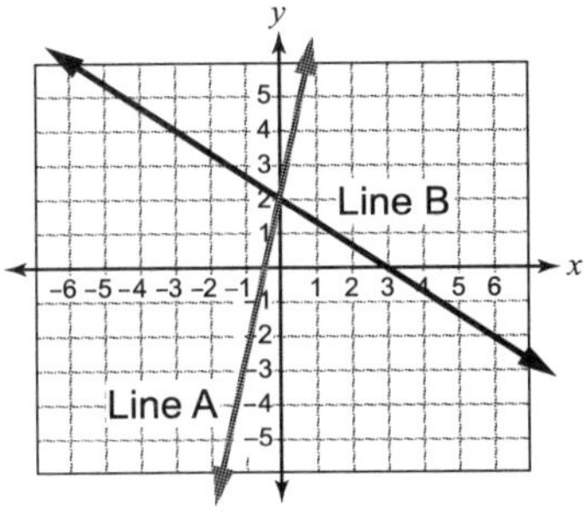

$|m_A| < |m_B|$ $\quad 0 > m_B$
$0 > m_A$ $\quad b_A < b_B$

$m_A > 0 > m_B$ $\quad b_A = b_B = 2$
$|m_A| > |m_B|$

$m_B > 0 > m_A$
$b_A = b_B = 2$

$|m_A| < |m_B|$ $\quad b_B = b_A = 2$
$m_B > 0 > m_A$

---

36. Compare the slopes and y-intercepts of the given graph.

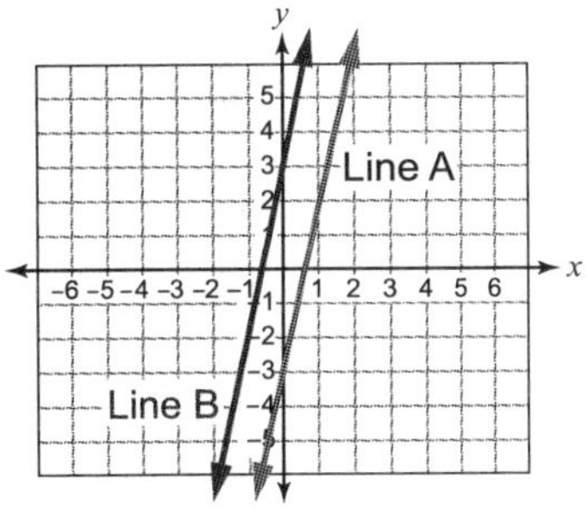

$m_A = m_B$ $\quad m_B < 0$
$m_A < 0$ $\quad b_B > b_A$

$m_A > 0$ $\quad m_A = m_B$
$m_B > 0$ $\quad b_B > b_A$

$m_A > m_B > 0$
$b_A > b_B$

$m_B > m_A > 0$
$b_B > b_A$

---

37. Compare the slopes and y-intercepts of the given graph.

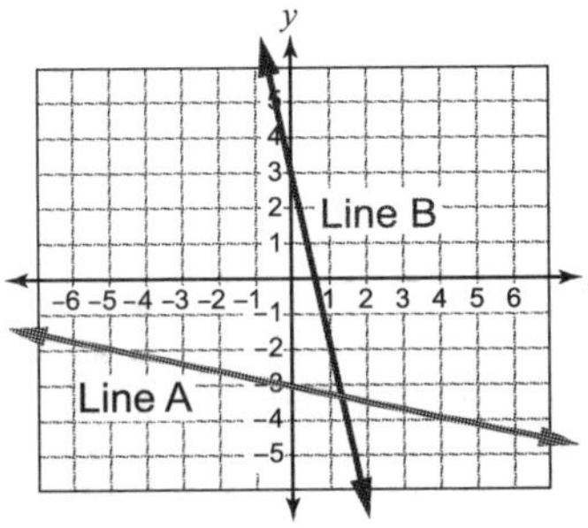

$m_A < m_B < 0$
$b_A < b_B$

$|m_B| > |m_A|$ $\quad m_B < 0$
$m_A < 0$ $\quad b_B > b_A$

$|m_A| < |m_B|$ $\quad b_A < b_B$
$m_A > 0 > m_B$

$m_B > m_A > 0$
$b_B > b_A$

---

38. Compare the slopes and y-intercepts of the given graph.

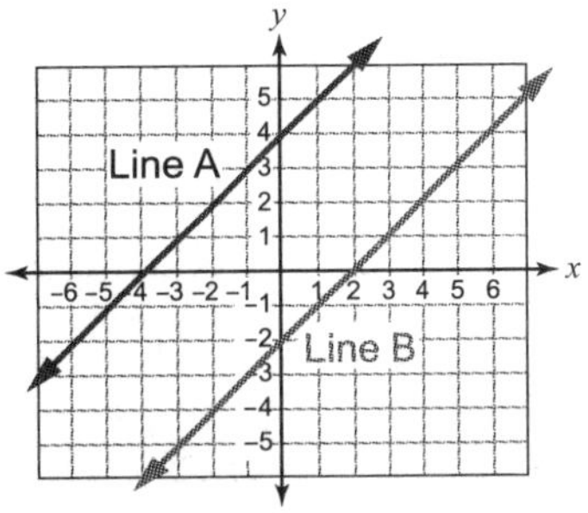

$m_A > 0$ $\quad m_A = m_B$
$m_B > 0$ $\quad b_A > b_B$

$m_A = m_B$ $\quad m_B > 0$
$m_A > 0$ $\quad b_B > b_A$

$m_B > m_A > 0$
$b_B > b_A$

$m_A > m_B > 0$
$b_A < b_B$

---

39. Compare the slopes and y-intercepts of the given graph.

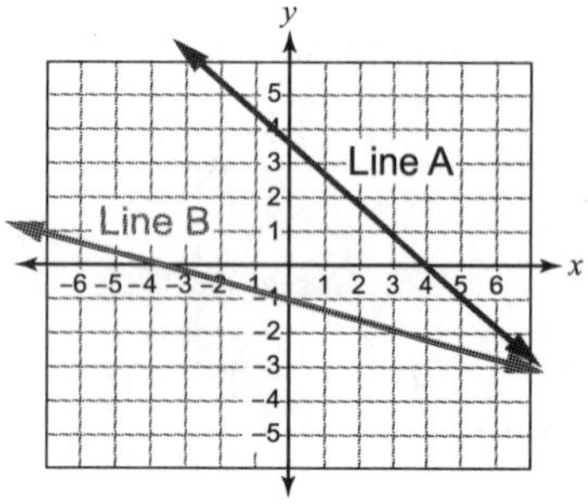

$\lvert m_A \rvert < \lvert m_B \rvert$	$0 > m_B$
$0 > m_A$	$b_A > b_B$

$m_B > m_A > 0$

$b_B > b_A$

$m_A < 0$	$\lvert m_A \rvert > \lvert m_B \rvert$
$m_B < 0$	$b_A > b_B$

$m_A > m_B > 0$

$b_A > b_B$

40. Compare the slopes and y-intercepts of the given graph.

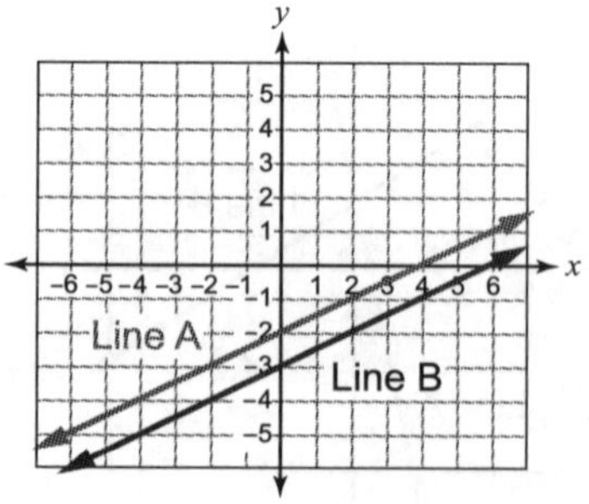

$m_A > 0$	$m_A = m_B$
$m_B < 0$	$b_B > b_A$

$m_A > 0$	$m_A = m_B$
$m_B > 0$	$b_A > b_B$

$m_B > m_A > 0$

$b_B < b_A$

$m_A = m_B$	$m_B < 0$
$m_A < 0$	$b_B < b_A$

HA1-405: Determining an Equation of a Line Given the Slope and Coordinates of One Point

You have already learned about slope, and you know that there are two ways to calculate the slope of a line: using the slope formula, or counting spaces with the $\frac{\text{rise}}{\text{run}}$ method. You have also seen in previous lessons that, when you are solving problems concerning lines, you may be given an equation for the line, or two points that are on the line, or the graph of the line, or maybe even a combination. In cases where you are given a graph or a pair of points, you may find it helpful to see the equation. In this lesson, you will learn how to find the equation of a line if you know a point on the line and the slope.

Before you begin, let's review some information about lines. Remember that there are many forms for the equation of a line. The form that you will be using in this lesson is slope-intercept form. The following definition will help you remember what this form looks like.

Slope-Intercept Form	The slope-intercept form for the equation of a line is $y = mx + b$, where m is the slope of the line, and b is the y-intercept.

Recall that the y-intercept of a line is the point where the graph of the line crosses the y-axis. Now, suppose that you want to find an equation for the line that goes through the point (5, 3) and has slope $-\frac{3}{5}$. Begin with the slope-intercept form. You know that m represents the slope of the line, so you can substitute the slope, $-\frac{3}{5}$, for m. The x and the y represent the coordinates of points on the line; therefore, you can substitute the coordinates of the point for these variables. You can use 5 for x and 3 for y. Carrying out these substitutions gives you:

$$y = mx + b$$
$$3 = \left(-\frac{3}{5}\right)(5) + b$$

The only thing that you don't know is b, but you can solve this equation for b as follows:

$$3 = \left(-\frac{3}{5}\right)(5) + b$$
$$3 = -3 + b$$
$$6 = b$$

The y-intercept, b, for this line is 6. To finish the problem, you must write down the actual equation for the line. You have all the information you need to do this. Simply go back to the slope-intercept form, and substitute only the slope and the y-intercept. Substituting $-\frac{3}{5}$ for m and 6 for b, and leaving x and y as they are, gives you the equation that you were looking for:

$$y = \left(-\frac{3}{5}\right)x + 6$$

This is the answer to the problem.

Look at the following examples.

Example 1 Find the equation for the line through the point (–6, 4) with slope = $\frac{1}{2}$.

$4 = \frac{1}{2}(-6) + b$

Step 1: Substitute the given information into the slope-intercept formula, $y = mx + b$.

$4 = \frac{(-6)}{2} + b$

$4 = -3 + b$

$4 + 3 = -3 + b + 3$

$7 = b$

Step 2: Solve for b.

$y = \frac{1}{2}x + 7$

Step 3: Rewrite the equation of the line, substituting only the slope and y-intercept.

Answer: $y = \frac{1}{2}x + 7$

Example 2 Find the equation for the line through the point (10, –4) with slope = 2.

$-4 = 2(10) + b$

Step 1: Substitute the given information into the slope-intercept formula, $y = mx + b$, and simplify.

$-4 = 20 + b$

$-4 - 20 = 20 + b - 20$

$-24 = b$

Step 2: Solve for b.

$y = 2x - 24$

Step 3: Rewrite the equation of the line, substituting only the slope and y-intercept.

Answer: $y = 2x - 24$

Before you finish this lesson, look at one more example.

Example 3 Write the equation of the line through point (6, 4) with slope $\frac{3}{2}$ in standard form.

$y = mx + b$

Step 1: Use the slope-intercept formula.

$4 = \frac{3}{2}(6) + b$

Step 2: Substitute $\frac{3}{2}$ for m, 4 for y, and 6 for x.

$4 = 9 + b$

Step 3: Simplify.

$4 - 9 = 9 + b - 9$

$-5 = b$

Step 4: Solve for b, the y-intercept.

$y = \frac{3}{2}x - 5$

Step 5: Substitute the slope and y-intercept into the slope-intercept form to write the equation of the line.

$\frac{-3}{2}x + y = -5$	**Step 6:**	Rearrange the equation into standard form.
$-2\left(\frac{-3}{2}x\right) + (-2)y = (-2)(-5)$ $3x - 2y = 10$	**Step 7:**	Multiply each side by –2 to make $A \geq 0$ and an integer.

Answer: $3x - 2y = 10$

Problem Set

1. Determine the equation of the line that has a slope of –2 and passes through the point (–3, 0). Write the equation in slope-intercept form.

2. Determine the equation of the line that has a slope of 1 and passes through the point (2, –6). Write the equation in slope-intercept form.

3. Determine the equation of the line that has a slope of 6 and passes through the point (4, –2). Write the equation in slope-intercept form.

4. Determine the equation of the line that has a slope of –3 and passes through the point (–3, 0). Write the equation in slope-intercept form.

5. Determine the equation of the line that has a slope of 2 and passes through the point (1, 4). Write the equation in slope-intercept form.

6. Determine the equation of the line that has a slope of 4 and passes through the point (2, 8). Write the equation in slope-intercept form.

7. Determine the equation of the line that has a slope of –3 and passes through the point (2, 7). Write the equation in slope—intercept form.

8. Determine the equation of the line that has a slope of 5 and passes through the point (1, 3). Write the equation in slope-intercept form.

9. Determine the equation of the line that has a slope of $-\frac{1}{4}$ and passes through the point (3, 1). Write the equation in slope-intercept form.

10. Determine the equation of the line that has a slope of $\frac{1}{2}$ and passes through the point (4, 2). Write the equation in slope-intercept form.

11. Determine the equation of the line that has a slope of 2 and passes through the point (5, 1). Write the equation in standard form.

12. Determine the equation of the line that has a slope of 3 and passes through the point (–2, –7). Write the equation in standard form.

13. Determine the equation of the line that has a slope of 3 and passes through the point (1, 2). Write the equation in standard form.

14. Determine the equation of the line that has a slope of 1 and passes through the point (–2, 1). Write the equation in standard form.

15. Determine the equation of the line that has a slope of –4 and passes through the point (0, 0). Write the equation in standard form.

16. Determine the equation of the line that has a slope of $\frac{4}{3}$ and passes through the point (0, –8). Write the equation in standard form.

17. Determine the equation of the line that has a slope of $-\frac{1}{2}$ and passes through the point (0, 4). Write the equation in standard form.

18. Raven charges a base price of $15.00 to baby-sit, as well as $3.00 per hour. Write an equation that will determine the total charge for Raven to baby-sit for *x* hours.

19. Taylor charges a base price of $10.00 to baby-sit, as well as $5.00 per hour. Write an equation that will determine the total charge for Taylor to baby-sit for *x* hours.

20. The population of the city of Mayfield is 55,000. If the population increases by 2,000 each year, what equation could be used to represent the population of the city in *x* years?

HA1-410: Determining an Equation of a Line Given the Coordinates of Two Points

By now, you know how to calculate slope, and how to find the equation of a line if you are given a point on the line and the slope. In this last lesson on slope and lines, you will learn how to put these two ideas together to find the equation of a line when you are given two points on the line. For this lesson, you need to remember two formulas: the slope formula and the slope-intercept form for the equation of a line. The following example demonstrates how to use the slope formula.

Example 1 Find the slope of the line through the points (3, 3) and (5, 6).

$$m = \frac{y_2 - y_1}{x_2 - x_1}$$

Step 1: Write the slope formula.

$$m = \frac{6-3}{5-3}$$

Step 2: Substitute the coordinates of the given points into the slope formula.

$$m = \frac{3}{2}$$

Step 3: Simplify the fraction.

Answer: $\frac{3}{2}$

The second formula you need to remember is the slope-intercept form for the equation of a line: $y = mx + b$, where m is the slope and b is the y-intercept. This is the formula you use to get the equation for a line once you know the slope and a point on the line. Now, suppose that you want to find an equation for a line, but the only information you have is a pair of points on the line. You must first use those two points to find the slope of the line, then use that slope and one of the two original points together with slope-intercept form to get the equation. Look at the following example:

Example 2 Find the equation of the line that passes through the points (2, –2) and (5, 3).

$$m = \frac{y_2 - y_1}{x_2 - x_1}$$

$$m = \frac{3-(-2)}{5-2}$$

$$m = \frac{5}{3}$$

Step 1: Find the slope of the line.

$$y = mx + b$$

$$(-2) = \left(\frac{5}{3}\right)(2) + b$$

$$-2 = \frac{10}{3} + b$$

Step 2: Substitute $\frac{5}{3}$ for the slope, 2 for x, and –2 for y. Once you know the slope, you can use the slope-intercept form to find the y-intercept. Do this by substituting the slope and one of the two original ordered pairs into the formula.

$$\frac{-10}{3} + (-2) = \left(\frac{10}{3}\right) + b - \frac{10}{3}$$

Step 3: Solve for b.

$$\frac{-10}{3} + \left(\frac{-2}{1}\right) = b$$

$$\frac{-10}{3} + \frac{-6}{3} = b$$

$$\frac{-16}{3} = -5\frac{1}{3} = b$$

$$y = \frac{5}{3}x + \left(-5\frac{1}{3}\right)$$

Step 4: Substitute the slope and y-intercept into slope-intercept form to write the equation of the line.

Answer: $y = \frac{5}{3}x + \left(-5\frac{1}{3}\right)$

Taking this process a step further, you can write the equation of the line in standard form. Remember that **standard form** is $ax + by = c$, where a, b and c are real numbers, $a \geq 0$, and a and b are not both zero. The next example demonstrates how to do this.

Example 3 Write the equation of the line through the points (3, 4) and (–3, –8) in standard form.

$$m = \frac{y_2 - y_1}{x_2 - x_1}$$

$$m = \frac{-8-4}{-3-3}$$

$$m = \frac{-12}{-6} = 2$$

Step 1: Find the slope of the line.

$$y = mx + b$$
$$4 = 2(3) + b$$

Step 2: Using slope-intercept form, substitute the values for slope and either ordered pair to find the y-intercept.

$$4 = 6 + b$$
$$4 - 6 = 6 + b - 6$$
$$-2 = b$$

Step 3: Solve for b.

$$y = 2x + (-2)$$

Step 4: Substitute the slope and y-intercept into the slope-intercept form.

$$y = 2x - 2$$
$$-2x + y = -2$$
$$-1(-2x + y) = (-1)(-2)$$
$$2x - y = 2$$

Step 5: Rearrange the equation into standard form.

Answer: $2x - y = 2$

Problem Set

Determine the equation of a line that contains the given points, and write the answer in slope-intercept form:

1. (–1, 1) and (1, 5)

2. (1, 0) and (2, 2)

3. (–2, 3) and (–1, 2)

4. (2, 3) and (3, 5)

5. (2, 2) and (1, 0)

6. (4, 3) and (2, 1)

7. (4, 2) and (6, 3)

8. (9, 3) and (4, 5)

9. (2, 3) and (8, 4)

10. (2, 3) and (1, 6)

Determine the equation of a line that contains the given points, and write the answer in standard form:

11. (–1, 5) and (2, 7)

12. (3, 1) and (1, –5)

13. (1, –3) and (5, 7)

14. (1, –1) and (3, 5)

15. (5, 1) and (3, 7)

16. (4, –3) and (–2, 3)

17. (3, –3) and (5, 3)

Solve:

18. A rental car company's rate (y) is a linear function of the number of days the car is rented (x). If the graph of the function includes the points (1, 75) and (6, 275), what is the equation of the line?

19. A blueprint of a wheelchair ramp shows two points: (0, 0) at the start of the ramp, and (60, 6) at the end of the ramp. What is the equation of the line between these two points?

20. Manuel wants to predict the cost of manufacturing tennis shoes (y). Two pairs cost $75 and ten pairs cost $303. If cost is expressed as a linear function of the number of pairs manufactured (x), what is the equation of the line?

HA1-415: Graphing Linear Inequalities with Two Variables

In this lesson, you will learn about graphing linear inequalities.

Linear Inequality	A linear statement that relates expressions to each other with inequality signs ($<$, $>$, $\leq$, or $\geq$)

A linear inequality looks very much like a linear equation, but there is an inequality symbol in place of the equal sign. For example, the equation $2x + y = 8$ is a linear equation, but $2x + y < 8$ is a linear inequality. Linear inequalities have many more solutions than linear equations. The reason is that inequality symbols relax the restrictions on what the variables can and cannot be. For example, (0, 8) is a solution to the equation $2x + y = 8$. You can easily verify that this ordered pair is also a solution to the inequality $2x + y \leq 8$:

$$2(0) + 8 \leq 8$$
$$8 \leq 8$$

Because it is true that $8 \leq 8$, (0,8) is a solution to the inequality. In fact, any solution to the linear equation $2x + y = 8$ will be a solution to the inequality $2x + y \leq 8$. Now consider the ordered pairs (0, 0) and (1, 2). Both of these ordered pairs are also solutions to the inequality $2x + y \leq 8$:

$$2(0) + 0 \leq 8 \qquad 2(1) + 2 \leq 8$$
$$0 + 0 \leq 8 \qquad 2 + 2 \leq 8$$
$$0 \leq 8 \qquad 4 \leq 8$$
$$\text{True} \qquad \text{True}$$

However, they are not solutions to the equation $2x + y = 8$:

$$2(0) + 0 = 8 \qquad 2(1) + 2 = 8$$
$$0 + 0 = 8 \qquad 2 + 2 = 8$$
$$0 = 8 \qquad 4 = 0$$
$$\text{False} \qquad \text{False}$$

The graph of a linear equation actually separates the coordinate plane into three sets of points: points on the line, points above the line, and points below the line. The line is called the boundary line and the regions on each side of the line are called half-planes.

Half-Plane	The regions formed when a boundary line divides the coordinate plane

The line that separates the two half-planes is called the **boundary line**, and it is the graph of the linear equation that would result if you changed the inequality symbol to an equal sign.

Boundary Line	The line that separates two half-planes

The graph of a linear inequality will include all points on one side of the boundary line and sometimes the line as well.

Shading is used to indicate which half-plane is indicated by the inequality.

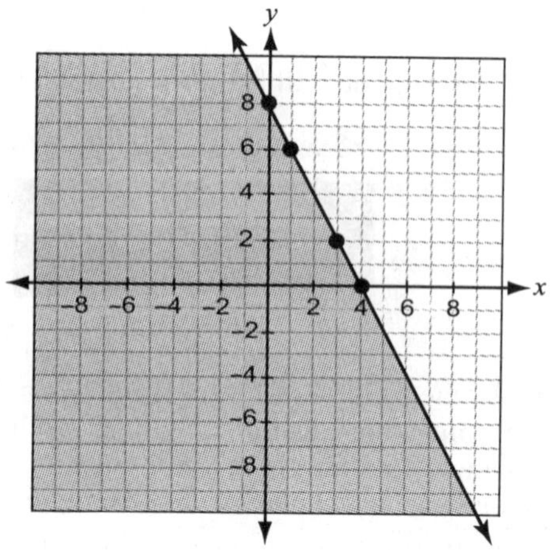

Step 1: Identify the related linear equation.
Step 2: Substitute values for x, and solve for y.
Step 3: Find the ordered pairs of the equation.
Step 4: Plot the ordered pairs.
Step 5: Graph the boundary line. If the inequality symbol is $<$ or $>$, use a dashed line for the boundary line.If the inequality symbol is greater than or equal to or less than or equal to, use a solid line, because then the line is part of the overall solution.
Step 6: Choose any point not on the boundary line. This point is called the **test point**. Substitute the coordinates of the test point into the inequality. If these numbers make the inequality true, then shade the half-plane that contains the test point. If these numbers make the inequality false, then shade the half-plane that does not contain the test point. The easiest test point to use is (0, 0) if it is not on the boundary line.

Example 1 Graph the inequality $3x + y \leq 10$.

x	y
1	7
2	4
3	1

Step 1: Identify the equation of the boundary line: $3x + y = 10$.
Step 2: Substitute values for x in the equation and solve for y.
Step 3: Find the ordered pairs: (1, 7), (2, 4) (3, 1).

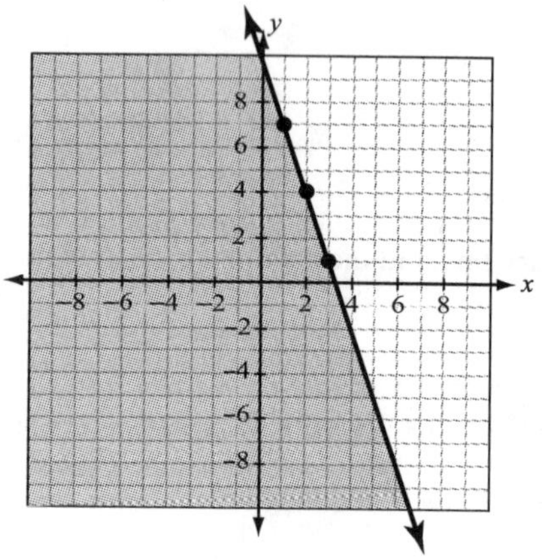

Step 4: Plot the ordered pairs: (1, 7), (2, 4) (3, 1).
Step 5: Graph the boundary line. Use a solid line because the inequality symbol is the less than or equal to symbol.
Step 6: Choose a test point (0, 0).

$$3x + y \leq 10$$
$$3(0) + 0 \leq 10$$
$$0 + 0 \leq 10$$
$$0 \leq 10$$

When you substitute the coordinates into the inequality, you will see that the numbers make the inequality true, so you will shade the side that contains the test point.

Example 2 Graph the inequality $y > 2x + 1$.

x	y
0	1
2	5
3	7

Step 1: Identify the equation of the boundary line: $y = 2x + 1$.
Step 2: Substitute values for x in the equation and solve for y.
Step 3: Find the ordered pairs: (0, 1); (2, 5); and (3, 7).

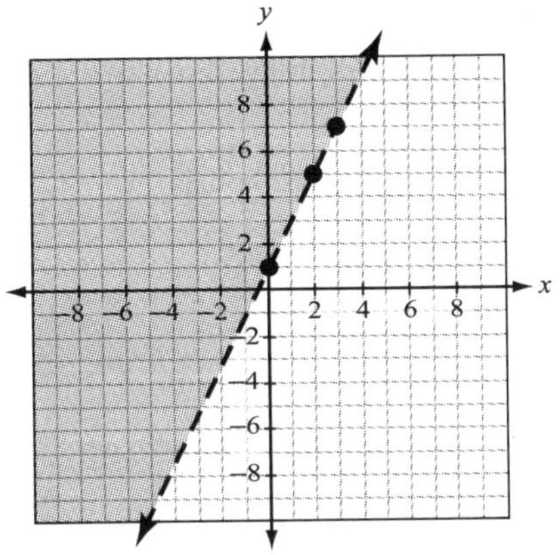

Step 4: Plot the ordered pairs.
Step 5: Graph the boundary line. Use a dashed line because the inequality symbol is the greater than symbol.
Step 6: Choose a test point: (0, 0). Substitute its coordinates into the inequality. $0 > 2(0) + 1$

$$0 > 2$$

Because these numbers make the inequality false, shade the side that does not contain the test point.

Example 3 Graph the inequality $x + y > 5$.

x	y
0	5
5	0
2	3

Step 1: Identify the equation of the boundary line: $x + y = 5$.
Step 2: Substitute values for x in the equation and solve for y.
Step 3: Find the ordered pairs: (0, 5), (5, 0), (2, 3).

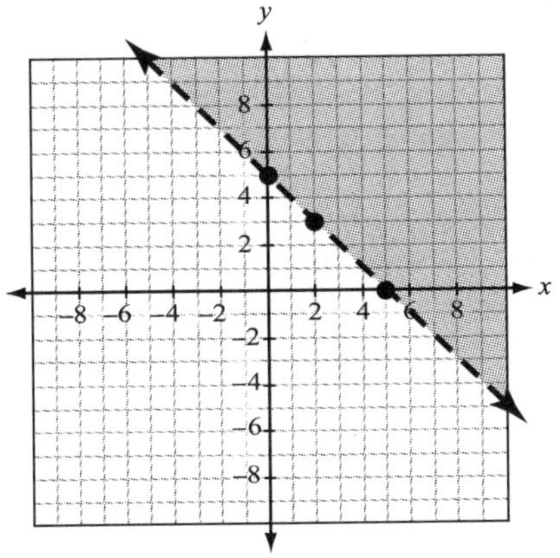

Step 4: Plot the ordered pairs.
Step 5: Graph the boundary line. Use a dashed line because the inequality symbol is the greater than symbol.
Step 6: Choose a test point (0, 0) and substitute it into the inequality. $0 + 0 > 5$

$$0 > 5$$

Because these numbers make the inequality false, shade the half of the side that does not contain the test point.

Example 4 Graph the inequality $x \leq 3$.

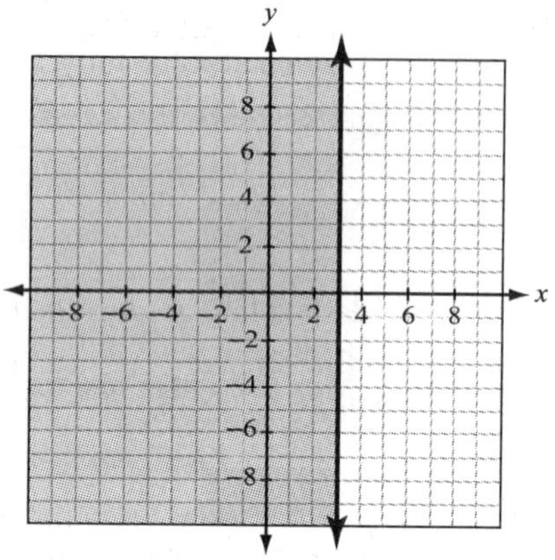

Step 1: Identify the equation of the boundary line: $x = 3$.

Step 2: Because this inequality contains only an x and no y, the graph is a vertical line. Draw a vertical line through the point (3, 0), since the equation is $x = 3$. Use a solid line because the inequality symbol is the less than or equal to symbol.

Step 3: Shade the half plane to the left of the line because the solution to the inequality includes all ordered pairs with x-values less than or equal to 3.

Problem Set

Graph the following linear inequalities:

1. $y > 3x$

2. $x \leq -2$

3. $y \leq 3x$

4. $y < 4$

5. $y \geq 3x$

6. $x - y > 2$

7. $x + y > 2$

8. $2x - y > -2$

9. $2x - y < -1$

10. $4x + 2y < 6$

HA1-416: Graphing Linear Inequalities with Two Variables Using the Graphing Calculator

In this lesson we will use the TI-83 Plus™ graphing calculator to describe the solutions of linear inequalities. The solutions to these inequalities are called half-planes. When we solve these inequalities, the linear equation defines a line or boundary that divides the coordinate plane into two half-planes. The coordinates that satisfy the inequality are the points on one of the half-planes.

If the line is included, the inequality symbols are $\geq$ or $\leq$ and the solution set is a closed half-plane. If the line is not included, the inequality symbols are $>$ or $<$ and the solution set is an open half-plane with the boundary represented by a dashed line.

The process has four steps:

- Step 0: (Settings) Initialize the calculator to a standard set of values so that we have a common starting point.
- Step 1: (Enter the expression) Enter the symbolic form of the inequality to be analyzed into the calculator.
- Step 2: (Graph) Obtain the graphs of the solution set from the symbolic form.
- Step 3: (Solve) Use information obtained graphically to describe the solution set or answer a specific question about the solutions of the inequality.

Before doing any of these examples you should check your calculator's settings to be sure that it will function as described above.

Press the [MODE] key and be sure that your settings appear as shown below.

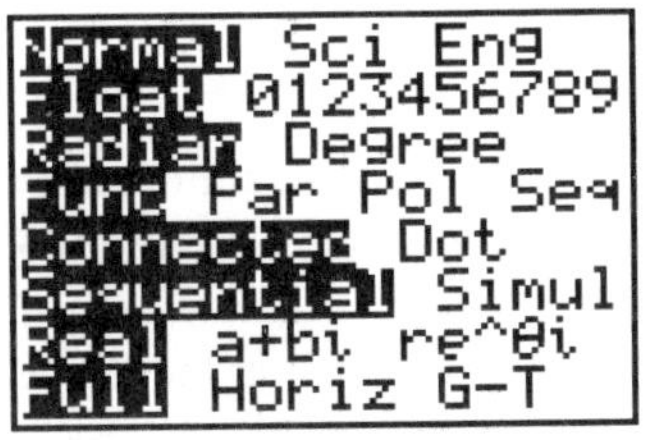

Next press the [2nd][ZOOM] keys and be sure that your settings appear as shown below.

Press the [ZOOM] key then the [6] key to select the standard window setting.

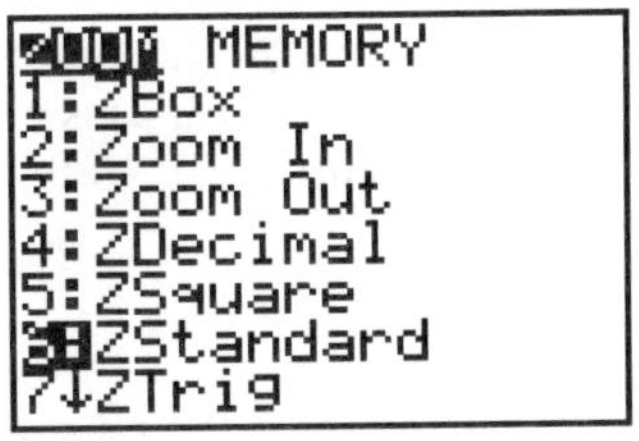

Press the WINDOW key to get the settings below..

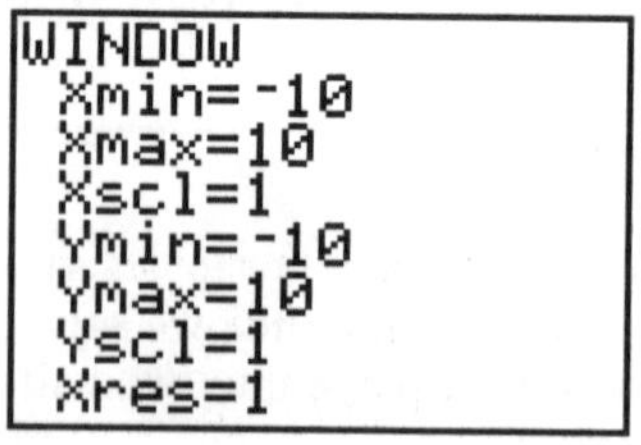

Example 1 Is the point (2, 2) a solution to the linear inequality $y < 3x + 4$?

Step 0: (Settings) Make sure you have the default settings as described above.

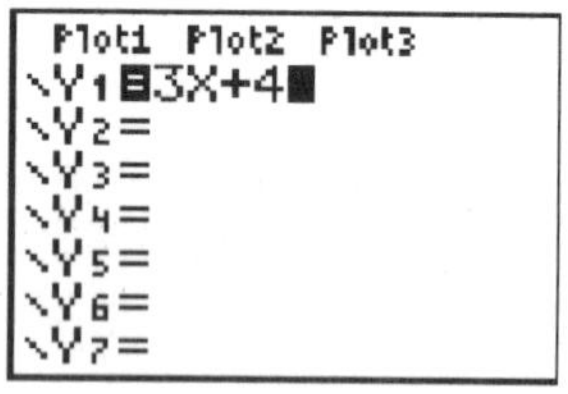

Step 1: (Enter the expression) Press the Y= key and enter the expression $3x + 4$ into the Y_1 slot.

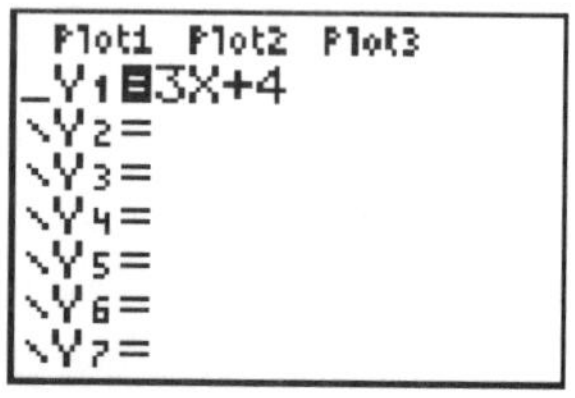

Next, use the ◄ key to put the cursor as far to the left on the screen as possible.

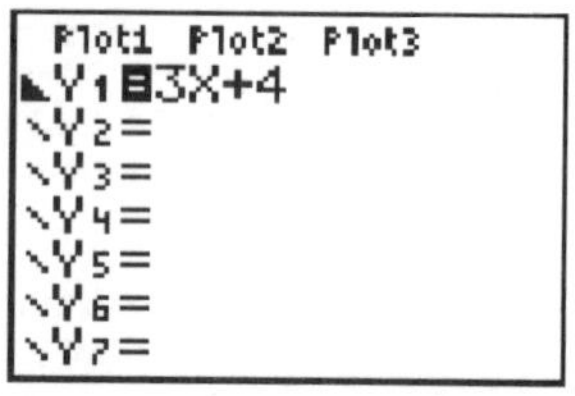

Note that the solid cursor becomes an underscore and it is blinking. Push ENTER ENTER ENTER to get the screen shown at left.

The symbol on the left tells the calculator to graph the solution to the "less than" or "less than or equal to" inequality type.

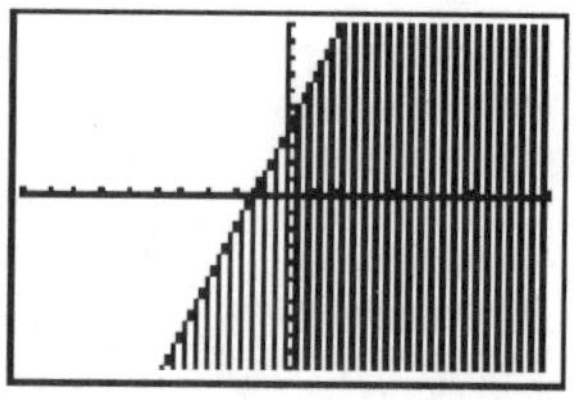

Step 2: (Graph) Press the GRAPH key.

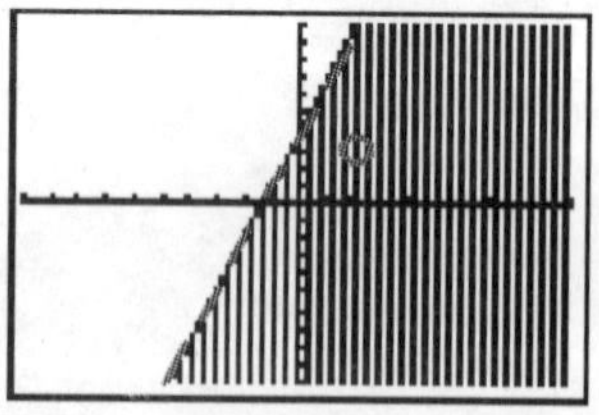

Step 3: (Solve) Notice that all of the points "below" and to the "right" of the line are shaded. These points represent the points that satisfy the inequality. The points on the line are NOT part of the solution set. The point (2, 2) is clearly in the shaded region so it belongs to the solution set.

Answer: The point (2, 2) is a solution to the inequality $y < 3x + 4$.

Example 2 Which of the following set of ordered pairs is in the solution set of $4x - 3y \le 21$: (1,–4), (9, 8), (6,–6), (0,–7)?

Step 0: (Settings) Make sure you have the default settings described at the beginning of the lesson.

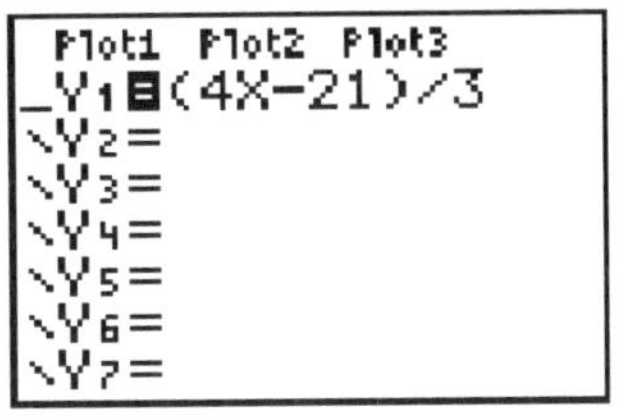

Step 1: (Enter the expression) The inequality must be solved for y in order to use the graphing calculator.

Adding $3y$ to both sides gives us:

$$4x \le 3y + 21.$$

Subtracting 21 from both sides gives us:

$$4x - 21 \le 3y.$$

Finally, dividing both sides by 3 gives us:

$$\frac{4x-21}{3} \le y \quad \text{or} \quad y \ge \frac{4x-21}{3}.$$

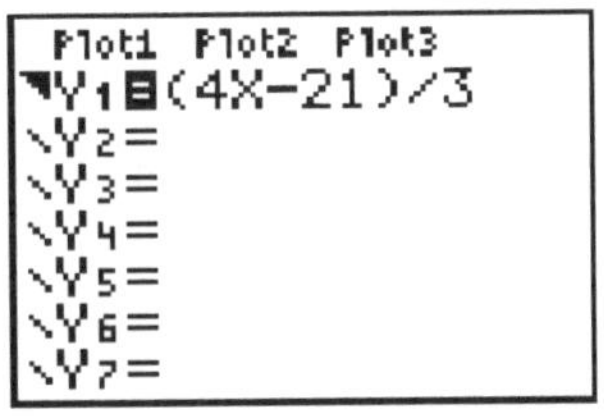

Press the [Y=] key and enter the expression $\frac{4x-21}{3}$ into the Y_1 slot.

Use the [◄] key to put the cursor as far to the left on the screen as possible. Note that the solid cursor becomes an underscore and it is blinking. Press [ENTER] [ENTER] so that the screen looks like the one to the left.

The symbol on the left tells the calculator to graph the solution to the "greater than" or "greater than or equal to" inequality type.

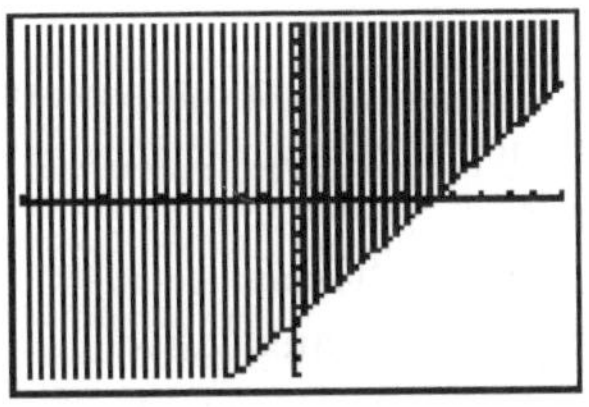

Step 2: (Graph) Press the [GRAPH] key.

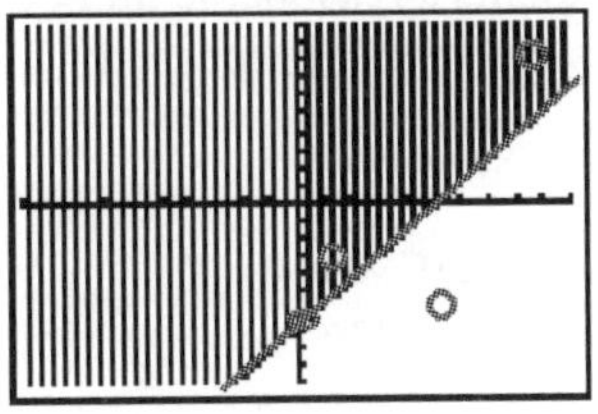

Step 3: (Solve) Notice that all of the points "above" and to the "left" of the line are shaded. These points represent the points that satisfy the inequality. The points on the line are part of the solution set. The points (1,–4), and (9, 8) are in the shaded region so they belong to the solution set. The point (0,–7) is on the line and is also included in the solution set. The point (6,–6) is not in the solution set.

Answer: The points (1, –4), (9, 8), and (0, –7) belong to the solution set, but the point (6, –6) does not.

Problem Set

1. Which of the following set of ordered pairs is in the solution set of $x - y \geq 2$?

(6, 3)	(–1, –1)
(–4, 0)	(–4, –2)

2. Which of the following set of ordered pairs is in the solution set of $3x - 6y > 15$?

(2, 1)	(3.5, –2.5)
(–5, 0)	(0, –2)

3. Which of the following set of ordered pairs is in the solution set of $-3x + 2y \geq 4$?

(–1, –1)	(2, 1)
(2, 5)	(0, 0)

4. Which of the following set of ordered pairs is in the solution set of $4x - 3y < 21$?

(9, 8)	(1, –4)
(6, –6)	(0, –7)

5. Sam bought x number of chocolate bars for \$1.50 each and y number of candies for 50 cents each. He spent less than \$10. Use the graphing calculator to graph the inequality $1.50x + 0.50y < 10$ and determine which of the following points represents a reasonable number of chocolate bars and candies bought.

(3, –2)	(–1, 4)
(4, 10)	(3, 7)

6. In her piggy bank, Keisha had x number of quarters and y number of nickels; together, the coins totaled more than \$1.50. Use the graphing calculator to graph the inequality $0.25x + 0.05y > 1.50$ and determine which of the following points represents a reasonable number of quarters and nickels Keisha collected.

(5, 3)	(9, –4)
(–2, 4)	(5, 8)

7. Jim bought x CDs for \$14 each and y DVDs for \$21 each. He spent no more than \$84. Use the graphing calculator to graph the inequality $14x + 21y \leq 84$ and determine which of the following points represents a reasonable possible total for the number of CDs and DVDs Sam bought.

(–1, 4)	(5, –2)
(3, 2)	(4, 3)

8. ABC school purchased new computer equipment for its computer labs. The school purchased x computers at \$1,850 each and y printers at \$545 each. The school spent no less than \$29,500. Use the graphing calculator to graph the inequality $1850x + 545y \geq 29,500$ and determine which of the following points represents a reasonable number of computers and printers the school purchased.

(–5, 36)	(18, 3)
(10, 4)	(20, –2)

9. Draw the graph that represents the solution of the inequality $y > \frac{1}{4}x - 3$.

10. Draw the graph that represents the solution of the inequality $8x + 4y \leq 6$.

HA1-442: Interpreting Graphs of Functions in Real-Life Situations

Have you ever read a newspaper article or magazine that included a line graph? You may have found that being able to interpret the graph helped you understand the article. Authors often use line graphs because they can clearly illustrate a relationship between two variables.

When interpreting a graph of a function, it is useful to determine the character of the function; that is, the intervals where the function is increasing, decreasing, and constant.

Increasing	A function is increasing when the graph is rising from the left to the right.
Decreasing	A function is decreasing when the graph is falling from the left to the right.
Constant	A function is constant when the graph is a horizontal line.

The graphs below illustrate several functions. The first function is increasing over its entire domain, whereas the second function is decreasing over its entire domain. The third graph is constant over its entire domain. Finally, the last graph is not increasing, decreasing, or constant over its entire domain; however, it is increasing, decreasing, and constant over certain intervals.

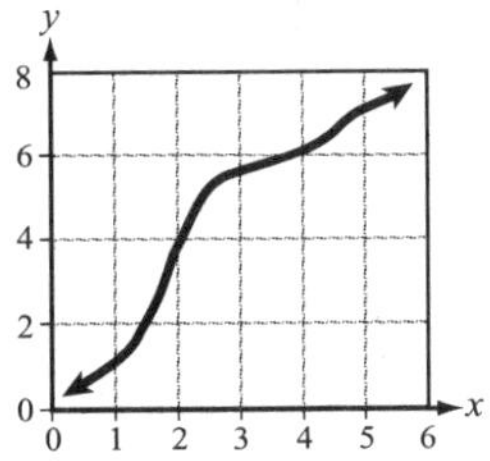

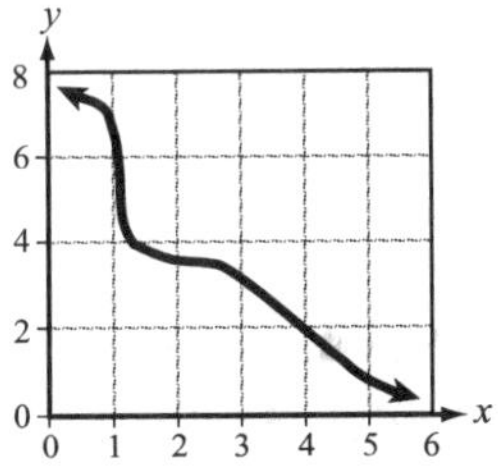

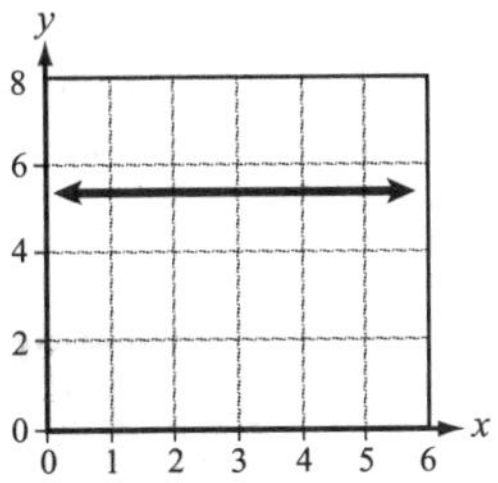

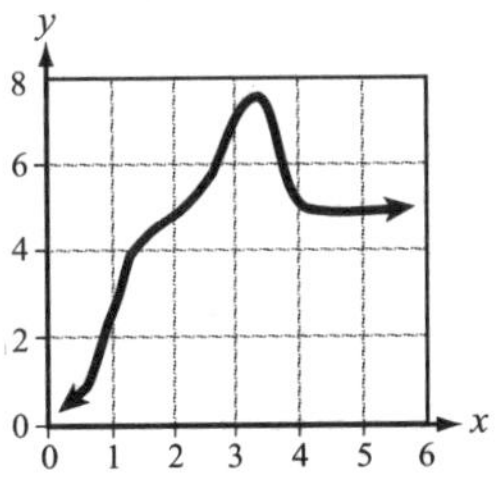

Example 1 The graph of the function f is given. What is the character of the function f on its entire domain?

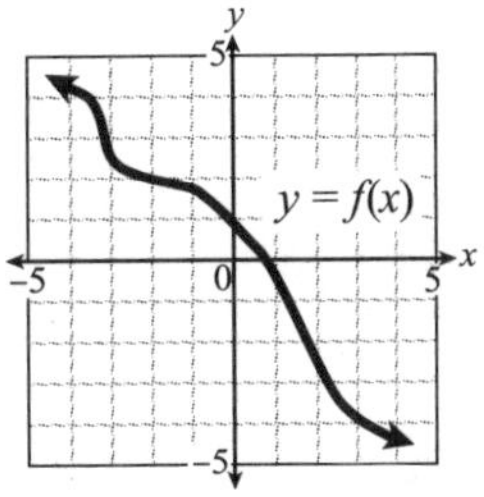

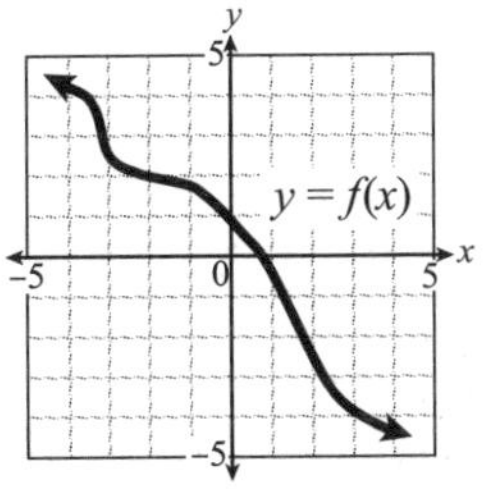

Step 1: The graph is falling over its entire domain, so the function f is decreasing.

Answer: The function f is decreasing over its entire domain.

Example 2 The graph of the function *f* is given. On which interval(s) is the function increasing?

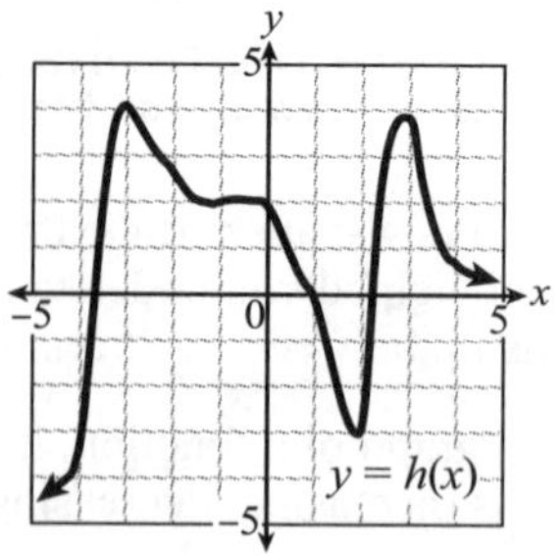

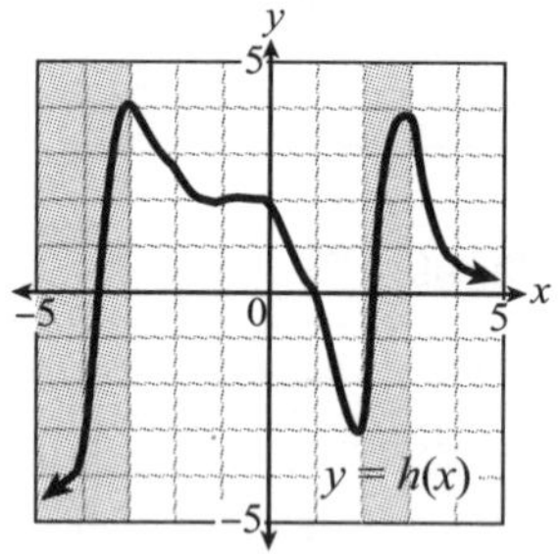

Step 1: Locate the intervals where the graph of the function is rising.
The graph is rising when $x < -3$ and $2 < x < 3$.

Answer: The function *h* is increasing when $x < -3$ and $2 < x < 3$.

Example 3 A retail manager recorded the number of customers, N, who entered the store each day, *d*, for a week. In the graph, $d = 1$ corresponds to Monday. What day of the week was the busiest, and what day was the slowest?

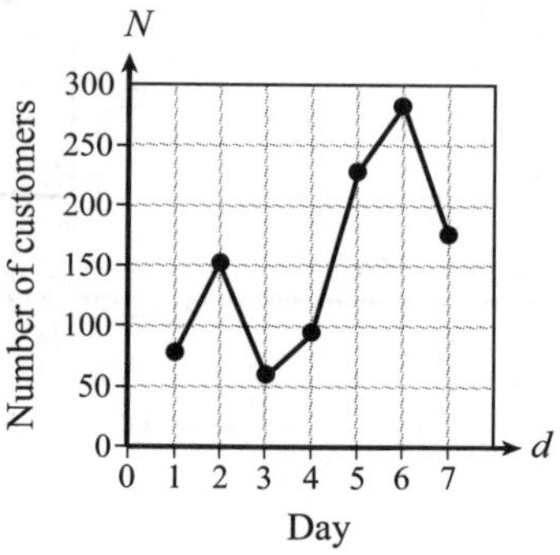

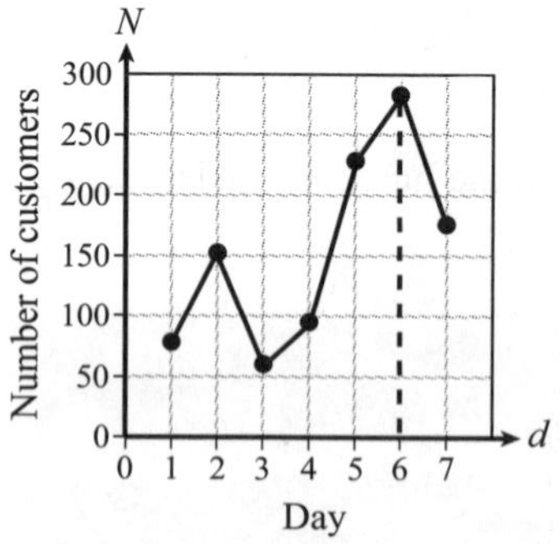

Step 1: The busiest day of the week occurs at the highest point (peak) on the graph. The highest point occurs when $d = 6$, which corresponds to Saturday.

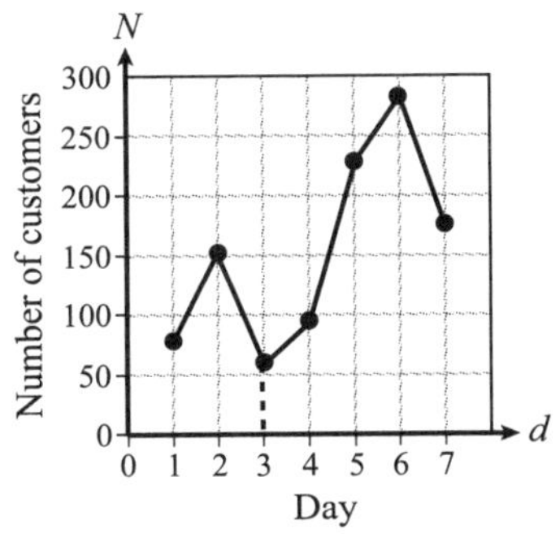

Step 2: The slowest day of the week occurs at the lowest point on the graph. The lowest point occurs when $d = 3$, which corresponds to Wednesday.

Answer: The busiest day was Saturday, and the slowest day was Wednesday.

Example 4 Which graph best represents the following report?

At the beginning of the month, the stock price slowly increased and then began to decrease. About midway through the month, the stock reached its lowest price. The stock price then slightly increased before remaining constant for a short period of time. The price then sharply increased to reach its peak before decreasing again.

Graph A

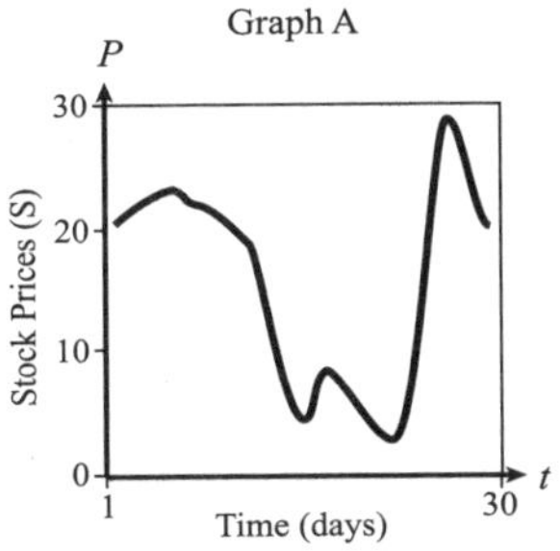

Graph B

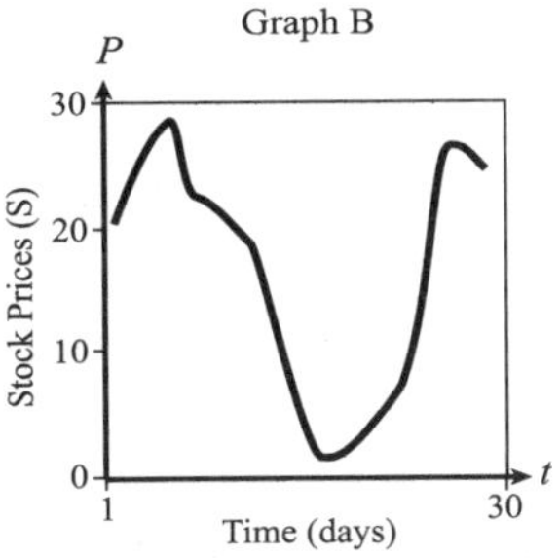

Graph C

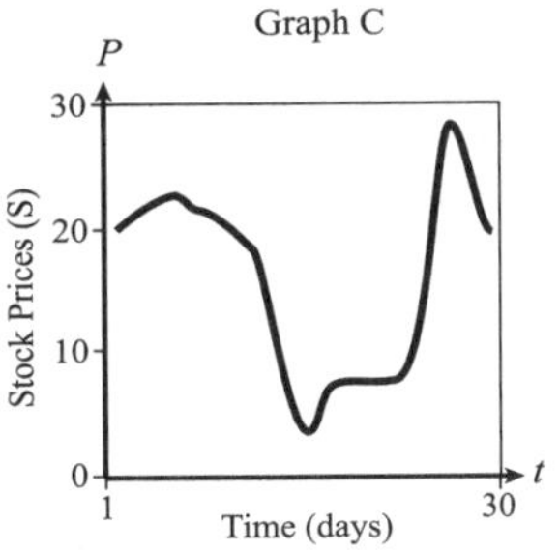

Graph D

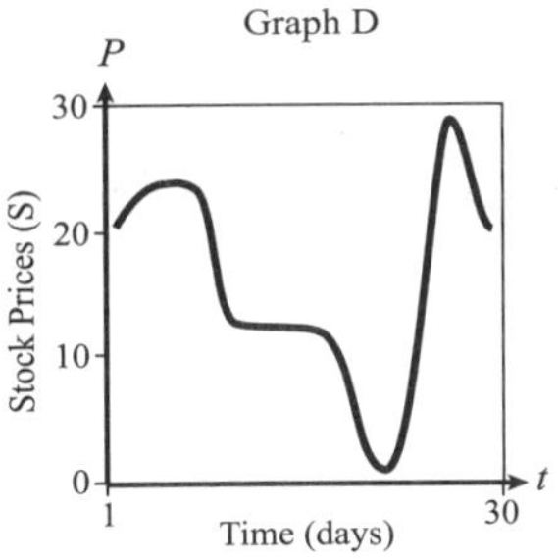

At the beginning of the month, the stock price slowly increased and then began to decrease.

Graphs A B, C, and D fit this description.

Step 1: Compare the first sentence in the report to each graph. Determine which graph fits that description.

About midway through the month, the stock reached its lowest price.

Graphs B, C, and D fit this description.

Step 2: Compare the second sentence in the report to the remaining graphs from Step 1. Determine which of the graphs fits that description.

The stock price then slightly increased before remaining constant for a short period of time.

Graph C fits this description.

Step 3: Compare the third sentence in the report to the remaining graphs from Step 2. Determine which of the graphs fits that description.

The price then sharply increased to reach its peak before decreasing again.

Graph C fits this description.

Step 4: Compare the fourth sentence in the report to the remaining graph from Step 3. Determine which graph fits that description.

Answer: Graph C best represents the report.

Example 5 The manager of a party supply store needs to write a report about the sales over the past year. In the graph, A represents the amount of sales in dollars, and m is the month, where $m = 1$ corresponds to January. Which report best represents the amount of sales?

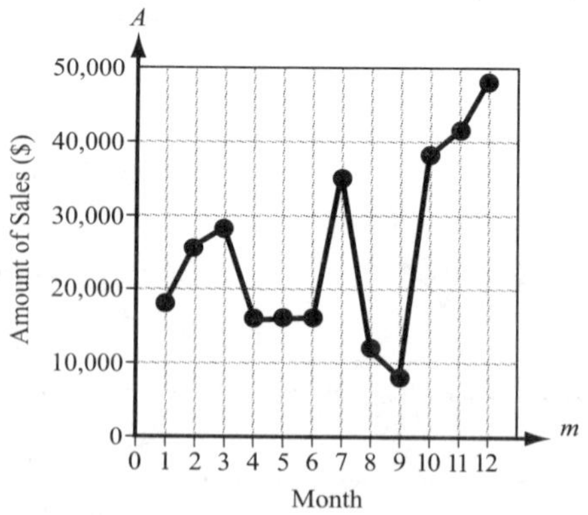

Report A: *At the beginning of the year, sales increased. From March to April, sales decreased and remained constant for the next three months. Sales picked up again in July to reach the peak, but then sharply decreased to reach the lowest amount in October. In November, sales increased and continued increasing through the end of the year.*

Report B: *At the beginning of the year, sales slowly increased. From March to April, sales decreased and remained constant through June. Sales picked up again in July, but then sharply decreased to reach the lowest amount in September. In October, sales sharply increased and continued increasing through the end of the year to reach the peak in December.*

Report C: *Sales increased for two months at the beginning of the year. From March to April, sales decreased and remained constant for the next two months. Sales increased in July and August, but then decreased to reach the lowest amount in September. In October, sales sharply increased and continued increasing until reaching the peak in December.*

Step 1: Identify the intervals where the graph is increasing, decreasing and constant. Also identify the highest point and lowest point.

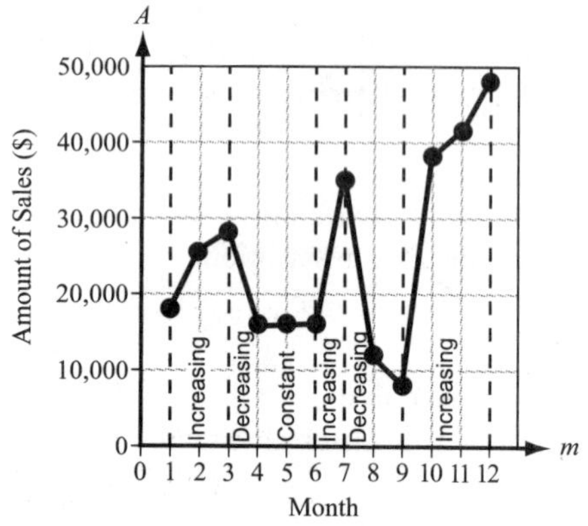

The highest amount of sales occurred in month 12, December.

The lowest amount of sales occurred in month 9, September.

Step 2: Choose the report that summarizes the observations in Step 1.

At the beginning of the year, sales slowly increased. From March to April, sales decreased and remained constant through June. Sales picked up again in July, but then sharply decreased to reach the lowest amount in September. In October, sales sharply increased and continued increasing through the end of the year to reach the peak in December.

Answer: Report B best represents the amount of sales.

Problem Set

1. The graph of the function f is given. Determine which of the statements is true about the function f.

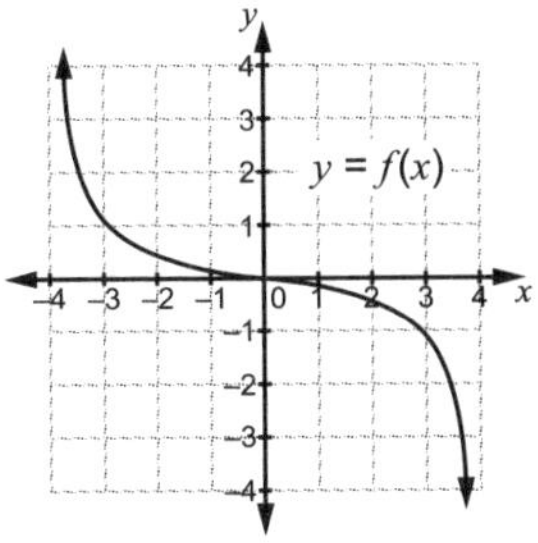

f is increasing and decreasing.

f is decreasing.

f is increasing.

f is constant.

2. The graph of the function f is given. Determine which of the statements is true about the function f.

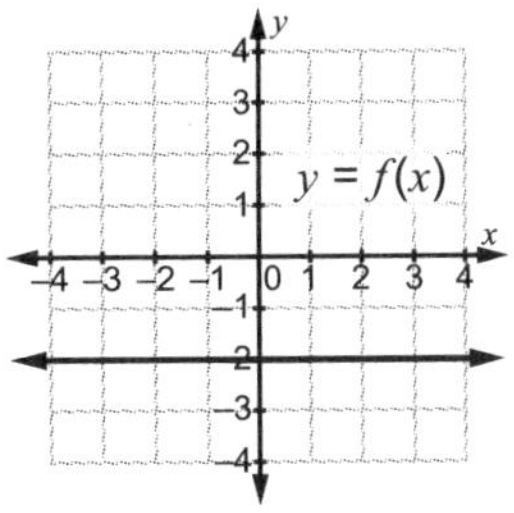

f is constant.

f is increasing and decreasing.

f is decreasing.

f is increasing.

3. The graph of the function f is given. Determine which of the statements is true about the function f on the interval $x > -2$.

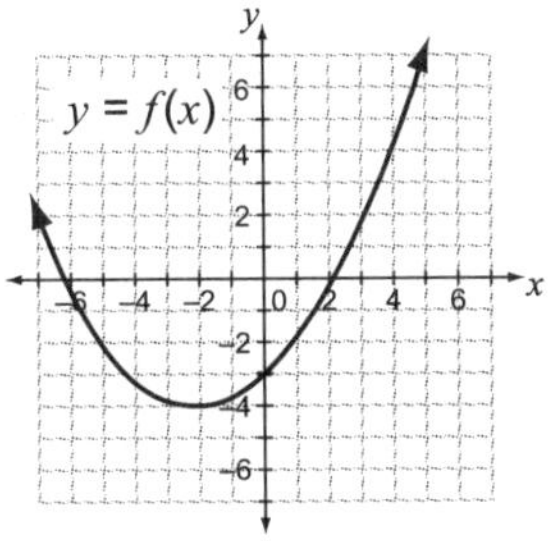

f is constant.

f is increasing and decreasing.

f is increasing.

f is decreasing.

4. The graph of the function f is given. Determine which of the statements is true about the function f on the interval $x > 2$.

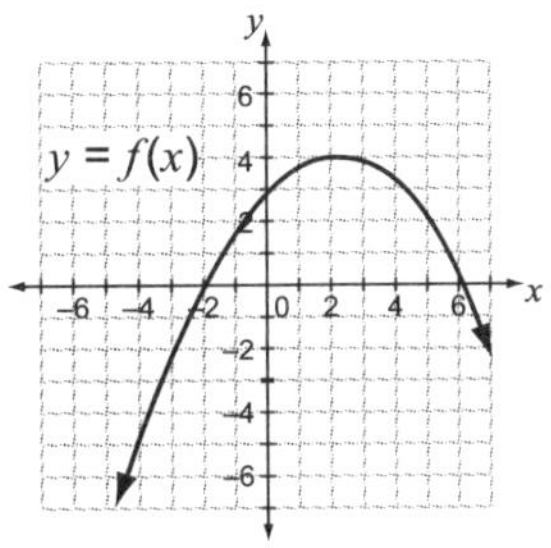

f is decreasing.

f is constant.

f is increasing and decreasing.

f is increasing.

5. The graph of the function f is given. Determine which of the statements is true about the function f.

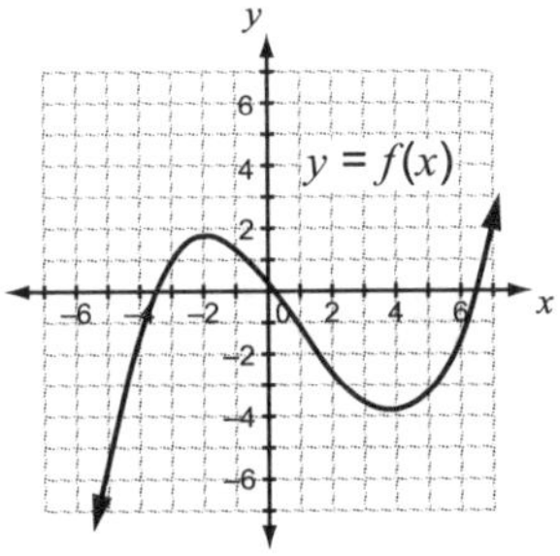

f is constant.

f is increasing.

f is increasing and decreasing.

f is decreasing.

6. The graph of the function f is given. Determine which of the statements is true about the function f on the interval $x < -2$.

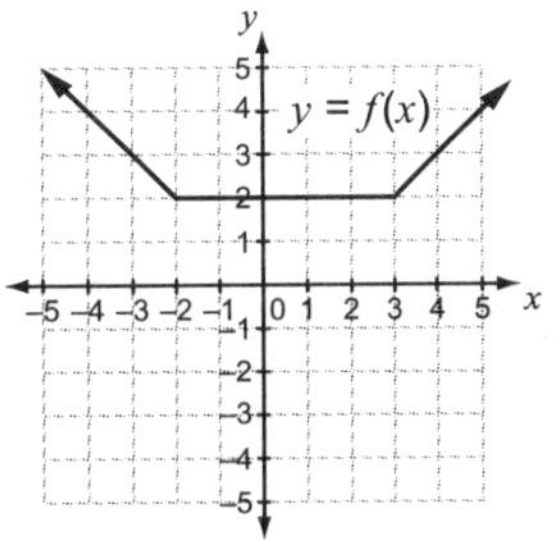

f is decreasing.

f is constant.

f is increasing and decreasing.

f is increasing.

7. The graph of the function f is given. Determine which of the statements is true about the function f.

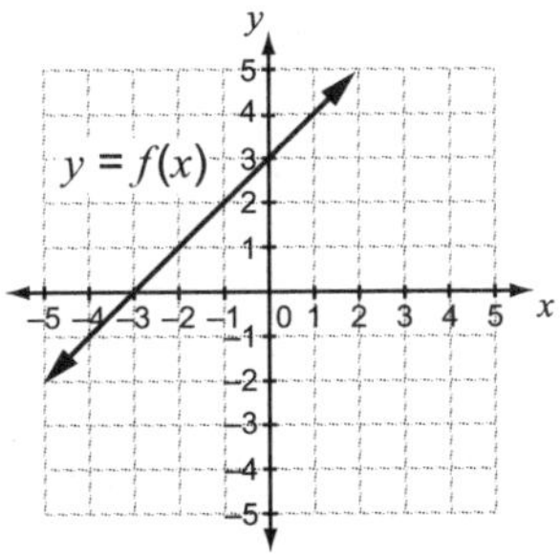

f is increasing and decreasing.

f is constant.

f is decreasing.

f is increasing.

8. The graph of the function f is given. Determine which of the statements is true about the function f on the interval $0 < x < 2$.

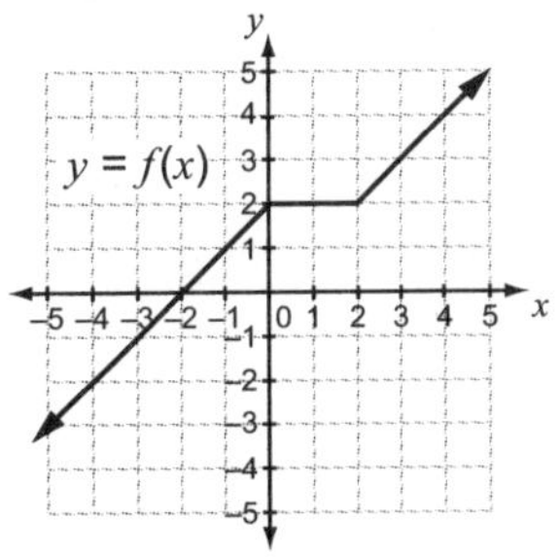

f is increasing and decreasing.

f is decreasing.

f is constant.

f is increasing.

9. The graph of the function g. is given. On which of the intervals is the function g constant?

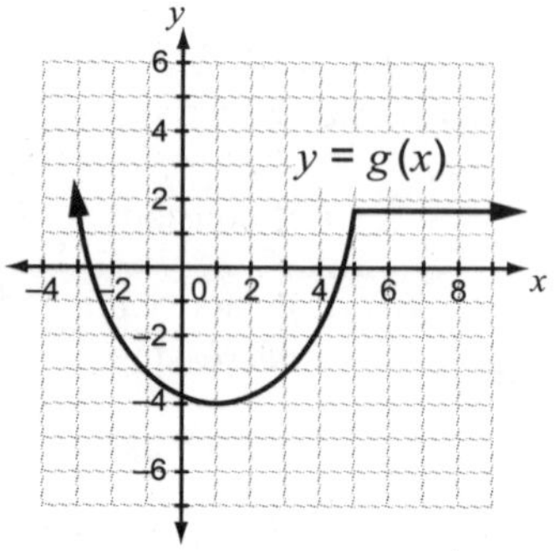

$x < 5$

$x > 5$

$1 < x < 5$

$x < 1$

10. The graph of the function g is given. On which of the intervals is the function g always increasing?

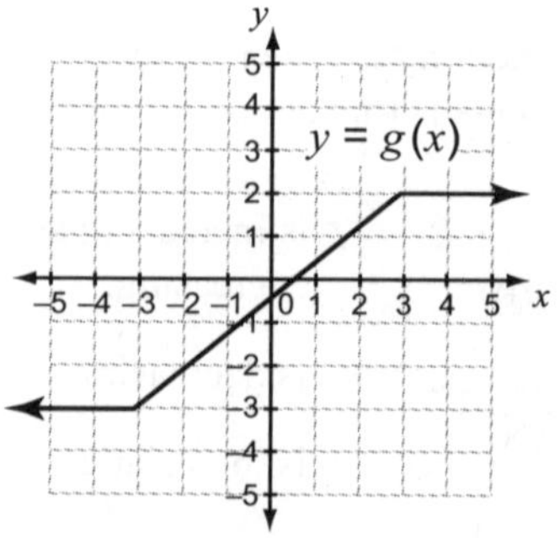

$x > 3$

$x < 3$

$-3 < x < 3$

$x < -3$

11. The graph of the function g. is given. On which of the intervals is the function g always increasing?

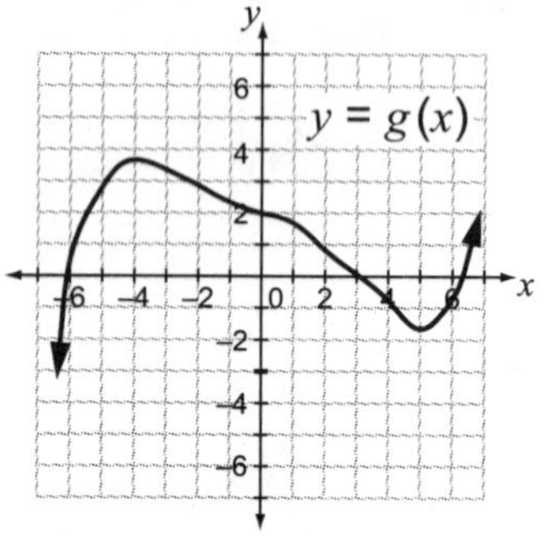

$x > -4$

$x < 5$

$-4 < x < 5$

$x < -4$

12. The graph of the function g. is given. On which of the intervals is the function g always decreasing?

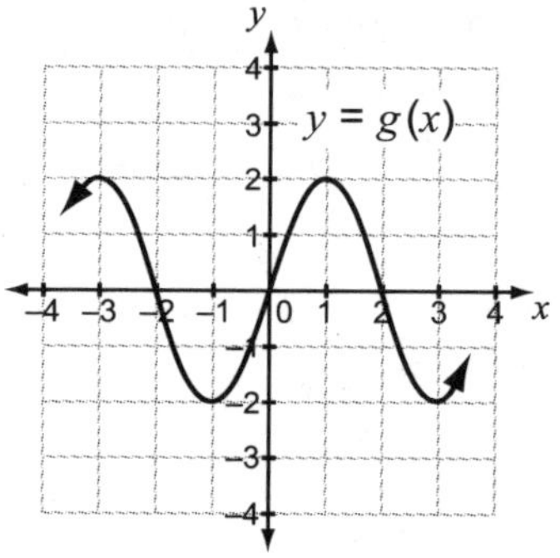

$x < 1$

$-1 < x < 1$

$1 < x < 3$

$x > -3$

13. The graph of the function g. is given. On which of the intervals is the function g always increasing?

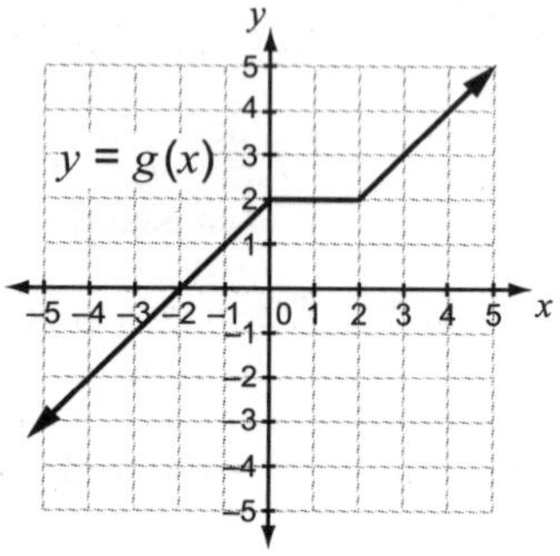

$x < 0$

$x < 2$

$x > 0$

$0 < x < 2$

14. The graph of the function g. is given. On which of the intervals is the function g always constant?

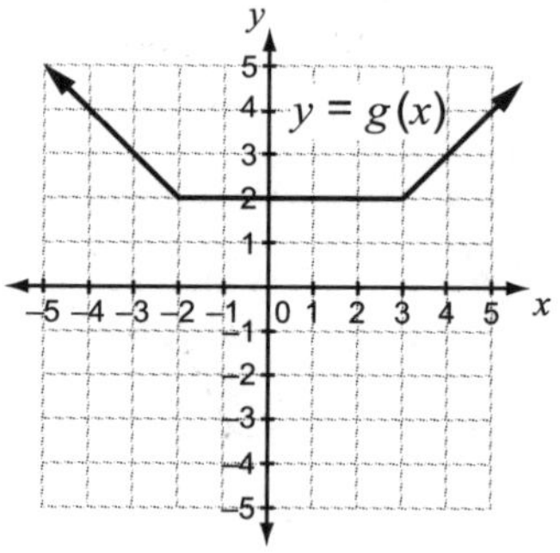

$-2 < x < 3$

$x < 3$

$x > -2$

$x < -3$

15. The graph of the function g. is given. On which of the intervals is the function g always decreasing?

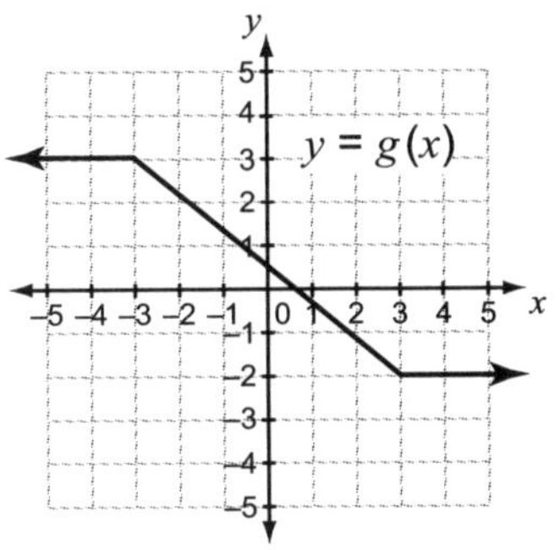

$x < -2$

$-3 < x < 3$

$x < 3$

$x > -3$

16. Bobby checked his pulse each minute after he began exercising. His pulse is recorded in the graph, where P is his pulse in beats per minute, and t is the time in minutes. When was his pulse constant?

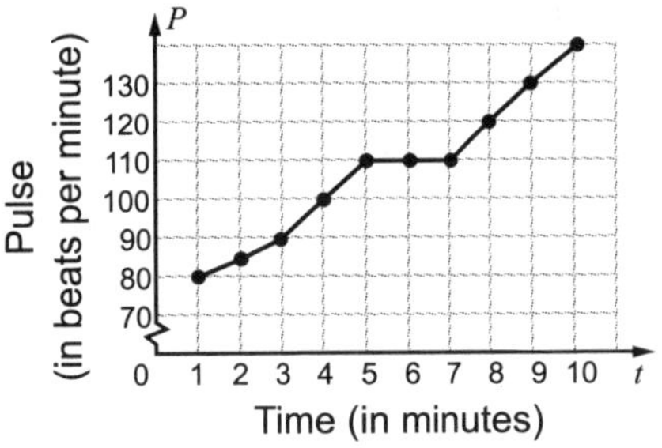

17. The monthly sales of a car company are represented in the graph, where n is the number of cars sold, and t is the month. Which of the following statements is incorrect?

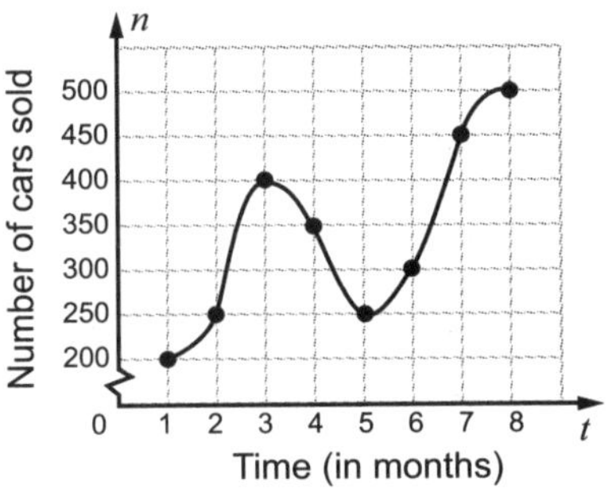

The number of cars sold was the greatest in month 3.

The number of cars sold decreased from month 3 to month 5.

The number of cars sold increased from month 5 to month 8.

The number of cars sold increased from month 1 to month 3.

18. The daily price of a share of a certain stock is represented in the graph, where P is the price in dollars and t is the time in days. Between which two consecutive days was there the smallest increase in the price of a share of this stock?

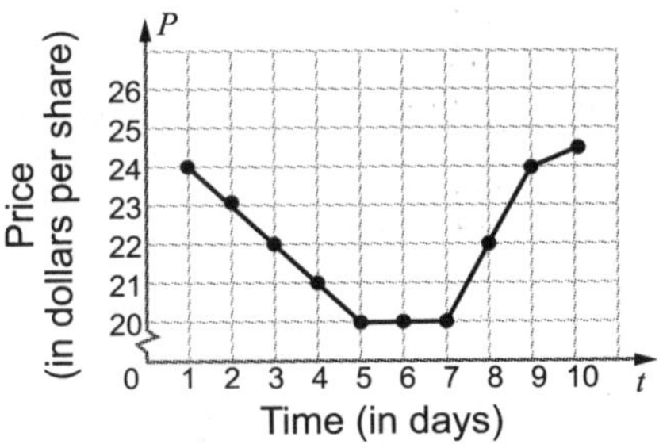

The smallest increase in the price of the stock occurred between day 7 and day 8.

The smallest increase in the price of the stock occurred between day 4 and day 5.

The smallest increase in the price of the stock occurred between day 8 and day 9.

The smallest increase in the price of the stock occurred between day 9 and day 10.

19. The population of a city between 1980 and 1990 is represented in the graph, where P is the number of people in thousands, and t is the year. On the horizontal axis, $t = 0$ corresponds to 1980. In what year was the population the smallest?

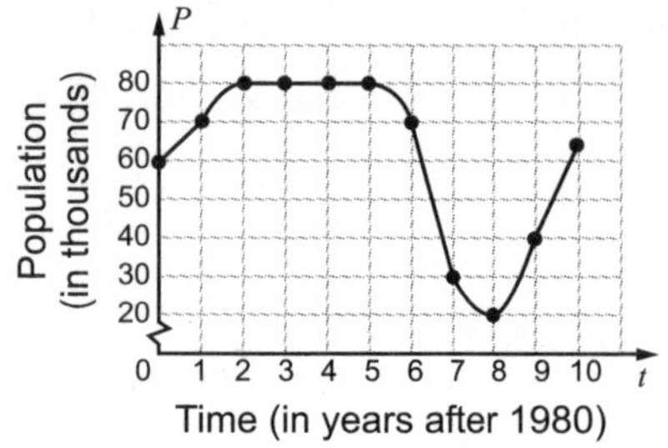

20. The temperatures in a certain city from noon until 10:00 p.m. are represented in the graph, where F is the temperature in degrees Fahrenheit, and t is the number of hours after noon. On the graph, $t = 0$ corresponds to noon. What happened to the temperature from 5:00 p.m. to 9:00 p.m.?

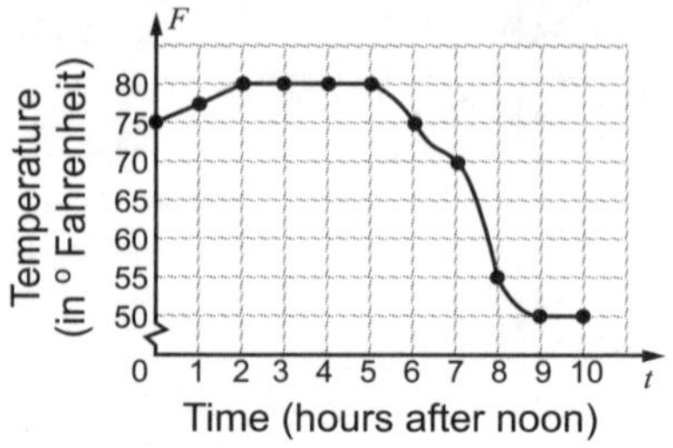

21. The profits of a company over a 10-month period are represented in the graph, where P is the profit in thousands of dollars, and t is the time in months. On the horizontal axis, $t = 1$ corresponds to January. Between which two consecutive months was the increase in profit the greatest?

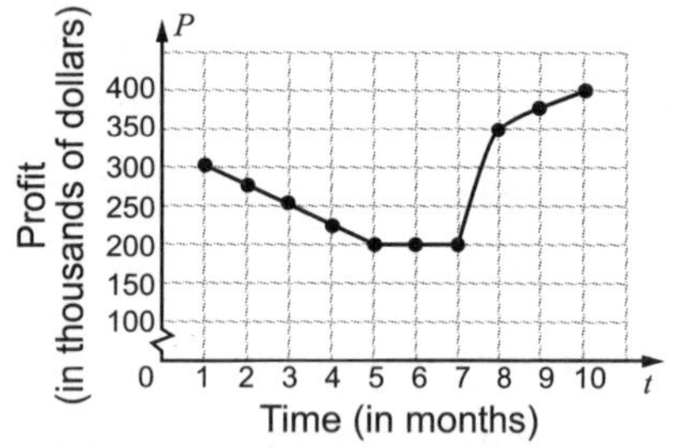

22. The average cost of a tank of gasoline over a period of 10 months is recorded in the graph, where C is the cost in dollars for a tank of gasoline, and t is the month the cost was recorded. What happened to the cost of a tank of gasoline from month 4 to month 7?

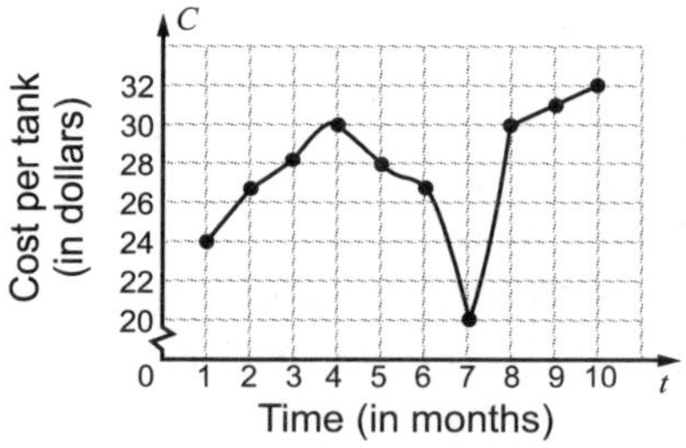

23. The amount of money in Maria's savings account is represented in the graph, where A is the amount saved in dollars, and t is the number of months since she started saving her money. After how many months did Maria have the most money in her savings account?

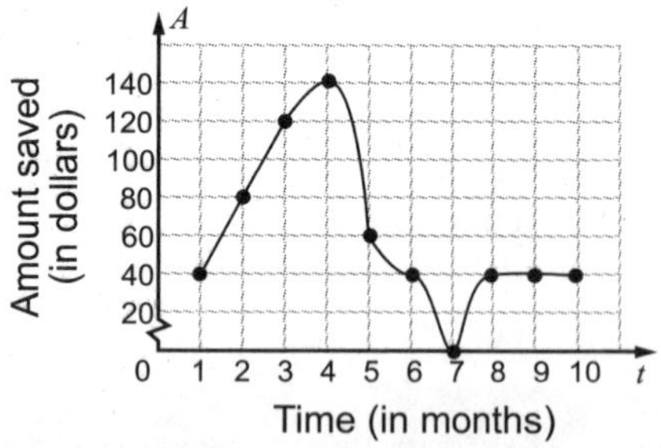

24. Determine which of the graphs best represents the average monthly rainfall, R, in inches, for a certain city over a period of 12 months. It is known that at the beginning of the year, the rainfall amount slowly increased, then remained constant, then increased, and finally decreased rapidly.

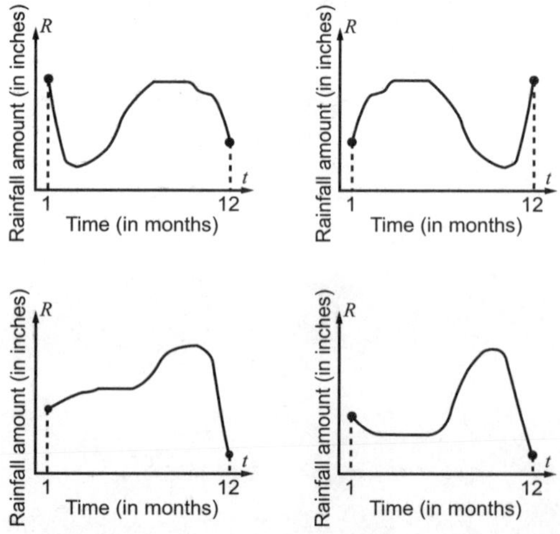

25. Determine which of the graphs best represents the interest rates, *R*, over a period of 12 months. It is known that at the beginning of the year, the rates were the highest. They then decreased rapidly to the lowest point, then increased, then remained constant, and finally decreased.

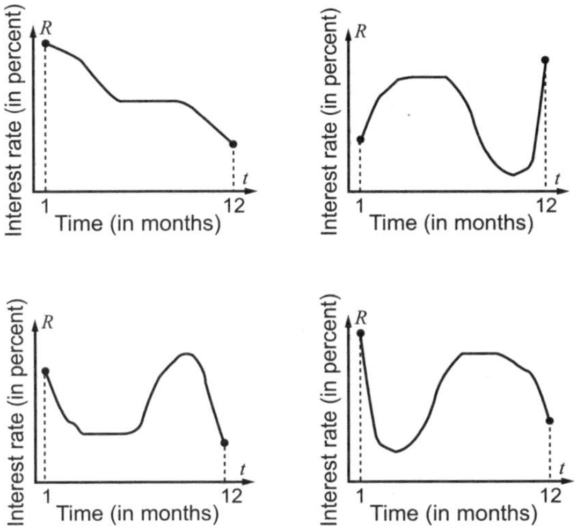

26. Determine which of the graphs best represents the average monthly temperatures, T, in degrees Fahrenheit, for a city during the first 6 months of the year. It is known that the temperature slowly increased, then remained constant, and then increased again to reach the highest temperature.

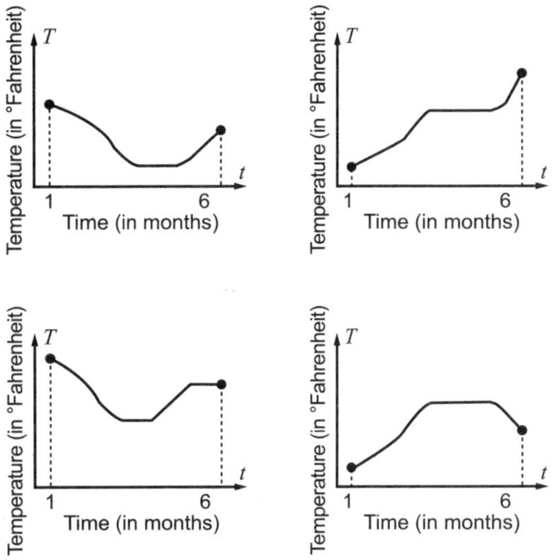

27. Luis recorded his average grade, G, in percent, in his math class for 9 weeks. His average increased rapidly at first, then it increased more slowly, and then remained constant. Which graph best represents his average grade over this time period?

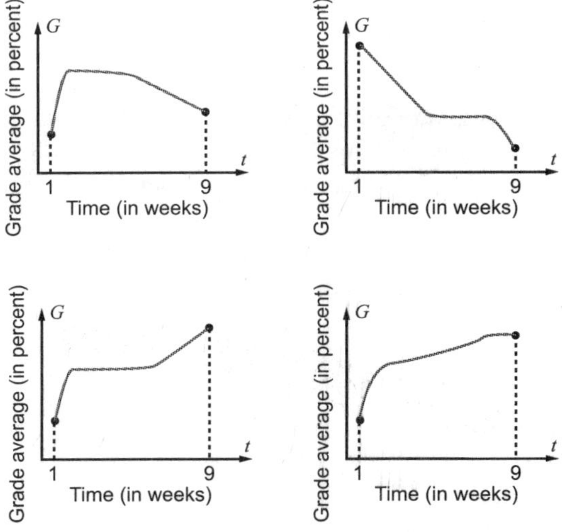

28. Determine which of the graphs best represents the profit, P, in thousands of dollars, of a company over a 12-month period. It is known that the profit was constant at first, then rapidly decreased, then remained constant again, and then finally slowly increased.

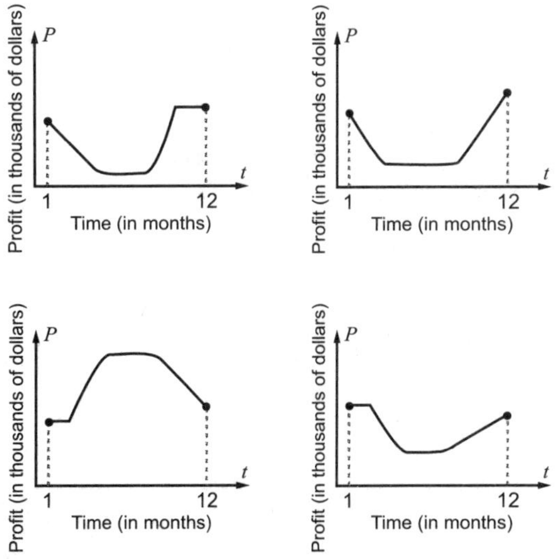

29. Determine which graph best represents the amount of money, A, in dollars, in a savings account over a 12-month period. It is known that the amount of money in the account increased rapidly at first to the greatest amount. Afterwards, it decreased and then slowly increased.

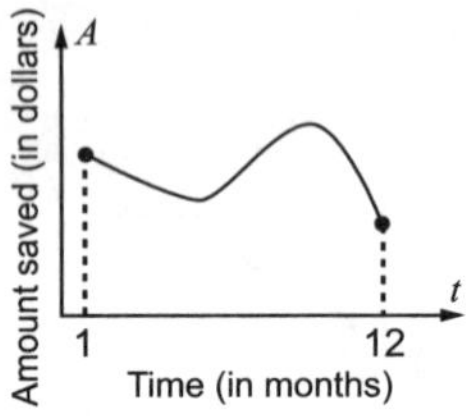

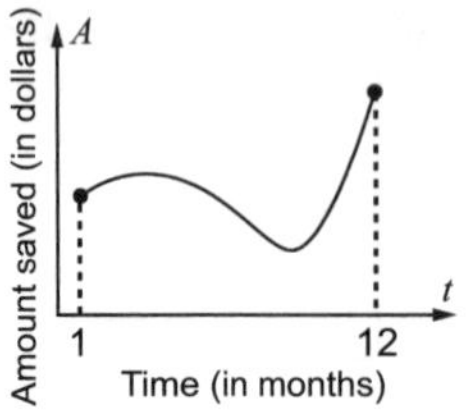

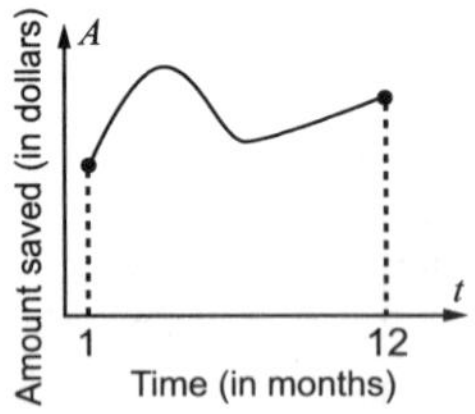

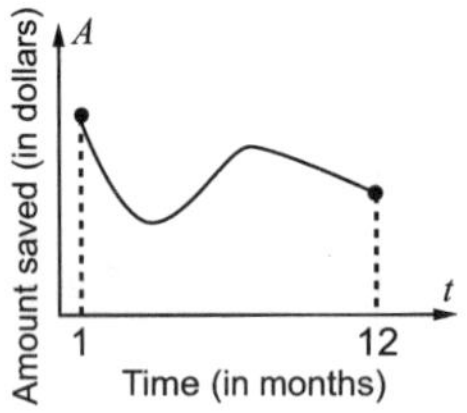

30. Manuel rode his bicycle for 5 hours. The graph represents his speed, S, in miles per hour, during his trip. It is known that his speed increased slowly at first, then remained constant, and then finally decreased rapidly. Which graph best represents his speed on this 5-hour trip?

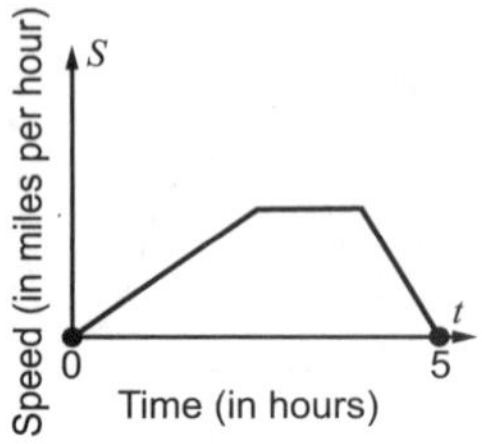

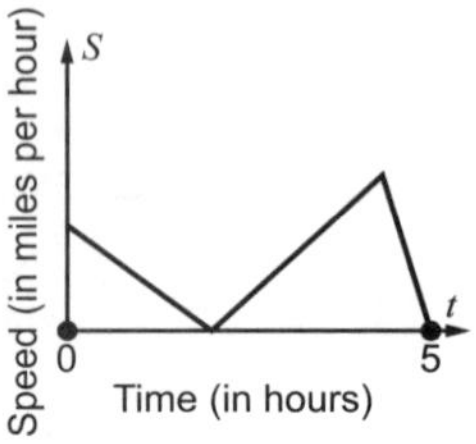

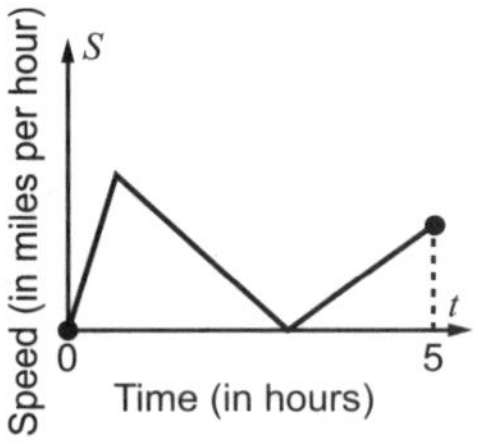

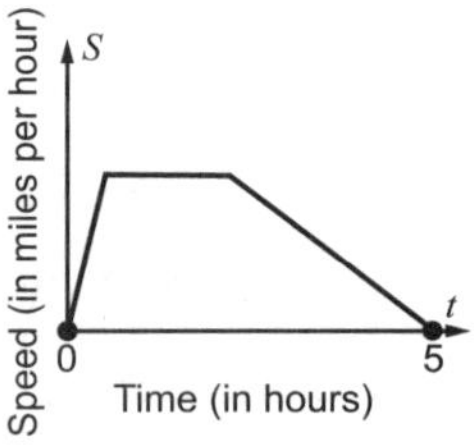

31. A used-car company wrote a report about the number of cars sold over an 8-month period. In the graph, N is the number of cars sold, and t is the time in months. Which report best represents the number of cars sold?

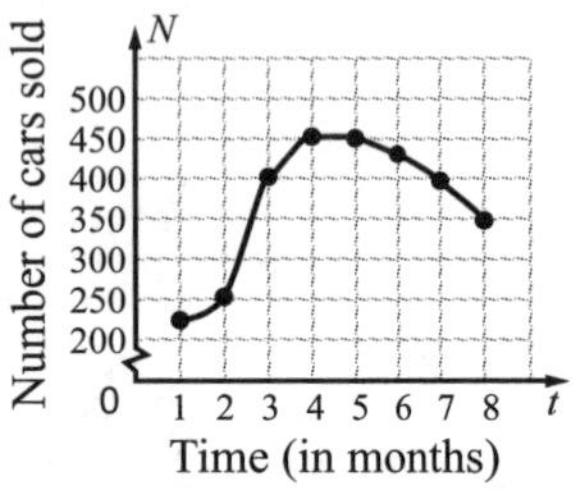

Sales increased for the first 3 months and were then constant for the remainder of the time period.

Sales increased very slowly for the first 3 months, then remained constant before decreasing rapidly.

Although the sales increased for the first 3 months, they increased rapidly between months 3 and 4. Sales were then constant for a month and then decreased slowly.

Although the sales increased for the first 3 months, they increased rapidly between months 2 and 3. Sales were then constant for a month and then decreased slowly.

32. A company wrote a report on its profit over a period of 10 months. In the graph, P is the profit in thousands of dollars, and t is the month. Which report best represents this company's profit over the 10-month period?

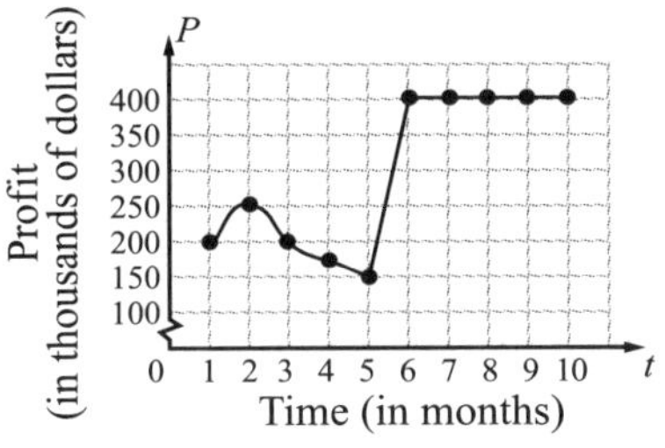

The profit increases, then decreases, and then increases rapidly for the remainder of the time period.

The profit increases slowly and then increases rapidly to its greatest value of $400,000. Then it remains constant for the remainder of the time period.

The profit decreases and then increases rapidly to its greatest value of $400. Then it remains constant for the remainder of the time period.

The profit increases, then decreases, and then increases rapidly to its greatest value of $400,000. Then it remains constant for the remainder of the time period.

33. A city published a report on its platoon over a 10-year period beginning in 1990. In the graph, P represents the population in thousands, and t represents the number of years since 1990. Which report best represents the population of this city over the given time period?

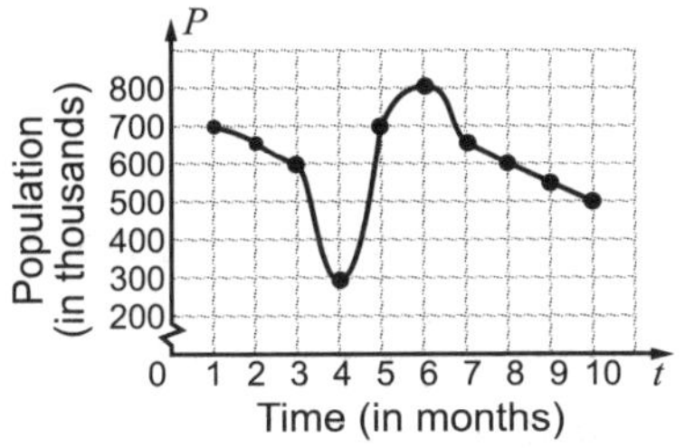

The population decreased rapidly for the first 3 years. Then it increased for 2 years but decreased again for the next 4 years.

The population decreased rapidly for the first 3 years. Then it remained constant for 2 years but decreased again for the next 4 years.

The population decreased for the first 3 years and then increased for the remainder of the 10-year period.

The population decreased for the first 3 years but decreased rapidly between 1993 and 1994. Then it increased for 2 years but decreased again for the next 4 years.

34. Nancy decided to analyze the amount of money in her savings account over the last 10 months. In the graph, A is the amount of money in her account, and t is the month. Which statement best represents the money in her account over the given time period?

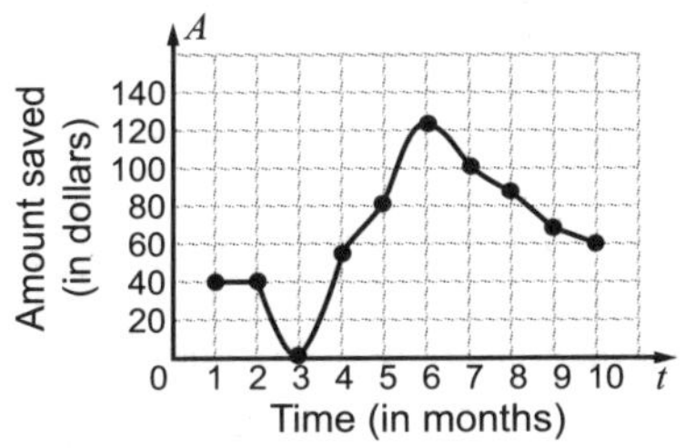

The amount of money in her account steadily increased to $120 and then decreased.

The money in her account increased to $40 for the first month, then decreased to $0. Then it increased to a high of $120 over the next 3 months, then decreased.

The money in her account was constant at $40 for the first month, then decreased to $0. Then it increased for the remainder of the time period.

The money in her account was constant at $40 for 1 month, then decreased to $0. Then it increased over the next 3 months to a high of $120 and then decreased.

35. John was considering buying a car which gets better as mileage, so he decided to record the cost of a tank of gas for a period of 10 months to determine any trends in the price. In the graph, C represents the cost of one tank of gasoline, and t is the month. Which statement best represents the gasoline costs over this time period?

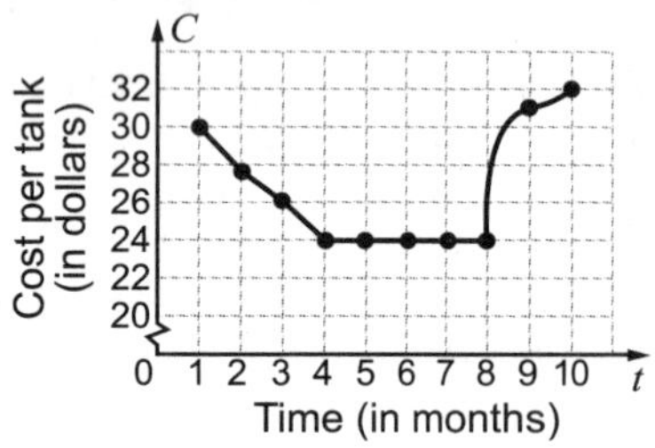

The cost of a tank of gasoline decreased for 3 months to $24 where it remained for 4 months. It rapidly increased between months 8 and 9, then increased more.

The cost of a tank of gasoline decreased for 3 months to $24 where it remained for 6 months. It rapidly increased between months 8 and 9, then increased more.

The cost of a tank of gasoline decreased for 6 months to $24, then rapidly increased between months 8 and 9, then increased to $32 by month 10.

The cost of a tank of gasoline decreased for 3 months to $24 where it remained for 4 months. It rapidly decreased between months 8 and 9 and then increased.

36. Roberto has a job such that his weekly pay depends on tips. He decided to observe his pay for 10 weeks to see if he wanted to keep his job. In the graph, S represents his weekly pay, and t represents the week. Which statement best describes his pay during this 10-week period?

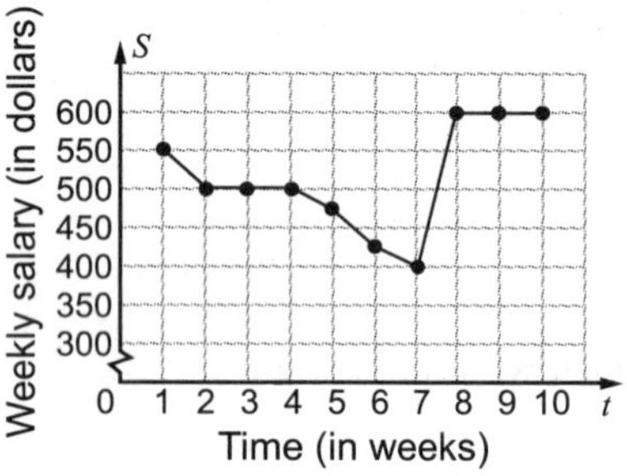

Roberto's pay increased at first, then remained constant. Afterwards, it decreased for 3 weeks and then rapidly increased to $600, where it remained constant.

Roberto's pay decreased at first, then remained constant. Afterwards, it decreased for 3 weeks and then slowly increased to $600.

Roberto's pay decreased at first, then remained constant. Afterwards, it decreased more for 3 weeks and then rapidly increased to $600, where it remained constant.

Roberto's pay increased at first, then remained constant. Afterwards, it decreased for 3 weeks and then remained constant.

37. A report on interest rates for home mortgage loans was released. In the graph, R represents the interest rate, and t is the time in months. Which statement best describes the interest rates during this time period of 10 months?

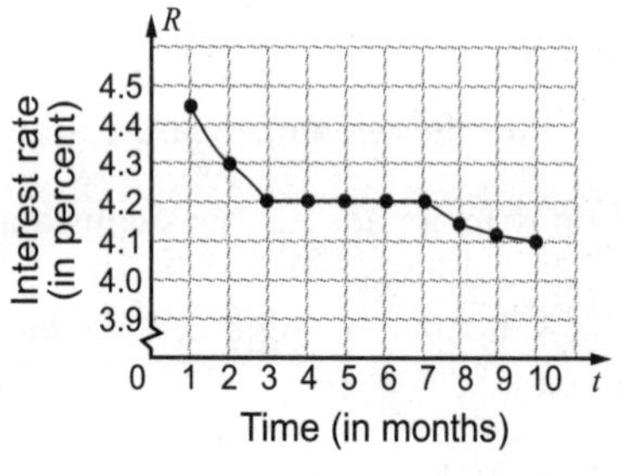

The interest rate increased for the first 2 months and then remained constant at 4.2% for 4 months. Then the rate increased for the final 3 months.

The interest rate decreased for the first 2 months and then remained constant at 4.2% for the final months.

The interest rate decreased for the first 2 months and then remained constant at 4.2% for 4 months. Then the rate decreased again for the final 3 months.

The interest rate decreased for the first 2 months and then increased for the remainder of the time.

38. Juanita decided that she wanted to invest her money in a particular stock. She watched its prices for a period of 10 days prior to purchasing it. In the graph, *P* represents the price of one share of the stock, and *t* is the day. Which statement best represents the price of this stock over the 10-day period of time?

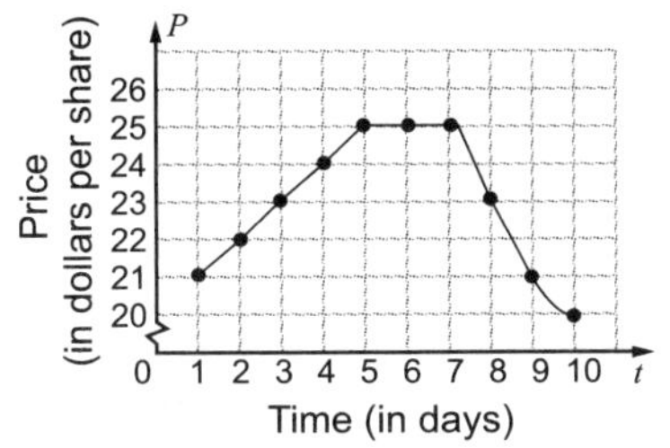

The price of one share of the stock increased steadily at first to #25 per share, where it remained for 2 days. Then the price decreased for the final 3 days.

The price of one share of the stock increased steadily at first to #25 per share, and then the price decreased for the remaining days.

The price of one share of the stock increased steadily at first to #20 per share, where it remained for 4 days. Then the price decreased for the final 3 days.

The price of one share of the stock increased steadily at first to #25 per share, where it remained for the remainder of the time.

39. A community hired a company to do a report on the water level in a nearby lake. In the graph, *W* represents the level of the water in inches above normal, and *t* represents the time in months. Which report best represents the water level in this lake over the given period of time?

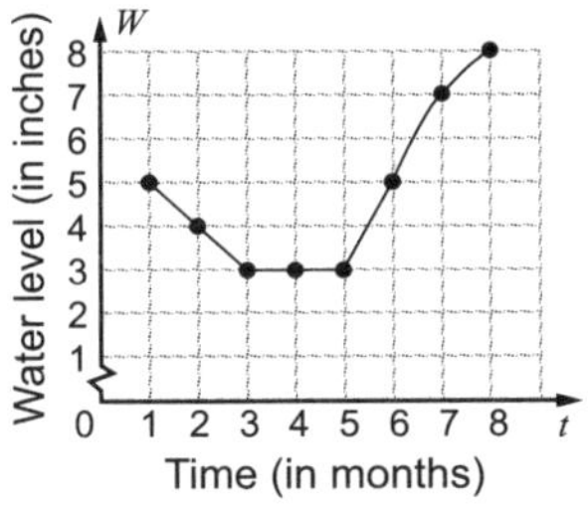

The water level in the lake fell from 5 inches to 3 inches above normal and stayed there for 5 months. Then it increased to its highest level of 8 inches above normal.

The water level in the lake fell from 5 inches to 3 inches above normal and stayed there for 2 months. Then it decreased for the remainder of the time.

The water level in the lake increased from 3 inches to 5 inches above normal and stayed there for 2 months. Then it increased to its highest level of 8 inches above normal.

The water level in the lake fell from 5 inches to 3 inches above normal and stayed there for 5 months. Then it increased to its highest level of 8 inches above normal.

40. Freddy wanted to visit a certain city, so he watched the average temperatures there for a period of one year. In the graph, *F* represents the average temperature in degrees Fahrenheit, and *t* represents the month. Which statement best describes the temperatures in this city during the year?

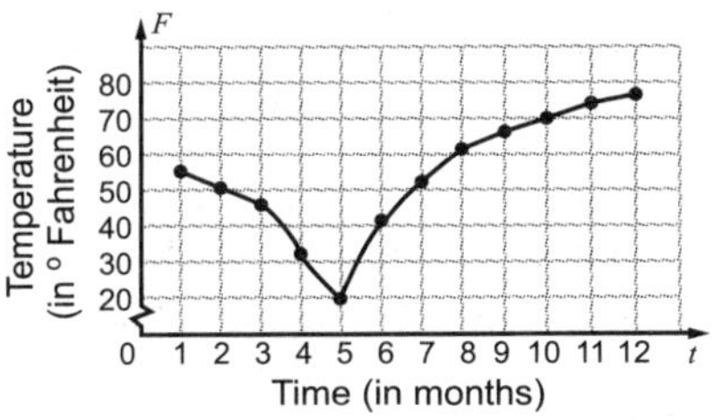

The temperatures decreased for 4 consecutive months to a low of 20 degrees and then increased for the next 7 months.

The temperatures decreased for 5 consecutive months to a low of 20 degrees and then increased for the next 6 months.

The temperatures decreased for 6 consecutive months to a low of 20 degrees and then increased for the next 6 months.

The temperatures decreased for 4 consecutive months to a high of 20 degrees and then decreased for the next 7 months.

HA1-449: Applying Inductive and Deductive Reasoning

In mathematics, there are two kinds of reasoning that are used to make conclusions. Both are very important to the mathematical process. These two kinds of reasoning are called deductive reasoning and inductive reasoning.

Deductive Reasoning	The process of generating a conclusion based on true statements

Inductive Reasoning	The process of making a generalization based on observations of specific examples

In ancient times, mathematicians measured the angles of many different triangles. They found that the sum of the angles of every triangle was 180 degrees, so they concluded that the sum of the angles of a triangle is always 180 degrees. They based their conclusion on observations of many different triangles. Their conclusion was based on inductive reasoning.

In later times, the fact that the sum of the angles of every triangle is always 180 degrees was proved based on other known true statements. The conclusion was then based on deductive reasoning. Deductive reasoning is the process of using known true statements to prove other true statements. Therefore, deductive reasoning is the basis of proof in mathematics. Inductive reasoning cannot be used as a proof because it is not possible to observe every example, but it is very important to the discovery process in mathematics.

Since conclusions obtained by inductive reasoning are based on observations, they are not proven statements and are often called conjectures.

Conjecture	An unproven statement based on observations

Deductive reasoning is used to prove that a conjecture is true, but a conjecture can be proved false by finding a counterexample.

Counterexample	An example that disproves a conjecture

Suppose the following observations are made: 19 is an odd number, 29 is an odd number, and 79 is an odd number. A possible conjecture is: If a number is an odd number, then its units digit is 9. This conjecture is actually false and can be proved false by a counterexample. One such counterexample is 15 because 15 is an odd number and its units digit is not 9. Other counterexamples are possible, but only one is needed to prove that a conjecture is false.

Conjectures are often written in "If-then" format. For example, "If a number is an odd number, then its units digit is 9." This statement is called a conditional statement. A conditional statement has two parts: the hypothesis, which is the statement after "If", and the conclusion, which is the statement after "then". .

Conditional Statement	A logical statement written in "If-then" format, where the clause after "If" is the hypothesis and the clause after the "then" portion is the conclusion

Hypothesis of a Conditional Statement	The clause that follows "If" in a conditional statement

Conclusion of a Conditional Statement	The clause that follows "then" in a conditional statement

Consider the above conditional statement once again: If a number is an odd number, then its units digit is 9.

Variables are sometimes used to denote the hypothesis and conclusion of a conditional statement. If p is used to denote the hypothesis of this conditional statement and q is used to denote the conclusion of this conditional statement, then p and q are as follows:

p: A number is an odd number.

q: Its units digit is 9.

Therefore, the conditional statement can be written in symbolic form as "If p, then q."

In order to show that a conditional statement is false, it is necessary to find a counterexample which shows that statement p, the hypothesis, is true, and that statement q, the conclusion, is false.

When a conclusion follows logically from given statements, it is called a valid conclusion. One method of obtaining a valid conclusion is to use the following argument. When "If p, then q" is true and p is true, then by deductive reasoning, q must also be true.

For example, use deductive reasoning to draw a valid conclusion using the given true statements. If a triangle is a scalene triangle, then it has no equal sides. Triangle ABC is a scalene triangle. Therefore, we can conclude that triangle ABC has no equal sides.

A conditional statement has three other conditional statements associated with it. They are called the converse, the inverse, and the contrapositive..

Converse	The statement formed by reversing the hypothesis and conclusion of the original conditional statement. The converse of *If* p*, then* q is *If* q*, then* p.

Inverse	The statement formed by negating the hypothesis and conclusion of the original conditional statement. The inverse of *If* p*, then* q is *If not* p*, then not* q.

Contrapositive	The statement formed by reversing and negating both the hypothesis and conclusion of the original conditional statement The contrapositive of *If* p*, then* q is *If not* q*, then not* p.

For example, given a conditional statement we can find the converse, inverse, and contrapositive.

Conditional Statement:	If x is greater than 5, then x is greater than 3.
Converse:	If x is greater than 3, then x is greater than 5.
Inverse:	If x is not greater than 5, then x is not greater than 3.
Contrapositive:	If x is not greater than 3, then x is not greater than 5.

A statement and its contrapositive always have the same truth value. If one is true, then the other is true. If one is false, then the other is false. The converse and inverse also have the same truth value. The statement and its converse may or may not have the same truth value. In the above example, the conditional statement and its contrapositive are both true statements. The converse and the inverse are both false statements. Since they are false, it is possible to produce a counterexample. There are many counterexamples, but only one is needed. One counterexample is $x = 4$. A conditional statement is false when the hypothesis is true and the conclusion is false. The converse is false because, if we replace x with 4, the hypothesis, "4 is greater than 3", is a true statement and the conclusion, "4 is greater than 5", is a false statement.

Deductive reasoning is used to solve equations. Consider the equation, $2(x-3)+1 = x-4$. When solving for x, as shown, true statements in the form of algebraic properties are used. In other words, when applying algebraic properties, such as the distributive property, deductive reasoning is used.

$$\begin{aligned} 2(x-3)+1 &= x-4 \\ 2x-6+1 &= x-4 \\ 2x-5 &= x-4 \\ 2x-x-5 &= x-x-4 \\ x-5 &= -4 \\ x-5+5 &= -4+5 \\ x &= 1 \end{aligned}$$

Therefore, by deductive reasoning, if $2(x-3)+1 = x-4$, then $x = 1$.

Example 1 Write a conditional statement with the given hypothesis p and conclusion q.

p: You play soccer for 2 hours.
q: You have fun.

If p, then q.
If you play soccer for 2 hours, then you have fun.

Step 1: Place the hypothesis, p, after "If" and the conclusion, q, after "then" in the statement.

Answer: If you play soccer for 2 hours, then you have fun.

Example 2 Find a counterexample for the given conditional statement: If x is divisible by 3, the x is divisible by 9.

p: x is divisible by 3.
q: x is divisible by 9.

Step 1: Identify the hypothesis, p, and the conclusion, q, in the given conditional statement.

Let $x = 6$.
6 is divisible by 3 and 6 is not divisible by 9.

Step 2: Find a value for x that makes the hypothesis true and the conclusion false.

Answer: 6
Note: Other counterexamples are 12, 15, 21, 24,...

Example 3 Which of the following arguments contains a conclusion that is based on deductive reasoning?

a) On Monday morning, you get stopped by a train on your way to school. On Tuesday morning, you get stopped by a train on your way to school. On Wednesday morning, you get stopped by a train on your way to school. You conclude that you will get stopped by a train every day on your way to school.

b) The newspaper contains a story saying that every Monday, a train goes through your town. Today is Monday. You conclude that a train will go through your town today.

When you observe a train on Monday, Tuesday, and Wednesday and conclude that you will get stopped by a train every day, you are basing your conclusion on observations. Inductive reasoning is based on observations. When you base your conclusion on true statements, you are using deductive reasoning.

Step 1: Examine the two arguments to see which one is based on deductive reasoning.

Answer: The newspaper contains a story saying that every Monday, a train goes through your town. Today is Monday. You conclude that a train will go through your town today.

Example 4 State the converse of the given conditional statement: If $2x + 1 = 5$, then $x = 2$.

p: $2x + 1 = 5$
q: $x = 2$

Step 1: Identify the hypothesis and conclusion of the original conditional statement.

The converse of *If p, then q* is *If q, then p*.

Step 2: Reverse the hypothesis and conclusion of the original conditional statement.

Answer: If $x = 2$, then $2x + 1 = 5$.

Example 5 Use deductive reasoning to draw a valid conclusion using the given statements.

Fred earned a higher grade on the test than Jackie.
Fred earned a lower grade on the test than Trevor.

Trevor
Fred
Jackie

Step 1: List the people in the order of their grades from highest to lowest (or lowest to highest). Fred is above Jackie because his grade is higher. Trevor is above Fred because Fred's grade is higher than Trevor's grade.

Trevor earned a higher grade than Jackie.	**Step 2:**	The person at the top of the list earned a higher grade than the person at the bottom of the list.
	Answer:	Trevor earned a higher grade on the test than Jackie.

Problem Set

Write a conditional statement for each with the given hypothesis *p* and conclusion *q*:

1. *p*: The storm is a hurricane.
q: The winds are more than 73 miles per hour.

2. *p*: An object is tossed up into the air.
q: The object will fall to the ground.

3. *p*: You do not wake up on time.
q: You will be late for school.

4. *p*: You have no money.
q: You cannot go to the movies.

5. *p*: You exceed the speed limit.
q: You will get a speeding ticket.

6. *p*: Two numbers are negative.
q: The sum of the numbers is negative.

7. *p*: n is an even integer.
q: $n + 2$ is an even integer.

8. *p*: $x \neq 3$
q: $2x - 4 \neq 2$

Solve:

9. Find a counterexample for the conditional statement "If an object is an appliance, then it is a refrigerator."

Shoe	Couch
Washing Machine	Broom

10. Find a counterexample for the conditional statement "If there is no school today, then today is Saturday."

Tuesday	Monday
Wednesday	Sunday

11. Find a counterexample for the conditional statement "If an animal lives in the ocean, then it is a whale."

Cat	Dog
Octopus	Pig

12. Find a counterexample for the conditional statement "If a number is prime, then it is an odd number."

2	9
10	3

13. Find a counterexample for the conditional statement "If x is any real number, then x^2 is greater than x."

4	1
2	3

14. Find a counterexample for the conditional statement "If x is greater than 0, then x is greater than 4."

–1	2
5	6

15. Find a counterexample for the conditional statement "If a person lives in Louisiana, then the person lives in New Orleans."

A person who lives in Dallas, Texas

A person who lives in Jacksonville, Florida

A person who lives in New York, New York

A person who lives in Baton Rouge, Louisiana

16. Which argument has a conclusion that is based on deductive reasoning?

If x is greater than 1, then x^2 is greater than x. You know that 3 is greater than 1. You conclude that 3^2 is greater than 3.

You noticed that your teacher did not give homework for three consecutive Fridays. You conclude that your teacher will never give homework on Friday.

You noticed that your garbage was picked up on three consecutive Wednesdays. You conclude that your garbage will be picked up every Wednesday.

You made the following observations: $2 \cdot 3$ is an even number, $2 \cdot 4$ is an even number, and $2 \cdot 5$ is an even number. You conclude that $2 \cdot x$ is always an even number.

17. Which argument has a conclusion that is based on inductive reasoning?

If x is an odd number, then x^2 is an odd number. 9 is an odd number. You conclude that 9^2 is an odd number.

You noticed that 3^2 is an odd number, 7^2 is an odd number, and 11^2 is an odd number. You conclude that x^2 is always an odd number.

If $x = 0$, then $xy = 0$. $xy \neq 0$. You conclude that $x \neq 0$.

If x is a negative number, then $-x$ is a positive number. -10 is a negative number. You conclude that $-(-10)$ is a positive number.

18. In the given argument, tell whether the conclusion is based on inductive or deductive reasoning.

You measured the angles of 3 triangles and made the following observations. The sum of the measures of the angles of triangle *ABC* was 180 degrees. The sum of the measures of the angles of triangle *DEF* was 180 degrees. The sum of the measures of the angles of triangle *XYZ* was 180 degrees. You conclude that the sum of the measures of the angles of every triangle is 180 degrees.

19. In the given argument, tell whether the conclusion is based on inductive or deductive reasoning.

In a certain school, the following observations were made: Room A has 32 desks, Room B has 34 desks, and Room C has 35 desks. You conclude that all classrooms in the school have more than 30 desks.

20. In the given argument, tell whether the conclusion is based on inductive or deductive reasoning.

You made the following observations in your algebra class: line l has a positive slope and it rises to the right, line m has a positive slope and it rises to the right, and line n has a positive slope and it rises to the right. You conclude that all lines that have a positive slope rise to the right.

21. In the given argument, tell whether the conclusion is based on inductive or deductive reasoning.

You learned in science that if an object is tossed into the air, then gravity will cause it to fall back to earth. You tossed a stone into the air. You conclude that the stone will fall back to earth.

22. In the given argument, tell whether the conclusion is based on inductive or deductive reasoning.

If an airplane is delayed, we will miss our connecting flight. The airplane is delayed. We conclude that we will miss our connecting flight.

23. In the given argument, tell whether the conclusion is based on inductive or deductive reasoning.

You learned that if $\frac{a}{b}$ is an integer, then b is a factor of a. You know that $\frac{144}{16}$ is an integer. You conclude that 16 is a factor of 144.

24. State the converse of the given statement.

"If you are industrious, then you are rich."

25. State the inverse of the given statement.

"If you live in Texas, then you live in the United States."

26. State the inverse of the given statement.

"If x is a rational number, then x is a real number."

27. State the converse of the given statement.

"If x is greater than 0, then $x + 2$ is greater than 0."

28. State the contrapositive of the given statement.

"If the weather is nice, then you will go to the baseball game."

29. State the contrapositive of the given statement.

"If the animal does not live in the sea, then the animal is not a fish."

30. State the contrapositive of the given statement.

"If the triangle is equilateral, then the sides of the triangle are equal in length."

Use deductive reasoning to draw a valid conclusion using the given statements:

31. Andy made more baskets than Joe, but fewer baskets than Jamie.

Andy made the most baskets.

Jamie made fewer baskets than Joe.

Jamie made more baskets than Joe.

Andy made the fewest baskets.

32. AJ spends less time studying than Sandra. Joey spends more time studying than Sandra.

Joey spends less time studying than AJ.

Sandra spends the least time studying.

Sandra spends the most time studying.

AJ spends less time studying than Joey.

33. Florida has warmer weather than Tennessee. Michigan has cooler weather than Tennessee.

Michigan has warmer weather than Florida.

Florida has warmer weather than Michigan.

Tennessee has the warmest weather.

Tennessee has the coolest weather.

34. Morgan is shorter than Kaitlin. Morgan is taller than Shelby.

Morgan is the tallest.

Kaitlin is shorter than Shelby.

Morgan is the shortest.

Shelby is the shortest.

35. $-(x+3) = 2(x+6)$

If $-(x+3) = 2(x+6)$, then $x = 3$.

If $-(x+3) = 2(x+6)$, then $x = 5$.

If $-(x+3) = 2(x+6)$, then $x = -3$.

If $-(x+3) = 2(x+6)$, then $x = -5$.

36. $-2x+3 = 7-(x+2)$

If $-2x+3 = 7-(x+2)$, then $x = -1$.

If $-2x+3 = 7-(x+2)$, then $x = 1$.

If $-2x+3 = 7-(x+2)$, then $x = -2$.

If $-2x+3 = 7-(x+2)$, then $x = 2$.

37. If you eat eggs for breakfast, then you do not get hungry in the morning. You ate eggs for breakfast.

You get hungry in the morning.

If you do not get hungry in the morning, then you ate eggs for breakfast.

If you do not eat eggs for breakfast, then you get hungry in the morning.

You do not get hungry in the morning.

38. If point A lies in Quadrant II on a coordinate plane, then the coordinates of point A do not have the same sign. Point A lies in Quadrant II on a coordinate plane.

The coordinates of point A do not have the same sign.

If the coordinates of point A do not have the same sign, then point A lies in Quadrant II on a coordinate plane.

The coordinates of point A have the same sign.

If point A does not lie in Quadrant II on a coordinate plane, then the coordinates of point A have the same sign.

39. If the sun is shining, the weather is nice for going to the beach. The weather is not nice for going to the beach.

If the sun is not shining, then the weather is not nice for going to the beach.

The sun is not shining.

If the weather is nice for going to the beach, then the sun is shining.

The sun is shining.

40. If x is an integer, then $2x$ is an even integer. $2x$ is not an even integer.

x is not an integer.

If x is not an integer, then $2x$ is not an even integer.

x is an integer.

If $2x$ is an even integer, then x is an integer.

HA1-492: Simplifying Square and Cube Roots

Square roots are used to find a number if its square is given.

Square Root	Let x be a real number and y be a non-negative real number such that $x^2 = y$ Then x **is the square root of** y. If $x \geq 0$, it is called the principal square root of y and it is denoted as $x = \sqrt{y}$. (radical sign: $\sqrt{\ }$; radicand: y) Square roots may also be written as $\sqrt{y} = y^{\frac{1}{2}}$

Remember, the square root of a negative number is not defined. Note that a positive number, y, has two square roots:

$$\text{Positive square root: } \sqrt{y} = y^{\frac{1}{2}}$$

$$\text{Negative square root: } -\sqrt{y} = -y^{\frac{1}{2}}$$

If asked to find a square root in words, give both the positive and negative square roots.

Properties of Square Roots	Let x and y be positive real numbers and n be a positive integer. Then $\sqrt{x^2} = x$ or $(x^2)^{\frac{1}{2}} = x$ $\sqrt{xy} = \sqrt{x} \cdot \sqrt{y}$ or $(xy)^{\frac{1}{2}} = (x)^{\frac{1}{2}} \cdot (y)^{\frac{1}{2}}$ $\sqrt{\frac{x}{y}} = \frac{\sqrt{x}}{\sqrt{y}}$ or $\left(\frac{x}{y}\right)^{\frac{1}{2}} = \frac{(x)^{\frac{1}{2}}}{(y)^{\frac{1}{2}}}$ $\sqrt{x^n} = (\sqrt{x})^n$ or $(x^n)^{\frac{1}{2}} = \left(x^{\frac{1}{2}}\right)^n = x^{\frac{n}{2}}$

Similarly, cube roots can be used to find a number when its cube is given.

Cube Root	Let x and y be real numbers such that $$x^3 = y.$$ Then x **is the cube root of** y and it is denoted as $$x = \sqrt[3]{y}.$$ (root or index: 3; radical sign: $\sqrt{\ }$; radicand: y) Cube roots can also be written as $$\sqrt[3]{y} = y^{\frac{1}{3}}.$$

There is only one cube root of a number. The radicand can be positive, negative, or zero.

Properties of Cube Roots	Let x and y be positive real numbers and n be a positive integer. Then $$\sqrt[3]{x^3} = x \quad \text{or} \quad (x^3)^{\frac{1}{3}} = x$$ $$\sqrt[3]{xy} = \sqrt[3]{x} \cdot \sqrt[3]{y} \quad \text{or} \quad (xy)^{\frac{1}{3}} = (x)^{\frac{1}{3}} \cdot (y)^{\frac{1}{3}}$$ $$\sqrt[3]{\frac{x}{y}} = \frac{\sqrt[3]{x}}{\sqrt[3]{y}} \quad \text{or} \quad \left(\frac{x}{y}\right)^{\frac{1}{3}} = \frac{(x)^{\frac{1}{3}}}{(y)^{\frac{1}{3}}}$$ $$\sqrt[3]{x^n} = (\sqrt[3]{x})^n \quad \text{or} \quad (x^n)^{\frac{1}{3}} = \left(x^{\frac{1}{3}}\right)^n = x^{\frac{n}{3}}$$

Example 1 Simplify: $\sqrt{144}$

$\sqrt{144} = \sqrt{(12)^2}$

Step 1: Express the given number as a perfect square of a non-negative number.

$\sqrt{(12)^2} = 12$

Step 2: Use the definition of square root to get the principal square root.

Answer: $\sqrt{144}$ simplifies to 12.

Example 2 Simplify: $\sqrt[3]{z^9}$

$\sqrt[3]{z^9} = (z^9)^{\frac{1}{3}}$ **Step 1:** Express the cube root as the $\frac{1}{3}$ power.

$= z^{\frac{9}{3}}$ **Step 2:** Use the property $(x^n)^{\frac{1}{3}} = \left(x^{\frac{1}{3}}\right)^n = x^{\frac{n}{3}}$

$= z^3$ **Step 3:** Simplify the exponent.

Answer: $\sqrt[3]{z^9}$ simplifies to z^3.

Example 3 Simplify $\sqrt{64a^6b^{14}}$, where a and b are positive real numbers.

$\sqrt{64a^6b^{14}} = (64a^6b^{14})^{\frac{1}{2}}$ **Step 1:** Express the square root as the $\frac{1}{2}$ power.

$= (64)^{\frac{1}{2}} \cdot (a^6)^{\frac{1}{2}} \cdot (b^{14})^{\frac{1}{2}}$ **Step 2:** Use the property $(xy)^{\frac{1}{2}} = (x)^{\frac{1}{2}} \cdot (y)^{\frac{1}{2}}$.

$= (8^2)^{\frac{1}{2}} \cdot (a^6)^{\frac{1}{2}} \cdot (b^{14})^{\frac{1}{2}}$ **Step 3:** Express 64 as a perfect square.

$= 8^{\frac{2}{2}} \cdot a^{\frac{6}{2}} \cdot b^{\frac{14}{2}}$ **Step 4:** Use the property $(x^n)^{\frac{1}{2}} = \left(x^{\frac{1}{2}}\right)^n = x^{\frac{n}{2}}$.

$= 8a^3b^7$ **Step 5:** Simplify the exponents.

Answer: $\sqrt{64a^6b^{14}}$ simplifies to $8a^3b^7$.

Example 4 Simplify: $\left(\frac{125x^9y^{15}}{27z^{12}}\right)^{\frac{1}{3}}$

$\left(\frac{125x^9y^{15}}{27z^{12}}\right)^{\frac{1}{3}} = \frac{(125x^9y^{15})^{\frac{1}{3}}}{(27z^{12})^{\frac{1}{3}}}$ **Step 1:** Use the property $\left(\frac{x}{y}\right)^{\frac{1}{3}} = \frac{(x)^{\frac{1}{3}}}{(y)^{\frac{1}{3}}}$.

$= \frac{(125)^{\frac{1}{3}} \cdot (x^9)^{\frac{1}{3}} \cdot (y^{15})^{\frac{1}{3}}}{(27)^{\frac{1}{3}} \cdot (z^{12})^{\frac{1}{3}}}$ **Step 2:** Use the property $(xy)^{\frac{1}{3}} = (x)^{\frac{1}{3}} \cdot (y)^{\frac{1}{3}}$ in both the numerator and the denominator.

$= \frac{(5^3)^{\frac{1}{3}} \cdot (x^9)^{\frac{1}{3}} \cdot (y^{15})^{\frac{1}{3}}}{(3^3)^{\frac{1}{3}} \cdot (z^{12})^{\frac{1}{3}}}$ **Step 3:** Express 125 and 27 as perfect cubes.

$$= \frac{5^{\frac{3}{3}} \cdot x^{\frac{9}{3}} \cdot y^{\frac{15}{3}}}{3^{\frac{3}{3}} \cdot z^{\frac{12}{3}}}$$

Step 4: Use the property $(x^n)^{\frac{1}{3}} = \left(x^{\frac{1}{3}}\right)^n = x^{\frac{n}{3}}$ in both the numerator and the denominator.

$$= \frac{5x^3y^5}{3z^4}$$

Step 5: Simplify the exponents.

Answer: $\left(\frac{125x^9y^{15}}{27z^{12}}\right)^{\frac{1}{3}}$ simplifies to $\frac{5x^3y^5}{3z^4}$.

Example 5 The volume, V, of a cube equals $64x^3y^9$ where x and y are positive real numbers. Find the area, A, of the top side of the cube.

$V = 64x^3y^9$

Step 1: The volume of the cube is given.

$V = l^3$

Step 2: Formula for the volume of the cube where l is the length of an edge of the cube.

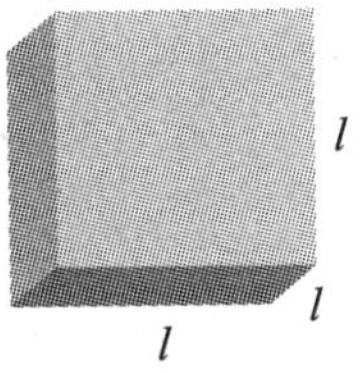

$l^3 = 64x^3y^9$

Step 3: Since both l^3 and $64x^3y^9$ are equal to V, they are equal to each other.

$l^3 = \sqrt[3]{64x^3y^9}$

Step 4: Apply the definition of a cube root.

$l = (64x^3y^9)^{\frac{1}{3}}$

Step 5: Express the cube root as the $\frac{1}{3}$ power.

$l = (64)^{\frac{1}{3}} \cdot (x^3)^{\frac{1}{3}} \cdot (y^9)^{\frac{1}{3}}$

Step 6: Use the property $(xy)^{\frac{1}{3}} = (x)^{\frac{1}{3}} \cdot (y)^{\frac{1}{3}}$.

$l = (4^3)^{\frac{1}{3}} \cdot (x^3)^{\frac{1}{3}} \cdot (y^9)^{\frac{1}{3}}$

Step 7: Express 64 as a perfect cube.

$l = 4^{\frac{3}{3}} \cdot x^{\frac{3}{3}} \cdot y^{\frac{9}{3}}$

Step 8: Use the property $(x^n)^{\frac{1}{3}} = \left(x^{\frac{1}{3}}\right)^n = x^{\frac{n}{3}}$.

$l = 4xy^3$

Step 9: Simplify the exponents to get the length, l, of an edge of the cube.

$A = l^2$

Step 10: Recall the formula for the area, A, of a square. All sides (and also the top side) of a cube are squares. The length of a side of the square equals the length, l, of an edge of the cube.

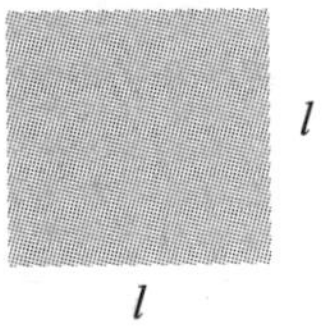

$A = (4xy^3)^2$	**Step 11:** Replace l with the value found in Step 9.
$A = 16x^2y^6$	**Step 12:** Simplify the expression.

Answer: The area of the top side of the given cube is $16x^2y^6$.

Problem Set

1. Simplify: $\sqrt[3]{-512}$

2. Simplify: $\sqrt[3]{729}$

3. Simplify: $\sqrt[3]{-1{,}000}$

4. Simplify: $\sqrt{256}$

5. Simplify: $\sqrt{196}$

6. Simplify: $(441)^{\frac{1}{2}}$

7. Simplify: $(512)^{\frac{1}{3}}$

8. Simplify $(x^{14})^{\frac{1}{2}}$ where x is a positive real number.

9. Simplify $(a^{24})^{\frac{1}{2}}$ where a is a positive real number.

10. Simplify $\sqrt{b^8}$ where b is a positive real number.

11. Simplify $\sqrt{z^{20}}$ where z is a positive real number.

12. Simplify $(x^{21})^{\frac{1}{3}}$ where x is a real number.

13. Simplify $(y^{24})^{\frac{1}{3}}$ where y is a real number.

14. Simplify $\sqrt[3]{x^{12}}$ where x is a real number.

15. Simplify $\sqrt[3]{c^{27}}$ where c is a real number.

16. Simplify $\sqrt{4a^6b^4}$ where a and b are positive real numbers.

17. Simplify $\sqrt{81x^8y^{14}}$ where x and y are positive real numbers.

18. Simplify $\sqrt{9a^{12}b^2c^{10}}$ where a, b, and c are positive real numbers.

19. Simplify $\sqrt{25p^{16}q^{14}r^8}$ where p, q, and r are positive real numbers.

20. Simplify $\sqrt[3]{64x^9y^{15}}$ where x and y are real numbers.

21. Simplify $\sqrt[3]{27a^{18}b^6}$ where a and b are real numbers.

22. Simplify $\sqrt[3]{125p^{21}q^{12}r^9}$ where p, q, and r are real numbers.

23. Simplify $\sqrt[3]{216x^{21}y^{24}z^{18}}$ where x, y, and z are real numbers.

24. Simplify $\left(\frac{25x^8}{49y^{10}}\right)^{\frac{1}{2}}$ where x and y are positive real numbers and $y \neq 0$.

25. Simplify $\left(\dfrac{36a^{16}}{121b^{12}}\right)^{\frac{1}{2}}$ where a and b are positive real numbers and $b \neq 0$.

26. Simplify $\left(\dfrac{16x^{12}}{25y^{6}z^{4}}\right)^{\frac{1}{2}}$ where x, y, and z are positive real numbers, $y \neq 0$, and $z \neq 0$.

27. Simplify $\left(\dfrac{144x^{6}y^{4}}{169z^{16}}\right)^{\frac{1}{2}}$ where x, y and z are positive real numbers and $z \neq 0$.

28. Simplify $\left(\dfrac{343x^{15}}{64y^{30}}\right)^{\frac{1}{3}}$ where x and y are real numbers and $y \neq 0$.

29. Simplify $\left(\dfrac{8a^{21}b^{15}}{125c^{27}}\right)^{\frac{1}{3}}$ where a, b and c are real numbers and $c \neq 0$.

30. Simplify $\left(\dfrac{216x^{12}}{27y^{6}z^{9}}\right)^{\frac{1}{3}}$ where x, y and z are real numbers, $y \neq 0$, and $z \neq 0$.

31. The volume of a cube is given by $V = 64x^{15}$. Find the length of each edge a.

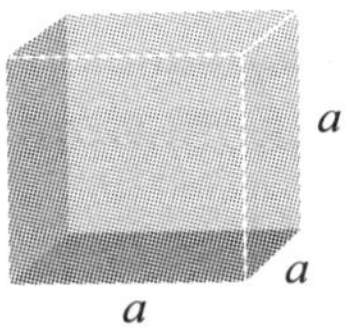

32. The volume of a cube is given by $V = 216y^{21}$. Find the length of each edge a.

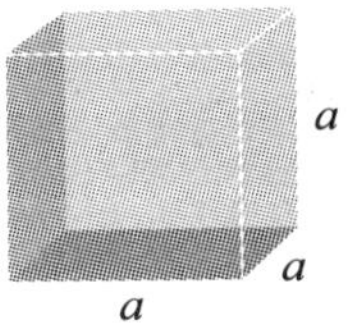

33. The volume of a cube is given by $V = 27p^{30}q^{18}$. Find the length of each edge a.

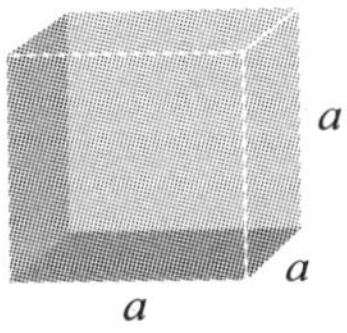

34. The volume of a cube is given by $V = 8u^{18}v^{21}w^{24}$. Find the length of each edge a.

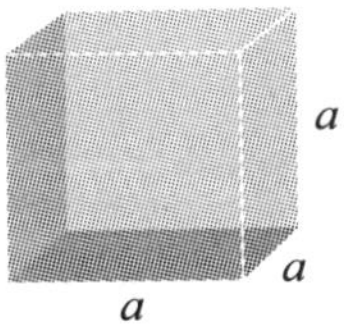

35. The volume of a cube is given by $V = 729x^{45}$. Find the length of each edge a.

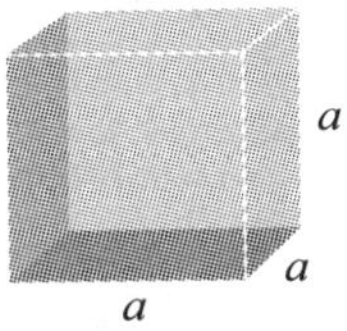

36. The area of a square is given by $A = 64x^{16}$. Find the length of each side a.

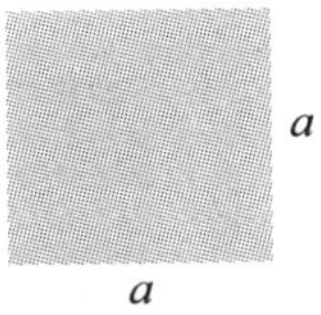

37. The area of a square is given by $A = 121c^{18}$. Find the length of each side a.

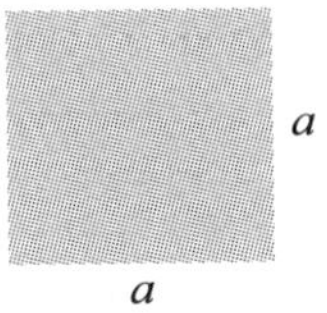

38. The area of a square is given by $A = 81x^{26}y^{22}$. Find the length of each side a.

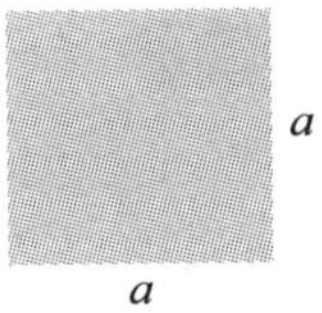

39. The area of a square is given by $A = 36x^{8}y^{10}z^{12}$. Find the length of each side a.

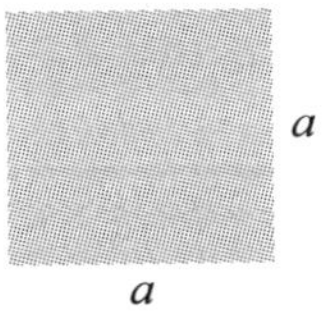

40. The area of a square is given by $A = 400x^{36}$. Find the length of each side a.

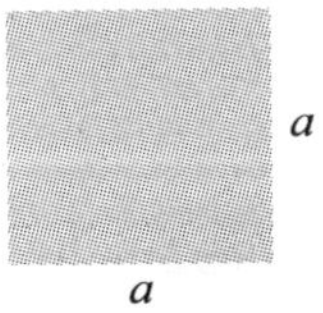

HA1-540: Finding the Mean, Median, and Mode from Data and Frequency Distribution Tables

Statistics can represent a set of data as a **measure of central tendency**.

Measure of Central Tendency	A number that represents the center of a set of data

Three such statistics are discussed in this lesson: mean, median, and mode.

Mean	The sum of the data divided by the number of items of data. The mean is also called the average or the arithmetic average.

Example 1 Determine the mean of the following set of data: {27, 23, 31, 12, 45, 28, 17, 11, 11, 21}
Note: A set of data is denoted by a list of numbers enclosed in braces.

$$\frac{27+23+31+12+45+28+17+11+11+21}{10} = \frac{226}{10}$$

$$\text{Mean} = 22.6$$

Step 1: Find the mean by adding the numbers and dividing by 10, the number of data items.

Answer: Mean = 22.6

Median	The middle value of an ordered set of data. When there are two middle values, the median is the average of those two values.

Example 2 Find the median of the following set of data: {27, 23, 31, 12, 45, 28, 17, 11, 21}

11, 12, 17, 21, 23, 27, 28, 31, 45

Step 1: List the data items in ascending order.

11, 12, 17, 21, **23**, 27, 28, 31, 45

Step 2: The middle value of the list is 23, so the median is 23.

Answer: Median = 23

Example 3 Find the median of the following set of data: {27, 23, 31, 19, 12, 45, 28, 17, 11, 22}

11, 12, 17, 19, 22, 23, 27, 28, 31, 45

Step 1: List the data in ascending order.

11, 12, 17, 19, **22**, **23**, 27, 28, 31, 45

Step 2: It is clear that 22 and 23 are the two values in the middle of the list.

$$\frac{22+23}{2} = \frac{45}{2} = 22.5$$

Step 3: Find the average of 22 and 23.

Answer: Median = 22.5

If there are n pieces of data then:

(a) If n is odd, the middle number is in position $\frac{n+1}{2}$.

(b) If n is even, the middle two numbers are in positions $\frac{n}{2}$ and $\frac{n}{2}+1$.

Mode	The number that occurs most often in the set of data

Example 4 Twenty people were asked to taste three cereals numbered 1, 2 and 3 then pick their favorite. The results were 1, 2, 3, 3, 1, 3, 3, 3, 1, 2, 3, 3, 3, 1, 3, 3, 3, 1, 3, 2. Find the mode.

Cereal	Frequency
1	5
2	3
3	12
Total	20

Step 1: It is often easier to interpret the data if it is arranged in a frequency table.

Note: A frequency table gives the number of times each item appears. It is not a frequency distribution because categories are not intervals. The number of the cereal only identifies the cereal, it is not a numerical value.

Answer: The mode is Cereal 3, because the frequency table shows that Cereal 3 occurs 12 times, which is more often than the other two cereals appear.

The mode is usually used as a measure of central tendency in situations where the mean and median make no sense. For example, in Example 4, the mean of the numbers is 2.35. It does not make sense to say that cereal number 2.35 is the mean. One problem with using the mode is that it is possible for two numbers to appear with equally great frequency.

Example 5 Find the mode of 4, 5, 6, 4, 5, 6, 7, 6, 4.

Number	Frequency
4	3
5	2
6	3
7	1
Total	9

Step 1: Make a frequency table.

Step 2: In the data, 4 and 6 each appear three times, 5 appears twice and 7 appears once. In this example, there are two modes, 4 and 6.

Answer: The modes are 4 and 6.

Problem Set

1. During one week in September, the high temperatures were:

   83°, 89°, 90°, 90°, 88°, 84°, 85°

   Find the mode of the temperatures.

2. During one week in September, the high temperatures were:

   80°, 88°, 87°, 84°, 82°, 86°, 88°

   Find the mode of the temperatures.

3. Jonathan surveyed 8 students and recorded the number of video games each played last weekend:

   1, 5, 4, 4, 0, 3, 5, 2

   What is the mean number of games played by these students?

4. Jerome surveyed 9 students and recorded the number of video games each played last weekend:

   0, 1, 4, 3, 4, 3, 2, 1, 0

   What is the mean number of games played by these students?

5. Jamal surveyed 8 students and recorded their weekly allowances:

   \$20, \$15, \$18, \$23.50, \$16.50, \$10, \$20, \$13

   What is the median of the allowances?

6. Jamal surveyed 8 students and recorded their weekly allowances:

   \$20, \$10, \$15, \$14.50, \$17.50, \$10, \$18, \$25

   What is the mean of the allowances?

7. Andrew surveyed 8 students and recorded their weekly allowances:

   \$10, \$16, \$10, \$16, \$22, \$23, \$15, \$8

   What is the median of the allowances?

8. Jamal surveyed 8 students and recorded their weekly allowances:

   \$10, \$20, \$17, \$12.50, \$15, \$19.50, \$15, \$25

   What is the mode of the allowances?

9. The following is a list of numbers of siblings for each of the 27 students in Mr. Mundy's algebra class:

   5, 0, 2, 3, 3, 4, 0, 6, 2, 1, 2, 4, 0, 6, 2, 6, 1, 2, 4, 3, 3, 2, 5, 3, 5, 1, 6

   The frequency table for this data is shown:

Siblings	Frequency
0	A
1	B
2	C
3	D
4	E
5	F
6	G

   What number will go in box A?

10. The following is a list of numbers of siblings for each of the 27 students in Mr. Mundy's algebra class:

   5, 0, 2, 3, 3, 4, 0, 6, 2, 1, 2, 4, 0, 6, 2, 6, 1, 2, 4, 3, 3, 2, 5, 3, 5, 1, 6

   The frequency table for this data is shown:

Siblings	Frequency
0	A
1	B
2	C
3	D
4	E
5	F
6	G

   What number will go in box B?

11. How many data values were used to make the table?

Test Scores	Number of Students
100	5
90	33
80	20
70	15
60	8

12. What is the mode of the data set represented by this frequency table?

Test Scores	Number of Students
100	5
90	33
80	20
70	15
60	8

13. What is the sum of the data?

Test Scores	Number of Students
100	3
80	5
60	12
40	11
20	6
0	3

14. The table shows a survey of the number of trips made to the mall during one week in July. What is the mean number of trips?

Trips	Frequency	Total
1	2	2
2	4	8
3	5	15
4	3	12
5	1	5
Sum	**15**	**42**

15. The table shows a survey of the number of trips made to the mall during one week in July. What is the median number of trips?

Trips	Frequency	Total
1	2	2
2	4	8
3	5	15
4	3	12
5	1	5
Sum	**15**	**42**

16. Below are two sets of scores for two different competitive divers. Higher scores are better.

Hal's Scores: 5.9, 5.4, 5.8, 5.5, 5.4, 5.6, 5.7
Bill's Scores: 5.2, 5.3, 5.9, 5.9, 5.8, 5.3, 5.3

One scoring method requires that the two lowest scores are dropped. Then, the mean of the remaining five scores is computed and used as the final score.
Who will win if this method is used?

17. Below are two sets of scores for two different competitive divers. Higher scores are better.

Hal's Scores: 5.9, 5.4, 5.8, 5.5, 5.4, 5.6, 5.7
Bill's Scores: 5.2, 5.3, 5.9, 5.9, 5.8, 5.3, 5.3

One scoring method requires that the highest score and the lowest score are dropped. Then, the median of the remaining five scores is computed and used as the final score.
Who will win if this method is used?

18. What is the median of the data set represented by the frequency table?

Test Scores	Frequency
100	17
75	4
50	3
25	2
0	1

19. What is the mean of the data set represented by the frequency table?

Test Scores	Number of Students
100	3
80	5
60	12
40	11
20	6
0	3

20. What is the median of the data set represented by the frequency table?

Number of Tardies this Week	Number of Students
0	125
1	70
2	5
3	2
4	1

HA1-541: Analyzing Data Using the Measures of Central Tendency and the Range

Statistics play an important role in mathematics by helping us analyze data to draw conclusions. In this lesson, we will discuss the two different types of data, the measures of central tendency, and the range.

There are two types of data, quantitative data and categorical data.

Quantitative Data	Numeric data
Categorical Data	Non-numeric data

Examples of quantitative data include the ages of students at your school or the temperatures of the classrooms. Examples of categorical data include the letter grades earned on your algebra tests or the types of vehicles students drive.

There are four measures that are useful when analyzing data. The first three measures are the mean, median, and mode. These measures are called the measures of central tendency. The fourth measure, the range, is a measure of variation.

Mean	The sum of the values in a data set divided by the number of values in the data set. The mean is often called the average.
Median	The middle value in a data set when the data is ordered from least to greatest. If there is an even number of data values, the median is the mean of the two middle values.
Mode	The value or category that occurs most frequently in a data set. It is possible for a data set to have more than one mode or no mode.
Range	The difference between the maximum and minimum values in a data set

Notice that the measure of the data that can be calculated depends on the type of data. If the data is quantitative, the mean, median, mode, and range can all be calculated. However, if the data is categorical, only the mode can be calculated.

Example 1 A quality control technician measured the amounts of ketchup in a sample of six 8-ounce bottles. If the weights (in ounces) of the ketchup are 7.8, 7.7, 7.9, 8.2, 8.0, and 8.1, determine the median amount of ketchup in these bottles.

{7.7, 7.8, 7.9, 8.0, 8.1, 8.2}	**Step 1:** Order the data set.
{7.7, 7.8, **7.9**, **8.0**, 8.1, 8.2}	**Step 2:** Locate the middle weights, 7.9 oz and 8.0 oz.

$$\text{Median} = \frac{7.9 + 8.0}{2} = \frac{15.9}{2} = 7.95 \text{ oz}$$

Step 3: Since there are an even number of weights, the median is equal to the mean of the two middle weights.

Answer: The median weight is 7.95 oz.

Example 2 The following data set contains the ages (in years) of nine people on a volleyball team: {32, 23, 40, 25, 25, 27, 31, 36, 28}. What is the range of the ages?

{23, 25, 27, 28, 31, 32, 36, 40}

Step 1: Order the data set.

{**23**, 25, 27, 28, 31, 32, 36, **40**}

Step 2: Identify the minimum age and the maximum age.

$$\text{Range} = 40 - 23 = 17 \text{ years}$$

Step 3: Calculate the difference between the maximum and minimum ages.

Answer: The range is 17 years.

Example 3 The number of typing errors made by a secretary in five different memos are 5, 2, 1, 0, and 3. Which measure of central tendency gives the lowest number of errors?

$$\text{Mean} = \frac{5 + 2 + 1 + 0 + 3}{2} = \frac{11}{5} = 2.2 \text{ errors}$$

Step 1: Calculate the mean number of typing errors.

{0, 1, **2**, 3, 5}

Median = 2 errors

Step 2: Calculate the median number of typing errors by first ordering the data set and then locating the middle value.

{0, 1, **2**, 3, 5}

There is no mode since each value occurs once.

Step 3: Identify the mode by first ordering the data set and then locating the value that occurs most frequently.

The median is smallest in value.

Step 4: Compare the values of the mean, median, and mode. Choose the measure of central tendency that is smallest in value.

Answer: The median, 2 errors, gives the lowest number of errors.

Example 4 A biologist recorded the life span of 100 butterflies. What measure of the data was used to determine that the average life span of the butterflies was 34 days?

The mean is also called the average.

Step 1: Identify the key word, “average”, in the question.

Answer: The mean was used to determine that the average life span of the butterflies was 34 days.

Example 5 Six people were asked about their daily TV viewing habits. The table below shows the number of hours each person spends watching TV. Which measure(s) of the data would be least affected if Steven watched 2 more hours of TV per day, and Kevin watched 2 hours less?

Person	Number of Hours per Day
Nicole	1
Steven	3
Sarah	3
Kevin	4
Carter	2
Todd	5

Step 1: Create a new column in the table for the adjusted TV viewing times. Add 2 hours to Steven's time and subtract 2 hours from Kevin's time.

Person	Number of Hours per Day	Adjusted Number of Hours per Day
Nicole	1	1
Steven	3	5
Sarah	3	3
Kevin	4	2
Carter	2	2
Todd	5	5

Step 2: Order the original data set and calculate the mean, median, mode, and range.

The original ordered data set is $\{1, 2, 3, 3, 4, 5\}$.

$$\text{Mean} = \frac{1+2+3+3+4+5}{6} = \frac{18}{6} = 3 \text{ hours}$$

$$\text{Median} = \frac{3+3}{2} = \frac{6}{2} = 3 \text{ hours}$$

Mode = 3 hours

Range = 5 – 1 = 4 hours

Step 3: Order the adjusted data set and calculate the mean, median, mode, and range.

The adjusted data set is $\{1, 2, 2, 3, 5, 5\}$.

$$\text{Mean} = \frac{1+2+2+3+5+5}{6} = \frac{18}{6} = 3 \text{ hours}$$

$$\text{Median} = \frac{2+3}{2} = \frac{5}{2} = 2.5 \text{ hours}$$

Mode = 2 hours and 5 hours

Range = 5 – 1 = 4 hours

Step 4: Compare the values of the mean, median, mode, and range. Choose the measure(s) of the data that remained the same.

The mean and range remained the same.

Answer: The mean and range would be least affected.

Problem Set

1. Calculate the mean of the following data set:
$\{2, 7, 4, 3, 2, 6\}$

2. Calculate the median of the following data set:
$\{2, 7, 4, 3, 2, 6\}$

3. Calculate the mode of the following data set:

$\{2, 7, 4, 3, 2, 6\}$

4. Calculate the range of the following data set:

$\{2, 7, 4, 3, 2, 6\}$

5. A convenience store sells five different sizes of fountain drinks. The prices of each size are \$0.59, \$0.79, \$0.99, \$1.09, and \$1.19. What is the average price of a fountain drink?

6. A convenience store sells five different sizes of fountain drinks. The prices of each size are \$0.59, \$0.79, \$0.99, \$1.09, and \$1.19. What is the median price of a fountain drink?

7. A convenience store sells five different sizes of fountain drinks. The prices of each size are \$0.59, \$0.79, \$0.99, \$1.09, and \$1.19. What is the mode(s) of the prices of the fountain drinks?

8. A convenience store sells five different sizes of fountain drinks. The prices of each size are \$0.59, \$0.79, \$0.99, \$1.09, and \$1.19. What is the range of the prices of the fountain drinks?

9. Sandy recorded the eye color of each student in her chemistry class. She found that 16 students had brown eyes, 7 had hazel eyes, 4 had green eyes, and 4 had blue eyes. Which eye color(s) is the mode?

10. The heights (in inches) of six players on a professional basketball team are given in the data set: $\{72, 75, 71, 79, 75, 72\}$. What is the mean height of these six players?

11. The heights (in inches) of six players on a professional basketball team are given in the data set: $\{72, 75, 71, 79, 75, 72\}$. What is the median height of these six players?

12. The heights (in inches) of six players on a professional basketball team are given in the data set: $\{72, 75, 71, 79, 75, 72\}$. What is the mode(s) of the heights of these six players?

13. The heights (in inches) of six players on a professional basketball team are given in the data set: $\{72, 75, 71, 79, 75, 72\}$. What is the range of the heights of these six players?

14. Jamal can take one of several routes to get to his office in the morning. The distances (in miles) of each of these routes are 5.2, 3.8, 5.9, 5.2, and 4.9. What is the average distance between Jamal's house and his office?

15. A car salesman sold 4 red cars, 2 blue cars, 6 black cars, and 6 white cars. Determine the mode(s).

16. Given the data set $\{8, 15, 12, 19, 18, 18\}$, which statement best interprets the data?

The mean and the median are not the same.
The range is 8.
The mean and mode are equal.
The mean, median, and mode are all 15.

17. The square footage of six homes is given in the following data set: {1,500, 1,750, 1,425, 1,850, 1,900}. Which statement best interprets the data?

The median is greater than the mean.
The median is less than the mode.
The mode is equal to the mean.
The median is less than the mean.

18. The number of students in each of Mr. Anderson's algebra classes is given in the data set: {23, 28, 21, 23, 23, 20}. Which statement best interprets the data?

The mean does not equal the median.
The range of the data is 23 students.
The mean, median, and mode are all 23 students.
Only the mean is 23 students.

19. The number of phone calls a secretary made each day last week are 19, 17, 27, 17, and 10. Which statement best interprets the data?

Only the median is 17 phone calls.
The mean and median are both 17 phone calls.
The range is less than the median.
The median, mode, and range are all 17 phone calls.

20. The number of viewers (in millions) who watched five different TV shows last week is given in the data set $\{5.8, 9.7, 28.8, 16.9, 5.8\}$. Which measure of central tendency will give the highest number of viewers?

Mode	Range
Mean	Median

21. The number of viewers (in millions) who watched five different TV shows last week is given in the data set $\{5.8, 9.7, 28.8, 16.9, 5.8\}$. Which measure of central tendency will give the lowest number of viewers?

Mean　　Median
Range　　Mode

22. The following data set contains the standardized test scores of six different students: {31, 24, 27, 34, 17, 17}. Which measure of central tendency will give the highest score?

Range　　Median
Mean　　Mode

23. The following data set contains the standardized test scores of six different students: {31, 24, 27, 34, 17, 17}. Which measure of central tendency will give the lowest score?

Range　　Mode
Median　　Mean

24. The ages (in years) of each person in a high school were recorded. What measure of the data was used to determine that the ages varied by 6 years?

25. A bank teller recorded the number of deposits she received each day for a month. What measure of the data was used to determine that the average number of deposits was 37 per day?

26. A doctor recorded the blood type of each of his patients. Which measure of the data was used to determine that the most common blood type was A-positive?

Median　　Mean
Mode　　Range

27. A basketball coach recorded each of his players' heights (in inches). Which measure of the data was used to determine that one-half of the players were shorter than 70 inches?

Range　　Median
Mean　　Mode

28. A survey of 75 college students was conducted. What measure of the data was used to determine that one-half of the students slept more than 7.1 hours per night?

29. The shoe sizes of twenty 4th-grade students were recorded. What measure of the data was used to determine that the most common shoe size was 3?

30. The weights (in pounds) of several newborn babies in a hospital nursery were recorded. What measure of the data was used to determine that the weights varied by 5.3 pounds?

31. The annual attendances (in millions) of several theme parks across the country are shown in the table below. Which measures of the data would be affected if the theme park in Florida had 6 million more people visit?

Location	Attendance (in millions)
Florida	31.2
California	24.7
Texas	17.1
New York	16.9
Ohio	6.8
Missouri	5.3

Mean and range

Median and mode

Mean, median, and mode

Mean, median, and range

32. The annual attendances (in millions) of several theme parks across the country are shown in the table below. Which measures of the data would be affected if the theme park in Texas had 4 million fewer people visit?

Location	Attendance (in millions)
Florida	31.2
California	24.7
Texas	17.1
New York	16.9
Ohio	6.8
Missouri	5.3

Mean, median, and mode

Mode and range

Mean and median

Mean, median, and range

33. The annual attendances (in millions) of several theme parks across the country are shown in the table below. Which measure of the data would be least affected if the theme park in New York had 7.8 million more people visit?

Location	Attendance (in millions)
Florida	31.2
California	24.7
Texas	17.1
New York	16.9
Ohio	6.8
Missouri	5.3

Mean

Mode

Median

Range

34. The annual attendances (in millions) of several theme parks across the country are shown in the table below. Which measures of the data would be least affected if a new park in Washington opened with an annual attendance of 13.5 million?

Location	Attendance (in millions)
Florida	31.2
California	24.7
Texas	17.1
New York	16.9
Ohio	6.8
Missouri	5.3

Mean and median

Mode and range

Range and mean

Mode and median

35. The annual attendances (in millions) of several theme parks across the country are shown in the table below. Which measures of the data would be affected if the theme parks in Ohio and Missouri closed?

Location	Attendance (in millions)
Florida	31.2
California	24.7
Texas	17.1
New York	16.9
Ohio	6.8
Missouri	5.3

Mean, median, and mode

Median and range only

Mean and median only

Mean, median, and range

36. The table below shows the number of hours Antonio worked each day last week. Which measures of the data would be affected if he worked 4 fewer hours on Friday?

Day	Hours Worked
Monday	6
Tuesday	5.5
Wednesday	8
Thursday	6.5
Friday	7

Mean and median only

Mean and mode

Median and range only

Mean, median, and range

37. The table below shows the number of hours Antonio worked each day last week. Which measures of the data would be affected if he worked 1 extra hour on Friday?

Day	Hours Worked
Monday	6
Tuesday	5.5
Wednesday	8
Thursday	6.5
Friday	7

Mean and median

Mean, median, and mode

Mean and mode

Mean, mode, and range

38. The table below shows the number of hours Antonio worked each day last week. Which measure(s) of the data would be least affected if he worked 3 fewer hours on Tuesday?

Day	Hours Worked
Monday	6
Tuesday	5.5
Wednesday	8
Thursday	6.5
Friday	7

Mode only

Mode and range

Mean and median

Median and mode

39. The table below shows the number of hours Antonio worked each day last week. Which measures of the data would be affected if he had not worked on Friday?

Day	Hours Worked
Monday	6
Tuesday	5.5
Wednesday	8
Thursday	6.5
Friday	7

Mean, median, and mode

Mean, mode, and range

Mean and mode

Mean and median

40. The table below shows the number of hours Antonio worked each day last week. Which measure of the data would be least affected if he worked 6 hours on Saturday?

Day	Hours Worked
Monday	6
Tuesday	5.5
Wednesday	8
Thursday	6.5
Friday	7

Median

Mean

Mode

Range

HA1-545: Making a Frequency Distribution Table

Data gives numerical information about a set of objects. For example, data may be a list of the weights of catfish in a pond or a list of salaries of workers in an appliance factory. A **statistic** is a number that gives information about a set of data. For example, the number of pieces of data is a statistic. The role of statistics is to collect, organize, and study data. This lesson focuses on organizing data.

Statistic	A number that gives information about a set of data.

Data falls into one of two categories, **continuous** or **discrete**.

Continuous Data	Data is called continuous if the data can be any real numbers.

Discrete Data	Data is called discrete if only certain fixed values are allowed for the data. Discrete data usually consists of integers.

A good way to organize data is in a frequency distribution table, which is made by grouping the data into intervals. The intervals should be constructed according to the following rules, which apply to both continuous and discrete data.

Step 1: The intervals must be of equal size.
Step 2: There must be no gaps between the intervals.
Step 3: The lower limits of the intervals must be multiples of a fixed integer.

This first example involves the ages of men on a basketball team. When an age is given as 29 years (for example), it means that the age is at least 29 years, but less than 30 years. For this reason, even if age is given as a whole number of years, it is really continuous data.

Example 1 The ages of 16 men on a basketball team are as follows:

29	32	27	22
38	27	35	38
22	23	29	32
38	31	30	30

Construct a frequency distribution table for this data using 4 intervals.

Intervals	Tally	Frequency
20 – 24		
25 – 29		
30 – 34		
35 – 39		

Step 1: First construct a table containing the appropriate intervals. Establish the intervals. There is no absolute rule for determining the size of the intervals. This data will fit nicely into 4 intervals if the width of the intervals is 5, and the lower limit of the first interval is 20.

Note: The lower limit of the first interval has to be a multiple of the width of one of the intervals.

You should check what the distribution would look like, using intervals with widths such as 4 and 6.

<table>
<tr><th>Intervals</th><th>Tally</th><th>Frequency</th></tr>
<tr><td>20 – 24</td><td>|||</td><td></td></tr>
<tr><td>25 – 29</td><td>||||</td><td></td></tr>
<tr><td>30 – 34</td><td>𝍸</td><td></td></tr>
<tr><td>35 – 39</td><td>||||</td><td></td></tr>
</table>

Step 2: The next step is to place a tally mark in the tally column for each piece of data.

<table>
<tr><th>Intervals</th><th>Tally</th><th>Frequency</th></tr>
<tr><td>20 – 24</td><td>|||</td><td>3</td></tr>
<tr><td>25 – 29</td><td>||||</td><td>4</td></tr>
<tr><td>30 – 34</td><td>𝍸</td><td>5</td></tr>
<tr><td>35 – 39</td><td>||||</td><td>4</td></tr>
</table>

Step 3: Count the tally marks and place the totals in the frequency column.

<table>
<tr><th>Intervals</th><th>Tally</th><th>Frequency</th></tr>
<tr><td>20 – 24</td><td>|||</td><td>3</td></tr>
<tr><td>25 – 29</td><td>||||</td><td>4</td></tr>
<tr><td>30 – 34</td><td>𝍸</td><td>5</td></tr>
<tr><td>35 – 39</td><td>||||</td><td>4</td></tr>
<tr><td>Total</td><td></td><td>16</td></tr>
</table>

Step 4: Finally, total the frequencies. This checks that all pieces of data have been counted. Since there were 16 ages, the total of the frequencies must be 16.

Here is the histogram for the above frequency distribution table:

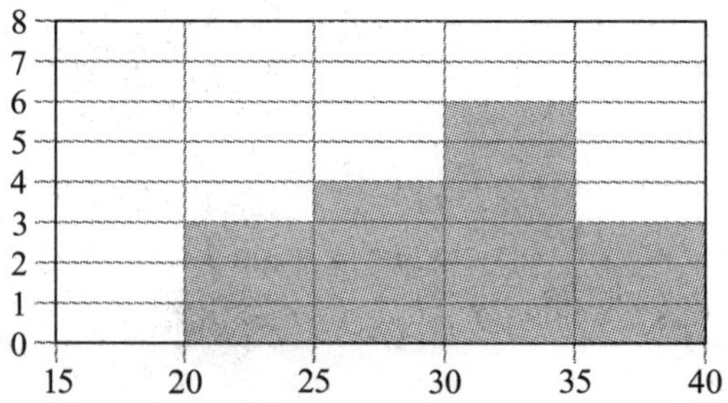

The next example involves discrete data.

Example 2 The shoe sizes for sixteen students are as follows:

7	8	8	9
6	5	5	4
8	7	8	10
11	11	6	4

Construct a frequency distribution for this data using intervals with a width of three.

Intervals	Tally	Frequency
3 – 5		
6 – 8		
9 – 11		

Step 1: Establish intervals for the data. The lower limit of the first interval must be a multiple of the width. To include all of the data, the lower limit of the first interval is three. Since shoe size must be a whole number, the only numbers that can appear in the first interval are three, four and five. Six, seven and eight are the numbers that can be in the second interval, and so on.

Intervals	Tally	Frequency
3 – 5	IIII	
6 – 8	卌 III	
9 – 11	IIII	

Step 2: Tally the data.

Intervals	Tally	Frequency
3 – 5	IIII	4
6 – 8	卌 III	8
9 – 11	IIII	4

Step 3: Count the tallies for each interval and record the sum in the frequency column.

Intervals	Tally	Frequency
3 – 5	IIII	4
6 – 8	卌 III	8
9 – 11	IIII	4
Total		16

Step 4: Total the frequencies.

The above table could also have been given following the rules for continuous data using intervals 3 – 6, 6 – 9, and 9 – 11. The advantage of the above table is that since there is nowhere to place a number such as 5.3; anyone looking at the table can see that the data is discrete. The histogram for the above table is:

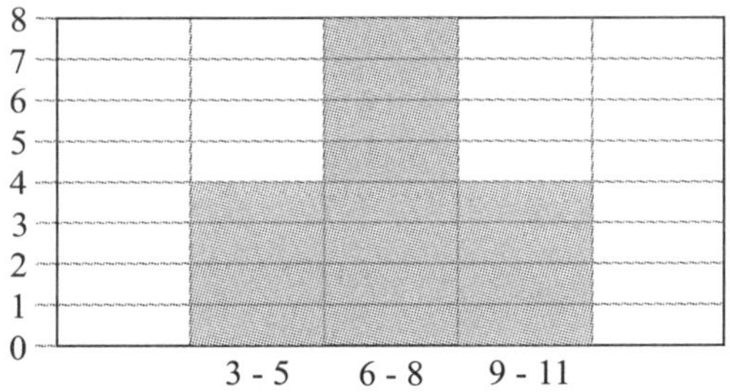

The intervals along the bottom could also have been labeled in the same manner that intervals are labeled for continuous data.

Problem Set

Solve:

1. Has the frequency distribution table been correctly completed for the given data? Answer yes or no.

Data: 8, 14, 12, 19, 16, 26, 24, 27, 20, 23, 23, 12, 22, 16, 26

Interval	Tally	Frequency
8 – 12	III	3
13 – 17	III	3
18 – 22	III	3
23 – 27	卌 I	6
	Total	15

2. Has the frequency distribution table been correctly completed for the given data? Answer yes or no.

Data: 7, 15, 18, 8, 23, 24, 29, 14, 23, 6, 7, 5, 8, 19, 14

Interval	Tally	Frequency
4 – 8	卌	5
9 – 13	I	1
14 – 18	卌	5
19 – 22	I	1
23 – 27	II	2
28 – 32	I	1
	Total	15

3. Has the frequency distribution table been correctly completed for the given data? Answer yes or no.

Data: 25, 19, 29, 30, 17, 36, 17, 27, 13, 15, 20, 31, 29, 19, 24

Interval	Tally	Frequency
10 – 19	ЖЖ I	6
20 – 29	ЖЖ I	6
30 – 39	III	3
	Total	15

4. Twenty people were surveyed and asked to give their average monthly cell phone bill. The frequency distribution is shown. What is the missing interval?

Interval	Tally	Frequency
0 – 49	X	1
50 – 99	XXXXXX	6
	XXXX	4
150 – 199	XXXX	4
200 – 249	XXXX	4
250 – 299	X	1

5. A frequency distribution table for the daily high temperatures is shown. What is the interval increment?

Interval	Tally	Frequency
50 – 59	XXXX	4
60 – 69	XXXXXXXXX	9
70 – 79	XXXXX	5
80 – 89	XXXXXXXX	8

6. A frequency distribution table for the age of children in a preschool is shown. What is the missing interval?

Interval	Tally	Frequency
1	XX	2
2	XXX	3
	XXXX	4
4	XXXXXXX	7
5	XXXX	4

7. The frequency distribution table for the time in minutes it takes 24 light bulbs to burn out is shown. What is the missing interval?

Interval	Tally	Frequency
850 – 899	XXXXX	5
	X	1
950 – 999	XX	2
1000 – 1049	XXX	3
1050 – 1099	XX	2
1100 – 1149	X	1
1150 – 1199	XXXXX	5
1200 – 1249	XX	2
1250 – 1299	XXX	3

8. The frequency distribution table for the test scores of an Algebra class is shown. How many people are in the class?

Interval	Tally	Frequency
1 – 10		0
11 – 19		0
21 – 30	X	1
31 – 40	XX	2
41 – 50	XXX	3
51 – 60	XX	2
61 – 70	XX	2
71 – 80	XXXX	4
81 – 90	XXX	3
91 – 100	X	1

9. A frequency distribution table for the daily high temperatures is shown. What is the interval increment?

Interval	Tally	Frequency
50 – 54	XXXX	4
55 – 59	XXXXXXX	7
60 – 64	XX	2
65 – 69	X	1

10. The frequency distribution table for the time in minutes it takes 22 light bulbs to burn out is shown. What is the missing interval?

Interval	Tally	Frequency
850 – 899	XXX	3
900 – 949	X	1
950 – 999	X	1
1000 – 1049	XXX	3
1050 – 1099	XX	2
	XXXX	4
1150 – 1199	XX	2
1200 – 1249	XXXX	4
1250 – 1299	XX	2

11. Construct a frequency distribution table from the data shown using 5-point intervals.

Data: 40, 66, 45, 45, 65, 67, 54, 48, 56, 62, 41, 66, 58, 60, 59

12. Construct a frequency distribution table from the data shown using 10-point intervals.

Data: 40, 34, 30, 35, 45, 39, 61, 37, 69, 47, 51, 61, 45, 47, 36

13. Construct a frequency distribution table from the data shown using 5-point intervals.

Data: 17, 29, 23, 22, 28, 15, 21, 15, 15, 24, 26, 25, 29

14. Construct a frequency distribution table from the data shown using 5-point intervals.

Data: 46, 57, 47, 41, 46, 43, 51, 55, 57, 57, 41, 46, 59, 52, 40

15. Construct a frequency distribution table from the data shown using 10-point intervals.

Data: 64, 93, 68, 82, 87, 79, 90, 80, 86, 71, 99, 93, 75, 81, 69

16. Construct a frequency distribution table from the data shown using 10-point intervals.

Data: 22, 24, 25, 29, 34, 35, 42, 23, 25, 36, 43, 34

17. Construct a frequency distribution table from the data shown using 10-point intervals.

Data: 32, 34, 35, 39, 44, 45, 52, 33, 35, 46, 53, 44

18. Construct a histogram using the frequency distribution table.

Interval	Tally	Frequency
1 – 9	~~IIII~~	5
10 – 19	III	3
20 – 29	IIII	4
30 – 39	~~IIII~~ III	8

19. Construct a histogram using the frequency distribution table.

Interval	Tally	Frequency
1 – 4	IIII	4
5 – 9	II	2
10 – 14	~~IIII~~ III	8
15 – 19	~~IIII~~ I	6

20. Construct a histogram using the frequency distribution table.

Interval	Tally	Frequency
1 – 4	~~IIII~~ III	8
5 – 9	~~IIII~~ ~~IIII~~	10
10 – 14	III	3
15 – 19	~~IIII~~	5

HA1-555: Computing the Range, Variance, and Standard Deviation of a Set of Data

When describing a set of data, it is useful to have a way to describe how widely the data is scattered. A **measure of dispersion** is a statistic that does just this.

Measure of Dispersion	A statistic that describes how widely data is scattered.

Three measures of dispersion are discussed in this lesson: range, variance, and standard deviation. The range of a set of data is the difference between the largest and smallest numbers in the set of data.

Range	The difference between the largest and smallest values in a set of data.

Example 1 Find the range of the following set of data: 12, 45, 33, 44, 21, 11, 17, 33, 36, 29, 18, 43

Step 1: We see that the largest value is 45 and the smallest value is 11. Thus the range is $45 - 11 = 34$.

Answer: 34

Before introducing the other measures of dispersion some notation is needed. If x is a set of data then $\sum x$ (read as "sigma x") is used to denote the sum of the set of data.

Note: $\sum$ **is the Greek letter sigma. It is the Greek equivalent of a capital S and is always used to denote a sum.**

Example 2 If $X = \{1, -2, 3, -4, 5, 6, 7\}$, find $\sum x$ and $\sum x^2$.

Step 1: To find $\sum x$, add the numbers in the set:

$$\begin{aligned}\sum x &= 1 + (-2) + 3 + (-4) + 5 + 6 + 7 \\ &= 16\end{aligned}$$

Step 2: To find $\sum x^2$, add the squares of the numbers in the set:

$$\begin{aligned}\sum x^2 &= 1^2 + (-2)^2 + 3^2 + (-4)^2 + 5^2 + 6^2 + 7^2 \\ &= 1 + 4 + 9 + 16 + 25 + 36 + 49 \\ &= 140\end{aligned}$$

Answer: $\sum x = 16, \sum x^2 = 140$

Given a set of data called X, the mean is denoted using $\bar{x}$ (read as "x bar") and n is used for the number of items of data. With this notation, the mean is given by the formula $\bar{x} = \frac{\sum x}{n}$.

Note: If a set is called Y, then y is used in place of x in the notation for sums and the mean. That is, the mean is $\bar{y} = \frac{\sum y}{n}$.

The **variance** measures how widely the data is scattered around the mean.

Variance	Variance is a measure of how widely data is scattered around the mean. It measures the accuracy of the mean. A high variance indicates a low accuracy, and a low variance indicates a high accuracy.

The variance is denoted by σ^2. Note that σ is the lower case form of Σ. The formula is $\sigma^2 = \frac{\sum(x-\bar{x})^2}{n}$.

The quantity $x - \bar{x}$ measures how far x is from the mean. It is called its deviation of x from the mean. The variance can be described as the mean of the squares of the deviations from the mean.

Example 3 Calculate the variance of the following set of data: $X = \{3, 10, 80, 15, 11, 9, 92, 63, 54, 3\}$

Step 1: Find the mean ($\bar{x}$).

$$\bar{x} = \frac{3 + 10 + 80 + 15 + 11 + 9 + 92 + 63 + 54 + 3}{10}$$
$$= \frac{340}{10}$$
$$= 34$$

Step 2: Calculate $\sum(x-\bar{x})^2$.

$$\begin{aligned}\sum(x-\bar{x})^2 &= (3-34)^2 + (10-34)^2 + (80-34)^2 + (15-34)^2 + (11-34)^2 \\ &\quad + (9-34)^2 + (92-34)^2 + (63-34)^2 + (54-34)^2 + (3-34)^2 \\ &= (-31)^2 + (-24)^2 + (46)^2 + (-19)^2 + (-23)^2 + (-25)^2 + (58)^2 + (29)^2 + (20)^2 + (-31)^2 \\ &= 961 + 576 + 2116 + 361 + 529 + 625 + 3364 + 841 + 400 + 961 \\ &= 10{,}734\end{aligned}$$

Step 3: Solve for the variance using the formula $\sigma^2 = \dfrac{\sum(x-\bar{x})^2}{n}$

$$\begin{aligned}\sigma^2 &= \frac{\sum(x-\bar{x})^2}{n}\\ &= \frac{10{,}734}{10}\\ &= 1{,}073.4\end{aligned}$$

Answer: The variance is 1,073.4.

The **standard deviation** of a set of data, denoted σ, is the positive square root of the variance. The standard deviation is found using the formula $\sigma = \sqrt{\sigma^2}$

Standard Deviation	The positive square root of the variance

Example 4 Calculate the standard deviation of the following set of data and round the answer to the nearest hundredth: $X = \{3, 10, 80, 15, 11, 9, 92, 63, 54, 3\}$

Step 1: Set up the equation, $\sigma = \sqrt{\sigma^2}$.

Step 2: The data is the same as the previous example, so you know that $\sigma^2 = 1073.4$.

$\sigma = \sqrt{1073.4}$

Step 3: Solve for σ.

$\sigma = 32.76$

Answer: The standard deviation is 32.76.

Problem Set

1. The following data represents ACT scores for a sample of ten students at Lakeview High:
20, 23, 15, 13, 30, 22, 10, 28, 16, 25
Compute the mean of the data.

2. The following data represents test scores for a sample of twelve students in Ms. Logan's class:
90, 93, 85, 93, 91, 98, 100, 88, 76, 85, 99, 89
Compute the mean of the data. Round your answer to the nearest hundredth.

3. The following data represents weekly temperatures for a city in Louisiana:
88°, 87°, 91°, 90°, 92°, 90°, 93°
Compute the mean of the data. Round your answer to the nearest hundredth.

4. On a recent trip, Mrs. Finley kept the following record of her gasoline purchases:
$15, $22, $21, $23, $17
Compute the mean of the data.

5. According to the weather bureau records, the total amount of rainfall in Genoa City during 2002 was as follows:

Month	Rainfall (in inches)
January	39
February	42
March	38
April	39
May	36
June	25
July	25
August	19
September	23
October	27
November	31
December	35

Compute the mean of the data. Round your answer to the nearest hundredth.

6. The following data represents ACT scores for a sample of ten students at Lakeview High:

20, 23, 15, 13, 30, 22, 10, 28, 16, 25

Compute the range of the data.

7. The following data represents test scores for a sample of twelve students in Ms. Logan's class:

90, 93, 85, 93, 91, 98, 100, 88, 76, 85, 99, 89

Compute the range of the data.

8. According to the weather bureau records, the total amount of rainfall in Genoa City during 2002 was as follows:

Month	Rainfall (in inches)
January	39
February	42
March	38
April	39
May	36
June	25
July	25
August	19
September	23
October	27
November	31
December	35

Compute the range of the data. Round your answer to the nearest hundredth.

9. On a recent trip, Mr. Freely kept the following record of his gasoline purchases:

\$15, \$22, \$21, \$23, \$17

Compute the range of the data.

10. The following data represents weekly temperatures for a city in Arizona:

88°, 87°, 91°, 90°, 92°, 90°, 93°

Compute the range of the data.

11. The following data represents ACT scores for a sample of ten students at Lakeview High:

20, 23, 15, 13, 30, 22, 10, 28, 16, 25

Compute the variance of the data. Round your answer to the nearest hundredth.

12. The following data represents test scores for a sample of twelve students in Ms. Logan's class:

90, 93, 85, 93, 91, 98, 100, 88, 76, 85, 99, 89

Compute the variance of the data. Round your answer to the nearest hundredth.

13. The following data represents weekly temperatures for a city in Louisiana:

88°, 87°, 91°, 90°, 92°, 90°, 93°

Compute the variance of the data. Round your answer to the nearest hundredth.

14. On a recent trip, Mrs. Finley kept the following record of her gasoline purchases:

\$15, \$22, \$21, \$23, \$17

Compute the variance of the data. Round your answer to the nearest hundredth.

15. According to the weather bureau records, the total amount of rainfall in Genoa City during 2002 was as follows:

Month	Rainfall (in inches)
January	39
February	42
March	38
April	39
May	36
June	25
July	25
August	19
September	23
October	27
November	31
December	35

Compute the variance of the data. Round your answer to the nearest hundredth.

16. Mr. Brown is analyzing his cell phone bills for the past five months. They are as follows:

Month	Amount of Bill
April	$42.23
May	$41.55
June	$52.32
July	$86.25
August	$42.55

Compute the variance of the data. Round your answer to the nearest hundredth.

17. The following data represents the number of pencils that seven students in the fifth grade have. The results are as follows:

Student	Number of Pencils
Janae	3
Raven	6
Sydney	12
Mia	14
Ann	8
Terri	5
Bryce	2

Compute the variance of the data. Round your answer to the nearest hundredth.

18. The following data represents the number of students that were tardy each day in the first week of school at Lakeview High:

Monday	8
Tuesday	4
Wednesday	3
Thursday	5
Friday	2

Compute the standard deviation of the data. Round your answer to the nearest hundredth.

19. The following data represents the number of pencils that seven students in the fifth grade have. The results are as follows:

Student	Number of Pencils
Janae	3
Raven	6
Sydney	12
Mia	14
Ann	8
Terri	5
Bryce	2

Compute the standard deviation of the data. Round your answer to the nearest hundredth.

20. Mr. Brown is analyzing his cell phone bills for the past five months. They are as follows:

Month	Amount of Bill
April	$42.23
May	$41.55
June	$52.32
July	$86.25
August	$42.55

Compute the standard deviation of the data. Round your answer to the nearest hundredth.

HA1-560: Determining Probability of an Event and Complementary Event from a Random Experiment

In statistics, an **experiment** is an activity to test the likelihood of an event occurring. Some examples are:

1. Tossing a coin
2. Rolling a die
3. Drawing a ball out of a bag containing 8 blue balls and 11 red balls

The result of an experiment is called an **outcome**. When performing an experiment, the possible outcomes are usually known ahead of time. The set of all possible outcomes is called the **sample space** and is denoted by S. Some examples are:

1. For tossing a coin, $S = \{\text{heads, tails}\}$
2. For rolling a die, $S = \{1, 2, 3, 4, 5, 6\}$
3. For drawing a ball out of a bag containing 8 blue balls and 11 red balls, $S = \{\text{red ball, blue ball}\}$

Outcomes are termed **equally likely** if they have the same chance of occurring. For example:

1. When tossing a balanced coin, heads and tails are equally likely.
2. When rolling a balanced die, each of the numbers one through six is equally likely.
3. When drawing a ball out of a bag containing 8 blue balls and 11 red balls, each ball has the same chance of being picked if color is the only difference, but the outcomes (red ball and blue ball) are not equally likely. It is more likely that a red ball will be selected because there are more of them.

An **event** is a set of outcomes. For example, when rolling a die, an event might be rolling an even number. This event is comprised of three outcomes, namely 2, 4 and 6. An event is a subset of the sample space. The **probability** of an event measures how likely an event is to occur and is denoted P(event). Providing all outcomes are equally likely, probability is given by

$$P(\text{event}) = \frac{\text{numbers of ways an event can occur}}{\text{total number of possible outcomes}}$$

Example 1 For the experiment of tossing a fair coin find $P(\text{heads})$.

$P(\text{heads}) = \frac{1}{2}$ **Step 1:** This event consists of only one outcome, so there is only one way it can occur. The total number of possible outcomes is two, either heads or tails.

Answer: $P(\text{heads}) = \frac{1}{2}$

The result of Example 1 should agree with your experiences of tossing a coin. When tossing a coin, you would expect to get heads about half of the time.

Example 2 For the experiment of rolling a fair die, find the probability of rolling a multiple of three.

$\frac{2}{6} = \frac{1}{3}$

Step 1: Rolling a multiple of three consists of two outcomes, 3 and 6. The total number of possible outcomes is six.

Answer: $P(\text{multiple of } 3) = \frac{1}{3}$

Example 3 For the experiment of picking a ball out of a bag containing 8 blue balls and 11red balls, find the probability of picking a red ball.

$P(\text{red ball}) = \frac{11}{19}$

Step 1: There are 11 ways to pick a red ball and 19 balls in all, each of which is equally likely to be selected

Answer: $P(\text{red ball}) = \frac{11}{19}$

In general, it is true that $\mathbf{0 \le P(event) \le 1}$. If an event is impossible, its probability is zero. For example, when rolling a die, $P(7) = 0$. If an event is certain to occur, its probability is one (1). For example, when tossing a coin, $P(\text{heads or tails}) = 1$. The closer the probability of an event is to one (1), the more likely the event is to occur. The closer the probability of an event is to zero (0), the less likely the event is to occur. Often the outcomes making up an event are termed **favorable outcomes** and all other outcomes are termed **unfavorable outcomes**. For a given event, the complementary event, denoted $\overline{\text{event}}$, consists of all of the unfavorable outcomes.

Example 4 In the experiment of rolling a fair die, what is $P(\bar{6})$?

$P(\bar{6}) = \frac{5}{6}$

Step 1: Since P(6) is the probability of rolling a 6, $P(\bar{6})$ is the probability of not rolling a 6. Therefore, "not six" consists of the five outcomes of rolling 1 through 5.

Answer: $P(\bar{6}) = \frac{5}{6}$

It is common practice to denote an event by a single letter. The following is an important fact in probability: For any event A it is true that $P(\bar{A}) = 1 - P(A)$ To see that this is true, suppose an experiment has n possible outcomes and A is an event consisting of k outcomes. Then $\bar{A}$ consists of $n - k$ outcomes, and:

$$P(\bar{A}) = \frac{n-k}{n} = \frac{n}{n} - \frac{k}{k} = 1 - P(A)$$

Example 5 Suppose A is the event of drawing a heart from a standard deck of 52 cards. (A standard deck contains four suits: clubs, diamonds, hearts and spades, each with 13 cards) Find $P(A)$ and $P(\bar{A})$.

$$P(A) = \frac{13}{52} = \frac{1}{4}$$

Step 1: Since A consists of 13 favorable outcomes, place 13 over the total number of possible outcomes, 52.

$$P(\bar{A}) = 1 - P(A) = 1 - \frac{1}{4} = \frac{3}{4}$$

Step 2: Because you have found $P(A)$, now you can find $P(\bar{A})$.

Answer: $P(A) = \frac{1}{4}$ and $P(\bar{A}) = \frac{3}{4}$

Sometimes the likelihood of an event occurring is expressed using odds. For example, if the odds in favor of an event are 10 to 19, it means that for each 10 favorable outcomes, there are 19 unfavorable outcomes. "The odds against an event are 5 to 3" means that for each 5 unfavorable outcomes, there are 3 favorable outcomes. The odds can also be expressed as a fraction, so for example "odds in favor of an event are $\frac{7}{11}$" means the same as "odds in favor of an event are 7 to 11."

Odds can be determined from the following formulas:

1. The odds in favor of event A are $\frac{P(A)}{P(\bar{A})}$
2. The odds against event A are $\frac{P(\bar{A})}{P(A)}$

To see that the first formula is true, suppose an experiment has n possible outcomes and A is an event consisting of k outcomes. Then $\bar{A}$ consists of $n-k$ outcomes. From the previous discussion, you know that the odds in favor of A are $P(A) = \frac{k}{n-k}$ The following calculation establishes the first formula.

$$\frac{P(A)}{P(\bar{A})} = \frac{\frac{k}{n}}{\frac{n-k}{n}} = \frac{k}{n} \cdot \frac{n}{n-k} = \frac{k}{n-k}$$

The second formula is established in a similar manner.

Example 6 Suppose that a single card is drawn from a standard 52-card deck. Define A to be the event of selecting a 3. Find: A) Find the odds in favor of A. B) The odds against A.

$$\frac{P(A)}{P(\bar{A})} = \frac{\frac{4}{52}}{1-\frac{4}{52}} = \frac{\frac{4}{52}}{\frac{48}{52}} = \frac{4}{48} = \frac{1}{12}$$

Step 1: To find part A), start with the formula for the odds in favor of A, then substitute $\frac{4}{52}$ into the formula and solve.

$$\frac{P(\bar{A})}{P(A)} = \frac{1 - \frac{4}{52}}{\frac{4}{52}} = \frac{\frac{48}{52}}{\frac{4}{52}} = \frac{48}{4} = \frac{12}{1}$$

Step 2: To find part B), start with the formula for the odds against A. Substitute $\frac{4}{52}$ into the formula and solve.

Answer: Odds in favor of $A = \frac{1}{12}$

Odds against $A = \frac{12}{1}$

In Example 6, the answers should be expressed as: "The odds in favor of A are 1 to 12 and the odds against A are 12 to 1." It is always true that the fractions of the odds for and against A are reciprocals.

Odds should always be expressed in lowest terms, as shown in the next example.

Example 7 Express the odds of 12 to 8 in lowest terms.

$$\frac{12}{8} = \frac{3}{2}$$

Step 1: Expressed as a fraction the odds are $\frac{12}{8}$, which reduces to $\frac{3}{2}$. So, "odds of 12 to 8" in lowest terms is "odds of 3 to 2."

Answer: $\frac{3}{2}$

Problem Set

1. A clown has a large bouquet of balloons. There are 12 white, 10 blue, 25 red, and 20 green balloons. If the clown pulls one string at random, calculate the probability of selecting a green balloon.

2. A letter is chosen at random from the word "Mississippi". Find the probability of selecting the letter "p".

3. A bag contains different colored pencils. There are 8 yellow, 5 blue, 6 green, and 9 red pencils. If one pencil is drawn at random, find the probability of selecting a blue pencil.

4. A letter is chosen at random from the word "VALEDICTORIAN". Find the probability of selecting the letter "N".

5. A bag contains different colored marbles. There are 6 yellow, 4 blue, 8 green, and 7 red marbles. If one marble is drawn at random, find the probability of selecting a yellow marble.

6. A letter is chosen at random from the word "MISSISSIPPI". Find the probability of selecting the letter "S".

7. A bag contains different colored marbles. There are 6 yellow, 4 blue, 8 green, and 7 red marbles. If one marble is drawn at random, find the probability of selecting a blue marble.

8. Each ninth grade student is enrolled in one elective. If one ninth grade student is selected at random, calculate the probability that the student is enrolled in French.

Woodworking	50
French	40
Food and Nutrition	60
Typing	80
Music	70

9. Each ninth grade student is enrolled in one elective. The tally is shown. If one ninth grade student is selected at random, calculate the probability that the student is enrolled in Music.

Woodworking	50
French	40
Food and Nutrition	60
Typing	80
Music	70

10. Each ninth grade student is enrolled in one elective. The tally is shown. If one ninth grade student is selected at random, calculate the probability that the student is enrolled in Woodworking.

Woodworking	50
French	40
Food and Nutrition	60
Typing	80
Music	70

11. A letter is chosen at random from the word "VALEDICTORIAN". Find the probability of selecting a consonant.

12. A letter is chosen at random from the word "SALUTATORIAN". Find the probability of selecting a consonant

13. A letter is chosen at random from the word "VALEDICTORIAN". Find the probability of selecting a vowel.

14. A bag contains different colored pencils. There are 8 yellow, 5 blue, 6 green, and 9 red pencils. If one pencil is drawn at random, find the probability of selecting a yellow or blue pencil.

15. A bag contains different colored pencils. There are 8 yellow, 5 blue, 6 green, and 9 red pencils. If one pencil is drawn at random, find the probability of selecting a green pencil.

16. A clown has a large bouquet of balloons. There are 12 yellow, 10 blue, 25 violet, and 20 orange balloons. If the clown pulls one string at random, calculate the probability of selecting a blue or orange balloon.

17. A clown has a large bouquet of balloons. There are 12 yellow, 10 blue, 25 violet, and 20 orange balloons. If the clown pulls one string at random, calculate the probability of selecting a violet or orange balloon.

18. Bob owns a used car dealership. The different types of vehicles on the lot are shown in the table below. If one vehicle is selected at random, find the odds that the vehicle will be an SUV.

Type of Vehicle	Number on Bob's Lot
Sports Cars	50
4-door Sedans	40
Minivans	25
SUVs	60

19. A college freshman must take one science course. The tally is shown. If one college freshman is selected at random, calculate the odds that the student is enrolled in Biology.

Biology	20
Chemistry	80
Physics	40
Computer Science	50

20. A college freshman must take one science course. The tally is shown. If one college freshman is selected at random, calculate the odds that the student is enrolled in Physics.

Biology	120
Chemistry	80
Physics	40
Computer Science	50

HA1-565: Solving Problems Involving Independent, Dependent, and Mutually Exclusive and Inclusive Events

Consider the experiment of tossing a coin and rolling a die. For this experiment, an event might be tossing heads and rolling a multiple of three. Such an event is called a **compound event**, because it is made up of other, simpler events.

Compound Event	An event that is made up of two or more events

The objective in this lesson is to study ways to derive the probability of a compound event from the probabilities of the events that make up the compound event. If A is the event of tossing heads and B is the event of rolling a multiple of three, then (A and B) is the event of tossing heads and rolling a multiple of three, and its probability is denoted by $P(A \text{ and } B)$.

All of the problems in this lesson could be solved using the direct approach. That is, write down the sample space and then count the number of favorable outcomes. It is easier to derive the result using rules of probability than it is to use the direct approach. Having said that, it can be useful to look at the direct approach because it makes it easier to understand why the rules work the way they do.

Example 1 Determine $P(A \text{ and } B)$ as described in the experiment above.

Step 1: Write down the sample space. Outcomes are described using ordered pairs. For example (T, 2) is the outcome of tossing tails and rolling a two. The sample space is:
{(T, 1), (T, 2), (T, 3), (T, 4), (T, 5), (T, 6), (H, 1), (H, 2), (H, 3), (H, 4), (H, 5), (H, 6)}

Step 2: There are two favorable outcomes, (H, 3) and (H, 6), and 12 outcomes in total. Therefore:

$$P(A \text{ and } B) = \frac{2}{12} = \frac{1}{6}.$$

Answer: $P(A \text{ and } B) = \frac{1}{6}$

Another way to determine $P(A \text{ and } B)$ is to argue as follows: Because $P(A) = \frac{1}{2}$, you would expect half of the tosses to result in heads. Each time you toss heads there is a $\frac{1}{3}$ chance that you will roll a multiple of three. It follows that you toss heads *and* roll a multiple of three in $\frac{1}{3}$ of $\frac{1}{2}$ of the attempts, or $\frac{1}{6}$ of the attempts. This means that $P(A \text{ and } B) = \frac{1}{6}$.

In the above argument, it was assumed that A and B were **independent events.** That is, event A has no effect on the probability of event B and vice versa. In other words, tossing heads does not make it more or less likely that you will roll a multiple of three.

Independent Event	An event B is said to be independent of event A if event A has no effect on the probability that event B occurs, or $P(B \mid A) = P(B)$.

Events that are not independent are called **dependent**.

Dependent Event	If $P(B\|A) \neq P(B)$, then event A has an effect on the probability that event B occurs, and event B is dependent on event A.

Using a similar argument, it is possible to establish the multiplication rule for independent events.

Multiplication Rule for Independent Events	If A and B are independent events, then $P(A \text{ and } B) = P(A) \cdot P(B)$.

Example 2 Suppose a card is drawn from a standard 52-card deck and a die is rolled. Find the probability of drawing a club and rolling an odd number.

$P(A \text{ and } B) = \frac{13}{52} \cdot \frac{3}{6} = \frac{1}{8}$

Step 1: Let A be the event of drawing a club and let B be the event of rolling an odd number. It is not possible that the card drawn can affect the roll of the die, so A and B are independent events. Note that there are 13 clubs and 3 odd numbers, so by the multiplication rule:

$$\frac{13}{52} \cdot \frac{3}{6} = \frac{1}{8}.$$

Answer: $P(A \text{ and } B) = \frac{1}{8}$

It would be very cumbersome in Example 2 to write down the sample space.

Now let's look at dependent events. Suppose you draw a card from a deck of cards, look at it and then draw a second card. Consider the probability that the second card is a diamond. The probability is affected by the card that was drawn first. If the first card was not a diamond, there are 51 cards left, 13 of which are diamonds, so the probability of drawing a diamond is $\frac{13}{51}$. If the first card is a diamond, there are 51 cards left, 12 of which are diamonds, so the probability of drawing a diamond is $\frac{12}{51}$, or $\frac{4}{17}$. The probability of B given A is the probability that event B occurs given that it is known that event A has already occurred. It is denoted $P(B \mid A)$.

Example 3 Let A be the event of drawing a red ball from a bag containing 10 red balls and 15 blue balls, and let B be the event of drawing a blue ball. Find $P(B \mid A)$, assuming the first ball is not placed back in the bag.

$P(B|A) = \frac{15}{24} = \frac{5}{8}$

Step 1: If a red ball has been drawn, there are 9 red balls left and 15 blue balls left, making 24 balls in all. The probability that the second ball is a blue ball, or $P(B|A)$, is $\frac{15}{24}$ or $\frac{5}{8}$.

Answer: $P(B|A) = \frac{5}{8}$

If a ball is drawn and not replaced, it is said to be drawn without replacement. If it is drawn and then replaced, it is said to be drawn with replacement.

Example 4 Let A be the event of drawing a red ball from a bag containing 10 red balls and 15 blue balls, and let B be the event of drawing a blue ball. Find the probability of drawing a red ball without replacement, followed by a blue ball.

$P(A) = \frac{10}{25} = \frac{2}{5}$

$P(B) = \frac{5}{8}$

Step 1: The probability of drawing a red ball is $\frac{10}{25}$, or $\frac{2}{5}$. From Example 3, you know the probability of drawing a blue ball is $\frac{5}{8}$. Using an argument like the one in Example 1, the desired probability is $\frac{2}{5} \cdot \frac{5}{8} = \frac{10}{40} = \frac{1}{4}$.

Answer: $\frac{1}{4}$

In Example 4, $P(A \text{ and } B) = \frac{1}{4}$. The probability turned out to be $P(A) \cdot P(B \mid A)$. In fact, the argument used in Example 4 works for any pair of dependent events and is stated in the following rule:

Multiplication Rule for Dependent Events	If A and B are dependent events, then $P(A \text{ and } B) = P(A) \cdot P(B \mid A)$.

When defining the event (A and B), it is important to state what the experiment is and what the events are. In Example 4, the compound event (A and B), a red ball followed by a blue ball, is not the same event as drawing two balls, one red and one blue. This is because you can get one red ball and one blue ball in two ways: by drawing a red ball followed by a blue ball, or by drawing a blue ball followed by a red ball.

If A and B are two events, (A and B) is the compound event in which one or both events A and B occur(s).

Suppose the experiment is to roll a die. Let A be the event that a number greater than four is rolled and let B be the event that an even number is rolled. In this situation, (A or B) is the event that 2, 4, 5 or 6 is rolled. This is not the same as (A and B), the event that both A and B occur. The event (A and B) occurs only if a 6 is rolled, because the number must be both greater than 4 and even. Events A and B are said to be **mutually exclusive** if they cannot occur at the same time; otherwise, they are **mutually inclusive.**

Mutually Exclusive Events	Events are said to be mutually exclusive if they cannot occur at the same time.

Mutually Inclusive Events	Events are said to be mutually inclusive if they can occur at the same time.

Examples of mutually exclusive events are: tossing heads and tossing tails; drawing a heart from a deck of cards and drawing a spade; or rolling an odd number on a die and rolling 4.

There are two important rules to learn.

Addition Rule for Mutually Exclusive Events	If events A and B are mutually exclusive, then $P(A \text{ or } B) = P(A) + P(B)$, where $P(A \text{ or } B)$ is the total probability of either event occurring.

To see why this is true, consider the case where A consists of r events and B consists of s events in a sample space with n equally likely outcomes. If A and B are mutually exclusive events, then (A or B) contains $r + s$ outcomes.

Then $P(A \text{ or } B) = \frac{r+s}{n} = \frac{r}{n} + \frac{s}{n} = P(A) + P(B)$

The next rule is:

Addition Rule for Mutually Inclusive Events	If events A and B are mutually inclusive events, then $P(A \text{ or } B) = P(A) + P(B) - P(A \text{ and } B)$.

To see why this is true, consider the case where A consists of r events and B consists of s events in a sample space with n equally likely outcomes. If A and B are mutually inclusive events, then the outcomes in (A and B) have been counted twice—once as part of A and once as part of B—so one set needs to be subtracted. This means that if (A and B) contains t outcomes, then (A or B) contains $r + s - t$ outcomes.

Then $P(A \text{ or } B) = P(A) + P(B) - P(A \text{ and } B)$.

Example 5 Consider the experiment of drawing a card from a standard deck. Let S be the event that a spade is drawn, C the event that a club is drawn, and A the event that an ace is drawn.

A) Find the probability that a club or a spade is drawn.

B) Find the probability that a club or an ace is drawn.

$$P(S \text{ or } C) = P(S) + P(C) = \frac{1}{4} + \frac{1}{4} = \frac{1}{2}$$

Step 1: To find Part A): The required probability is $P(S \text{ or } C)$. It is not possible to draw a card that is both a spade and a club, so S and C are mutually exclusive.

$$P(C \text{ and } A) = \frac{1}{52}$$

$$P(C \text{ or } A) = P(C) + P(A) - P(C \text{ and } A) = \frac{1}{4} + \frac{1}{13} - \frac{1}{52} = \frac{4}{13}$$

Step 2: To find Part B): The required probability is $P(C \text{ or } A)$. It is possible to draw a card that is both a club and an ace, namely the ace of clubs. This means that (C and A) is the event of drawing the ace of clubs.

Answer: A) $P(S \text{ or } C) = \frac{1}{2}$

B) $P(C \text{ or } A) = \frac{4}{13}$

Example 6 Suppose two dice are rolled. Find the probability of rolling a one and a six. Caution: If A is the event of rolling a one, and B is the event of rolling a six, then (A and B) is not the event of rolling a one and a six. The reason is that for event (A and B), A and B are events involving the roll of only one die, and therefore do not belong in a problem where the experiment is to roll two dice.

$$P(1 \text{ or } 6) = \frac{1}{36} + \frac{1}{36} = \frac{1}{18}$$

Step 1: There are two outcomes that give rise to one and six: one on the first die and six on the second, and six on the first die and one on the second. These two events are mutually exclusive and each has probability $\frac{1}{6}$ or $\frac{1}{6} \cdot \frac{1}{6} = \frac{1}{36}$.

Answer: $P(1 \text{ or } 6) = \frac{1}{18}$

If two coins are tossed, then using an argument similar to the one in Example 6, you can show that $P(\text{Heads and Tails}) = \frac{1}{2}$. Try tossing two coins 50 times and record the results. You will find that approximately half of the attempts will result in heads and tails.

Problem Set

1. Mike rolls a die. What is the probability that he gets a 3 or a 4?

2. Mrs. Jacob assigned projects to her class by placing various topics into a box. Each student then selected one project at random from the box. Six of the cards indicated writing a report, 3 indicted building a model, and 3 indicated making a poster. What is the probability that the first student had to write a report or make a poster?

3. A spinner has five equal parts like the one shown here. If the spinner is spun once, what is the probability that it will land on yellow or green?

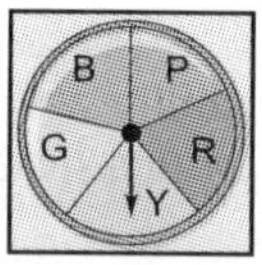

B = blue
G = green
Y = yellow
R = red
P = purple

4. Mr. Tran rewards his best students with stickers. A pack of 20 stickers contains 3 "Excellent" stickers, 5 "Good Job" stickers, 10 "Super Student" stickers, and 2 "Awesome" stickers. If each student is only allowed to pick one sticker at random, what is the probability that the first student will pick an "Excellent" sticker or a "Super Student" sticker?

5. A bag contains 6 red marbles, 2 blue marbles, and 8 white marbles. A marble is randomly drawn from the bag, It is then replaced, and a second marble is drawn. What is the probability of drawing a red marble and a white marble?

6. A box contains 8 red ribbons, 7 white ribbons, and 5 blue ribbons. A ribbon is selected at random from the box. After replacing it, a second ribbon is selected. What is the probability of selecting a red ribbon and a blue ribbon?

7. A coin is tossed and a die is rolled at the same time. What is the probability of getting heads on the coin and rolling a 5 on the die?

8. A coin is tossed and a die is rolled at the same time. What is the probability of getting tails on the coin and rolling a 2 on the die?

9. Mrs. Smith's class has 15 girls and 17 boys signing up for the spelling bee. If two of the students are selected at random without replacement, what is the probability that both students will be boys?

10. A basket contains 3 bananas, 2 oranges, 1 apple, and 3 pears. If two pieces of fruit are randomly selected from the basket without replacement, what is the probability of getting an apple first and an orange second?

11. A cube has six equal sides labeled 3, 6, 9, 12, 15, and 18. If the cube is rolled once, what is the probability of getting an odd number or a number greater than 9?

12. Cali has 7 cards labeled A, L, G, E, B, R, and A. She places them into a box and randomly selects one card from the box. What is the probability that she selects a card labeled with the letter A, or a vowel?

HA1-800: Solving Equations

To solve an equation, the variable in the equation must be isolated. The variable is isolated by using the inverse operations in order to "undo" or solve the equation. When applying the inverse operations, the same step must be performed to both sides of the equation to maintain equality. Listed below are properties used in solving equations.

Addition Property for Equations	For all real numbers, a, b, and c, if $a = b$, then $a + c = b + c$.
Subtraction Property for Equations	For all real numbers, a, b, and c, if $a = b$, then $a - c = b - c$.
Multiplication Property for Equations	For all real numbers, a, b, and c, if $a = b$, then $a \cdot c = b \cdot c$.
Division Property for Equations	For all real numbers, a, b, and c, if $a = b$ and $c \neq 0$, then $\frac{a}{c} = \frac{b}{c}$.
Distributive Property	$a(b + c) = a(b) + a(c)$

The following steps are useful in solving equations that contain fractions, parentheses, and/or variables on both sides.

1. Clear the equation of fractions.
2. Use the Distributive Property to remove grouping symbols.
3. Simplify each side of the equation by combining like terms.
4. Use the Addition Property and Subtraction Property for equations to isolate the term that contains the variable.
5. Use the Multiplication Property and Division Property for equations to isolate the variable.
6. Check your answer.

To check the solution of any equation, substitute the solution into the original equation for the variable and simplify. If you get a true statement, the solution is correct.

Example 1 Solve: $4x = 20$

$$\frac{4x}{4} = \frac{20}{4}$$

Step 1: Divide both sides by 4.

$$x = 5$$

Step 2: Simplify.

$4x = 20$

$4(5) = 20$

$20 = 20$ ✓

Step 3: Check your answer.

Answer: $x = 5$

Example 2 Solve: $-3x + 7 = -14$

$-3x + 7 - 7 = -14 - 7$	**Step 1:** Subtract 7 from both sides.
$-3x = -21$	**Step 2:** Simplify.
$\frac{-3x}{-3} = \frac{-21}{-3}$	**Step 3:** Divide both sides by –3.
$x = 7$	**Step 4:** Simplify.
$-3x + 7 = -14$ $-3(7) + 7 \stackrel{?}{=} -14$ $-21 + 7 \stackrel{?}{=} -14$ $-14 = -14$ ✓	**Step 5:** Check your answer.

Answer: $x = 7$

Example 3 Solve: $2(13 - g) = -10$

$26 - 2g = -10$	**Step 1:** Distribute the 2.
$26 - 26 - 2g = -10 - 26$	**Step 2:** Subtract 26 from both sides.
$-2g = -36$	**Step 3:** Combine like terms.
$\frac{-2g}{-2} = \frac{-36}{-2}$	**Step 4:** Divide both sides by –2.
$g = 18$	**Step 5:** Simplify.
$2(13 - g) = -10$ $2(13 - 18) \stackrel{?}{=} -10$ $2(-5) \stackrel{?}{=} -10$ $-10 = -10$ ✓	**Step 6:** Check your answer.

Answer: $g = 18$

Example 4 Solve: $\frac{2}{5}m + 10 = -4$

$5\left(\frac{2}{5}m + 10\right) = 5(-4)$	**Step 1:** Multiply both sides by 5 (LCD) to clear the equation of fractions.
$\frac{10}{5}m + 50 = -20$	**Step 2:** Distribute the 5.
$2m + 50 = -20$	**Step 3:** Simplify.
$2m + 50 - 50 = -20 - 50$	**Step 4:** Subtract 50 from both sides.
$2m = -70$	**Step 5:** Simplify.
$\frac{2m}{2} = \frac{-70}{2}$	**Step 6:** Divide both sides by 2.
$m = -35$	**Step 7:** Simplify.
$\frac{2}{5}m + 10 = -4$ $\frac{2}{5}(-35) + 10 \stackrel{?}{=} -4$ $\frac{2}{{}_1\cancel{5}}(-\cancel{35}^{-7}) + 10 \stackrel{?}{=} -4$ $2(-7) + 10 \stackrel{?}{=} -4$ $-4 = -4$ ✓	**Step 8:** Check your answer.

Answer: $m = -35$

Example 5 Solve: $6x + 3(x - 2) = 2x - 13$

$6x + 3x - 6 = 2x - 13$	**Step 1:** Distribute the 3.
$9x - 6 = 2x - 13$	**Step 2:** Combine like terms.
$9x - 2x - 6 = 2x - 2x - 13$	**Step 3:** Subtract $2x$ from both sides.
$7x - 6 = -13$	**Step 4:** Combine like terms.
$7x - 6 + 6 = -13 + 6$	**Step 5:** Add 6 to both sides.
$7x = -7$	**Step 6:** Simplify.
$\frac{7x}{7} = \frac{-7}{7}$	**Step 7:** Divide both sides by 7.
$x = -1$	**Step 8:** Simplify.

$$6x + 3(x-2) = 2x - 13$$
$$6(-1) + 3(-1-2) \stackrel{?}{=} 2(-1) - 13$$
$$-6 + 3(-3) \stackrel{?}{=} -2 - 13$$
$$-6 - 9 \stackrel{?}{=} -15$$
$$-15 = -15 \quad \checkmark$$

Step 9: Check your answer.

Answer: $m = -1$

Problem Set

Solve the following:

1. $2h + 7 = 15$

2. $2h - 7 = 15$

3. $4 - h = -8$

4. $-5 - 2x = 10$

5. $3f + 15 = 51$

6. $3f - 15 = 51$

7. $\frac{1}{6}f + 15 = 57$

8. $\frac{1}{6}f - 15 = 57$

9. $8d + 24 = 56$

10. $8d - 24 = 56$

11. $-4x + 26 = 54$

12. $-4x - 26 = 54$

13. $12 - \frac{3}{4}m = 36$

14. $\frac{1}{4}(b + 6) = 12$

15. $\frac{1}{4}(f - 12) = 32$

16. $\frac{1}{4}(12 + f) = 48$

17. $15 - 6g + 4g = -8$

18. $-8x - 3(x - 4) = 3x + 26$

19. $-6 + \frac{3}{5}x = \frac{24}{5}$

20. $\frac{26}{5} = 4 - \frac{8}{5}x$

HA1-801: Solving Inequalities

To solve inequalities, use the same properties that were used in solving equations: addition, subtraction, multiplication, and division properties. It is important to remember that when multiplying or dividing by a negative number the direction of the inequality sign must be reversed.

Addition Property for Inequalities	$a < b$ and $a + c < b + c$ are equivalent inequalities.

Subtraction Property for Inequalities	$a < b$ and $a - c < b - c$ are equivalent inequalities.

Multiplication Property for Inequalities	If c is positive, the inequalities $a < b$ and $ac < bc$ are equivalent inequalities. If c is negative, the inequalities $a < b$ and $ac > bc$ are equivalent inequalities.

Division Property for Inequalities	If c is positive, the inequalities $a < b$ and $\frac{a}{c} < \frac{b}{c}$ are equivalent inequalities. If c is negative, the inequalities $a < b$ and $\frac{a}{c} > \frac{b}{c}$ are equivalent inequalities.

Use the information in the table below to determine when to use an open or closed circle when graphing.

Open Circle	Closed Circle
$>$ greater than	$\geq$ greater than or equal to
$<$ less than	$\leq$ less than or equal to

Example 1 Graph: $x \le -2$

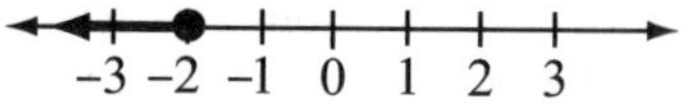

Step 1: Graph the inequality. Since –2 is a part of the solution, use a closed circle.

Answer:

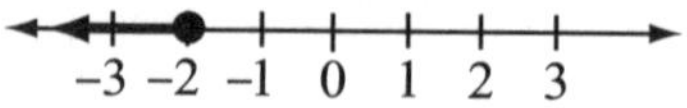

Example 2 Solve the inequality and graph the solution set for $h-(-4)>1$.

$h-(-4)>1$
$h+4>1$

Step 1: Simplify.

$h+4-4>1-4$

Step 2: Subtract 4 from both sides.

$h>-3$

Step 3: Simplify.

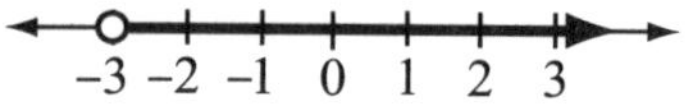

Step 4: Graph the inequality. Since –3 is not a solution, use an open circle.

Answer: $h>-3$

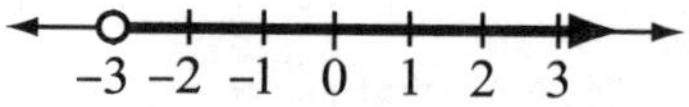

Example 3 Solve the inequality and graph the solution set for $-2h+6\le 4$.

$-2h+6\le 4$
$-2h+6-6\le 4-6$

Step 1: Subtract 6 from both sides.

$-2h\le -2$

Step 2: Simplify.

$\frac{-2h}{-2}\ge\frac{-2}{-2}$

Step 3: Divide both sides by –2 and reverse the direction of the inequality.

$h\ge 1$

Step 4: Simplify.

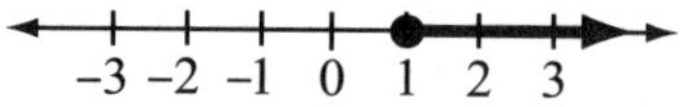

Step 5: Graph the inequality. Since 1 is part of the solution, use a closed circle.

Answer: $h\ge 1$

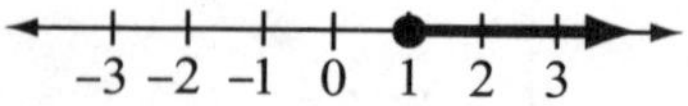

Example 4 Solve the inequality and graph the solution set for $3x + 15 + 2x - 8 > -18$.

$3x + 15 + 2x - 8 > -18$
$5x + 7 > -18$

Step 1: Combine like terms.

$5x + 7 - 7 > -18 - 7$

Step 2: Subtract 7 from both sides.

$5x > -25$

Step 3: Simplify.

$\frac{5x}{5} > \frac{-25}{5}$

Step 4: Divide both sides by 5.

$x > -5$

Step 5: Simplify.

Step 6: Graph the inequality. Since –5 is not part of the solution, use an open circle.

Answer: $x > -5$

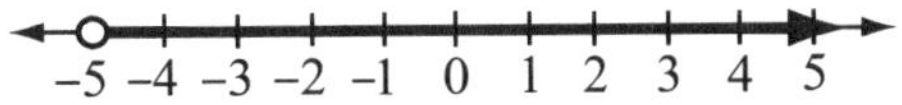

Example 5 Solve the inequality and graph the solution set for $-11m + 3(2m + 5) \leq 1 - 3(m - 6)$.

$-11m + 3(2m + 5) \leq 1 - 3(m - 6)$
$-11m + 6m + 15 \leq 1 - 3m + 18$

Step 1: Distribute the 3 to $(2m + 5)$ and distribute the –3 to $m - 6$.

$-5m + 15 \leq 19 - 3m$

Step 2: Combine like terms.

$-5m + 3m + 15 \leq 19 - 3m + 3m$

Step 3: Add $3m$ to both sides.

$-2m + 15 \leq 19$

Step 4: Combine like terms.

$-2m + 15 - 15 \leq 19 - 15$

Step 5: Subtract 15 from both sides.

$-2m \leq 4$

Step 6: Simplify.

$\frac{-2m}{-2} \geq \frac{4}{-2}$

Step 7: Divide both sides by –2 and reverse the direction of the inequality sign.

$m \geq -2$

Step 8: Simplify.

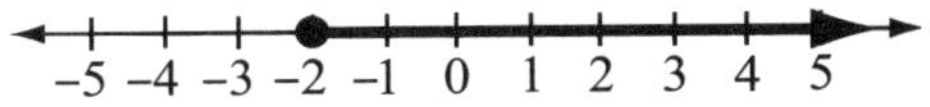

Step 9: Graph the inequality. Since –2 is part of the solution, use a closed circle.

Answer: $m \geq -2$

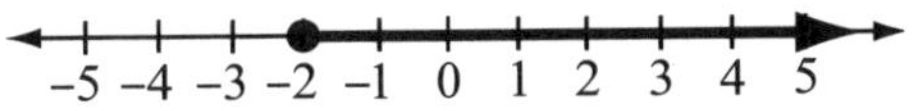

Problem Set

Solve the inequalities:

1. $a + (-9) < -2$

2. $k - 4 < 6$

3. $b - (-7) < 5$

4. $75 > p + 52$

5. $-43 > h + 16$

6. $h + 5 \geq 7$; Graph the solution set.

7. $h - 5 > 7$; Graph the solution set.

8. $f - 4 \leq 5$; Graph the solution set.

9. $f + 4 \leq 9$; Graph the solution set.

10. $g + (-2) \leq -6$; Graph the solution set.

11. $16h - 18 - 14h + 17 > 29$

12. $-11h + 18 + 12h - 17 \geq 25$

13. $-20f + 25 + 21f - 31 < 37$

14. $\frac{1}{3}p - 4 < -7$

15. $3x + 7 > -5$

16. $11 > 6h - 7$

17. $4k + 20 + 6k - 12 \leq 48$

18. $9f - 4(2f - 5) \leq 63$

19. $-13w + 4(2w + 5) - 12 \leq 63$

20. $5(1 - x) < 4(3 - x)$

HA1-802: Factoring

In this lesson, you will factor trinomials of the form $x^2 + bx + c$, where b is the coefficient of the x term and c is a constant. When factoring a trinomial, you are asked to find the binomial factors that equal the trinomial. The product of the factors of the trinomial must equal c, and the sum of the factors must equal b. When the coefficient of the x^2 term is one, there are some useful rules to determine which factors to use.

1. The product of the first terms in the binomial is the first term in the trinomial.
2. The product of the last terms in the binomial is the last term in the trinomial.
3. The sum of the last terms in the binomials is equal to the coefficient of x, or b.

The following table will help determine the appropriate sign to use for each of the factors.

	$c > 0$	$c < 0$
$b > 0$	Both factors positive	Larger factor positive
$b < 0$	Both factors negative	Larger factor negative

In order to check the answer, we can use the FOIL method. Recall that the FOIL method is used to multiply two binomials.

FOIL Method	A method used to multiply two binomials, in the following order:		
	F	First	Multiply the first terms of each binomial.
	O	Outer	Multiply the outer terms: the first term of the first binomial and the second term of the second binomial.
	I	Inner	Multiply the inner terms: the second term of the first binomial and the first term of the second binomial.
	L	Last	Multiply the last terms of each binomial.

Example 1 Find the missing term: $x^2 + 9x + 20 = (x + 4)(x + ?)$

Factors of 20	Sum of Factors
1, 20	21
2, 10	12
4, 5	**9**

Step 1: Determine which two numbers will have a product of 20 and a sum of 9.

$(x + 4)(x + \underline{5})$

Step 2: Use the factors from Step 1 to determine the missing term.

Answer: The missing term is 5.

Example 2 Factor: $x^2 + 11x + 24$

$x \cdot x$

Step 1: Determine which factors are multiplied to get x^2.

Factors of 24	Sum of Factors
1, 24	25
2, 12	14
3, 8	**11**
4, 6	10

Step 2: Determine which two numbers will have a product of 24 and a sum of 11.

$(x + 3)(x + 8)$

Step 3: Show the factors in binomial form.

Step 4: Use the FOIL method to check.

F $x \cdot x = x^2$

O $x \cdot 8 = 8x$

I $3 \cdot x = 3x$

L $3 \cdot 8 = 24$

$x^2 + 8x + 3x + 24 = x^2 + 11x + 24$

Answer: $(x + 3)(x + 8)$

Example 3 Factor: $x^2 - 8x + 16$

$x \cdot x$

Step 1: Determine which factors are multiplied to get x^2.

Factors of 16	Sum of Factors
−1, −16	−17
−2, −8	−10
−4, −4	**−8**

Step 2: Determine which two numbers will have a product of 16 and a sum of −8.

$(x-4)(x-4)$

Step 3: Show the factors in binomial form.

Step 4: Use the FOIL method to check.

F $x \cdot x = x^2$

O $x \cdot (-4) = -4x$

I $-4 \cdot x = -4x$

L $-4 \cdot (-4) = 16$

$x^2 - 4x - 4x + 16 = x^2 - 8x + 16$

Answer: $(x-4)(x-4)$

Example 4 If the area of a rectangle is $x^2 + 27x + 50$ and the length is $(x+25)$, find the width.

$$\text{Area} = \text{length} \times \text{width}$$
$$= (x+25)(\quad)$$

Step 1: Recall the formula for the area of a rectangle and substitute the given information into the formula.

Factors of 50	Sum of Factors
1, 50	51
2, 25	**27**
5, 10	15

Step 2: Determine which two numbers will have a product of 50 and a sum of 27.

$(x+25)(x+2)$

Step 3: Show the factors in binomial form.

$x^2 + 27 + 50 = (x+25)(\underline{x+2})$

Step 4: Determine the width from the factors in the binomial, since the length was given as $(x+25)$.

Answer: The width of the rectangle is $(x+2)$.

Example 5 Factor: $x^2 - 14x - 32$

$x \cdot x$

Step 1: Determine which factors are multiplied to get x^2.

Factors of 32	Sum of Factors
1, –32	–31
2, –16	**–14**
4, –8	–4

Step 2: Determine which two numbers will have a product of –32 and a sum of –14.

$(x+2)(x-16)$

Step 3: Show the factors in binomial form.

Step 4: Use the FOIL method to check.

F $x \cdot x = x^2$

O $x \cdot (-16) = -16x$

I $2 \cdot x = 2x$

L $2 \cdot (-16) = -32$

$x^2 - 16x + 2x - 32 = x^2 - 14x - 32$

Answer: $(x+2)(x-16)$

Problem Set

Factor:

1. $x^2 + 4x + 4$
2. $x^2 + 11x + 10$
3. $x^2 + 11x + 18$
4. $x^2 + 10x + 24$
5. $x^2 + 12x + 32$
6. $x^2 + 13x + 30$
7. $x^2 + 13x + 40$
8. $z^2 + 20z + 64$
9. $x^2 + 12x + 20$
10. $x^2 + 20x + 99$
11. $x^2 - 6x + 8$
12. $w^2 - 5w + 6$
13. $y^2 - 11y + 28$
14. $x^2 - 8x + 7$
15. $a^2 - 9a + 18$
16. $r^2 - 4r + 4$
17. $z^2 - 12z + 35$
18. $x^2 + 5x - 36$
19. $x^2 - 4x - 96$
20. $x^2 + 11x - 60$

HA1-803: Solving Systems of Equations

We have learned how to solve systems of equations using graphing and the substitution method. Another way to solve a system of equations is by using the elimination method. When using this method, the coefficients of one of the variables in each equation of the system must be opposites so that one of the variables can be eliminated when adding. Also, both of the equations need to be written in standard form.

$$\begin{aligned} 3x - 2y &= 4 \\ 2x + 2y &= 6 \end{aligned}$$

In the system of equations above, the coefficient of the y in the first equation is -2. The coefficient of the y in the second equation is 2. Since the coefficients are opposites, we can add the two equations to eliminate the y variable and solve for x.

$$\begin{aligned} 3x - 2y &= 4 \\ \underline{2x + 2y} &\underline{= 6} \\ 5x &= 10 \\ \frac{5x}{5} &= 10 \\ x &= 2 \end{aligned}$$

Elimination Method	The use of addition to eliminate one variable and solve a system of equations

If the coefficients of one of the variables in the system of equations are not the same use equivalent equations. Equivalent equations are found by multiplying one or both equations by a number so the coefficients of one of the variables in the system of equations are the same.

When solving a system of equations, the objective is to find at least one ordered pair that makes both equations true. In solving, you may get one solution, no solution, or infinitely many solutions. In the example above, substitute $x = 2$ into one of the equations in the system to solve for y.

$$\begin{aligned} 3(2) - 2y &= 4 \\ 6 - 2y &= 4 \\ 6 - 6 - 2y &= 4 - 6 \\ -2y &= -2 \\ y &= 1 \end{aligned}$$

This system has one solution, (2, 1).

Use the chart below to help determine what type of solution you have when using the elimination method.

one solution	When solving the equation, one ordered pair (x, y) will satisfy the system.
no solution	When solving the equation, the result will be a false statement.
infinitely many solutions	When solving the equation, the result will be a true statement.

Example 1 Solve the following system of equations:
$$x + y = -2$$
$$x - y = 6$$

$$\begin{aligned} x + y &= -2 \\ x - y &= 6 \\ \hline 2x &= 4 \end{aligned}$$

Step 1: Add the two equations to eliminate y.

$$\begin{aligned} \frac{2x}{2} &= \frac{4}{2} \\ x &= 2 \end{aligned}$$

Step 2: Solve the equation for x.

$$\begin{aligned} x + y &= -2 \\ 2 + y &= -2 \\ 2 - 2 - y &= -2 - 2 \\ y &= -4 \end{aligned}$$

Step 3: Substitute $x = 2$ into one of the original equations and solve for y.

Answer: The solution to the system of equations is (2,–4).

Example 2 Given the following system, what must be done to the second equation to get equivalent equations:
$$3x + 2y = 5$$
$$2x - y = 8$$

$$\begin{aligned} 3x + 2y &= 5 \\ 2x - y &= 8 \end{aligned}$$

Step 1: Determine what number could be multiplied by either the first or second equation so that one of the variables has opposite coefficients.

$$\begin{aligned} 3x + 2y &= 5 &\rightarrow 3x + 2y &= 5 \\ 2(2x - y) &= 2(8) &\rightarrow 4x - 2y &= 16 \end{aligned}$$

Step 2: Multiply the second equation by 2 so the y terms are opposites.

Answer: To get equivalent equations, multiply the second equation by 2.

Example 3 Solve the following system of equations:
$$2x + 3y = -7$$
$$2x = y + 1$$

$$\begin{aligned} 2x &= y + 1 \\ 2x - y &= y - y + 1 \\ 2x - y &= 1 \end{aligned}$$

Step 1: Change the second equation into standard form.

$$\begin{aligned} 2x + 3y &= -7 &\rightarrow 2x + 3y &= -7 \\ -1(2x - y) &= -1(1) &\rightarrow -2x + y &= -1 \end{aligned}$$

Step 2: Multiply the second equation by –1 so the x terms are opposites.

$$\begin{aligned} 2x + 3y &= -7 \\ -2x + y &= -1 \\ \hline 4y &= -8 \end{aligned}$$

Step 3: Add the two equations to eliminate x.

$$\begin{aligned} \frac{4y}{4} &= \frac{-8}{4} \\ y &= -2 \end{aligned}$$

Step 4: Solve the equation for y.

$$\begin{aligned} 2x + 3y &= -7 \\ 2x + 3(-2) &= -7 \\ 2x - 6 &= -7 \\ 2x - 6 + 6 &= -7 + 6 \\ 2x &= -1 \\ \frac{2x}{2} &= \frac{-1}{2} \\ x &= -\frac{1}{2} \end{aligned}$$

Step 5: Substitute $y = -2$ into one of the original equations and solve for x.

Answer: The solution to the system of equations is $\left(-\frac{1}{2}, -2\right)$.

Example 4 Solve the following system of equations:
$$\begin{aligned} 6x + 4y &= 2 \\ -3x - 2y &= 8 \end{aligned}$$

$$\begin{aligned} 6x + 4y &= 2 &\rightarrow\quad 6x + 4y &= 2 \\ 2(-3x - 2y) &= 2(8) &\rightarrow\quad -6x - 4y &= 16 \end{aligned}$$

Step 1: Multiply the second equation by 2 so the x terms are opposites.

$$\begin{aligned} 6x + 4y &= 2 \\ -6x - 4y &= 16 \\ \hline 0 &= 18 \end{aligned}$$

Step 2: Add the two equations to eliminate x.

Answer: This is a false statement; therefore, there is no solution to the system of equations.

Example 5 Solve the following system of equations:
$$\begin{aligned} 2x - 5y &= -4 \\ -3x + 3y &= -12 \end{aligned}$$

$$\begin{aligned} 3(2x - 5y) &= 3(-4) &\rightarrow\quad 6x - 15y &= -12 \\ 2(-3x + 3y) &= 2(-12) &\rightarrow\quad -6x + 6y &= -24 \end{aligned}$$

Step 1: Multiply the first equation by 3 and the second equation by 2 so the x terms are opposites.

$$\begin{aligned} 6x - 15y &= -12 \\ -6x + 6y &= -24 \\ -9y &= -36 \end{aligned}$$

Step 2: Add the two new equations.

$\frac{-9y}{-9} = \frac{-36}{-9}$ $y = 4$	**Step 3:** Solve the equation for *y*.
$2x - 5y = -4$ $2x - 5(4) = -4$ $2x - 20 = -4$ $2x - 20 + 20 = -4 + 20$ $2x = 16$ $\frac{2x}{2} = \frac{16}{2}$ $x = 8$	**Step 4:** Substitute $y = 4$ into one of the original equations and solve for *x*.
	Answer: The solution to the system of equations is (8, 4).

Problem Set

Solve the following systems of equations:

1. $x + y = 2$; $7x - 4 = 6$

2. $x + 2y = 11$; $-x + 3y = 14$

3. $2x - y = 14$; $9x + y = 63$

4. $-x + 5y = 10$; $x + y = 2$

5. $4x + y = 19$; $-4x + y = -5$

6. $3x + y = 4$; $2x - y = -4$

7. $-4x - 3y = 7$; $4x = 6y + 2$

8. $4x + 2y = -2$; $4x - 2y = -6$

9. $x + y = -5$; $x - y = 1$

10. $x - y = -3$; $x + y = 5$

11. $3a + 2b = -7$; $a - b = 1$

12. $-3a + 2b = -4$; $a - b = -3$

13. $x - 3y = -13$; $2x + y = -5$

14. $-x - 3y = -13$; $2x + y = 6$

15. $2x + 3y = 1$; $-x + y = 2$

16. $5x + 3y = -2$; $x + y = 2$

17. $2m + 5n = 6$; $3m - n = 9$

18. $3a + 2b = -3$; $2a - 3b = 11$

19. $6x - 3y = 21$; $4x - 2y = 14$

20. $6x - 3y = 21$; $4x - 2y = 7$

HA1-804: Solving Systems of Equations by Graphing

We have previously learned how to graph linear equations in two variables. Recall that a linear equation in two variables is an equation of the form $Ax + By = C$, with A and B not both zero. Many real-world problems can be solved using one or more linear equations. Two or more linear equations form a system of linear equations where the equations each use the same variables.

In this lesson, we will use the method of graphing to solve systems of linear equations. When solving a system of equations, each equation is graphed on the same coordinate plane. The point where the lines intersect, if it exists, is the solution. The table below shows the three possible types of solutions.

Graphs of Equations	Number of Solutions or Points of Intersection
Intersecting Lines 	One
Parallel Lines 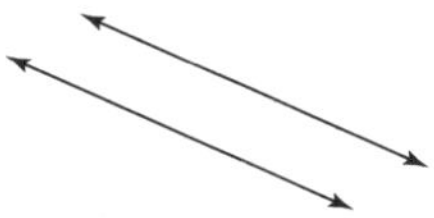	None
Same Line 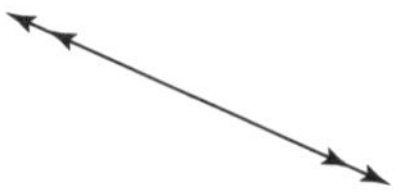	Infinitely Many

To check a solution, substitute the x-value and the y-value into each equation. If the point satisfies both equations, the solution is correct.

Example 1 Determine whether (3,–1) is a solution to the following system of equations.

$$x + y = 2$$
$$2x - 3y = 9$$

Step 1: Substitute (3,–1) into each equation and simplify.

First Equation	**Second Equation**
$x + y = 2$	$2x - 3y = 9$
$3 + (-1) \stackrel{?}{=} 2$	$2(3) - 3(-1) \stackrel{?}{=} 9$
$2 = 2$ √	$6 + 3 \stackrel{?}{=} 9$
	$9 = 9$ √

Answer: Since (3,–1) satisfies both equations, it is a solution to the system.

Example 2 Find the solution to the system of equations by graphing.

$$x = 2$$
$$2x - 4y = 8$$

x	y
0	–2
4	0

Step 1: Determine the x- and y-intercepts for the second equation using a table, since the first equation is a vertical line through $x = 2$.

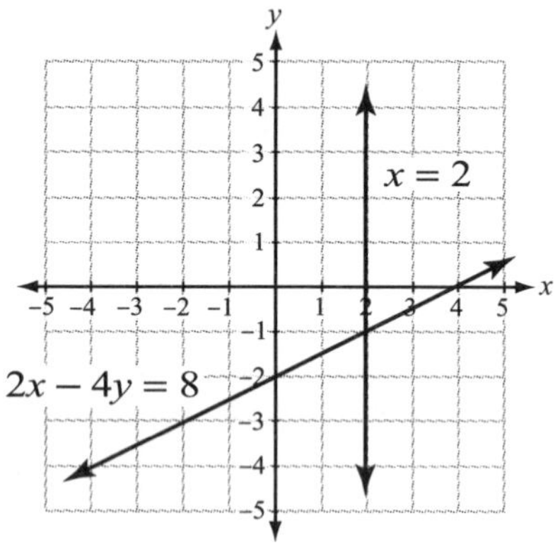

Step 2: Draw the graphs on the same coordinate plane.

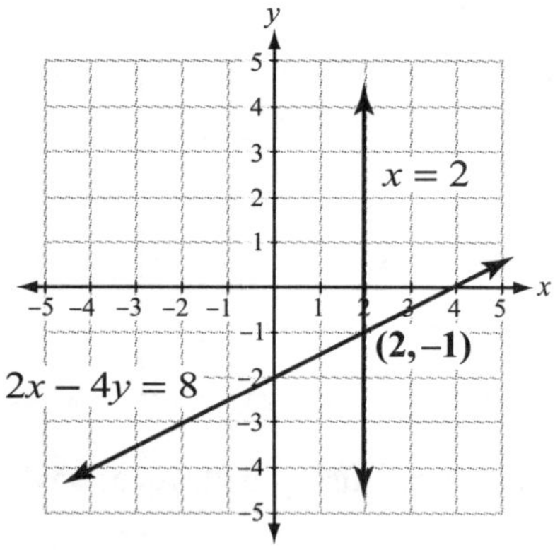

Step 3: Find the point(s) of intersection of the two lines.

First Equation

$x = 2$

$2 = 2$ √

Second Equation

$2x - 4y = 8$

$2(2) - 4(-1) \stackrel{?}{=} 8$

$4 + 4 \stackrel{?}{=} 8$

$8 = 8$ √

Step 4: Verify that the ordered pair (2,–1) is the solution to the given system by substituting (2,–1) into each equation of the system.

Answer: Since the graphs intersect at (2,–1), it is the solution to the system.

Example 3 Find the solution to the system of equations by graphing.

$$2x - y = 3$$
$$x + 2y = 4$$

Step 1: Determine the x- and y-intercepts for both equations.

$2x - y = 3$

x	y
0	−3
1.5	0

$x + 2y = 4$

x	y
0	2
4	0

Step 2: Draw the graphs on the same coordinate plane.

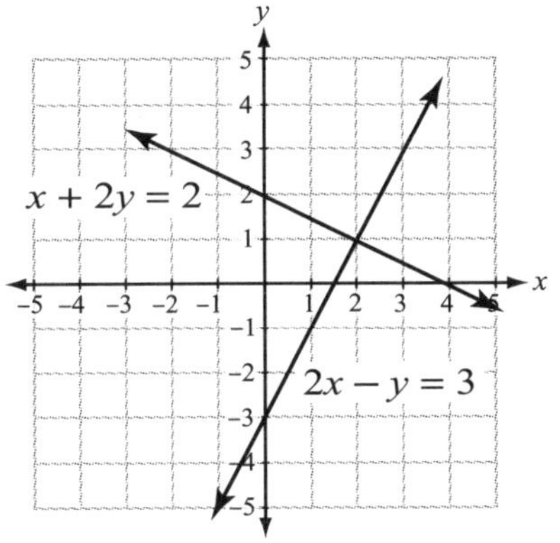

Step 3: Find the point(s) of intersection of the two lines.

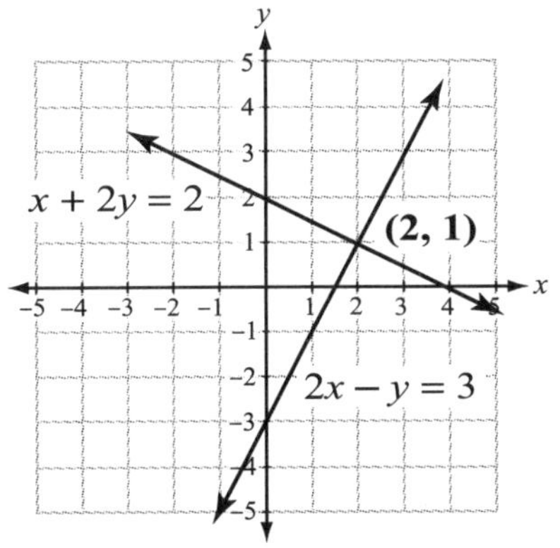

Step 4: Verify that the ordered pair (2, 1) is the solution to the given system by substituting (2, 1) into each equation of the system.

First Equation

$2x - y = 3$

$2(2) - 1 \stackrel{?}{=} 3$

$4 - 1 \stackrel{?}{=} 3$

$3 = 3$ √

Second Equation

$x + 2y = 4$

$2 + 2(1) \stackrel{?}{=} 4$

$2 + 2 \stackrel{?}{=} 4$

$4 = 4$ √

Answer: Since the graphs intersect at (2, 1), it is the solution to the system.

Example 4 Find the solution to the system of equations by graphing.

$$4x + 2y = 8$$
$$-2x - y = -4$$

$4x + 2y = 8$ $\quad -2x - y = -4$

x	y
0	4
2	0

x	y
0	4
2	0

Step 1: Determine the x- and y-intercepts for both equations.

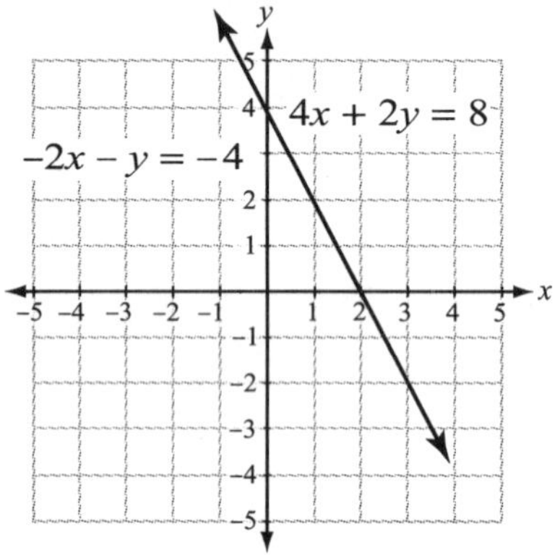

Step 2: Draw the graphs on the same coordinate plane. Since the intercepts are the same, both equations yield the same line.

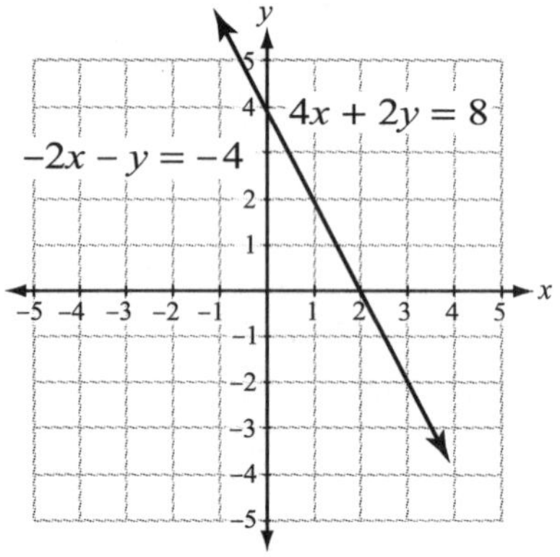

Step 3: Find the point(s) of intersection of the two lines. Every point on the line is a solution to both equations. Therefore, there are infinitely many solutions.

Answer: Since the graphs are the same line, the system has infinitely many solutions.

Example 5 Find the solution to the system of equations by graphing.

$$x + 3y = 5$$
$$y = x + 1$$

Step 1: Determine the x- and y-intercepts for both equations.

$x + 3y = 5$

x	y
0	$\frac{5}{3}$
5	0

$y = x + 1$

x	y
0	1
−1	0

Step 2: Draw the graphs on the same coordinate plane.

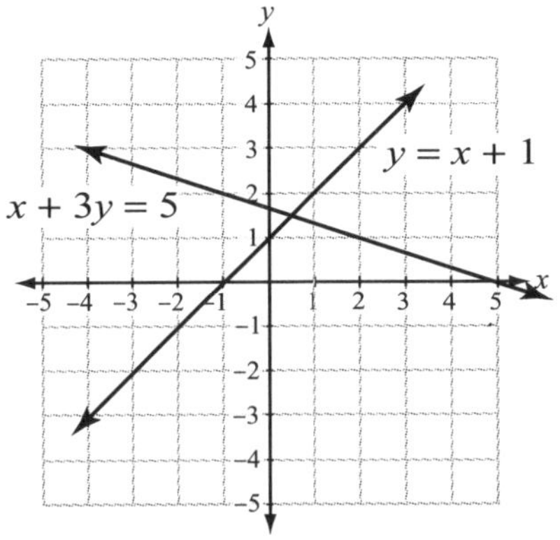

Step 3: Find the point(s) of intersection of the two lines.

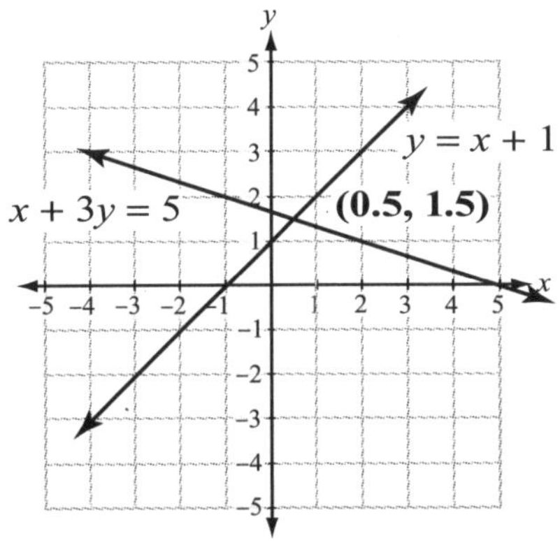

Step 4: Verify that the ordered pair (0.5, 1.5) is the solution to the given system by substituting (0.5, 1.5) into each equation of the system.

First Equation

$$x + 3y = 5$$
$$0.5 + 3(1.5) \stackrel{?}{=} 5$$
$$0.5 + 3(1.5) \stackrel{?}{=} 5$$
$$5 = 5 \ \surd$$

Second Equation

$$y = x + 1$$
$$1.5 \stackrel{?}{=} 0.5 + 1$$
$$1.5 = 0.5 \ \surd$$

Answer: Since the graphs intersect at (0.5, 1.5), it is the solution to the system.

Problem Set

Solve:

1. Determine whether (0, 2) is a solution to the following system of equations:

$$x = y - 2$$
$$3x = 3y - 6$$

No Yes

2. Determine whether (1, 3) is a solution to the following system of equations:

$$2x + y = -1$$
$$x - y = 4$$

Yes No

3. Determine whether (–1, 2) is a solution to the following system of equations:

$$x - y = -3$$
$$2x + 4y = 5$$

No Yes

4. Find the solution to the following system of equations:

$$x + y = -1$$
$$2x - y = 4$$

5. Find the solution to the following system of equations:

$$2x + y - 6 = 0$$
$$2x + 2y = 4$$

6. Solve the following system of equations by graphing:

$$x - y = -2$$
$$4x + y = -6$$

7. Solve the following system of equations by graphing:

$$y = 1$$
$$y + x = -3$$

8. Solve the following system of equations by graphing:

$$x - y = -4$$
$$x - y = 3$$

9. Graph the system of equations and determine which of the following statements is true about the system:

$$2x + y = 5$$
$$6x + 3y = 15$$

Since the graphs do not intersect, the system has no solution.

Since the graphs are the same, the system has an infinite number of solutions.

Since the graphs do not intersect, the system has an infinite number of solutions.

Since the graphs only intersect at one point, the system has exactly one solution.

10. Graph the system of equations and determine which of the following statements is true about the system:

$$3x + y = -5$$
$$9x + 3y = 15$$

Since the graphs only intersect at one point, the system has two solutions.

Since the graphs do not intersect, the system has no solution.

Since the graphs are parallel, the system has exactly one solution.

Since the graphs are the same, the system has an infinite number of solutions.

HA1-805: Applying Algebra Concepts

You've learned how to solve word problems involving equations and inequalities, and we've looked at some applications of linear functions. Now it's time to put those skills together and work on applying algebra concepts to word problems that involve systems, tables, linear functions, and quadratic equations.

Example 1 Kayla likes to buy snacks with her allowance. Last week, she bought two bags of chips and one bag of cookies for $2. This week, she bought one bag of chips and two bags of cookies for $2.50. Find the price for each bag of chips and each bag of cookies.

x = price for each bag of chips
y = price for each bag of cookies

$$\begin{cases} 2x + y = 2 \\ x + 2y = 2.50 \end{cases}$$

Step 1: Set up the equations. Let x be the price for each bag of chips and y be the price for each bag of cookies. On the first week, she bought two bags of chips and one bag cookies for $2. Therefore, the equation is $2x + y = 2$. On the next week, she bought one bag of chips and two bag of cookies for $2.50. Therefore, the equation is $x + 2y = 2.50$.

$$\begin{cases} 2x + y = 2 \\ -2(x + 2y = 2.50) \end{cases}$$

$$\begin{aligned} 2x + y &= 2 \\ +(-2x) - 4y &= -5 \\ \hline -3y &= -3 \\ y &= 1 \end{aligned}$$

Price for a bag of cookies

$$\begin{aligned} 2x + (1) &= 2 \\ 2x &= 1 \\ x &= \frac{1}{2} = 0.50 \end{aligned}$$

Price for a bag of chips

Step 2: Solve the equation using the Elimination Method.

Answer: The price for each bag of chips is $0.50 and the price for each bag of cookies is $1.

Example 2 A total of 500 seats were sold for a play. The floor seats cost $25.00. The balcony seats cost $15.00. The play brought in a total of $10,000. Find the number of tickets sold on the floor and the number of seats tickets on the balcony.

$$\begin{aligned} x + y &= 500 \\ 25x + 15y &= 10{,}000 \end{aligned}$$

Step 1: Set up the equations.

$y = 500 - x$

Step 2: Solve the equations using the Substitution Method.

$$
\begin{aligned}
25x + 15(500 - x) &= 10,000 \\
25x + 7,500 - 15x &= 10,000 \\
10x + 7,500 &= 10,000 \\
10x &= 2,500 \\
x &= 250
\end{aligned}
$$

$$
\begin{aligned}
y &= 500 - 250 \\
&= 250
\end{aligned}
$$

Answer: There were 250 tickets sold for the first floor and 250 tickets sold for the balcony.

Example 3 What are the dimensions of a rectangular box, if its perimeter is 34 inches and its area is 60 square inches?

$$
\begin{aligned}
P &= 2l + 2w \\
A &= lw
\end{aligned}
$$

Step 1: Set up the equations. Recall that the Perimeter = $2l + 2w$ and Area = lw.

Let **length** $= x$

width $= y$

Therefore,

$$
\begin{aligned}
2x + 2y &= 34 \\
xy &= 60
\end{aligned}
$$

$$
\begin{aligned}
2y &= 34 - 2x \\
y &= 17 - x
\end{aligned}
$$

Step 2: Solve the equations using the Substitution Method.

$$
\begin{aligned}
x(17 - x) &= 60 \\
17x - x^2 &= 60 \\
x^2 - 17x + 60 &= 0 \\
(x - 12)(x - 5) &= 0
\end{aligned}
$$

$x = 5$ or $x = 12$

Since length is greater than width, $x = 12$ and $y = 5$

Answer: Therefore, the length of the box is 12 inches and the width of the box is 5 inches.

Problem Set

1. A carpenter braces a 6×8 foot wall by nailing a board diagonally across the wall. How long is the bracing board?

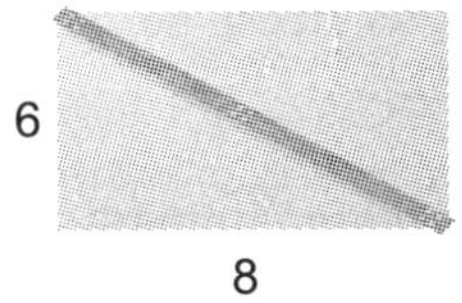

2. Sizes of TV screens and computer monitors are given according to the length of the diagonal. If a monitor screen is a rectangle which is 12 inches by 5 inches, how long is the diagonal?

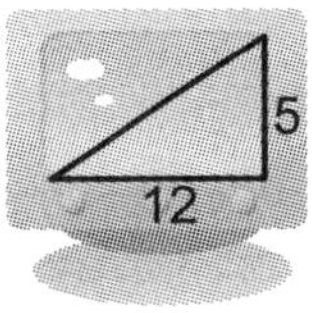

3. If a piece of square tile is 16 inches on a side, how long is the diagonal of the tile? Round your answer to the nearest hundredth.

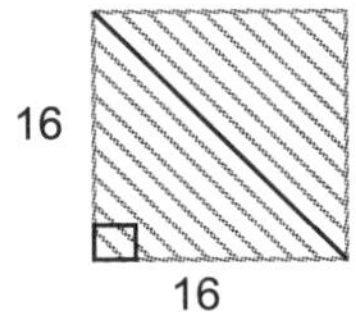

4. A surveyor needs to find the distance across a lake from point C to point B. He measures the distance from A to C to be 170 meters, and the distance from A to B to be 80 meters. Find the distance across the lake.

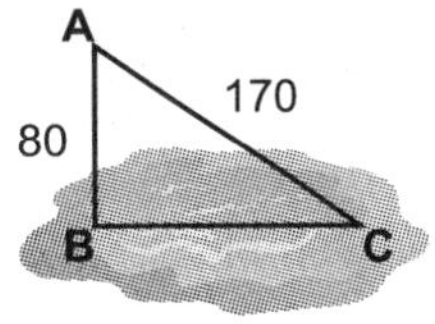

5. A hot-air balloon is rising vertically but it is attached to a 34-meter cord. When the balloon reaches its maximum height of 34 meters, an observer uses a laser to determine that the distance between him and the balloon is 36 meters. How far is he from the point of liftoff of the balloon? Round the answer to the nearest hundredth.

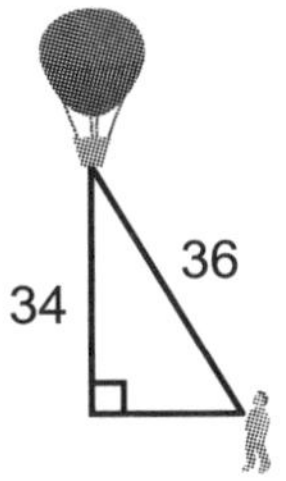

6. The area of a rectangle is 84 square feet. If the length is 17 feet more than the width, find the length of the rectangle.

7. The width of a rectangle rug is 3 feet less than its length. If the area of the rug is 80 feet, find the width of the rug. Round your answer to the nearest tenth.

8. The length of a rectangular swimming pool is 15 feet longer than its width. If the area of the pool is 127 square feet, what are the length and width of the pool?

9. Alvin sold 80 tickets for the school play. He gave the treasurer \$460. If adult tickets cost \$8 each and child tickets cost \$4 each, how many of each did he sell?

10. A large bag of flour weighs 3 pounds more than three small bags of flour. Find the weight of each if 4 large bags and 4 small bags weigh 60 pounds.

HA1-815: Simplifying Expressions with Negative and Zero Exponents

Recall the exponential expression a^n, where the variable a represents the base and the variable n represents the exponent. This expression is read as "a to the nth power". For example, look at, 4^2. It can be simplified to $4 \cdot 4$ or 16, where 4 is the base, 2 is the exponent, and is read as "four to the second power."

Exponent	a number or variable that expresses how many times the base is multiplied

Base	the repeated factor

Power	a term that is written as a product of equal factors that includes a base and an exponent

Previously, we worked problems involving positive exponents. For example:

1. $x^2 \cdot x^3 = x^{2+3} = x^5$
2. $(yz)^2 = y^2 \cdot z^2 = y^2z^2$
3. $(2^2)^3 = 2^{2 \cdot 3} = 2^6 = 64$

These three examples were simplified using the following properties: the Product of Powers Property, the Power of a Product Property, and the Power of a Power Property, respectively.

For all real numbers a and b, and any positive integer m and n,

Product of Powers Property: $a^m \cdot a^n = a^{m+n}$

Power of a Product Property: $(ab)^m = a^m \cdot b^m$

Power of a Power Property: $(a^m)^n = a^{m \cdot n}$

We will now work problems involving negative or zero exponents. For example, how would 4^{-2} and 4^0 be simplified? There are two new properties that will help us to simplify these problems. They are the **Zero Exponent Property** and **Negative Exponent Property**.

Zero Exponent Property	For any nonzero real number b and any positive integer n, $b^0 = 1$.

Negative Exponent Property	For any nonzero real number b and any positive integer n, $b^{-n} = \frac{1}{b^n}$ and $\frac{1}{b^{-n}} = b^n$.

Using these two properties, we can simplify 4^{-2} and 4^0. We can rewrite 4^{-2} as $\frac{1}{4^2}$ using the Negative Exponent Property and then simplify.

$$4^{-2} = \frac{1}{4^2}$$
$$= \frac{1}{16}$$

Using the Zero Exponent Property, we can say that $4^0 = 1$. Let's look at the following problems and evaluate them using these properties along with the properties we previously learned.

Example 1 Evaluate the expression: 7^0

$7^0 = 1$

Step 1: Use the Zero Exponent Property.

Answer: 1

Example 2 Are the expressions 6^{-3} and $\frac{1}{18}$ the same?

$6^{-3} = \frac{1}{6^{-3}}$

Step 1: Use the Negative Exponent Property to rewrite 6^{-3}.

$= \frac{1}{216}$

Step 2: Simplify.

$\frac{1}{18} \neq \frac{1}{216}$

Step 3: Compare the two expressions.

Answer: No

Example 3 Evaluate the expression: $2 \cdot 2^{-6} \cdot 2^4$

$2 \cdot 2^{-6} \cdot 2^4 = 2^1 \cdot 2^{-6} \cdot 2^4$

Step 1: Replace 2 with 2^1, since $2 = 2^1$.

$= 2^{1 + (-6) + 4}$ **Step 2:** Use the Product of Powers Property to add the exponents.

$= 2^{-1}$ **Step 3:** Simplify.

$= \frac{1}{2^1}$ **Step 4:** Use the Negative Exponent Property.

$= \frac{1}{2}$ **Step 5:** Simplify.

Answer: $\frac{1}{2}$

Example 4 Rewrite the expression with positive exponents: $\frac{x^{-3}}{(2z)^{-2}}$

$\frac{x^{-3}}{(2z)^{-2}} = \frac{(2z)^2}{x^3}$ **Step 1:** Use the Negative Exponent Property.

$= \frac{2^2 z^2}{x^3}$ **Step 2:** Use the Power of a Product Property.

$= \frac{4z^2}{x^3}$ **Step 3:** Simplify.

Answer: $\frac{4z^2}{x^3}$

Example 5 The population growth of a small town in Alabama is modeled by the equation $P = 10,651(1.035)^t$ where t represents the time in years and P represents the population. What is the population "now"?

$P = 10,651(1.035)^t$
$= 10,651(1.035)^0$ **Step 1:** Since the time "now" is $t = 0$, substitute 0 in the equation for t.

$= 10,651(1)$ **Step 2:** Use the Zero Exponent Property.

$= 10,651$ **Step 3:** Simplify.

Answer: The population "now" is 10,651 people.

Problem Set

1. Evaluate the expression: $\frac{1}{11^{-2}}$

2. Evaluate the expression: $\frac{1}{2^{-3}}$

3. Evaluate the expression: 15^0

4. Evaluate the expression: $\left(-\frac{1}{8}\right)^0$

5. Which expression simplifies to $\frac{1}{25}$?

-5^2 $\quad 5^{-5}$

1^{-25} $\quad 5^{-2}$

6. Which expression simplifies to $\frac{1}{16}$?

8^{-2} $\quad 1^{-16}$

2^{-4} $\quad \frac{1}{2^{-4}}$

7. Which expression simplifies to 64?

1^{-64} $\quad \frac{1}{32^{-2}}$

$\frac{1}{8^{-2}}$ $\quad 8^{-2}$

8. Which expression simplifies to 32?

$\frac{1}{8^{-4}}$ $\quad 16^{-2}$

1^{-32} $\quad \frac{1}{2^{-5}}$

9. Evaluate the expression: $8^{-2} \cdot 8^4 \cdot 8^{-2}$

10. Evaluate the expression: $6^{-4} \cdot 6^8 \cdot 6^{-4}$

11. Evaluate the expression: $5^{-2} \cdot 5^2 \cdot 5^{-2}$

12. Evaluate the expression: $3^{-3} \cdot 3^4 \cdot 3^{-3}$

13. Rewrite the expression with positive exponents:

$(3y)^{-4}$

14. Rewrite the expression with positive exponents:

$(2xy)^{-5}$

15. Rewrite the expression with positive exponents:

$\frac{(2y)^{-4}}{(3x)^{-3}}$

16. Rewrite the expression with positive exponents:

$\frac{(5x)^{-2}}{(2y)^{-5}}$

17. James invested money in a local stock. The money can be modeled by using the expression $P = 6{,}000(1.2)^t$ where t is the time in years and P is the amount of money, in dollars. How much money did he have 5 years ago? Round your answer to the nearest dollar.

18. Emma invested money in a local stock. The money can be modeled by using the expression $P = 10{,}000(1.1)^t$ where t is the time in years and P is the amount of money, in dollars. How much money did she have three years ago? Round your answer to the nearest dollar.

19. The number of deer in a state park can be modeled by using the expression $D = 800(2)^t$ where t is the time in years and D is the number of deer in the park. How many deer were in the park 3 years ago?

20. The number of hawks in the state can be modeled by using the expression $H = 12{,}500(2)^t$ where t is the time in years and H is the number of hawks in the state. How many hawks were in the state 2 years ago?

HA1- 818: Simplifying Expressions Using the Division Properties of Exponents

Remember that in the expression a^n the "a" represents the base, the "n" represents the exponent, and the whole expression is the power. These three concepts that we previously learned will be important to remember when working with the exponents.

Previously, we worked problems that involved multiplying expressions containing positive, negative, and zero exponents. For example:

$$1.\ (ab)^{-3} = \frac{1}{(ab)^3} = \frac{1}{a^3b^3}$$

$$2.\ 3^{-3} \cdot 3^4 \cdot 3^{-1} = 3^{-3+4+(-1)} = 3^0 = 1$$

$$3.\ \frac{x}{(2y)^{-2}} = \frac{x}{2^{-2} \cdot y^{-2}} = 2^2 \cdot x \cdot y^2 = 4xy^2$$

These examples were simplified using the properties listed below:

For all real numbers a and b, and any positive integer m and n,

Product of Powers Property: $a^m \cdot a^n = a^{m+n}$

Power of a Product Property: $(ab)^m = a^m \cdot b^m$

Power of a Power Property: $(a^m)^n = a^{m \cdot n}$

For any nonzero real number b and any positive integer n,

Zero Exponent Property: $b^0 = 1$

Negative Exponent Property: $b^{-n} = \frac{1}{b^n}$, and $\frac{1}{b^{-n}} = b^n$

Now that we have reviewed the properties that help simplify exponents containing positive, negative, and zero exponents, let's look at expressions involving division. For example, how would we simplify the following expressions: $\left(\frac{2}{3}\right)^2$ and $\frac{x^3}{x^2}$? The first expression is a quotient raised to an exponent. The second expression is a quotient where the numerator and denominator have the same base but different exponents. The **Quotient of Powers Property** and the **Power of a Quotient Property** will allow us to simplify these expressions.

Quotient of Powers Property	For all real numbers a and any positive integer m and n, $\frac{a^m}{a^n} = a^{m-n},\ a \neq 0$

Power of a Quotient Property	For all real numbers a and b, and any positive integer m and n, $$\left(\frac{a}{b}\right)^m = \frac{a^m}{b^m},\ b \neq 0$$

Let's use these properties to simplify the expressions. For $\left(\frac{2}{3}\right)^2$, we apply the Power of a Quotient Property to rewrite the expression as $\frac{2^2}{3^2}$, which is $\frac{4}{9}$ when simplified. We apply the Quotient of Powers Property to simplify the expression $\frac{x^3}{x^2}$ as x^{3-2}, which is x^1 or x. The next two examples will use these division properties to simplify expressions with exponents.

Example 1 Simplify the quotient: $\frac{y^5}{y^3}$

$\frac{y^5}{y^3} = y^{5-3}$ **Step 1:** Use the Quotient of Powers Property.

$= y^2$ **Step 2:** Simplify.

Answer: y^2

Example 2 Simplify the quotient: $\left(\frac{4}{5}\right)^2$

$\left(\frac{4}{5}\right)^2 = \frac{4^2}{5^2}$ **Step 1:** Use the Power of a Quotient Property.

$= \frac{16}{25}$ **Step 2:** Simplify.

Answer: $\frac{16}{25}$

Now let's look at some examples that will require the application of more than one property to simplify the expression.

Example 3 Simplify the expression: $\left(\frac{2a}{b^2}\right)^4$

$\left(\frac{2a}{b^2}\right)^4 = \frac{(2a)^4}{(b^2)^4}$ **Step 1:** Use the Power of a Quotient Property.

$= \dfrac{2^4 \cdot a^4}{(b^2)^4}$ **Step 2:** Use the Power of a Product property.

$= \dfrac{2^4 \cdot a^4}{b^8}$ **Step 3:** Use the Power of a Power Property.

$= \dfrac{16a^4}{b^8}$ **Step 4:** Simplify.

Answer: $\dfrac{16a^4}{b^8}$

Example 4 Simplify the expression: $\left(\dfrac{x^3}{4y^2}\right)^{-2}$

$\left(\dfrac{x^3}{4y^2}\right)^{-2} = \dfrac{(x^3)^{-2}}{(4y^2)^{-2}}$ **Step 1:** Use the Power of a Quotient Property.

$= \dfrac{(x^3)^{-2}}{4^{-2} \cdot (y^2)^{-2}}$ **Step 2:** Use the Power of a Product Property.

$= \dfrac{x^{-6}}{4^{-2} \cdot y^{-4}}$ **Step 3:** Use the Power of a Power Property.

$= \dfrac{4^2 \cdot y^4}{x^6}$ **Step 4:** Use the Negative Exponent Property.

$= \dfrac{16y^4}{x^6}$ **Step 5:** Simplify.

Answer: $\dfrac{16y^4}{x^6}$

Example 5 Simplify the expression: $\left(\dfrac{y}{z^{-4}}\right)^{-3} \cdot \left(\dfrac{y^2}{z}\right)^{-1}$

$\left(\dfrac{y}{z^{-4}}\right)^{-3} \cdot \left(\dfrac{y^2}{z}\right)^{-1} = \dfrac{y^{-3}}{(z^{-4})^{-3}} \cdot \dfrac{(y^2)^{-1}}{z^{-1}}$ **Step 1:** Use the Power of a Quotient Property.

$= \dfrac{y^{-3}}{z^{12}} \cdot \dfrac{y^{-2}}{z^{-1}}$ **Step 2:** Use the Power of a Power Property.

$= \frac{y^{-3} \cdot y^{-2}}{z^{12} \cdot z^{-1}}$ **Step 3:** Combine the numerators and denominators.

$= \frac{y^{-3+(-2)}}{z^{12+(-1)}}$ **Step 4:** Use the Product of Powers Property.

$= \frac{y^{-5}}{z^{11}}$ **Step 5:** Simplify.

$= \frac{1}{y^5 z^{11}}$ **Step 6:** Use the Negative Exponent Property.

Answer: $\frac{1}{y^5 z^{11}}$

Problem Set

Simplify each quotient:

1. $\frac{a^9}{a^{11}}$
2. $\frac{y^{16}}{y^4}$
3. $\frac{7^{15}}{7^5}$
4. $\frac{y^2}{y^{18}}$
5. $\left(\frac{4}{5}\right)^3$
6. $\left(\frac{3}{2}\right)^6$
7. $\left(\frac{5}{6}\right)^3$
8. $\left(\frac{x}{y^2}\right)^3$

Simplify each expression:

9. $\left(\frac{2a^4}{5b^7}\right)^2$
10. $\left(\frac{3x^6}{2y^2}\right)^3$
11. $\left(\frac{3a^5}{2b^8}\right)^4$
12. $\left(\frac{13x}{11y^9}\right)^2$
13. $\left(\frac{2x^4}{7x^2}\right)^{-2}$
14. $\left(\frac{3x^3}{4x^5}\right)^{-3}$
15. $\left(\frac{9a^5}{4a^2}\right)^{-3}$
16. $\left(\frac{3b^6}{2b^3}\right)^{-4}$
17. $\left(\frac{x^5}{y^{-3}}\right)^{-2} \cdot \left(\frac{x^2}{y}\right)$
18. $\left(\frac{y^{-6}}{x^{-5}}\right)^{-2} \cdot \left(\frac{x^8}{y^7}\right)$
19. $\left(\frac{2a^3}{3b^2}\right)^{-3} \cdot \left(\frac{a^4}{b^3}\right)$
20. $\left(\frac{2a^3}{b^{-2}}\right)^{-5} \cdot \left(\frac{5b^{10}}{a^4}\right)$

HA1-840: Introduction to Matrices

It is often convenient to arrange numbers in rows and columns for keeping records, comparing data, and other everyday situations. Rectangular arrangements of numbers can be expressed mathematically using matrices. This lesson introduces matrices and the terminology associated with matrices.

Matrix (matrices)	A rectangular arrangement of elements in rows and columns that is enclosed in brackets

$$
\begin{array}{ccc}
 & & \overbrace{\text{columns}}^{\text{2 columns}} \\
 & & \begin{array}{cc} 1 & 2 \end{array} \\
\text{2 rows} & \left\{\begin{array}{l} \text{Row 1} \\ \text{Row 2} \end{array}\right. & \begin{bmatrix} 8 & 3 \\ 4 & 9 \end{bmatrix}
\end{array}
$$

For example, consider the matrix below:

Quiz grades

$$
\begin{array}{cc}
 & \text{Weeks} \\
 & \begin{array}{ccc} 1 & 2 & 3 \end{array} \\
\begin{array}{c} \text{Lang. Arts} \\ \text{Math} \\ \text{Science} \\ \text{History} \\ \text{Health} \end{array} & \begin{bmatrix} 9 & 8 & 7 \\ 9 & 9 & 6 \\ 8 & 7 & 9 \\ 10 & 9 & 9 \\ 6 & 7 & 8 \end{bmatrix}
\end{array}
$$

In the example above, each number inside the brackets represents a quiz score. A number in a matrix is also called an **element.**

Elements	Entries in a matrix

Each row in the matrix above represents one of five courses: language arts, mathematics, science, history, and health. For example, the science quiz grades in the third row are 8, 7, and 9.

Row	A horizontal line of numbers

Each column represents the quiz grades in a week. For example, the quiz grades during the second week are in the second column and are 8, 9, 7, 9, and 7.

Column	A vertical line of numbers

The dimensions of a matrix give the numbers of rows and columns.

Dimensions of a Matrix	The number of rows and columns. A matrix with m rows and n columns is an "$m \times n$" or "m-by-n" matrix.

The matrix has dimensions of 5×3.

Example 1 What are the dimensions of the following matrix? $\begin{bmatrix} 1 & 4 & 2 \\ 3 & 4 & 9 \\ 5 & 7 & 10 \\ 7 & 3 & 6 \end{bmatrix}$

$$\text{4 rows}\left\{\begin{matrix} \text{Row 1} \\ \text{Row 2} \\ \text{Row 3} \\ \text{Row 4} \end{matrix}\right. \begin{bmatrix} 1 & 4 & 2 \\ 3 & 4 & 9 \\ 5 & 7 & 10 \\ 7 & 3 & 6 \end{bmatrix}$$

Step 1: The dimensions of a matrix give the number of rows and columns in the matrix. First, determine the number of rows. There are four horizontal lines of elements, therefore, this matrix has four rows.

$$\overbrace{\begin{matrix} \text{Column 1} & \text{Column 2} & \text{Column 3} \end{matrix}}^{\text{3 columns}}$$

$$\text{4 rows}\left\{\begin{matrix} \text{Row 1} \\ \text{Row 2} \\ \text{Row 3} \\ \text{Row 4} \end{matrix}\right. \begin{bmatrix} 1 & 4 & 2 \\ 3 & 4 & 9 \\ 5 & 7 & 10 \\ 7 & 3 & 6 \end{bmatrix}$$

Step 2: Next, determine the number of columns in the matrix. A column is a vertical line of elements, so this matrix has three columns.

Answer: Therefore, the dimensions of the matrix are 4×3.

Example 2 What is the position of the number 3 in the following matrix? $\begin{bmatrix} 2 & 5 \\ 4 & 4 \\ 5 & 0 \\ 7 & \mathbf{3} \\ 0 & 6 \\ 9 & 9 \end{bmatrix}$

$$\begin{matrix} \text{Row 1} \\ \text{Row 2} \\ \text{Row 3} \\ \textbf{Row 4} \\ \text{Row 5} \\ \text{Row 6} \end{matrix} \begin{bmatrix} 2 & 5 \\ 4 & 4 \\ 5 & 0 \\ 7 & \mathbf{3} \\ 0 & 6 \\ 9 & 9 \end{bmatrix}$$

Step 1: First, determine the row position of the number. The number 3 is in Row 4.

$$\begin{matrix} \text{Column} & \textbf{Column} \\ 1 & \mathbf{2} \end{matrix}$$

$$\begin{bmatrix} 2 & 5 \\ 4 & 4 \\ 5 & 0 \\ 7 & \mathbf{3} \\ 0 & 6 \\ 9 & 9 \end{bmatrix}$$

Step 2: Next, determine the column position of the number. The number 3 is in Column 2.

Answer: Therefore, the number 3 is in row 4, Column 2.

Example 3 On four consecutive days, a track coach timed and ranked his four sprinters to determine how he would order them in a relay.

	Jason	Michael	Alex	Jesse
Monday	1st	3rd	2nd	4th
Tuesday	2nd	4th	1st	3rd
Wednesday	1st	2nd	4th	3rd
Thursday	2nd	3rd	1st	4th

Construct a matrix that shows how many times each person placed in each position.

$$\begin{matrix} & \text{Jason} & \text{Michael} & \text{Alex} & \text{Jesse} \\ \text{1st} & 2 & & & \\ \text{2nd} & 2 & & & \\ \text{3rd} & 0 & & & \\ \text{4th} & 0 & & & \end{matrix}$$

Step 1: To solve this problem, find the number of times each person placed 1st, 2nd, 3rd, and 4th, and write the results in the matrix. Jason placed 1st twice, so write 2 in the first row, first column. He placed 2nd twice, so write 2 in the second row, first column. He did not place either 3rd or 4th, so write 0 in the third and fourth rows of the first column.

$$\begin{matrix} & \text{Jason} & \text{Michael} & \text{Alex} & \text{Jesse} \\ \text{1st} & 2 & 0 & & \\ \text{2nd} & 2 & 1 & & \\ \text{3rd} & 0 & 2 & & \\ \text{4th} & 0 & 1 & & \end{matrix}$$

Step 2: Michael did not place 1st on any day, so write 0 in the first row, second column. He placed 2nd once, so write 1 in the second row, second column. He placed 3rd twice, so write 2 in the third row, second column. He placed 4th once, so write 1 in the fourth row, second column.

$$\begin{matrix} & \text{Jason} & \text{Michael} & \text{Alex} & \text{Jesse} \\ \text{1st} & 2 & 0 & 2 & \\ \text{2nd} & 2 & 1 & 1 & \\ \text{3rd} & 0 & 2 & 0 & \\ \text{4th} & 0 & 1 & 1 & \end{matrix}$$

Step 3: Repeat the process for Alex.

	Jason	Michael	Alex	Jesse
1st	2	0	2	0
2nd	2	1	1	0
3rd	0	2	0	2
4th	0	1	1	2

Step 4: Repeat the process for Jesse.

Answer: Therefore, this is the desired matrix:

	Jason	Michael	Alex	Jesse
1st	2	0	2	0
2nd	2	1	1	0
3rd	0	2	0	2
4th	0	1	1	2

Example 4 Four pre-K students are assessed each day on their behavior and given a summary report at the end of each week. The grades for the first two weeks are shown in the table below.

	Week 1					Week 2				
	M	T	W	T	F	M	T	W	T	F
Jane	A	B	B	A	C	A	B	C	B	A
Michelle	B	A	A	C	B	B	B	B	C	B
Randy	C	C	B	B	B	B	B	C	B	A
Jackson	B	B	B	B	A	B	B	B	C	A

Complete the following 4×6 matrix showing the students' reports. The column headings refer to the number of times a student received that letter grade during the week.

	Week 1			Week 2		
	A	B	C	A	B	C
Jane						
Michelle						
Randy						
Jackson						

	Week 1			Week 2		
	A	B	C	A	B	C
Jane	2	2	1	2	2	1
Michelle						
Randy						
Jackson						

Step 1: Start with Jane's grades. Find the total number of A's, B's, and C's Jane received the first week. She received two A's, two B's, and one C, so write 2, 2, and 1 in Jane's row of the matrix for the first week. For the second week, Jane again received two A's, two B's, and one C, so write 2, 2, and 1 in the matrix for the second week.

Step 2: Complete Michelle's row in the same manner. Michelle received two A's, two B's, and one C in the first week, so write 2, 2, and 1 in Michelle's row of the matrix for the first week. For the second week, Michelle received no A's, 4 B's, and one C, so write 0, 4, and 1 in the matrix for the second week.

	Week 1			Week 2		
	A	B	C	A	B	C
Jane	2	2	1	2	2	1
Michelle	2	2	1	0	4	1
Randy						
Jackson						

Step 3: Randy received no A's, three B's, and two C's in the first week, so write 0, 3, and 2 in Randy's row of the matrix for the first week. For the second week, Randy received one A, three B's, and one C, so write 1, 3, and 1 in the matrix for the second week.

	Week 1			Week 2		
	A	B	C	A	B	C
Jane	2	2	1	2	2	1
Michelle	2	2	1	0	4	1
Randy	0	3	2	1	3	1
Jackson						

Step 4: Jackson received one A, four B's, and no C's in the first week, so write 1, 4, and 0 in Jackson's row of the matrix for the first week. For the second week, Jackson received one A, three B's, and one C, so write 1, 3, and 1 in the matrix for the second week.

	Week 1			Week 2		
	A	B	C	A	B	C
Jane	2	2	1	2	2	1
Michelle	2	2	1	0	4	1
Randy	0	3	2	1	3	1
Jackson	1	4	0	1	3	1

Answer: Therefore, this is the desired matrix:

	Week 1			Week 2		
	A	B	C	A	B	C
Jane	2	2	1	2	2	1
Michelle	2	2	1	0	4	1
Randy	0	3	2	1	3	1
Jackson	1	4	0	1	3	1

Example 4 Last Saturday, the track team had a fund-raiser for an end-of-season party at a water park. Each athlete solicited sponsors to donate $1 for each lap. The three highest money-raisers were Jimmy, with six sponsors and sixteen laps; Caroline, with five sponsors and fourteen laps; and Jasmine, with four sponsors and eleven laps. Construct a matrix to organize this information, including the total amount earned for each athlete. Also, find the total amount earned by the three athletes.

Step 1: Set up a matrix with rows that represent each athlete. Let the columns represent the number of sponsors, the number of laps, and the amount earned in dollars.

	Sponsors	Laps	Earnings
Jimmy			
Caroline			
Jasmine			

	Sponsors	Laps	Earnings
Jimmy	6	16	96
Caroline			
Jasmine			

Step 2: Fill in Jimmy's row with 6 sponsors and 16 laps. Jimmy earned 6 × 16, or 96 dollars.

	Sponsors	Laps	Earnings
Jimmy	6	16	96
Caroline	5	14	70
Jasmine			

Step 3: Fill in Caroline's row with 5 sponsors and 14 laps. Caroline earned 5 × 14, or 70 dollars.

	Sponsors	Laps	Earnings
Jimmy	6	16	96
Caroline	5	14	70
Jasmine	4	11	44

Step 4: Fill in Jasmine's row with 4 sponsors and 11 laps. Jasmine earned 4 × 11, or 44 dollars.

The total earnings for these athletes is $96 + 70 + 44$, or 210 dollars.

Step 5: The total amount earned by the three athletes is the sum of the three elements in the "Earnings" column.

Answer: The desired matrix is shown in Step 4, and the total amount earned by the three athletes is $210.

Problem Set

Solve:

1. What are the dimensions of the following matrix?

$$\begin{bmatrix} 3 & 2 & 7 & 4 & 19 \\ 10 & 13 & 15 & 3 & 6 \\ 11 & 8 & 12 & 9 & 18 \\ 17 & 16 & 14 & 0 & 5 \end{bmatrix}$$

2. What are the dimensions of the following matrix?

$$\begin{bmatrix} 4 & 8 & 0 \\ 3 & 7 & 1 \end{bmatrix}$$

3. What are the dimensions of the following matrix?

$$\begin{bmatrix} 12 & 5 & 6 & 4 \\ 2 & 1 & 7 & 3 \\ 11 & 18 & 8 & 20 \\ 27 & 23 & 0 & 25 \\ 17 & 19 & 9 & 21 \end{bmatrix}$$

4. What are the dimensions of the following matrix?

$$\begin{bmatrix} 10 & 17 & 8 & 1 & 9 & 2 & 11 \\ 14 & 7 & 16 & 15 & 0 & 12 & 4 \\ 1 & 6 & 13 & 0 & 5 & 3 & 2 \\ 7 & 5 & 8 & 4 & 7 & 32 & 20 \\ 22 & 0 & 9 & 2 & 25 & 9 & 16 \end{bmatrix}$$

5. What are the dimensions of the following matrix?

$$\begin{bmatrix} 3 & 8 & 6 & 5 \\ 11 & 7 & 9 & 1 \end{bmatrix}$$

6. What are the dimensions of the following matrix?

$$\begin{bmatrix} 2 & 7 & 3 & 1 & 5 \\ 4 & 8 & 9 & 9 & 12 \\ 6 & 7 & 5 & 11 & 3 \end{bmatrix}$$

7. What are the dimensions of the following matrix?

$$\begin{bmatrix} 2 & 1 & 8 & 7 \\ 5 & 10 & 9 & 4 \\ 3 & 5 & 22 & 1 \\ 6 & 16 & 19 & 3 \end{bmatrix}$$

8. What is the position of the number "9" in the following matrix?

$$\begin{bmatrix} 5 & 3 & 6 & 2 \\ 4 & 5 & 5 & 7 \\ 6 & 9 & 8 & 1 \end{bmatrix}$$

9. What is the position of the number "8" in the following matrix?

$$\begin{bmatrix} 6 & 11 \\ 2 & 8 \\ 9 & 3 \\ 5 & 4 \\ 1 & 2 \\ 7 & 4 \end{bmatrix}$$

10. What is the position of the number "13" in the following matrix?

$$\begin{bmatrix} 14 & 2 & 3 & 9 \\ 8 & 1 & 10 & 5 \\ 17 & 6 & 11 & 13 \end{bmatrix}$$

11. What is the position of the number "1" in the following matrix?

$$\begin{bmatrix} 2 & 13 & 5 & 12 & 8 \\ 7 & 6 & 14 & 1 & 19 \end{bmatrix}$$

12. What is the position of the number "3" in the following matrix?

$$\begin{bmatrix} 2 & 3 & 4 \\ 4 & 5 & 2 \\ 0 & 7 & 8 \end{bmatrix}$$

13. What is the position of the number "23" in the following matrix?

$$\begin{bmatrix} 2 & 5 & 6 & 4 \\ 2 & 1 & 7 & 3 \\ 13 & 8 & 8 & 20 \\ 7 & 23 & 0 & 5 \\ 17 & 19 & 9 & 21 \end{bmatrix}$$

14. What is the position of the number "4" in the following matrix?

$$\begin{bmatrix} 5 & 9 & 0 \\ 2 & 7 & 4 \end{bmatrix}$$

15. What is the position of the number "4" in the following matrix?

$$\begin{bmatrix} 3 & 11 & 4 \\ 16 & 5 & 12 \\ 9 & 17 & 8 \end{bmatrix}$$

16. Mr. Bearden gave his top three students a prize based on their overall participation each day of the week.

Daily Prize Table

	Student A	Student B	Student C
M	balloon	pen	balloon
T	pen	pen	pen
W	balloon	balloon	balloon
TH	pencil	balloon	pencil
F	pencil	pen	pencil

Using the rows and columns defined below, write a matrix that describes how many times each student received a pen, pencil, and balloon.

Student A Student B Student C

pen
pencil
balloon
$\begin{bmatrix} \quad & \quad & \quad \\ & & \\ & & \end{bmatrix}$

17. Rochelle and Myra compared grades on their English quizzes for each day of the week.

Daily English Quiz Table

	Rochelle	Myra
M	B	A
T	A	A
W	A	C
TH	B	B
F	C	C

Using the rows and columns defined below, write a matrix that describes how many times each student received an A, B, and C.

Rochelle Myra

A
B
C
$\begin{bmatrix} \quad & \quad \\ & \\ & \end{bmatrix}$

18. A weather forecaster displayed the weekly weather conditions for two cities over a 6-week period.

	Genoa City	Pine Valley
Week 1	snow	sunny
Week 2	snow	rain
Week 3	rain	sunny
Week 4	rain	rain
Week 5	sunny	rain
Week 6	sunny	snow

Using the rows and columns defined below, write a matrix that describes how many times each city received rain, snow, and sunny weather.

	Genoa City	Pine Valley
rain		
snow		
sunny		

19. Daryl and Phoebe compared stamps that they collected each day.

Daily Stamps Collected

	Daryl	Phoebe
M	Heritage	Heritage
T	Breast cancer	American flag
W	American flag	American flag
TH	Heritage	Breast cancer
F	Heritage	Heritage

Using the rows and columns defined below, write a matrix that describes how many times Daryl and Phoebe received a heritage, breast cancer, and American flag stamp.

	Daryl	Phoebe
Heritage		
Breast cancer		
American flag		

20. A local hospital's daily report showed the following information for patients A, B, and C.

Daily Medicine Distribution

	Patient A	Patient B	Patient C
day 1	aspirin	multi-vitamin	multi-vitamin
day 2	aspirin	iron	aspirin
day 3	iron	iron	iron
day 4	multi-vitamin	multi-vitamin	aspirin
day 5	iron	aspirin	aspirin

Using the rows and columns defined below, write a matrix that describes how many times Patient A, Patient B, and Patient C received aspirin, iron, and a multi-vitamin.

	Patient A	Patient B	Patient C
aspirin			
iron			
mulit-vitamin			

21. Kelly and Michelle displayed the type of vehicles they leased each year.

Leased Vehicle Table

	Kelly	Michelle
Year 1	SUV	car
Year 2	truck	car
Year 3	car	truck
Year 4	SUV	truck
Year 5	truck	SUV

Using the rows and columns defined below, write a matrix that describes how many of the leased vehicles were SUVs, trucks, and cars.

	Kelly	Michelle
SUV		
truck		
car		

22. At a major electronics store, the four month sales report showed the following information for Group A and Group B.

	Group A	Group B
November	profit loss	profit gain
December	profit loss	profit gain
January	profit loss	profit loss
February	profit gain	profit loss

Using the rows and columns defined below, write a matrix that describes how many times Group A and Group B received a profit gain and a profit loss.

$$\begin{array}{c} \\ \text{profit gain} \\ \text{profit loss} \end{array} \begin{array}{c} \text{Group A} \quad \text{Group B} \\ \begin{bmatrix} \quad & \quad \\ \quad & \quad \end{bmatrix} \end{array}$$

23. The weekly on-off chart for two network computers is given below. Create a 2×2 matrix showing the total on-off usage for the week. Let the first column represent Computer A and the second column represent Computer B. Let the first row represent the number of computers "on" and the second row represent the number of computers "off".

	Computer A	Computer B
Week 1	on	on
Week 2	on	off
Week 3	off	on
Week 4	off	on
Week 5	off	on

24. At Simply Tea, Shelby is responsible for testing the temperature of three types of teas. The hourly temperature chart for each type of tea is given below. Create a 3×3 matrix showing the different temperatures. Let the first column represent Tea 1, the second column represent Tea 2, and the third column represent Tea 3. Let the first row represent the tea that is hot, the second row represent the tea that is warm, and the third row represent the tea that is cold.

	Tea 1	Tea 2	Tea 3
7:00	hot	hot	hot
8:00	hot	warm	hot
9:00	warm	warm	warm
10:00	warm	cold	cold
11:00	cold	cold	cold

25. The top two candidates running for governor are given below. Create a 2×2 matrix showing the responses of six cities. Let the first column represent Republican and the second column represent Democrat. Let the first row represent the "for" votes, and the second row represent the "against" votes.

	Republican	Democrat
City A	for	against
City B	against	for
City C	against	for
City D	against	for
City E	against	for
City F	for	against

26. The answers for two students' history quizzes are given below. Create a 2×2 matrix showing the responses for questions one through five. Let the first column represent Jason's quiz responses and the second column represent Nicholas' quiz responses. Let the first row represent the true responses and the second row represent the false responses.

	Jason	Nicholas
Question 1	TRUE	TRUE
Question 2	TRUE	TRUE
Question 3	FALSE	FALSE
Question 4	TRUE	TRUE
Question 5	FALSE	TRUE

27. The results from a light bulb testing company are listed below. Create a 2×2 matrix showing the results of good and bad light bulbs.
Let the first column represent Batch A and the second column represent Batch B. Let the first row represent the good light bulbs and the second row represent the bad light bulbs.

	Batch A	Batch B
Bulb 1	Good	Bad
Bulb 2	Bad	Bad
Bulb 3	Good	Bad
Bulb 4	Bad	Good
Bulb 5	Bad	Bad
Bulb 6	Good	Good

28. The plays by quarter that Thomas and Bryce contributed to the basketball game are listed below. Create a 3×2 matrix showing the different plays of the basketball game. Let the first column represent Thomas' plays and the second column represent Bryce's plays. Let the first row represent the number of free throws, the second row represent the number of jump shots, and the third row represent the number of lay-ups.

	Thomas	Bryce
1st quarter	Free throw	Lay-up
2nd quarter	Jump shot	Free throw
3rd quarter	Lay-up	Jump shot
4th quarter	Jump shot	Free throw

29. The results from a loan department are given below. Create a 2×3 matrix showing the outcome of the three customers. Let the first column represent Customer A, the second column represent Customer B, and the third column represent Customer C. Let the first row represent the approved results and the second row represent the denied results.

	Customer A	Customer B	Customer C
Officer 1	approved	denied	denied
Officer 2	approved	approved	approved
Officer 3	denied	denied	denied
Officer 4	approved	approved	approved
Officer 5	approved	denied	approved

30. The report cards for three students are given below. Create a 3×3 matrix showing the grades of the students. Let the first column represent Ann, the second column represent Cameron, and the third column represent Darren. Let the first row represent the A's, the second row represent the B's, and the third row represent the C's.

	Ann	Cameron	Darren
Spanish	A	A	B
History	B	A	A
Reading	C	C	B

31. Leslie and Michelle are stock market investors. The total number of stocks they purchased each year is shown below.

What is the total number of stocks each one purchased for these five years?

$$\begin{array}{c} \\ 2000 \\ 2001 \\ 2002 \\ 2003 \\ 2004 \end{array} \begin{array}{c} \text{Leslie} \quad \text{Michelle} \\ \begin{bmatrix} 13 & 14 \\ 16 & 15 \\ 13 & 16 \\ 14 & 17 \\ 15 & 14 \end{bmatrix} \end{array}$$

32. Nancy and Gerald are training for the Breast Cancer Awareness Marathon. Listed in the matrix are the miles they ran each day.

What was the average distance in miles Nancy and Gerald each ran?

	Nancy	Gerald
Monday	4.5	2.5
Tuesday	3.5	3.8
Wednesday	4.2	5.2
Thursday	4.8	3.5
Friday	3.0	2.0

33. Keisha works part-time at a boutique. Her wages for the past two weeks are shown in the matrix below.

What is her total wages earned for each week?

	Week 1	Week 2
Monday	$45	$54
Tuesday	$48	$52
Wednesday	$35	$56
Thursday	$54	$59
Friday	$58	$59
Saturday	$68	$52
Sunday	$65	$70

34. The Candy Company receives numerous orders for Valentine's Day. Listed below are the totals from the past four years.

What is the total number of each type of dessert ordered?

	Cakes	Cookies	Pies
2002	50	65	80
2003	45	70	64
2004	65	76	50
2005	40	60	65

35. Judy and her sister Jasmine clipped coupons each day from the newspaper. Curious to see who saved the most money this month, they decided to compare grocery bills. The following information is shown in the matrix below.

What is the total amount of money Judy and her sister each saved?

	Judy	Jasmine
Bill 1	$4.50	$9.90
Bill 2	$11.80	$12.50
Bill 3	$7.50	$8.25
Bill 4	$6.50	$6.00
Bill 5	$9.25	$10.30

36. Betty and Tyler collected money for their class trip. At the end of eight hours, they compared the total amount they collected. The information is shown in the matrix below.
What is the total amount of money Betty and Tyler each collected?

	Betty	Tyler
Hour 1	$10	$12
Hour 2	$18	$13
Hour 3	$16	$15
Hour 4	$15	$14
Hour 5	$12	$18
Hour 6	$18	$16
Hour 7	$19	$12
Hour 8	$11	$17

37. At summer camp, Trisha, Leah, Marc, Nathan, and Troy were playing a game. At the end of the day, they compared the total number of wins and losses for each round. The information is shown in the matrix below.

What is the total win-loss record?

	Win	Loss
Trisha	4	2
Leah	3	3
Marc	1	5
Nathan	5	1
Troy	6	0

38. Dairy Mart displayed their milk prices over a seven-week period.

If Maya bought a gallon of whole milk and a gallon of skim milk each week for seven weeks, what is the total amount of money she spent on each type of milk?

	Whole	Skim
Week 1	\$2.14	\$2.28
Week 2	\$2.09	\$2.29
Week 3	\$2.11	\$2.30
Week 4	\$2.15	\$2.30
Week 5	\$2.11	\$2.39
Week 6	\$2.20	\$2.27
Week 7	\$2.29	\$2.25

39. A weather forecaster displayed the seven day forecast for a town.

What are the average high-low temperatures for these seven days?

	High	Low
Thurs.	89°	72°
Fri.	88°	71°
Sat.	90°	73°
Sun.	91°	73°
Mon.	77°	58°
Tues.	76°	56°
Wed.	77°	59°

40. Katie and Andrea compared Spanish test scores at the end of the semester. Their test scores for the semester are shown in a matrix below.

What is the average Spanish test score for each student?

	Katie	Andrea
Test 1	88	90
Test 2	92	87
Test 3	85	79
Test 4	99	94

HA1-845: Operations with Matrices

This lesson describes how to perform basic operations with matrices.

Operation	A mathematical process of deriving one expression from other expressions according to a rule

Two matrix operations are addition and subtraction.

Matrix Addition (or Subtraction)	An operation in which corresponding elements of two matrices with the same dimensions are added (or subtracted) to form a matrix with the same dimensions as the original matrices

For example, to add $\begin{bmatrix} 2 & -4 & 6 \\ -8 & 10 & -12 \end{bmatrix}$ and $\begin{bmatrix} 0 & 1 & 3 \\ -5 & -7 & -9 \end{bmatrix}$, add corresponding elements. As shown in the equations below, the element 2 in the first matrix and the element 0 in the second matrix are both in the first row and first column of their respective matrices, so 2 + 0, or 2, is in the first row and first column of the matrix that represents the sum. Find the other elements in the sum by adding the other five pairs of corresponding elements, as shown below:

$$\begin{bmatrix} \mathbf{2} & -4 & 6 \\ -8 & 10 & -12 \end{bmatrix} + \begin{bmatrix} \mathbf{0} & 1 & 3 \\ -5 & -7 & -9 \end{bmatrix} = \begin{bmatrix} \mathbf{2+0} & -4+1 & 6+3 \\ -8+(-5) & 10+(-7) & -12+(-9) \end{bmatrix}$$

$$= \begin{bmatrix} \mathbf{2} & -3 & 9 \\ -13 & 3 & -21 \end{bmatrix}$$

Another matrix operation is scalar multiplication.

Scalar	A number or constant (as opposed to a 1-by-1 matrix)

Scalar Multiplication	An operation in which each element of a matrix is multiplied by a scalar to form a matrix with the same dimensions as the original matrix

For example, to multiply $\begin{bmatrix} 1 & -1 \\ 2 & 3 \end{bmatrix}$ by the scalar 5, multiply each element of the matrix by 5, as follows:

$$5\begin{bmatrix} 1 & -1 \\ 2 & 3 \end{bmatrix} = \begin{bmatrix} 5(1) & 5(-1) \\ 5(2) & 5(3) \end{bmatrix}$$

$$= \begin{bmatrix} 5 & -5 \\ 10 & 15 \end{bmatrix}$$

Example 1 Given matrices C and B, find $C + B$.

$$C = \begin{bmatrix} 2 & 5 \\ 3 & 7 \\ 6 & 4 \end{bmatrix} \quad B = \begin{bmatrix} 1 & 2 \\ 4 & 6 \\ 8 & 5 \end{bmatrix}$$

$$C + B = \begin{bmatrix} 2 & 5 \\ 3 & 7 \\ 6 & 4 \end{bmatrix} + \begin{bmatrix} 1 & 2 \\ 4 & 6 \\ 8 & 5 \end{bmatrix}$$

Step 1: Replace the variables in $C + B$ with the matrices represented by C and B.

$$= \begin{bmatrix} 2+1 & 5+2 \\ 3+4 & 7+6 \\ 6+8 & 4+5 \end{bmatrix}$$

Step 2: Add the corresponding elements of the matrices.

$$= \begin{bmatrix} 3 & 7 \\ 7 & 13 \\ 14 & 9 \end{bmatrix}$$

Step 3: Find the sums.

Answer:

$$C + B = \begin{bmatrix} 3 & 7 \\ 7 & 13 \\ 14 & 9 \end{bmatrix}$$

Example 2 Given the matrix A, find $-4A$.

$$A = \begin{bmatrix} 3 & -2 \\ 5 & 4 \\ -9 & 6 \\ 7 & 8 \end{bmatrix}$$

$$A = -4\begin{bmatrix} 3 & -2 \\ 5 & 4 \\ -9 & 6 \\ 7 & 8 \end{bmatrix}$$

Step 1: Replace the variable in $-4A$ with the matrix represented by A.

$$= \begin{bmatrix} -4(3) & -4(-2) \\ -4(5) & -4(4) \\ -4(-9) & -4(6) \\ -4(7) & -4(8) \end{bmatrix}$$

Step 2: Multiply each element in the matrix by the scalar -4.

$$= \begin{bmatrix} -12 & 8 \\ -20 & -16 \\ 36 & -24 \\ -28 & -32 \end{bmatrix}$$

Step 3: Find the products.

Answer:

$$-4A = \begin{bmatrix} -12 & 8 \\ -20 & -16 \\ 36 & -24 \\ -28 & -32 \end{bmatrix}$$

Example 3 Given the matrices A and B, find $4A - B$.

$$A = \begin{bmatrix} -2 & 3 & 1 \\ -8 & 6 & -2 \end{bmatrix} \qquad B = \begin{bmatrix} 2 & 4 & 2 \\ 4 & -7 & -1 \end{bmatrix}$$

$$4A - B = 4\begin{bmatrix} -2 & 3 & 1 \\ -8 & 6 & -2 \end{bmatrix} - \begin{bmatrix} 2 & 4 & 2 \\ 4 & -7 & -1 \end{bmatrix}$$

Step 1: Replace the variables in $4A - B$ with the matrices represented by A and B.

$$= 4\begin{bmatrix} -2 & 3 & 1 \\ -8 & 6 & -2 \end{bmatrix} + \begin{bmatrix} -2 & -4 & -2 \\ -4 & 7 & 1 \end{bmatrix}$$

Step 2: Subtract matrix B by changing the subtraction sign to an addition sign and changing each element in B to its opposite.

$$= \begin{bmatrix} 4(-2) & 4(3) & 4(1) \\ 4(-8) & 4(6) & 4(-2) \end{bmatrix} + \begin{bmatrix} -2 & -4 & -2 \\ -4 & +7 & +1 \end{bmatrix}$$

Step 3: Multiply each element in matrix A by the scalar 4.

$$= \begin{bmatrix} -8 & 12 & 4 \\ -32 & 24 & -8 \end{bmatrix} + \begin{bmatrix} -2 & -4 & -2 \\ -4 & +7 & +1 \end{bmatrix}$$

Step 4: Find the products in the first matrix.

$$= \begin{bmatrix} -8+(-2) & 12+(-4) & 4+(-2) \\ -32+(-4) & 24+7 & -8+1 \end{bmatrix}$$

Step 5: Add the corresponding elements of the matrices.

$$= \begin{bmatrix} -10 & 8 & 2 \\ -36 & 31 & -7 \end{bmatrix}$$

Step 6: Find the sums.

Answer:

$$4A - B = \begin{bmatrix} -10 & 8 & 2 \\ -36 & 31 & -7 \end{bmatrix}$$

Example 4 Given matrices D, E, and F, find $D + 3E - 2F$.

$$D = \begin{bmatrix} 2 & 5 \\ -3 & 6 \end{bmatrix} \quad E = \begin{bmatrix} 4 & 8 \\ 5 & 9 \end{bmatrix} \quad F = \begin{bmatrix} -1 & 2 \\ -3 & -5 \end{bmatrix}$$

$$D + 3E - 2F = \begin{bmatrix} 2 & 5 \\ -3 & 6 \end{bmatrix} + 3\begin{bmatrix} 4 & 8 \\ 5 & 9 \end{bmatrix} - 2\begin{bmatrix} -1 & 2 \\ -3 & -5 \end{bmatrix}$$

Step 1: Replace the variables in $D + 3E - 2F$ with the matrices represented by D, E, and F.

$$= \begin{bmatrix} 2 & 5 \\ -3 & 6 \end{bmatrix} + 3\begin{bmatrix} 4 & 8 \\ 5 & 9 \end{bmatrix} + 2\begin{bmatrix} 1 & -2 \\ 3 & 5 \end{bmatrix}$$

Step 2: Subtract $2F$ by changing the subtraction sign to an addition sign and changing each element in F to its opposite.

$$= \begin{bmatrix} 2 & 5 \\ -3 & 6 \end{bmatrix} + \begin{bmatrix} 3(4) & 3(8) \\ 3(5) & 3(9) \end{bmatrix} + \begin{bmatrix} 2(1) & 2(-2) \\ 2(3) & 2(5) \end{bmatrix}$$

Step 3: Multiply each element in matrix E by the scalar 3, and multiply each element in matrix F by the scalar 2.

$$= \begin{bmatrix} 2 & 5 \\ -3 & 6 \end{bmatrix} + \begin{bmatrix} 12 & 24 \\ 15 & 27 \end{bmatrix} + \begin{bmatrix} 2 & -4 \\ 6 & 10 \end{bmatrix}$$

Step 4: Find the products in the second and third matrices.

$$= \begin{bmatrix} 2 + 12 + 2 & 5 + 24 + (-4) \\ -3 + 15 + 6 & 6 + 27 + 10 \end{bmatrix}$$

Step 5: Add the corresponding elements of the three matrices.

$$= \begin{bmatrix} 16 & 25 \\ 18 & 43 \end{bmatrix}$$

Step 6: Find the sums.

Answer: $D + 3E - 2F = \begin{bmatrix} 16 & 25 \\ 18 & 43 \end{bmatrix}$

Example 5 This table shows the number of vehicles sold last month by three salesmen at a local dealership:

Number of Vehicles Sold

Name	Cars	SUVs	Trucks
Abbott	8	11	7
Buddy	11	12	9
Charles	9	8	8

If a salesman receives a \$100 commission for a car, a \$200 commission for a SUV, and a \$300 commission for a truck, find the total commission each salesman received last month.

$$C = \begin{bmatrix} 8 \\ 11 \\ 9 \end{bmatrix} \quad S = \begin{bmatrix} 11 \\ 12 \\ 8 \end{bmatrix} \quad T = \begin{bmatrix} 7 \\ 9 \\ 8 \end{bmatrix}$$

Step 1: Define three 3-by-1 matrices whose elements are the number of each type of vehicle sold by the salesmen. Matrix C is the number of cars, matrix S is the number of SUVs, and matrix T is the number of trucks sold.

Total commissions $= \$100C + \$200S + \$300T$	**Step 2:** Find the total commissions by multiplying the number of cars sold by \$100, the number of SUVs sold by \$200, and the number of trucks sold by \$300, and then add the results.
$= 100\begin{bmatrix} 8 \\ 11 \\ 9 \end{bmatrix} + 200\begin{bmatrix} 11 \\ 12 \\ 8 \end{bmatrix} + 300\begin{bmatrix} 7 \\ 9 \\ 8 \end{bmatrix}$	**Step 3:** Replace the variables in $100C + 200S + 300T$ with the matrices represented by C, S, and T.
$= \begin{bmatrix} 100(8) \\ 100(11) \\ 100(9) \end{bmatrix} + \begin{bmatrix} 200(11) \\ 200(12) \\ 200(8) \end{bmatrix} + \begin{bmatrix} 300(7) \\ 300(9) \\ 300(8) \end{bmatrix}$	**Step 4:** Multiply each element of C by the scalar 100, each element of S by the scalar 200, and each element of T by the scalar 300.
$= \begin{bmatrix} 800 \\ 1{,}100 \\ 900 \end{bmatrix} + \begin{bmatrix} 2{,}200 \\ 2{,}400 \\ 1{,}600 \end{bmatrix} + \begin{bmatrix} 2{,}100 \\ 2{,}700 \\ 2{,}400 \end{bmatrix}$	**Step 5:** Find the products in each matrix.
$= \begin{bmatrix} 800 + 2{,}200 + 2{,}100 \\ 1{,}100 + 2{,}400 + 2{,}700 \\ 900 + 1{,}600 + 2{,}400 \end{bmatrix}$	**Step 6:** Add the corresponding elements of the three matrices.
$= \begin{bmatrix} 5{,}100 \\ 6{,}200 \\ 4{,}900 \end{bmatrix}$	**Step 7:** Find the sums.
	Answer: The total commission for each salesman is shown in the matrix below.

$$\begin{matrix} \text{Abbott} \\ \text{Buddy} \\ \text{Charles} \end{matrix} \begin{bmatrix} \$5{,}100 \\ \$6{,}200 \\ \$4{,}900 \end{bmatrix}$$

Problem Set

Solve:

1. Given matrices A and B, find $A + B$.

$$A = \begin{bmatrix} 3 & 9 \\ 4 & 8 \end{bmatrix} \quad B = \begin{bmatrix} 3 & 4 \\ 9 & 5 \end{bmatrix}$$

2. Given matrices A and B, find $A + B$.

$$A = \begin{bmatrix} 1 & 2 & 1 \\ 6 & 3 & 4 \\ 9 & 8 & 7 \end{bmatrix} \quad B = \begin{bmatrix} 4 & 6 & 4 \\ 1 & 3 & 2 \\ 0 & 4 & 5 \end{bmatrix}$$

3. Given matrices C and D, find $C + D$.

$$C = \begin{bmatrix} 2 & 5 \\ 8 & 7 \end{bmatrix} \quad D = \begin{bmatrix} 6 & 2 \\ 3 & 5 \end{bmatrix}$$

4. Given matrices A and B, find $A + B$.

$$A = \begin{bmatrix} 3 & 5 & 8 \\ 0 & 8 & 7 \\ 4 & 3 & 1 \end{bmatrix} \quad B = \begin{bmatrix} 1 & 2 & 8 \\ 8 & 1 & 4 \\ 5 & 3 & 4 \end{bmatrix}$$

5. Given matrices R and S, find $R + S$.

$$R = \begin{bmatrix} 1 & 2 & 1 & 2 \\ 1 & 2 & 1 & 2 \\ 1 & 2 & 1 & 2 \\ 1 & 2 & 1 & 2 \end{bmatrix} \quad S = \begin{bmatrix} 2 & 1 & 2 & 1 \\ 2 & 1 & 2 & 1 \\ 2 & 1 & 2 & 1 \\ 2 & 1 & 2 & 1 \end{bmatrix}$$

6. Given matrices M and N, find $M + N$.

$$M = \begin{bmatrix} 3 & 7 \\ 8 & 8 \end{bmatrix} \quad N = \begin{bmatrix} 1 & 9 \\ 7 & 2 \end{bmatrix}$$

7. Given matrices A and B, find $A + B$.

$$A = \begin{bmatrix} 3 & 2 & 1 \\ 1 & 2 & 3 \\ 2 & 4 & 2 \end{bmatrix} \quad B = \begin{bmatrix} 1 & 2 & 3 \\ 4 & 5 & 6 \\ 7 & 8 & 9 \end{bmatrix}$$

8. Given matrices F and G, find $F + G$.

$$F = \begin{bmatrix} 1 & 2 & 3 & 4 \\ 1 & 2 & 3 & 4 \\ 1 & 2 & 3 & 4 \\ 1 & 2 & 3 & 4 \end{bmatrix} \quad G = \begin{bmatrix} 5 & 6 & 7 & 8 \\ 5 & 6 & 7 & 8 \\ 5 & 6 & 7 & 8 \\ 5 & 6 & 7 & 8 \end{bmatrix}$$

9. Given matrix H, find $3H$.

$$H = \begin{bmatrix} 3 & 3 \\ 1 & 1 \\ 1 & 3 \\ 4 & 1 \\ 3 & 3 \\ 1 & 2 \end{bmatrix}$$

10. Given matrix C, find $6C$.

$$C = \begin{bmatrix} 2 & 6 & 3 & 1 & 2 & 6 \\ 4 & 5 & 2 & 1 & 0 & 1 \end{bmatrix}$$

11. Given matrix D, find $2D$.

$$D = \begin{bmatrix} 4 & 1 \\ 3 & 2 \\ 4 & 1 \\ 2 & 5 \end{bmatrix}$$

12. Given matrix D, find $3D$.

$$D = \begin{bmatrix} 4 & 2 \\ 3 & 8 \\ 1 & 1 \\ 2 & 0 \end{bmatrix}$$

13. Given matrix S, find $4S$.

$$S = \begin{bmatrix} 3 & 9 & 0 \\ 7 & 5 & 2 \\ 0 & 6 & 4 \end{bmatrix}$$

14. Given matrix J, find $2J$.

$$J = \begin{bmatrix} 6 & 4 \\ 2 & 1 \\ 3 & 5 \\ 4 & 3 \\ 2 & 2 \end{bmatrix}$$

15. Given matrix X, find $3X$.

$$X = \begin{bmatrix} 5 & 4 \\ 3 & 2 \end{bmatrix}$$

16. Given matrices A and G, find $A + 2G$.

$$A = \begin{bmatrix} 3 & 2 \\ 5 & -2 \end{bmatrix} \quad G = \begin{bmatrix} 2 & -6 \\ 3 & 2 \end{bmatrix}$$

17. Given matrices C and E, find $2C + 2E$.

$$C = \begin{bmatrix} 6 & 4 & 1 \\ 6 & 2 & 3 \\ 6 & 5 & 1 \end{bmatrix} \quad E = \begin{bmatrix} 1 & 3 & 5 \\ 7 & 2 & 3 \\ 8 & 3 & 4 \end{bmatrix}$$

18. Given matrices U and V, find $3U + 2V$.

$$U = \begin{bmatrix} 2 & 4 \\ 3 & 2 \end{bmatrix} \quad V = \begin{bmatrix} 2 & 5 \\ 3 & 4 \end{bmatrix}$$

19. Given matrices G and J, find $2G + J$.

$$G = \begin{bmatrix} -4 & -4 \\ 8 & -6 \end{bmatrix} \quad J = \begin{bmatrix} -3 & 2 \\ -2 & -1 \end{bmatrix}$$

20. Given matrices J and K, find $J + 2K$.

$$J = \begin{bmatrix} -4 & 3 \\ -9 & 6 \end{bmatrix} \quad K = \begin{bmatrix} 3 & 2 \\ 2 & 1 \end{bmatrix}$$

21. Given matrices R and Q, find $2R + 2Q$.

$$R = \begin{bmatrix} 5 & 4 & 1 \\ 5 & 2 & 3 \\ 5 & 5 & 1 \end{bmatrix} \quad Q = \begin{bmatrix} -1 & 3 & 2 \\ 3 & 2 & -3 \\ 5 & -2 & 4 \end{bmatrix}$$

22. Given matrices N and K, find $N + 2K$.

$$N = \begin{bmatrix} 2 & 3 \\ 3 & -4 \end{bmatrix} \quad K = \begin{bmatrix} 4 & 2 \\ 3 & 4 \end{bmatrix}$$

23. Given matrices U and V, find $3U + V$.

$$U = \begin{bmatrix} -6 & 4 & 1 \\ 2 & -4 & 3 \end{bmatrix} \quad V = \begin{bmatrix} 3 & 2 & 7 \\ 6 & 9 & 8 \end{bmatrix}$$

24. Given matrices F and T, find $F + 2T$.

$$F = \begin{bmatrix} 3 & 4 & 1 \\ 2 & 2 & 3 \\ 6 & 5 & 1 \end{bmatrix} \quad T = \begin{bmatrix} 1 & 3 & 5 \\ 7 & 2 & 3 \\ 9 & 3 & 2 \end{bmatrix}$$

25. Given matrices F, G, and H, find $F - 3G + 2H$.

$$F = \begin{bmatrix} 2 & 2 \\ 1 & 1 \end{bmatrix} \quad G = \begin{bmatrix} 3 & 2 \\ 5 & 4 \end{bmatrix} \quad H = \begin{bmatrix} 2 & 1 \\ 4 & 3 \end{bmatrix}$$

26. Given matrices D, E, and F, find $D + 2E - F$.

$$D = \begin{bmatrix} 2 & 4 \\ 1 & 3 \end{bmatrix} \quad E = \begin{bmatrix} 3 & 1 \\ 6 & 4 \end{bmatrix} \quad F = \begin{bmatrix} 9 & 1 \\ 0 & 4 \end{bmatrix}$$

27. Given matrices B, A, and D, find $3B + 2A + D$.

$$B = \begin{bmatrix} 2 & 5 \\ 2 & 3 \end{bmatrix} \quad A = \begin{bmatrix} 3 & 1 \\ 5 & 4 \end{bmatrix} \quad D = \begin{bmatrix} 1 & 2 \\ 2 & 4 \end{bmatrix}$$

28. Given matrices S, E, and T, find $S - 3E + T$.

$$S = \begin{bmatrix} 4 & 1 \\ 1 & 3 \end{bmatrix} \quad E = \begin{bmatrix} 1 & 1 \\ 1 & 1 \end{bmatrix} \quad T = \begin{bmatrix} 3 & 1 \\ 5 & 4 \end{bmatrix}$$

29. Given matrices D, B, and C, find $2D - B + 2C$.

$$D = \begin{bmatrix} 5 & 4 & 2 \\ 3 & 2 & 1 \\ 6 & 3 & 1 \end{bmatrix} \quad B = \begin{bmatrix} 1 & 9 & 7 \\ 6 & 8 & 2 \\ 2 & 5 & 6 \end{bmatrix} \quad C = \begin{bmatrix} 2 & 4 & 2 \\ 1 & 2 & 1 \\ 3 & 1 & 2 \end{bmatrix}$$

30. Given matrices C, E, and T, find $C + E + T$.

$$C = \begin{bmatrix} 8 & 9 \\ 6 & 7 \end{bmatrix} \quad E = \begin{bmatrix} -1 & -1 \\ -1 & -1 \end{bmatrix} \quad T = \begin{bmatrix} 2 & 4 \\ 3 & 2 \end{bmatrix}$$

31. At Crimson University, the total number of male and female students in the math department for two weeks is given below.

Week 1 — Class

	A	B
male	25	30
female	10	7

Week 2 — Class

	A	B
male	13	5
female	15	23

To estimate the amount of students present in the department over the next fourteen weeks, the department head adds the two matrices above, and then multiplies the sum by seven. Find the resulting matrix.

32. Two neighboring towns are getting ready for their mayoral election. The total number of Democratic and Republican voters in Town A and Town B is given below.

Year 1 — Town

	A	B
Dem.	109	200
Rep.	75	318

Year 2 — Town

	A	B
Dem.	135	117
Rep.	211	313

To estimate the amount of registered voters over ten years, the mayor adds the two matrices above, and then multiplies the sum by five. Find the resulting matrix.

33. At Big Al's new and used car lot, the total number of cars and SUVs sold is given below.

Week 1

	new	used
cars	32	34
SUVs	26	22

Week 2

	new	used
cars	16	31
SUVs	36	27

To estimate the amount of vehicles sold over 18 weeks, Al adds the two matrices above, and then multiplies the sum by nine. Find the resulting matrix.

34. The receptionist at the university ordered pencils and pens for her department. The following shows the amount for two months

Month 1

	pens	pencils
blue	234	251
black	228	248

Month 2

	pens	pencils
blue	224	271
black	251	220

To estimate the amount of pens and pencils purchased over 12 months, the receptionist adds the two matrices above, and then multiplies the sum by six. Find the resulting matrix.

35. Miss Tyson has three stocks, HTTD, IRON, and ACE. The annual report shows the total stocks broken down over two business weeks.

Week 1

	HTTD	IRON	ACE
Mon.	52	62	29
Tues.	52	63	29
Wed.	50	61	30
Thur.	52	60	32
Fri.	53	68	31

Week 2

	HTTD	IRON	ACE
Mon.	54	62	29
Tues.	54	64	32
Wed.	55	65	33
Thur.	56	66	33
Fri.	56	66	35

To estimate the price of the stocks over the next eight weeks, Miss Tyson must add the two matrices above, and then multiply the sum by four. Find the resulting matrix.

36. The student teacher ratio between two local high schools for two years is given below.

Year 1

	School A	School B
teachers	119	86
students	875	918

Year 2

	School A	School B
teachers	123	88
students	675	723

To estimate the amount of teachers and students over twelve years, add the two matrices above, and then multiply the sum by six. Find the resulting matrix.

37. The Computer Store is having their annual clearance sale. Two stores sold the following computers over a two-day period.

Day 1

	Store A	Store B
laptop	50	27
desktop	66	59
tablet pc	20	43

Day 2

	Store A	Store B
laptop	48	45
desktop	89	68
tablet pc	48	39

To estimate the number of computers sold over the next twelve days, the store owner adds the two matrices above, and then multiplies the sum by six. Find the resulting matrix.

38. The Music Factor is having a clearance sale. The following table shows the inventory sold of two types of music.

Week 1

	Cassettes	LPs	CDs
Hiphop Music	234	105	359
Country Music	356	87	206

Week 2

	Cassettes	LPs	CDs
Hiphop Music	204	105	350
Country Music	306	75	222

To estimate the amount of music to be sold over the next four weeks, the store manager adds the two matrices above, and then multiplies the sum by two. Find the resulting matrix.

39. The total number of dogs and cats adopted for the first two days of Adopt-a-Pet week are listed below.

Day 1

	pound A	pound B
dogs	23	34
cats	43	32

Day 2

	pound A	pound B
dogs	32	31
cats	48	39

To estimate the number of animals to be adopted over the next four days, add the two matrices above, and then multiply the sum by two. Find the resulting matrix.

40. Two local sporting goods stores report shoe sales for December and January as follows:

December

	Store A	Store B
running	34	65
walking	43	23
basketball	54	34
golf	24	14

January

	Store A	Store B
running	24	34
walking	35	27
basketball	34	32
golf	29	24

To estimate the number of shoes by type to purchase for the next eight months, the store manager adds the two matrices above, and then multiplies the sum by four. Find the resulting matrix.

HA1-860: Using the Laws of Exponents

If you want to add $3 + 3 + 3 + 3 + 3$, you can use multiplication as a shortcut, and write $3 \cdot 5$. What if you want to multiply $3 \cdot 3 \cdot 3 \cdot 3 \cdot 3$, what kind of shortcut could you use?

Just as you can use multiplication as a shortcut to addition, you can use exponential form as a shortcut to multiplication.

$$3 \cdot 3 \cdot 3 \cdot 3 \cdot 3 = \overbrace{3^{5}}^{\text{power}} \leftarrow \text{exponent}$$

$$\underset{\text{base}}{\uparrow}$$

Base	The repeated factor
Exponent	A number or variable that expresses how many times the base is multiplied
Power	A term that is written as a product of equal factors that includes a base and an exponent

How do you read 3^5? "Three to the fifth power."

How do you read 2^7? "Two to the seventh power."

How do you read a^n? "*a* to the *n*th power."

What would you do if a problem asked you to multiply two powers, such as $2^3 \cdot 2^5$?

Think about what $2^3 \cdot 2^5$ means:

$$2^3 = 2 \cdot 2 \cdot 2$$
$$2^5 = 2 \cdot 2 \cdot 2 \cdot 2 \cdot 2$$
$$2^3 \cdot 2^5 = 2 \cdot 2 \cdot 2 \cdot 2 \cdot 2 \cdot 2 \cdot 2 \cdot 2$$
$$2^3 \cdot 2^5 = 2^8$$
$$2^8 = 256$$

What about $c^2 \cdot c^3$?

$$c^2 = c \cdot c$$
$$c^3 = c \cdot c \cdot c$$
$$c^2 \cdot c^3 = c \cdot c \cdot c \cdot c \cdot c$$
$$c^2 \cdot c^3 = c^5$$

Rule for Multiplying Exponents	When multiplying two or more monomials **with the same base**, add the exponents. $a^x \cdot a^y = a^{x+y}$

Note: If the bases are not the same, you cannot use this rule.

What would you do if you were asked to divide $\frac{8^3}{8^2}$?

What does 8^3 mean? $8 \cdot 8 \cdot 8$

What does 8^2 mean? $8 \cdot 8$

So, $\frac{8^3}{8^2}$ means $\frac{8 \cdot 8 \cdot 8}{8 \cdot 8}$; if you cancel the common factors repeated in the numerator and denominator, you are left with 8, or 8^1.

What about $\frac{g^7}{g^3}$?

$$g^7 = g \cdot g \cdot g \cdot g \cdot g \cdot g \cdot g$$
$$g^3 = g \cdot g \cdot g$$

Therefore, $\frac{g^7}{g^3} = \frac{g \cdot g \cdot g \cdot g \cdot g \cdot g \cdot g}{g \cdot g \cdot g}$; Cancel: $\frac{g \cdot g \cdot g \cdot \cancel{g} \cdot \cancel{g} \cdot \cancel{g} \cdot g}{\cancel{g} \cdot \cancel{g} \cdot \cancel{g}} = g \cdot g \cdot g \cdot g = g^4$.

Rule for Dividing Exponents	When dividing two monomials **with the same base**, subtract the exponents. $\frac{a^x}{a^y} = a^{x-y}$

Finally, what is $(3^2)^4$?

What does $(3^2)^4$ mean? $3^2 \cdot 3^2 \cdot 3^2 \cdot 3^2$

Now, using the rule for multiplying powers, you can add the exponents because the bases are the same: $3^2 \cdot 3^2 \cdot 3^2 \cdot 3^2 = 3^8$. If you look at the original problem, $(3^2)^4$, the same result can be obtained by multiplying the two exponents..

Power Rule of Exponents	When you raise a power to a power, **keep the base** and multiply the exponents. $(a^x)^y = a^{xy}$

Try these examples:

Example 1 Simplify: $\frac{d^6}{d^3}$

$\frac{d^6}{d^3}$

d^{6-3}

Step 1: Because you are dividing two powers with the same base, subtract the exponents. The answer is d^3.

d^3

Answer: d^3

Example 2 Evaluate: $5^2 \cdot 5$

$5^2 \cdot 5^1$

Step 1: Because you are multiplying two powers with the same base, add the exponents.

$5^{2+1} = 5^3 = 125$

Answer: 125

Example 3 Simplify: $(m^4)^7$

$m^{4 \cdot 7} = m^{28}$

Step 1: Because a power, m^4, is being raised to the seventh power, multiply the exponents.

Answer: m^{28}

Problem Set

Simplify:

1. 10^3
2. 7^4
3. 10^1
4. 2^5
5. 5^4
6. 6^4
7. 1^7
8. 11^3
9. $d^5 \cdot d^{10}$
10. $m \cdot m^3 \cdot m^6$
11. $b^4 \cdot b^6$
12. $r^{21} \cdot r^7$
13. $\frac{x^5}{x}$
14. $\frac{a^{20}}{a^4}$
15. $\frac{e^{100}}{e^{25}}$
16. $\frac{d^{12}}{d^7}$
17. $\frac{x^9}{x}$
18. $(x^9)^9$
19. $(m^6)^5$
20. $(rs^3)^{12}$

HA1-861: Simplifying Expressions with Negative and Zero Exponents

Recall that an exponent tells you how many times to multiply the base. For example, $3^2 = 3 \cdot 3$ and $x^4 = x \cdot x \cdot x \cdot x$. However, this definition does not hold for exponents less than or equal to zero. Therefore, you must consider another rule for zero and negative exponents: the **quotient rule** for exponents states that if $x \neq 0$, and n and m are real numbers, then:

$$\frac{x^n}{x^m} = x^{n-m}$$

For example: $\frac{x^2}{x^5} = x^{2-5} = x^{-3}$ and $\frac{x^2}{x^5} = \frac{x \cdot x}{x \cdot x \cdot x \cdot x \cdot x} = \frac{\cancel{x} \cdot \cancel{x}}{\cancel{x} \cdot \cancel{x} \cdot x \cdot x \cdot x} = \frac{1}{x \cdot x \cdot x} = \frac{1}{x^3}$

Therefore, $x^{-3} = \frac{1}{x^3}$

You can see that the base with a negative exponent is equal to its reciprocal with the positive exponent.

Remember that 1 raised to any exponent is 1, so $\left(\frac{1}{x}\right)^n = \frac{1^n}{x^n} = \frac{1}{x^n}$.

Thus the rule is as follows:

Negative Exponent Rule	For all real numbers x and n, where $x \neq 0$, $x^{-n} = \left(\frac{1}{x}\right)^n = \frac{1}{x^n}$.

Example 1 Simplify 4^{-2}.

$4^{-2} = \left(\frac{1}{4}\right)^2$

Step 1: Use the rule for negative exponents to rewrite as the reciprocal with a positive exponent.

$= \frac{1^2}{4^2}$

Step 2: Square the numerator and the denominator.

$= \frac{1}{16}$

Step 3: Use the exponents to evaluate the powers.

Answer: $\frac{1}{16}$

Example 2 Simplify: $\left(\frac{3}{5}\right)^{-3}$

$\left(\frac{3}{5}\right)^{-3} = \left(\frac{1}{\frac{3}{5}}\right)^3$

Step 1: Rewrite as the reciprocal with a positive exponent.

$= \left(\frac{5}{3}\right)^3$ **Step 2:** Eliminate the complex fraction by changing it to a simple fraction; that is, take the reciprocal of the reciprocal.

$= \frac{5^3}{3^3}$ **Step 3:** Simplify using the rules of exponents.

$= \frac{125}{27}$ **Step 4:** Evaluate the powers.

Answer: $\frac{125}{27}$

Example 3 Simplify: $(x^4y^3z)^{-2}$.

$(x^4y^3z)^{-2} = \left(\frac{1}{x^4y^3z}\right)^2$ **Step 1:** Rewrite as the reciprocal with a positive exponent.

$= \frac{1^2}{(x^4y^3z)^2}$ **Step 2:** Square both the numerator and the denominator.

$= \frac{1}{x^8y^6z^2}$ **Step 3:** Simplify by multiplying the exponent of each term in the denominator by two.

Answer: $\frac{1}{x^8y^6z^2}$

You now have rules for positive and negative exponents; zero, however, is neither positive nor negative. Therefore, you need a rule for the zero exponents. Go back to the quotient rule for exponents, which states that if $x \neq 0$ and n and m are real numbers, $\frac{x^n}{x^m} = x^{n-m}$. Consider $\frac{y^3}{y^3}$. According to the quotient rule for exponents, $\frac{y^3}{y^3} = y^{3-3} = y^0$.

Also remember that $\frac{y^3}{y^3} = \frac{y \cdot y \cdot y}{y \cdot y \cdot y} = 1$

$$y^0 = 1$$

Therefore, the rule is as follows:

Zero Exponents Rule	For all real numbers x, where $x \neq 0$, $x^0 = 1$.

Example 4 Simplify: 6^0.

$6^0 = 1$ **Step 1:** Use the zero exponents rule.

Answer: 1

Example 5 Simplify: $(-4)^0$.

$(-4)^0 = 1$	**Step 1:** Use the zero exponents rule.
	Answer: 1

Example 6 Simplify: $8z^0$.

$8z^0 = 8 \cdot 1$	**Step 1:** Use the zero exponents rule to simplify the z term.
$= 8$	**Step 2:** Multiply.
	Answer: 8

Example 7 Simplify: $\dfrac{x^{-3}y^2}{x^5y^{-7}}$

$\dfrac{y^2y^7}{x^5x^3}$	**Step 1:** Use the rule for negative exponents to rewrite those factors with negative exponents.
$\dfrac{y^9}{x^8}$	**Step 2:** Use the rule for multiplying exponents to combine like bases.
	Answer: $\dfrac{y^9}{x^8}$

Problem Set

Simplify the following:

1. 8^0
2. 4^0
3. -8^0
4. -4^0
5. -5^{-2}
6. -8^{-2}
7. $(-2)^{-2}$
8. $(-3)^{-2}$
9. x^7x^{-12}
10. x^4x^{-12}
11. $x^5x^{-8}x^{-2}x^9$
12. $x^{-6}x^3x^{-2}x^{12}$
13. $\dfrac{x^{-4}}{x^{-12}}$
14. $\dfrac{x^{-2}}{x^{-10}}$
15. $x^{-5}y^{-2}x^5y^4$
16. $x^{-4}y^8x^4y^{-5}$
17. $(x^{-4}y^3)^{-2}$
18. $\dfrac{x^{-12}y^{-7}}{x^{-9}y^{-2}}$
19. $\left(\dfrac{2^4x^{-3}y^{-4}}{x^{-6}}\right)^{-3}$
20. $\left(\dfrac{x^{-8}y^{-10}z^{-12}}{x^{-6}y^{-5}z^{-4}}\right)^3$

HA1-862: Dividing Polynomials Using Factoring

Just as you can simplify a numerical fraction by canceling out common factors, you can simplify a fraction with polynomials in the same way. For example, to simplify a fraction like $\frac{15}{25}$, you would factor a 5 out of the numerator and denominator to get $\frac{3 \cdot 5}{5 \cdot 5}$, then cancel the fives to get $\frac{3}{5}$.

To treat division with polynomials in the same way, you must set up your division problem in fraction form. The **Fundamental Property of Rational Expressions** states that if *x* divided by *y* is a rational expression and if *z* represents any real number not equal to zero, then *x* times *z* divided by *y* times *z* is equal to *x* divided by *y*.

$$\frac{x \cdot z}{y \cdot z} = \frac{x}{y}$$

Let's see how this property can be used with factoring to divide polynomials.

Example 1 Divide $x^2 + 10x + 24$ by $(x + 4)$.

$\frac{x^2 + 10x + 24}{x + 4}$ **Step 1:** Set the dividend as the numerator and the divisor as the denominator.

$\frac{(x + 6)(x + 4)}{x + 4}$ **Step 2:** Factor the numerator.

$\frac{(x + 6)\cancel{(x + 4)}}{\cancel{x + 4}}$ **Step 3:** Cancel the common factor $x + 4$ from the numerator and denominator.

Answer: $x + 6$

Example 2 Divide $x^2 - 5x + 6$ by $(x - 3)$

$\frac{x^2 - 5x + 6}{x - 3}$ **Step 1:** Set the dividend as the numerator and the divisor as the denominator.

$\frac{(x - 2)(x - 3)}{x - 3}$ **Step 2:** Factor the numerator.

$\frac{(x - 2)\cancel{(x - 3)}}{\cancel{x - 3}}$ **Step 3:** Cancel the common factor $x - 3$ from the numerator and denominator.

Answer: $x - 2$

Example 3 Divide $x^3 - 4x$ by $(x^2 + 2x)$.

$= \frac{x^3 - 4x}{x^2 + 2x}$ **Step 1:** Set the dividend as the numerator and the divisor as the denominator.

$= \dfrac{x(x^2 - 4)}{x(x + 2)}$

Step 2: Factor the numerator and denominator.

$= \dfrac{x(x - 2)(x + 2)}{x(x + 2)}$

Step 3: Factor the numerator further.

$= \dfrac{\cancel{x}(x - 2)\cancel{(x + 2)}}{\cancel{x}\cancel{(x + 2)}}$

Step 4: Cancel the common factors $x + 2$ and x from the numerator and denominator.

Answer: $x - 2$

Example 4 Divide $\dfrac{x^4 - 10x^2 + 9}{x^2 - 2x - 3}$

$= \dfrac{(x^2 - 9)(x^2 - 1)}{(x - 3)(x + 1)}$

Step 1: Factor the numerator and denominator.

$= \dfrac{(x - 3)(x + 3)(x - 1)(x + 1)}{(x - 3)(x + 1)}$

Step 2: Factor the numerator further.

$= \dfrac{\cancel{(x - 3)}(x + 3)(x - 1)\cancel{(x + 1)}}{\cancel{(x - 3)}\cancel{(x + 1)}}$

Step 3: Cancel the common factor $x - 3$ and $x + 1$ from the numerator and denominator.

$= (x + 3)(x - 1)$

$= (x^2 + 2x - 3)$

Step 4: Multiply resulting factors.

Answer: $x^2 + 2x - 3$

Problem Set

Simplify the following:

1. $\dfrac{3x^2}{3x}$
2. $\dfrac{12h^2}{2h^3}$
3. $\dfrac{x^3}{x^2}$
4. $\dfrac{8x^3}{16x^5}$
5. $\dfrac{8x^4}{16x^3}$
6. $\dfrac{a^4}{a}$
7. $\dfrac{-4x^2}{20x^3}$
8. $\dfrac{48x^2}{24x}$
9. $\dfrac{64x^3}{8x}$
10. $\dfrac{12x^4}{120x^2}$
11. $\dfrac{5k^3 - k^4}{k}$
12. $\dfrac{c^9 - c^8 + c^6 - c}{c}$
13. $\dfrac{84x^5 + 100x^2}{2x}$
14. $\dfrac{120y^5 - 75y^4}{3y}$
15. $\dfrac{14b^2 - 21}{7}$
16. $\dfrac{5x^5 + x^3}{x}$
17. $\dfrac{20y^3 - 15y}{5y}$
18. $\dfrac{x^2 + 3x - 10}{x - 2}$
19. $\dfrac{x^2 + 10x + 21}{x + 3}$
20. $\dfrac{y^2 - 8y + 16}{y - 4}$

HA1-863: Dividing Polynomials Using Long Division

Factoring does not always help you divide polynomials. In such cases, you can use long division. Long division for polynomials is similar to long division with whole numbers; you must be careful to include all variable terms in the dividend. Make sure all terms are written in descending exponential order. Use zero as the coefficient of any missing terms. Set the problem up like a normal long division problem. First, look at the first term of the divisor and the first term of the dividend. The first terms must match, so you need to know what number multiplied by the variable will result in the first term. Next, multiply this by the whole divisor and place the answer under the dividend. Now, subtract as you would with normal long division and bring the next term in the dividend down. Now, repeat the same steps as before, finding a value that can be multiplied by the first term of the divisor to result in the same value of the first term of the "new" dividend, and subtract it from the dividend. If there are more terms, continue dividing until you have reached the end of the polynomial. If there are no more terms left to bring down and you cannot divide anymore, look for a remainder. If there is none, the division is complete. If there is a remainder, you can add it to the quotient in the form of remainder over divisor.

Example 1 Divide $(x^3 + x^2 - 3x + 1) \div (x - 1)$.

$$x-1\overline{)x^3+x^2-3x+1}$$

Step 1: Rewrite the problem as a long division problem.

$$\begin{array}{r} x^2 \\ x-1\overline{)x^3+x^2-3x+1} \end{array}$$

Step 2: Find the term that will equal x^3 when multiplied by x. Place that term above x^2.

$$\begin{array}{r} x^2 \\ x-1\overline{)x^3+x^2-3x+1} \\ x^3-x^2 \end{array}$$

Step 3: Multiply x^2 by each term of the divisor, aligning terms with the same exponents.

$$\begin{array}{r} x^2 \\ x-1\overline{)x^3+x^2-3x+1} \\ \underline{-(x^3-x^2)} \\ 2x^2-3x \end{array}$$

Step 4: Subtract the partial product from the dividend and bring down the next term. Remember that $x^2 - (-x^2)$ is $2x^2$.

$$\begin{array}{r} x^2+2x-1 \\ x-1\overline{)x^3+x^2-3x+1} \\ \underline{-(x^3-x^2)} \\ 2x^2-3x \\ \underline{-(2x^2-2x)} \\ -x+1 \\ \underline{-(-x+1)} \\ 0 \end{array}$$

Step 5: Now, find the term that will equal $2x^2$ when multiplied by x. Continue the process until all terms are divided. If there is a remainder, place it over the divisor and add it to your quotient.

Answer: $x^2 + 2x - 1$

Example 2 Divide: $\dfrac{4x^2 - 9x + 15}{x - 6}$

$x - 6\,\overline{)4x^2 - 9x + 15}$

Step 1: In order to solve this problem we must perform long division.

$$\frac{4x^2 - 9x + 15}{x - 6} = x - 6\,\overline{)4x^2 - 9x + 15} \quad \text{quotient: } 4x + 15 + \frac{105}{x - 6}$$

$$\begin{array}{r} 4x^2 - 9x + 15 \\ -(4x^2 - 24x) \\ \hline 15x + 15 \\ -(15x - 90) \\ \hline 105 \end{array}$$

$$\frac{4x^2 - 9x + 15}{x - 6} = 4x + 15 + \frac{105}{x - 6}$$

Step 2: We perform the long division and simplify.

Answer: $4x + 15 + \dfrac{105}{x - 6}$

Example 3 Divide: $(9c^2 + 2c + 4) \div (c - 3)$

$= c - 3\,\overline{)9c^2 + 2c + 4}$

Step 1: We must use long division to solve this problem.

$= c - 3\,\overline{)9c^2 + 2c + 4}$

$c - 3\,\overline{)9c^2 + 2c + 4} \quad \text{quotient: } 9c + 29 + \dfrac{91}{c - 3}$

$$\begin{array}{r} 9c^2 + 2c + 4 \\ -(9c^2 - 27c) \\ \hline 29c + 4 \\ -(29c - 87) \\ \hline 91 \end{array}$$

$= 9c + 29 + \dfrac{91}{c - 3}$

Step 2: We perform the long division and simplify.

Answer: $9c + 29 + \dfrac{91}{c - 3}$

Example 4 Divide: $(8y^3 - 24y - 5) \div (2y + 3)$

$= 2y + 3\overline{)8y^3 - 24y - 5}$

Step 1: We must use long division to solve this problem.

$= 2y + 3\overline{)8y^3 - 24y - 5}$

Step 2: We perform the long division and simplify.

$$
\begin{array}{r}
4y^2 - 6y - 3 + \dfrac{4}{2y+3} \\
= 2y + 3\overline{)8y^3 + 0y^2 - 24y - 5} \\
-(8y^3 + 12y^2) \\
\hline
-12y^2 - 24y \\
-(-12y^2 - 18y) \\
\hline
-6y - 5 \\
-(-6y - 9) \\
\hline
4
\end{array}
$$

$= 4y^2 - 6y - 3 + \dfrac{4}{2y + 3}$

Answer: $4y^2 - 6y - 3 + \dfrac{4}{2y + 3}$

Problem Set

Divide the following using long division:

1. $(y^2 - 2y + 3) \div (y + 6)$
2. $(y^2 + 4y - 25) \div (y + 5)$
3. $(k^2 + 4k - 30) \div (k - 4)$
4. $(x^2 + 2x + 1) \div (x + 1)$
5. $(x^2 + 3x - 12) \div (x + 2)$
6. $(a^2 + 3a - 3) \div (a - 3)$
7. $(x^2 + 5x + 8) \div (x + 3)$
8. $(c^2 + 3c - 9) \div (c + 2)$
9. $(y^2 + 3y + 8) \div (y - 5)$
10. $(a^2 - 12a + 8) \div (a - 6)$
11. $(3y^2 - 3y - 2) \div (y - 6)$
12. $(2m^2 + 8m - 12) \div (m - 4)$
13. $(4x^2 - 4x - 20) \div (x - 2)$
14. $(3x^2 + 3x + 4) \div (x - 3)$
15. $(4y^2 + 2y + 3) \div (y + 1)$
16. $(a^2 - 144) \div (a - 14)$
17. $(a^2 - 81) \div (a - 10)$
18. $(y^3 - 4y^2 + 3y - 2) \div (y - 2)$
19. $(a^3 - a + 12) \div (a - 4)$
20. $(12x^3 - 41x + 7) \div (2x - 3)$

HA1-864: Dividing Polynomials Using Synthetic Division

Because dividing polynomials can be a long process, mathematicians developed a shortcut called synthetic division. This type of division is much faster than long division, and requires less space. However, synthetic division only works if the divisor is a binomial in the form $x + b$ or $x - b$. Synthetic division usually works if the power of x is one. If the power of $x \neq 1$, then use long division.

Synthetic division also requires that the coefficient of x is 1. If the coefficient of x is not 1, divide the binomial divisor by the coefficient of x to make it 1, then use synthetic division. The last step is to divide your answer by the same coefficient.

Example 1 Divide using synthetic division: $(5x^3 + x^2 - 20x - 4) \div (x - 2)$

$5x^3$	$+x^2$	$-20x$	-4
↓	↓	↓	↓
5	1	−20	−4

Step 1: Set up the problem for synthetic division, write the terms of the polynomial so that the degrees of the terms are in decreasing order. Then write the coefficient of the dividend.

2	5	1	−20	−4

Step 2: Write the opposite of b of the divisor $x - b$ to the left. In this case, the divisor is $x - 2$, so b is -2 and the opposite of b is 2.

2	5	1	−20	−4
	↓			
	5			

Step 3: Bring down the first coefficient, 5. Then Multiply $-b$, which is 2, by the first coefficient, 5, and place the result under the second coefficient.

2	5	1	−20	−4
	↓	10		
	5			

2	5	1	−20	−4
	↓	10	22	
	5	11		

Step 4: Add 1 and 10. Write the sum, then multiply it by 2 and write the product under the next term.

2	5	1	−20	−4
	↓	10	22	4
	5	11	2	0

Step 5: Continue to add then multiply, repeating the process until you have a remainder, under the last term. The remainder in this case is 0.

2	5	1	–20	–4	
	↓	10	22	4	
	5	11	2	0	
	↓	↓	↓		
	$5x^2$	$+ 11x$	$+ 2$		

Step 6: Write out the quotient. The numbers along the bottom row are the coefficients of the terms of the polynomial in descending order. Start with the degree of the term that is one less than that of the dividend. In this case, the degree of the dividend was 3, so we start with a degree of 2.

Answer: The quotient is $5x^2 + 11x + 2$.

Example 2 Divide using synthetic division: $x+3\overline{)3x^4 - 2x^3 + 24x - 228}$

$3x^4$	$- 2x^3$	$+ 0x^2$	$+ 24x$	$- 228$
↓	↓	↓	↓	↓
3	–2	0	24	–228

Step 1: Set up this problem just as we did in Example 1. But in this case, since the dividend does not contain a term with degree x^2, we add that term with a coefficient of zero. This serves as a placeholder in our problem.

–3	3	–2	0	24	–228

Step 2: Write the opposite of b of the divisor $x – b$ to the left. In this case, the divisor is $x+3$, so b is 3 and the opposite of 3 is –3.

–3	3	–2	0	24	–228
	↓	–9			
	3				

Step 3: Bring down the first coefficient, 3. Then multiply –3 by the first coefficient, 3, and place the product under the next coefficient.

–3	3	–2	0	24	–228
	↓	–9	33		
	3	–11			

Step 4: Add –2 and –9. Write the sum, then multiply that sum by –3 and write the product under the next coefficient.

–3	3	–2	0	24	–228
	↓	–9	33	–99	225
	3	–11	33	–75	–3

Step 5: Continue to add then multiply, repeating the process until you have a remainder under the last term. The remainder in this case is –3.

-3	3	-2	0	24	-228	
	↓	-9	33	-99	225	
	3	-11	33	-75	3	
	↓	↓	↓	↓		

$$3x^3 - 11x^2 + 33x - 75 \; \frac{-3}{x+3}$$

Step 6: Write out the quotient. Remember to use the result of our division as the coefficients, and start with a degree one less than the degree of the original dividend. We start with a degree of 3.

Answer: The quotient is $3x^3 - 11x^2 + 33x - 75 - \dfrac{3}{x+3}$.

Example 3 Divide $6x^3 + 5x^2 + 7x + 2$ by $3x + 1$.

$$\frac{3x+1}{3} = x + \frac{1}{3}$$

$$b = \frac{1}{3}$$

Step 1: The coefficient of x in the divisor is not 1, so divide $3x + 1$ by 3.

$-\frac{1}{3}$	6	5	7	2	
		-2	-1	-2	
	6	3	6	0	
	↓	↓	↓		

$$6x^2 + 3x + 6$$

Step 2: The new b term is $\frac{1}{3}$, so use its opposite, $-\frac{1}{3}$, as the divisor in synthetic division and proceed as usual.

$$= \frac{6x^2 + 3x + 6}{3} = 2x^2 + x + 2$$

Step 3: Because you divided the original divisor $3x + 1$ by 3, you must divide the partial answer by 3 to get the correct answer.

Answer: The quotient is $2x^2 + x + 2$.

1. In the synthetic division shown, what is the quotient?

$-\frac{1}{2}$	6	0	$\frac{1}{2}$	7	3
		-3	$\frac{3}{2}$	-1	-3
	6	-3	2	6	0

2. In the synthetic division shown, what is the divisor?

-5	1	2	-15
		-5	15
	1	-3	0

3. Can we use $x^2 + 5$ as a divisor in synthetic division? Answer yes or no.

4. If $x - 3$ is the divisor, what term would be used to multiply by the dividend in synthetic division?

5. If $x + 2$ is the divisor, what term would be used to multiply by the dividend in synthetic division?

6. Can we use $2x^3 + 3x^2 - 2x - 23$ as a divisor in synthetic division? Answer yes or no.

7. Can we use $2x + 3$ as a divisor in synthetic division? Answer yes or no.

8. In the synthetic division shown, what is the dividend?

1	1	0	2	−2	−1
		1	1	3	1
	1	1	3	1	0

9. Divide using synthetic division:

$$x + 5 \overline{) x^2 + 2x - 15}$$

10. Divide using synthetic division:

$$x + 1 \overline{) 2x^3 + 3x^2 - x - 2}$$

11. Divide using synthetic division. What is the coefficient of the x-term in the quotient?

$$x + 2 \overline{) x^3 - 2x^2 - 5x + 6}$$

12. Divide using synthetic division. What is the coefficient of the x-term in the quotient?

$$x + 1 \overline{) 3x^3 - x + 2}$$

13. Divide using synthetic division. What is the remainder?

$$x - 4 \overline{) 3x^5 - 7x^4 - 76x^2 - 15x - 4}$$

14. Divide using synthetic division. What is the coefficient of the x-squared term in the quotient?

$$x + 3 \overline{) 4x^4 - 38x^2 - 9x - 9}$$

15. Divide using synthetic division:

$$x + 3 \overline{) 2x^3 - 20x - 6}$$

16. Divide using synthetic division:

$$x - 2 \overline{) 3x^5 - 13x^3 + 6x^2 + 1}$$

17. Divide using synthetic division:

$$x + 6 \overline{) 2x^4 + 10x^3 - 13x^2 + 36}$$

18. Divide using synthetic division:

$$x - \frac{1}{2} \overline{) 4x^3 - 6x^2 - 2x + 2}$$

19. Divide using synthetic division:

$$x - \frac{1}{3} \overline{) 9x^3 - 3x^2 + 6x - 2}$$

20. Divide using synthetic division:

$$x - 6 \overline{) \frac{1}{3}x^3 - 2x^2 + x - 6}$$

HA1-866: Drawing a Line Using Slope-Intercept Form and Determining if Two Lines are Parallel or Perpendicular

You have previously learned about the slope of a line, how to graph a line, and the different forms for the equation of a line. Recall that the slope of a line is the slant of the line and is defined as the ratio of the change in y-coordinates to the corresponding change in x-coordinates. The formula for determining the slope is

$m = \frac{y_2 - y_1}{x_2 - x_1} = \frac{\text{rise}}{\text{run}}$.

The graph below shows a line that passes through the points (–2, 0) and (2, 2).

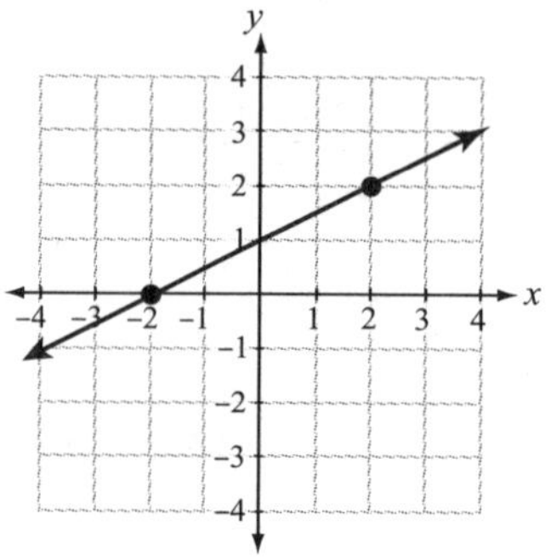

Using the points plotted on the graph, we can determine the slope of the line using the slope formula.

$$m = \frac{y_2 - y_1}{x_2 - x_1} = \frac{2 - 0}{2 - (-2)} = \frac{2}{2 + 2} = \frac{2}{4} = \frac{1}{2}$$

We can also determine the slope of the line by looking at the graph and counting the length of the rise and run.

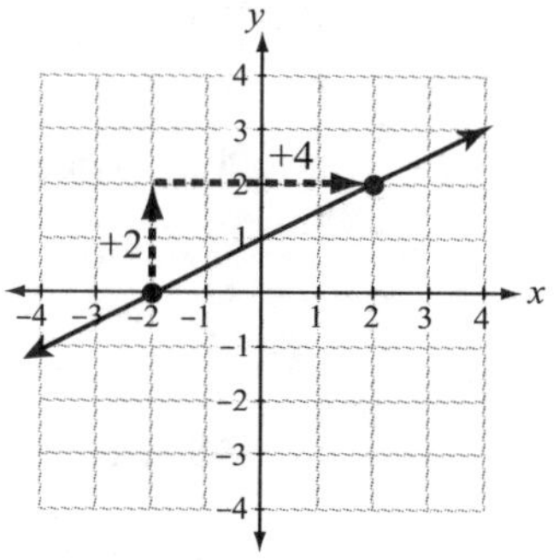

You can see that the rise is positive 2 and the run is positive 4. Therefore, the slope is $\frac{2}{4}$ or $\frac{1}{2}$. This is the same result we found using the slope formula. When using the rise over run method, the direction in which you count is very important. For example, if you count down instead of up, the rise is negative. If you count to the left instead of to the right, the run is also negative.

Therefore, whenever you have a positive slope, you always count up and to the right, or down and to the left. For a negative slope, you always count down and to the right, or up and to the left.

Now let's review the standard form and slope-intercept form for the equation of a line. Recall that when graphing equations it is easier to have the equation in slope-intercept form in order to easily identify the slope and y-intercept of t0he line.

Standard Form for the Equation of a Line	$Ax + By = C$, where A, B, and C are real numbers (written in integer form when possible). $A \geq 0$, A and B are not both zero.

Slope-intercept Form for the Equation of a Line	$y = mx + b$, where m is the slope and b is the y-intercept.

Example 1 Identify the slope and y-intercept. $y = 2x + 1$

$y = mx + b$
$y = \underline{2}x + \underline{1}$

Step 1: Recall that the slope-intercept form for the equation of a line is $y = mx + b$, where m is the slope and b is the y-intercept.

Answer: The slope is 2 and the y-intercept is 1.

Example 2 Identify the slope and y-intercept. $2x + 3y = 6$

$$2x + 3y = 6$$
$$3y = -2x + 6$$
$$y = -\frac{2}{3}x + 2$$

Step 1: Rewrite the equation in slope-intercept form.

$$y = mx + b$$
$$y = -\frac{2}{3}x + 2$$

Step 2: Recall that the slope-intercept form for the equation of a line is $y = mx + b$, where m is the slope and b is the y-intercept.

Answer: The slope is $-\frac{2}{3}$ and the y-intercept is 2.

Example 3 Graph the equation $y = x + 3$ using the slope and y-intercept.

$y = mx + b$
$y = x + 3$
$y = \underline{1}x + \underline{3}$
$m = 1,\ b = 3$

Step 1: Identify the slope and the y-intercept.

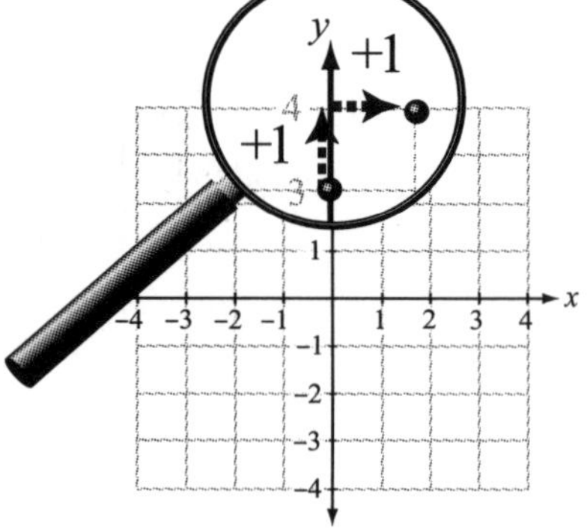

Step 2: Plot the point (0, 3) for the y-intercept and use the slope, 1, to obtain another point on the graph.

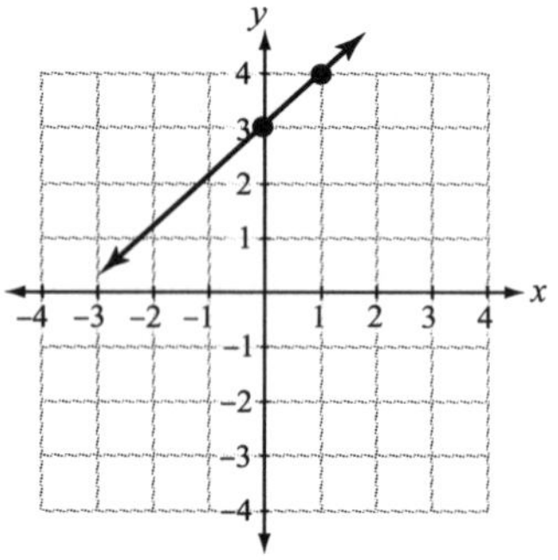

Step 3: Draw the line through the points.

Answer: The graph of $y = x + 3$ is given in step 3.

Example 4 Given the graph, find the equation of the line in slope-intercept form.

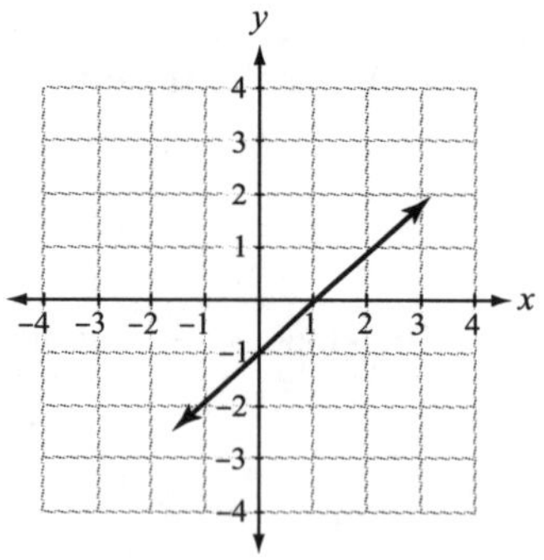

Step 1: Determine the y-intercept.

Since the graph intersects the y-axis at $(0, -1)$, $b = -1$.

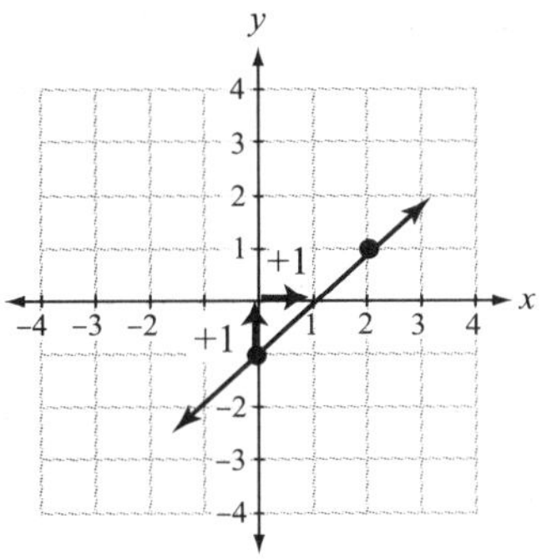

Step 2: Determine the slope.

Since the rise is +1 and the run is +1,

$m = \frac{1}{1} = 1.$

$y = mx + b$

$y = 1x - 1$

$y = x - 1$

Step 3: Substitute $m = 1$ and $b = -1$ into the slope-intercept form to find the equation.

Answer: $y = x - 1$

Slopes and the slope-intercept form of an equation are important in determining if two lines are parallel or perpendicular. If two lines have the same slope they are said to be parallel and if the slopes of two lines are opposite reciprocals of each other they are said to be perpendicular.

Parallel Lines	Two lines are parallel if and only if their slopes are equal.

Perpendicular Lines	Two lines are perpendicular if and only if their slopes are opposite reciprocals of each other.

Example 5 Determine whether the two lines are parallel, perpendicular, or neither.

$$4x + y = 1$$
$$y = \frac{1}{4}x + 3$$

$$4x + y = 1$$
$$4x - 4x + y = -4x + 1$$
$$y = -4x + 1$$

Step 1: Rewrite the first equation into the slope-intercept form.

The slope of the first line is –4.

The slope of the second line is $\frac{1}{4}$.

Step 2: Identify the slopes of the two lines.

Since the slopes are opposite reciprocals of each other, they are perpendicular.	**Step 3:** Compare the slopes of the two equations.
	Answer: The two lines are perpendicular.

Problem Set

Identify the slope and the *y*-intercept for the following:

1. $2x + 5y = 1$

2. $2x + y = 4$

3. $3x + y = 6$

4. $-7x + y = 8$

5. $7y = 14x - 21$

6. $-y = 3x + 7$

7. $-y = 12x + 14$

8. $-4x + y = 6$

9. $3y - 9 = 5x$

10. $3x - y = 3$

Graph the following equations using the slope and *y*-intercept:

11. $x - 2y = -4$

12. $12x - 3y = -3$

13. $y = 2x + 3$

14. $y = 3x - 2$

15. $x + 2y = 6$

16. $y - 4x = 1$

17. $3x + y = -3$

Determine if the two lines in the following questions are parallel, perpendicular, or neither:

18. $3y = 2x + 14$
$2x - 3y = 2$

19. $y = -3x + 4$
$6x + 2y = -10$

20. $4y = -12x + 16$
$y = 3x - 5$

HA1-877: Drawing Inferences and Making Predictions from Tables and Graphs

In many business and scientific situations, it is extremely important to be able to predict what is likely to happen in the future based on past data. Examples vary from predicting the weather to deciding what price to charge for merchandise, or deciding which product a company chooses to market. While producing a weather forecast is beyond the scope of this lesson, the concept can be used to illustrate the importance of predictions. Everyone knows that weather forecasts can be unreliable, yet enormous numbers of people rely on them every day. The point is that a method of prediction that tends to be right most of the time leads people to make decisions that are usually beneficial. The first example is intended to illustrate that it is not always possible to use data to make predictions.

Example 1 The chart shows the value of a mutual fund in dollars. The numbers along the bottom give the number of days since the beginning of the year. Can you predict what the value will be on day 250?

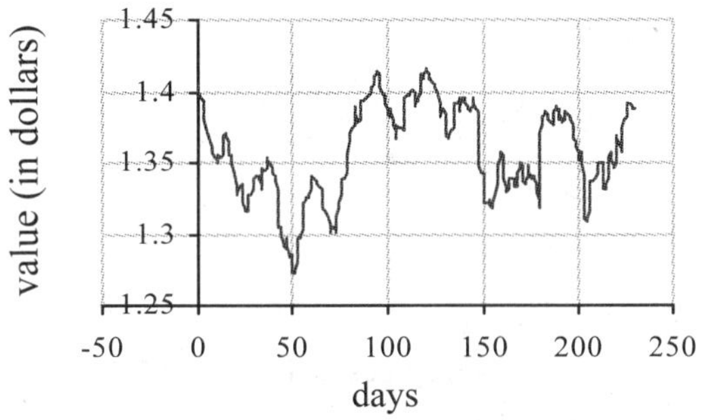

Step 1: There is no clear pattern to the ups and downs on the graph, and so, based solely on the data, there is no way to determine what is likely to happen next.

Answer: No, you cannot predict the value based on this chart.

Making predictions from data is known as **statistical inference**.

Statistical Inference	The process of making predictions from data

It is true that a market analyst may predict what is going to happen to the value of the mutual fund in Example 1 based on economic and financial market forces. That would not be statistical inference, because it uses factors that do not involve the data. The next example illustrates statistical inference.

Example 2 In a laboratory experiment, the following data was collected relating the temperature of a liquid in degrees Celsius (°C) and its volume in cubic centimeters (cc). Predict the volume:
A) when the temperature is 43°C. B) when the temperature is 124°C.

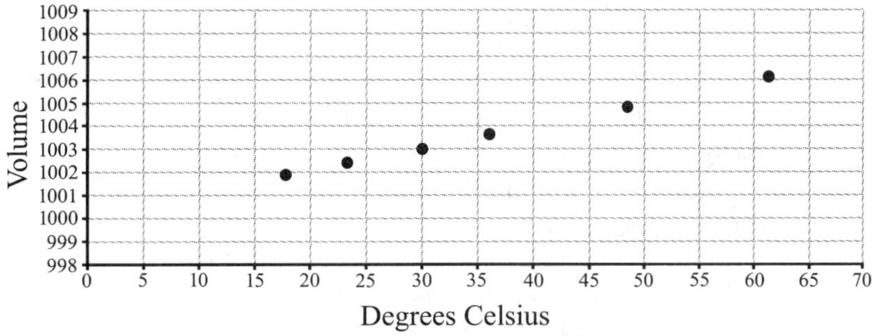

Step 1: The data appears to lie on a straight line, so draw a line through the points.

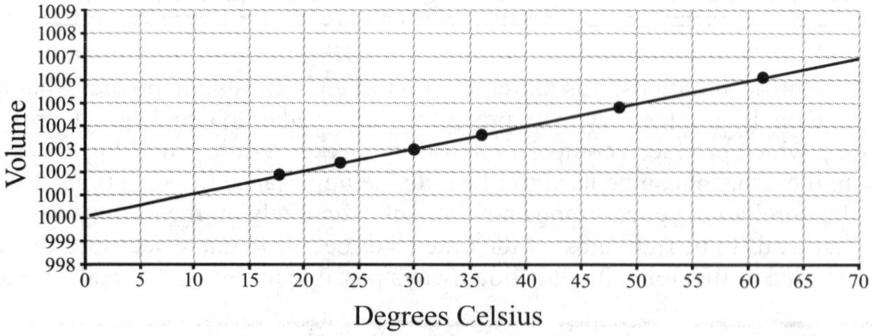

Step 2: Part (A) can now be answered by inspecting the graph. You can see that when the temperature is 43°C, the volume is about 1004.3 cc. The graph as it is drawn does not yield an answer to part (B).

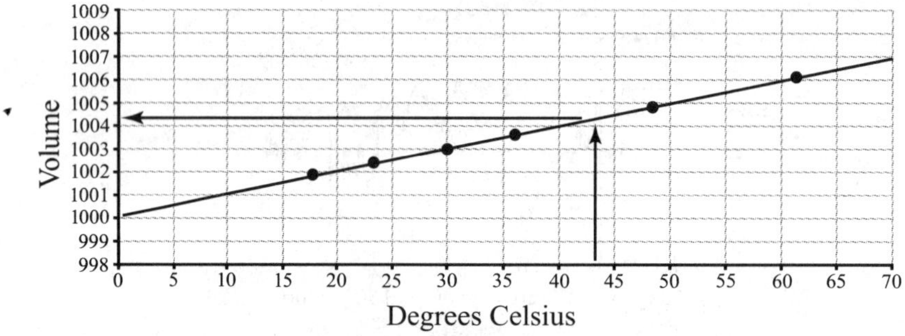

Step 3: One way to answer part (B) would be to extend the graph. A better approach is to find an equation for the line. Recall that the equation of a line can be given as $y = mx + b$, where m is the slope of the line and b is the y-intercept.

Step 4: One way to find the slope of a line is to identify two points on the line, call them (x_1, y_1) and (x_2, y_2), and then use the equation $m = \frac{y_2 - y_1}{x_2 - x_1}$. Note that $b = 1000$. The y-intercept is a good choice for (x_1, y_1), so $(x_1, y_1) = (0, 1000)$. For (x_2, y_2), it is best to pick a point whose coordinates are easy to identify, such as (70, 1007), so let $(x_2, y_2) = (70, 1007)$. Using these points:

$$m = \frac{1007 - 1000}{70 - 0}$$

$$= \frac{7}{70}$$

$$= 0.1$$

An equation of the line is $y = 0.1x + 1000$.

Step 5: Having found the equation of the line, it is now easy to answer both parts of the question.
A) If the temperature is 43°C then $x = 43$ and $y = 0.1(43) + 1000$, or $y = 1004.3$. This answer agrees with the answer found by inspecting the graph.
B) If the temperature is 124°C then $x = 124$ and $y = 0.1(124) + 1000$, or $y = 1012.4$. So, when the temperature is 124°C the volume is 1012.4cc

Answer: A) 1004.3cc B)1012.4cc

Note: In the graph of Example 2, there is a clear trend for temperatures up to 61°C, so it seems likely that the prediction for 43°C is accurate. The prediction for 124°C is another matter. It may happen that the liquid boils at a temperature lower than 124°C, in which case the prediction would be meaningless. However, based solely on the data it is the best that can be done. Issues of reliability are not considered in this lesson, what is considered is how to obtain predictions.

Example 3

Using the following graph:

A) Draw a straight line as close as possible to the points;
B) Find the equation of the line drawn in part (A);
C) Predict the value of y when $x = 8$ and comment on how reliable the prediction might be.

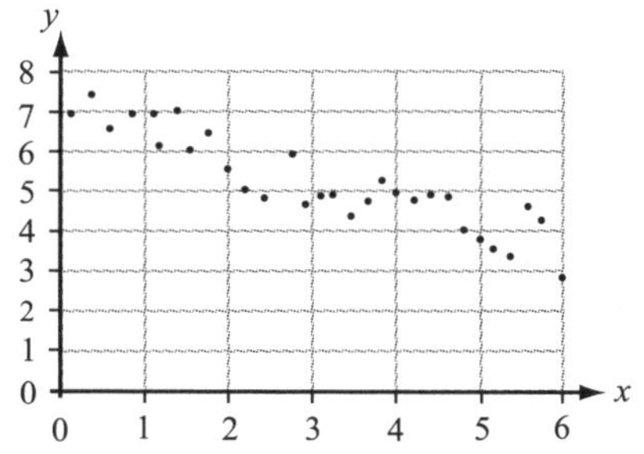

Step 1: Draw a line as close as possible to the points.

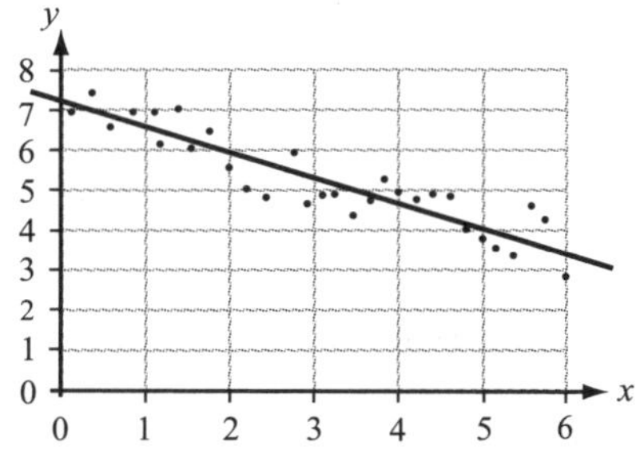

Step 2: The equation can be found using the same process you used in Example 2. The y-intercept is about 7.3, so let $(x_1, y_1) = (0, 7.3)$. It appears that the point (5, 4) lies on the line; this point is picked because its coordinates are easy to identify. So, let $(x_2, y_2) = (5, 4)$. Using these points:

$$\begin{aligned} m &= \frac{4-7.3}{5-0} \\ &= \frac{-3.3}{5} \\ &= -0.66 \end{aligned}$$

The equation of the line is $y = -0.66(x) + 7.3$.

Step 3: When $x = 8$, the equation gives $y = -0.66(8) + 7.3$, or $y = 2.02$.

Answer: A) Draw a straight line as close as possible to the points;

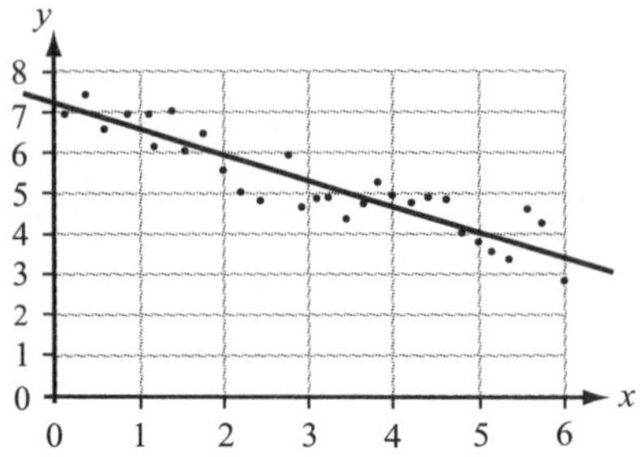

Answer: B) Find the equation of the line drawn in part (A);

$y = -0.66(x) + 7.3$

Answer: C) Predict the value of y when $x = 8$ and comment on how reliable the prediction might be. $y = -0.66(8) + 7.3 = 2.02$

The prediction is reasonably reliable since the dots fit fairly close to the line.

Data can often be easier to understand when it is presented in a histogram or bar graph. A bar graph is similar to a histogram, the difference being that in a bar graph there may be gaps between intervals or that the categories do not need to be intervals. The next example involves comparing two sets of data.

Example 4 A movie was shown to two groups of 50 people, who were then asked to rate the movie as poor, average, good, very good or excellent. The first group consisted of people under 25 years of age and the second group consisted of senior citizens. The results were used to construct the following two bar graphs.

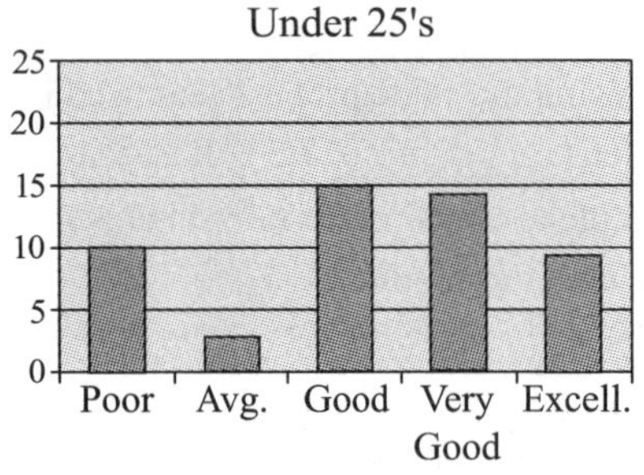

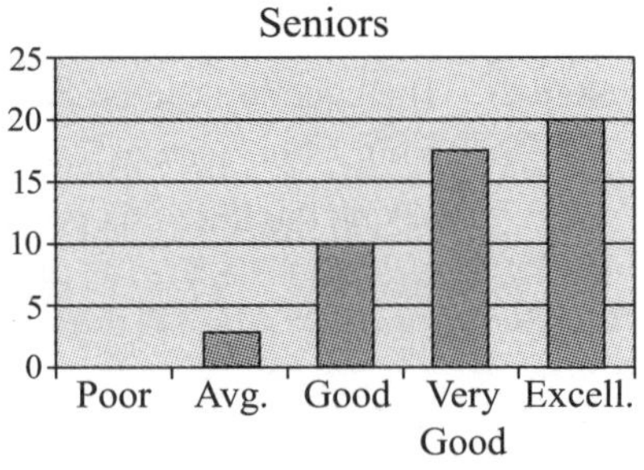

Answer the following questions based on the graphs:

A) Which group liked the movie better?

B) How many in each group thought the movie poor?

C) Why might this information be useful?

Step 1: To answer part A), look at the bars for "Good," "Very Good," and "Excell." For the seniors, the bars are taller in the categories for people who liked the movie the best, and shorter in categories for people who liked the movie the least. Overall, seniors liked the movie more than the "Under 25" group did.

Answer: A) People in the "Seniors" group liked the movie more than people in the "Under 25" group did.

Step 2: To answer part B), look at the "Poor" column. Ten people in the under 25 graph thought it was poor. In the "Seniors" graph, there is no bar for the "Poor" column, which indicates that none of the seniors thought the movie was poor.

Answer: B) In the "Under 25" group, 10 people thought the movie was poor; in the "Seniors" group, no one thought the movie was poor.

Step 3: To answer part C), look at both graphs again. The graphs are comparing two groups of people in different age categories. This graph could be used to help you decide which audiences would find the movie appealing. As an example, even though overall the movie may be popular, it would not be a good choice for a group of college students.

Answer: C) This information can help you decide which audiences would find the movie appealing.

Problem Set

1. The following table shows the weekly car sales and revenue of A & C Car Town. How many cars were sold on Wednesday?

Day	Number of Cars Sold	Money Made in Sales
Monday	35	420,000
Tuesday	20	240,000
Wednesday	17	204,000
Thursday	23	276,000
Friday	25	300,000
Saturday	43	516,000

2. The following table shows the weekly car sales and revenue of A & C Car Town. Determine what day of the week people would most likely purchase a new car.

Day	Number of Cars Sold	Money Made in Sales
Monday	35	420,000
Tuesday	20	240,000
Wednesday	17	204,000
Thursday	23	276,000
Friday	25	300,000
Saturday	43	516,000

3. Kellie's fee for teaching acting is shown on the graph below. The x-axis represents the number of students and the y-axis represents how much money Kellie makes per hour of work. How many students would Kellie be teaching if she made \$10.00 per hour?

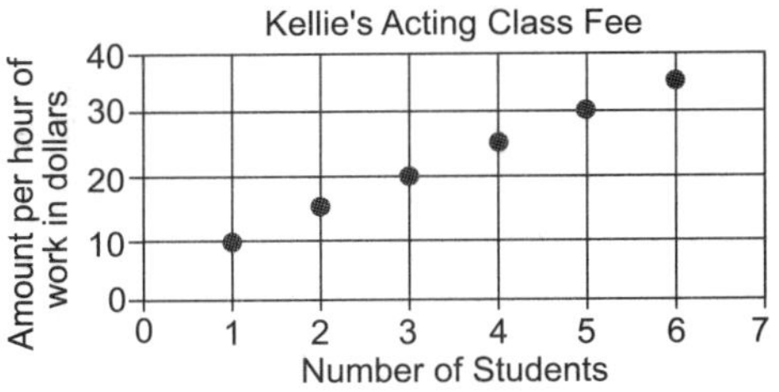

4. Kellie's fee for teaching acting is shown on the graph below. The x-axis represents the number of students and the y-axis represents how much money Kellie makes per hour of work. How much would Kellie make per hour of work if there were 5 students in her class?

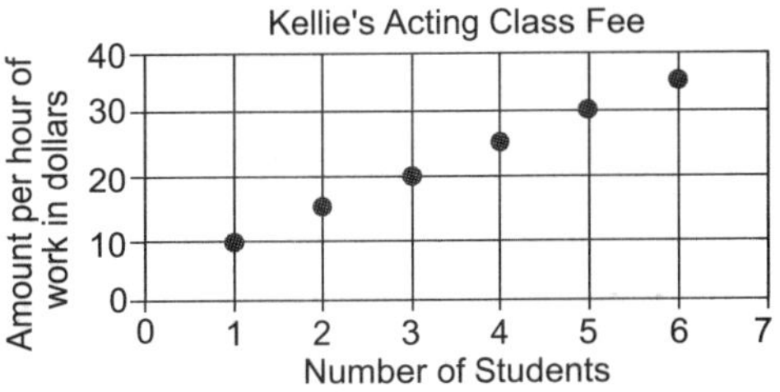

5. Mandy earns $4.00 per hour babysitting each child. The table shows how much she earns if she babysits for two hours?

Number of Hours(x)	Amount earned ($)
1	4
2	8
3	12
4	16
5	20

6. This graph relates the cost of mailing to the weight of the letter for the year 1991. How much did it cost to send a 4-ounce letter?

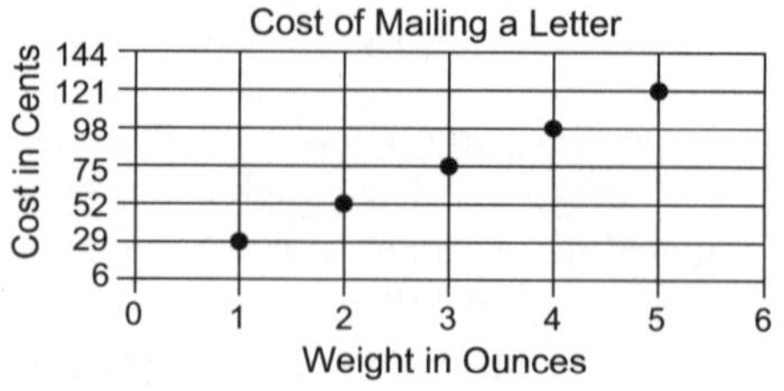

7. Below is a bar graph of the number of people who watch prime-time television in a given week. The x-axis represents the days of the week and the y-axis represents the number of people, in millions. On what night of the week is television watched by the greatest number of people?

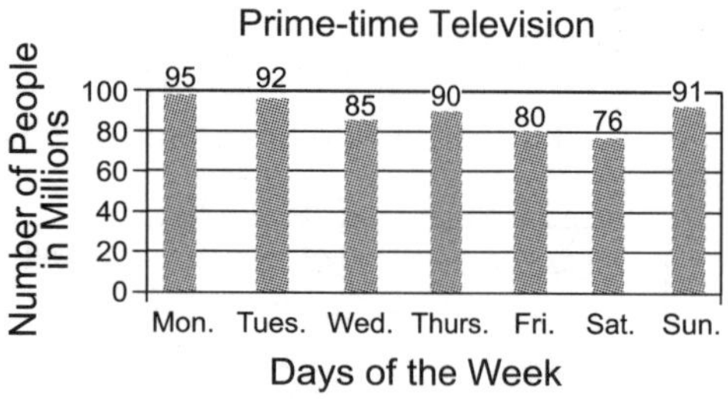

8. A sports drink company developed a number of new flavors. They tested these flavors with 97 testers. The graph shows the new flavors and how many people like each of them. How many new flavors did the company test?

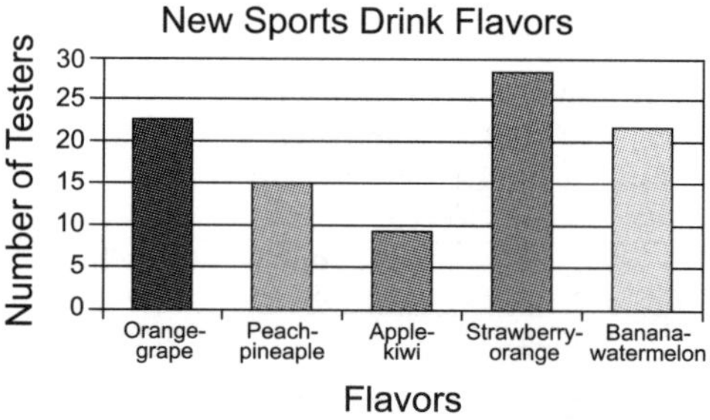

9. The following table shows the cooling rate of boiled water. How many minutes would a person have to wait for the temperature to be 73 degrees?

Time (min.)	0	5	10	15	20	25	30
Temp (°C)	100	90	81	73	66	60	55

10. The graph shows the value of a stock in a new company after the company was founded. What was the value of the stock when it was in the sixth month of operation?

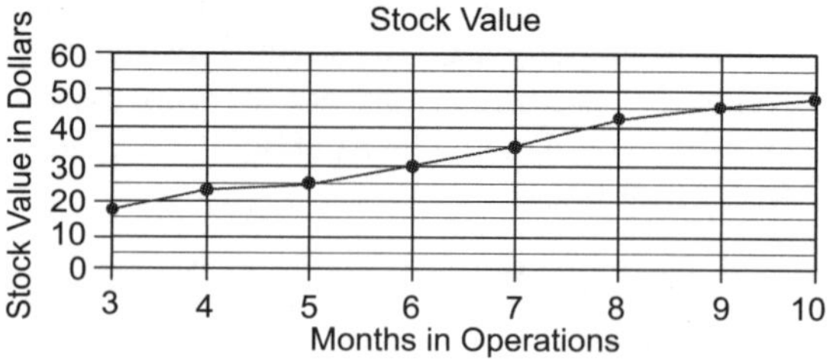

11. The table shows the weekly car sales and revenue of A & C Car Town. How many cars were sold for the week?

Day	Number of Cars Sold	Money Made in Sales
Monday	35	420,000
Tuesday	20	240,000
Wednesday	17	204,000
Thursday	23	276,000
Friday	25	300,000
Saturday	43	516,000

12. The table shows the weekly car sales and revenue of A & C Car Town. How much money did the company make in sales for Monday, Tuesday, and Thursday?

Day	Number of Cars Sold	Money Made in Sales
Monday	35	420,000
Tuesday	20	240,000
Wednesday	17	204,000
Thursday	23	276,000
Friday	25	300,000
Saturday	43	516,000

13. Mandy earns $4.00 per hour babysitting each child. The table shows how much she earns if she babysits one child. Find the equation for the data given by the table.

Number of Hours(x)	Amount earned ($)
1	4
2	8
3	12
4	16
5	20

14. The graph shows the cost of mailing, relative to the weight of the letter, for the year 1991. How much money would James need to send a 5-ounce letter and a 2-ounce letter?

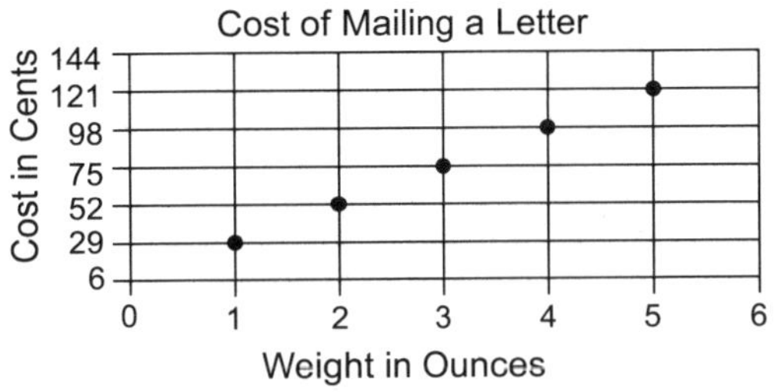

15. The following table shows the cooling rate of boiled water. What is the temperature difference that occurs between 10 and 20 minutes?

Time (min.)	0	5	10	15	20	25	30
Temp (°C)	100	90	81	73	66	60	55

16. If you bought stock on the third day and you sold it on the seventeenth day, by how much would you have gained or lost?

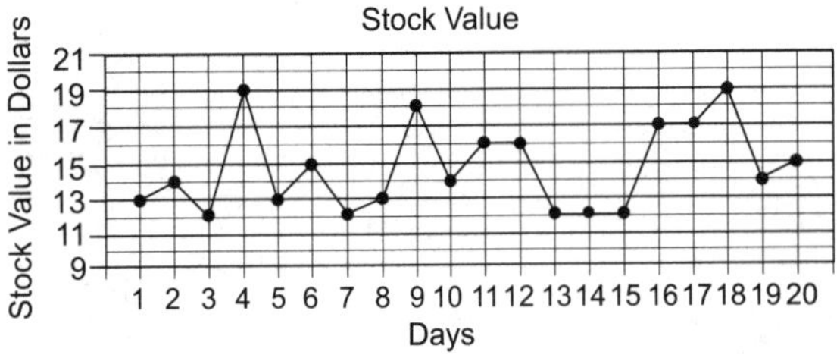

17. What is the equation of the li ne shown?

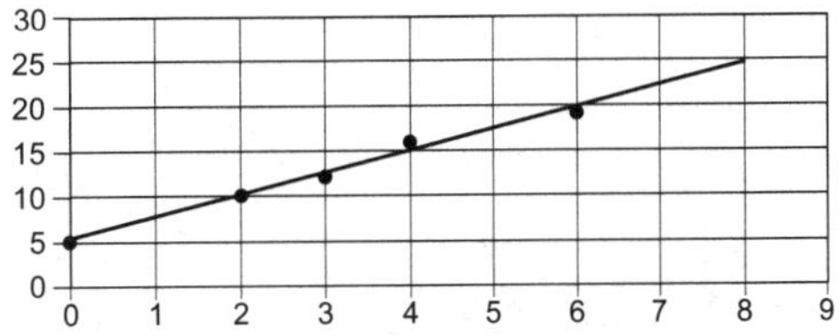

18. Which graph depicts the line of best fit?

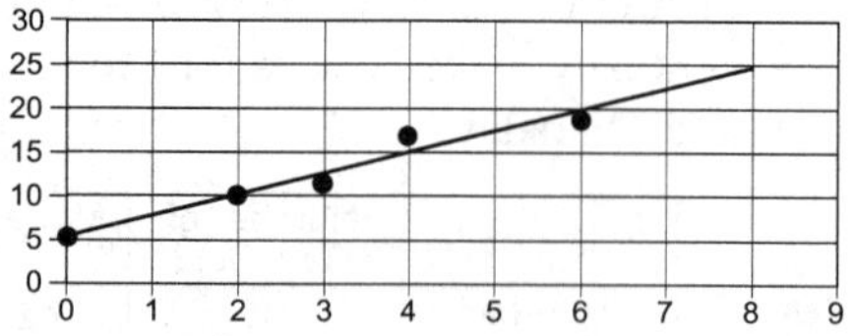

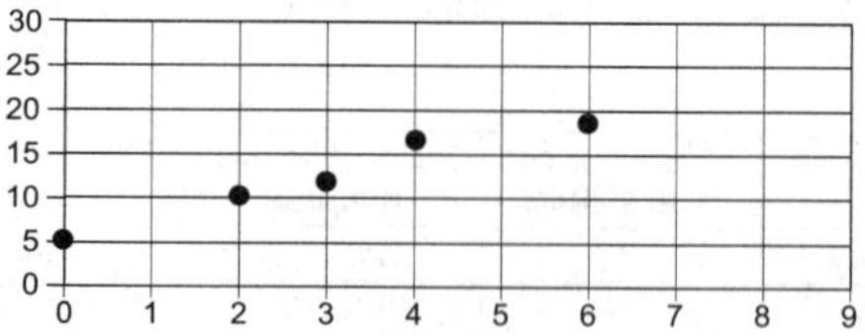

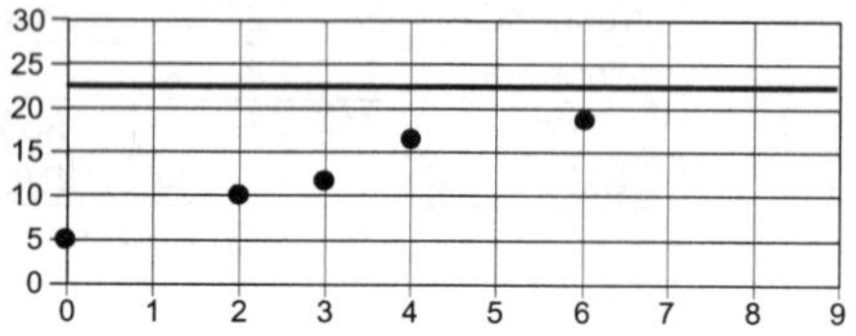

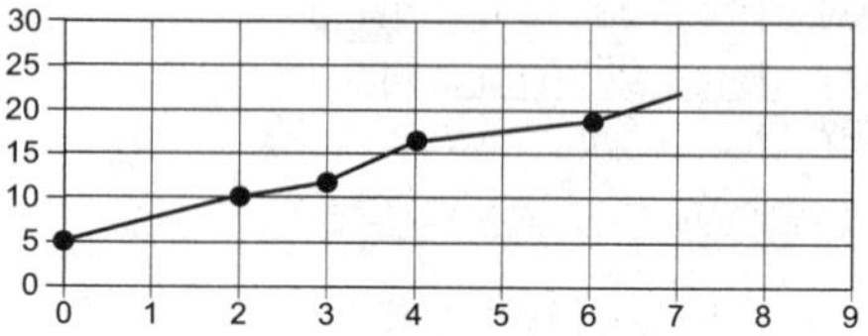

19. Predict the value of y when $x = 10$.

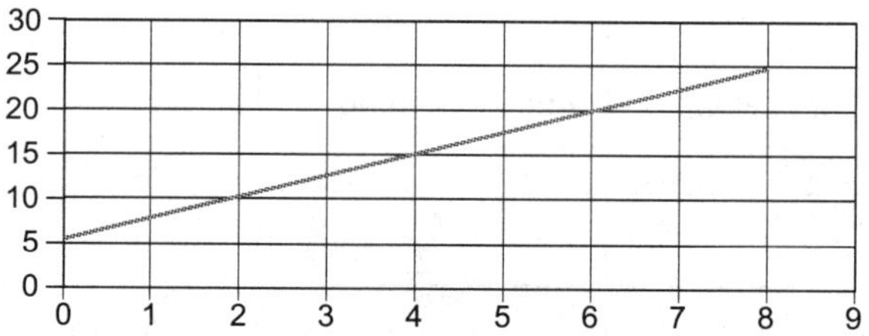

20. Predict the value of x when $y = 27$.

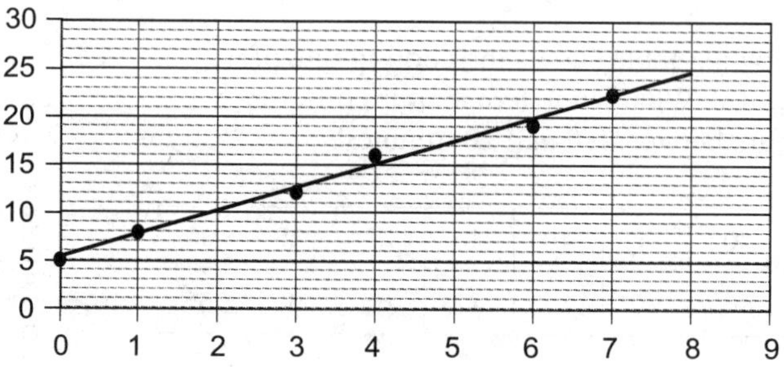

HA1-879: Applying Counting Techniques to Permutations and Combinations

Determining the size of the sample space (how many outcomes an event contains) is a fundamental concept for any probability problem. A **counting technique** is any technique that allows you to determine the size of a set. A **permutation** of a set of objects is the objects written in particular order.

Counting Technique	A technique that allows you to determine the size of a set

Permutation	A permutation of a set of objects is the objects written in a particular order.

Example 1 List the permutations of the letters *a*, *b*, and *c*.

abc, bac, cab, acb, bca, cba

Step 1: List all the possible ways to combine the letters.

Answer: *abc, bac, cab, acb, bca, cba*

Example 2 List the permutations of the letters *a*, *b*, *c*, and *d*.

abcd	*bacd*	*cabd*	*dabc*
abdc	*badc*	*cadb*	*dacb*
acbd	*bcad*	*cbad*	*dbac*
acdb	*bcda*	*cbda*	*dbca*
adbc	*bdac*	*cdab*	*dcab*
adcb	*bdca*	*cdba*	*dcba*

Step 1: List all the possible ways to combine the letters.

Answer: abcd, abdc, acbd, acdb, adbc, adcb, bacd, badc, bcad, bcda, bdac, bdca, cabd, cadb, cbad, cbda, cdab, cdba, dabc, dacb, dbac, dbca, dcab, dcba

Many counting problems can be solved using the **fundamental counting principle**.

Fundamental Counting Principle	If one task, A, can be performed in m different ways and a second task, B, can be performed in n different ways, then task A followed by task B can be performed in $m \cdot n$ ways. This concept can be extended in a similar manner if there are additional tasks: $m \cdot n \cdot r \ldots \cdot t$.

It states that the number of ways to choose two objects is the number of ways to choose the first object times the number of ways left to choose the second object after the first object has been chosen. In symbols, if the first object can be chosen in n_1 ways and the second object can then be chosen in n_2 ways, then both objects can be chosen in $n_1 \cdot n_2$ ways. This idea extends to choosing any number of objects. For example, the number of ways to choose five objects is: $n_1 \cdot n_2 \cdot n_3 \cdot n_4 \cdot n_5$. Examples 1 and 2 can be used to show that the fundamental counting principle is correct. In Example 1, the first letter can be chosen in three ways, which then leaves two choices for the second letter, leaving only one choice left for the third letter. By the fundamental counting principle, there are $3 \cdot 2 \cdot 1$, or 6 ways to choose the three letters. Similarly, the number of ways to choose a permutation of four letters is $4 \cdot 3 \cdot 2 \cdot 1$, or 24. The product $4 \cdot 3 \cdot 2 \cdot 1$ has a special name. It is four factorial and its symbol is 4!.

In general, if n is a whole number, then ***n*-factorial** ($n!$) is the product of the number n down to one. For example $7! = 7 \cdot 6 \cdot 5 \cdot 4 \cdot 3 \cdot 2 \cdot 1$, or $7! = 5040$. The fundamental counting principle gives the following important fact: The number of permutations of n objects is $n!$. This is true because the number of ways to choose the first object is n, the number of ways to choose the second object is one less than n, or $n - 1$ and so on, making the total number $n(n-1)(n-2)\ldots1$. This is just another way to write $n!$.

Example 3 In how many different ways can 10 people stand in line?

$10! = 10 \cdot 9 \cdot 8 \cdot 7 \cdot 6 \cdot 5 \cdot 4 \cdot 3 \cdot 2 \cdot 1$
$= 3,628,800$

Step 1: The answer is the number of permutations of 10 objects, or 10!

Answer: 3,628,800 different ways

Example 4 How many four-letter combinations are possible from a 26-letter alphabet?

$26 \cdot 26 \cdot 26 \cdot 26 = 456,976.$

Step 1: By the fundamental counting principle, the first letter can be chosen in 26 ways. Since letters can be repeated in a combination, the second letter can also be chosen in 26 ways, and so on. The total number is then 456,976.

Answer: 456,976 four-letter combinations

Example 5 How many four-letter combinations are possible from a 26-letter alphabet if no letter is to be repeated?

$26 \cdot 25 \cdot 24 \cdot 23 = 358,800$

Step 1: By the fundamental counting principle, the first letter can be chosen in 26 ways. Since letters can be repeated in a combination, the second letter can also be chosen in 26 ways, and so on. The total number is then 358,800.

Answer: 358,800 four-letter combinations

The quantity in Example 5 can be referred to as the *number of permutations of 26 objects four at a time*. Its symbol is ${}_{26}P_4$. In general, the number of different ways to order r objects out of a set of n objects is called the **number of permutations** of n objects r at a time. Its symbol is ${}_nP_r$ and is given by the formula ${}_nP_r = \dfrac{n!}{(n-r)!}$.

Verify that the formula ${}_nP_r$ is correct by working the following example using the fundamental counting principle and then by evaluating the formula.

Example 6 In how many ways can gold, silver and bronze medals be awarded in a race with six runners?

$6 \cdot 5 \cdot 4 = 120$

Step 1: By the fundamental counting principle, gold can be awarded in six ways. There are then five ways to award silver, followed by four ways to award bronze. The total number of ways is 120.

$$\begin{aligned} {}_6P_3 &= \frac{6!}{(6-3)!} \\ &= \frac{6!}{3!} \\ &= \frac{6 \cdot 5 \cdot 4 \cdot 3 \cdot 2 \cdot 1}{3 \cdot 2 \cdot 1} \\ &= \frac{720}{6} \\ &= 120 \end{aligned}$$

Step 2: Use the formula ${}_nP_r$ to find the number of permutations of six objects three at a time, or ${}_6P_3$.

Answer: 120 ways.

Example 7 In how many ways can seven people sit around a round table?

$6 \cdot 5 \cdot 4 \cdot 3 \cdot 2 \cdot 1 = 720$

Step 1: To see the answer, you need to imagine that you are one of the seven people. From your seat, you would be able to see the other six people seated in 6! ways, because the order could be any permutation of six people. If you chose to sit in another seat, even though people are seated in different positions, you would see the same 6! arrangements of the other people. So the total number of orders seven people can sit around a round table is 6!, or 720.

Answer: 720 ways

The problem in Example 7 leads to the idea of a **circular permutation**. The number of circular permutations of *n* objects is the number of different ways that the objects can be ordered around a circle. It is given by: The number of circular permutations of *n* objects is $(n-1)!$. The next example introduces a different type of problem.

Example 8 How many committees of three people can be formed from a pool of five people?

a b c	a b d	a b e	a c d	a c e
a d e	b c d	b c e	b d e	c d e

Step 1: Let the people be designated by *a, b, c, d,* and *e.* List the possible committees and count them. You will find 10 committees.

Answer: 10 committees

Example 8 is not a permutation problem, because different permutations lead to the same committee. For example, *a b c* and *b a c* are two different permutations, but they are the same committee. The quantity in Example 8 is called the number of *combinations of five objects three at a time*. Its symbol is ${}_5C_3$. In general ${}_nC_r$ is the number of combinations of *n* objects *r* at a time. You can get a formula for ${}_nC_r$ by looking at how it is related to ${}_nP_r$.
The number of permutations of *r* objects is r!. This means that ${}_nP_r$ counts the same committee *r*! times. Therefore, to get ${}_nC_r$, divide ${}_nP_r$ by *r*! as shown:

$${}_nC_r = \frac{{}_nP_r}{r!}$$

Another way to write this formula is:

$$ {}_nC_r = \frac{{}_nP_r}{r!} = \frac{n!}{(n-r)!} \cdot \frac{1}{r!} = \frac{n!}{r!(n-r)!} $$

Example 9 A poker hand consists of five cards drawn from a 52-card deck. How many different poker hands are there?

$$ {}_{52}C_5 = \frac{52 \cdot 51 \cdot 50 \cdot 49 \cdot 48}{5 \cdot 4 \cdot 3 \cdot 2 \cdot 1} = 2,598,960 $$

Step 1: In counting problems, you must decide if you are dealing with a combination or a permutation problem. Since the order in which the cards are drawn is irrelevant, the number of different poker hands is ${}_{52}C_5$. It is best to evaluate this using ${}_nC_r = \frac{{}_nP_r}{r!}$

Answer: 2,598,960 different poker hands

Most calculators have keys for evaluating ${}_nC_r$ *and* ${}_nP_r$. Refer to your calculator's instruction manual.

Problem Set

1. How many permutations are there for the letters W, S, and Y?

2. In how many ways can 9 people stand in line?

3. In how many ways can 7 people stand in line?

4. In how many ways can 5 people stand in line?

5. In how many ways can 3 people stand in line?

6. In how many ways can 8 people stand in line?

7. How many permutations are there for the letters F, G, and H?

8. How many permutations are there for the letters P, and O?

9. How many permutations are there for the letters A, E, I, O, and U?

10. How many permutations are there for the letters M, A, T, and H?

11. Calculate: ${}_{12}P_2$

12. Calculate: ${}_9P_1$

13. There are five candidates for a job and you have to select a first, second, and third choice from among the candidates. In how many ways can you do this?

14. There are three candidates for a job and you have to select a first, second, and third choice from among the candidates. In how many ways can you do this?

15. How many 2 letter words can be formed from a 28 letter alphabet if no letter is repeated?

16. How many 3 letter words can be formed from a 34 letter alphabet if no letter is repeated?

17. There are eight candidates for a job and you have to select a first, second, and third choice from among the candidates. In how many ways can you do this?

18. How many 6 member committees can you form from a pool of 8 people?

19. How many different 3-player teams can be formed from 10 people?

20. How many ways can 8 lucky charms be evenly arranged around a bracelet?

HA1-885: Histograms and the Normal Distribution

Representing data graphically is an important tool for understanding information that can be obtained from the data. One tool for representing data is called a **histogram**.

Histogram	A bar chart for quantitative variables in which the bars are depicted as touching each other.

A histogram is constructed from a frequency distribution table. A frequency distribution table depicts the frequency (number of times) with which a value occurs in a given interval. The intervals are placed on the horizontal axis and the values are placed on the vertical axis. The height of each bar illustrates how frequently a value occurs. You can see this in the example below.

1. A store owner wants to know more about the ways customers spend money in her store. She records all of the transaction totals for a single day.

Amount Spent	Frequency
$0.01 to $1.00	5
$1.01 to $2.00	8
$2.01 to $3.00	6
$3.01 to $4.00	12
$4.01 to $5.00	7
$5.01 to $6.00	5
$6.01 to $7.00	2
$7.01 to $8.00	0
$8.01 to $9.00	1
$9.01 to $10.00	1

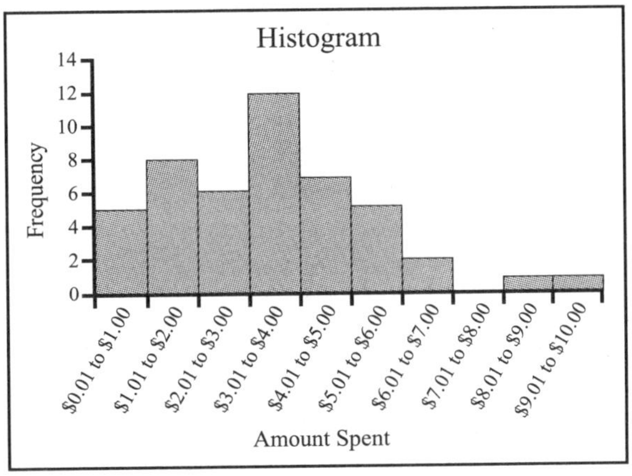

Since we do not have the original data (we only have data organized into continuous categories) it is not possible to compute the **measures of central tendency** (mean, median, and mode) directly.

Measure of Central Tendency	A number that represents the center of a set of data

There are several measures of central tendency for a histogram. They are called the **mean of a histogram**, the **modal class**, the **mode of a histogram**, and the **median of a histogram**.

Mean of a Histogram	The weighted average of the midpoints of the intervals of a histogram. It approximates the mean of the data.

To calculate the mean of a histogram:

Step 1: Find the midpoint for each interval to use as the data values.

$$\text{midpoint} = \frac{\text{low value} + \text{high value}}{2}$$

Step 2: Find the frequency of each interval.

Step 3: Find the sum of the frequencies.

Step 4: Find the product of the midpoint and frequency.

Step 5: Find the sum of the products.

Step 6: The mean is the sum of the products divided by the sum of the frequencies.

Modal Class	The interval with the largest frequency. On a histogram, this is the tallest bar.
Mode of a Histogram	The midpoint of the modal class
Median of a Histogram	The middle value of the data set

To find the median of a histogram, list all the midpoints of the intervals of the histogram. Use these as the data values to find the median.

Types of Histograms

The distribution of the data in a frequency table specifies the shape of the histogram. We can classify histograms by their basic shapes. The common shapes have special names: uniform (or rectangular), U-shaped, bell-shaped, and skewed.

- **Uniform Distribution**: Almost all intervals have the same frequency, and there is no mode.

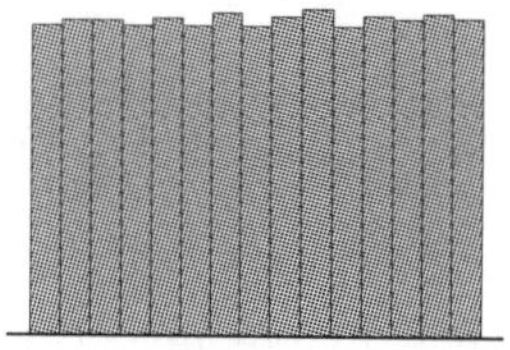

- **U-Shaped Distribution**: more frequencies at the ends and fewer in the middle, the mean and median are located near the center, and the mode is at the ends.

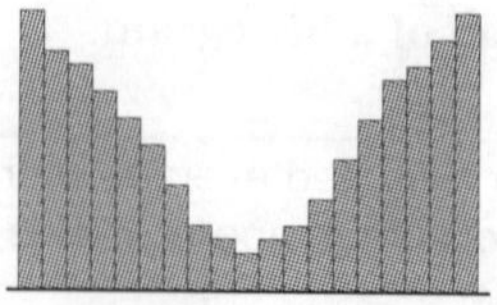

- **Bell-Shaped Distribution:** more frequencies in the middle and fewer on the ends; the mean, median, and mode are approximately equal.

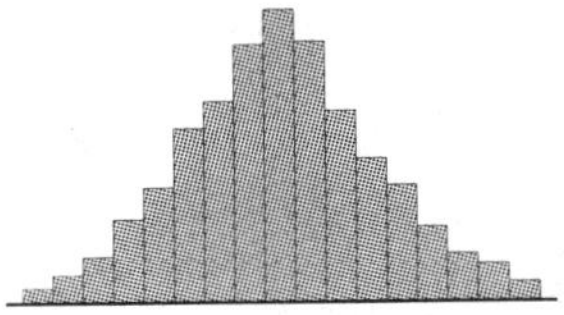

- **Skewed Distribution**: frequencies spread out in one direction from a central cluster.

Skewed Right

(the mean is located to the right of the median)

Skewed Left

(the mean is located to the left of the median)

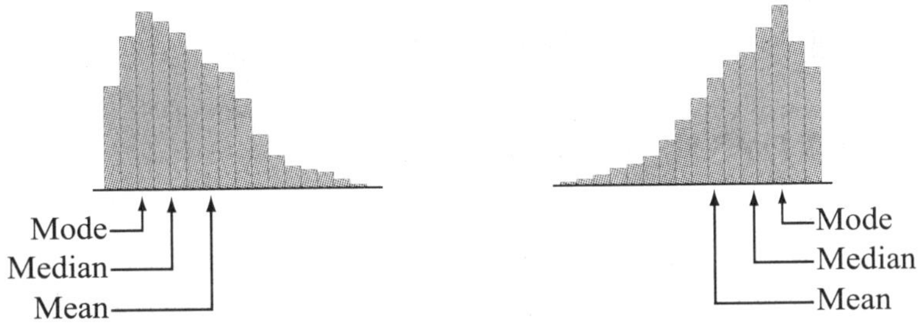

There is an ideal bell-shaped distribution that quite commonly appears in statistics. It is called the **normal distribution**. It has many uses and some very definite qualities. A normal distribution is a continuous, smooth, bell-shaped curve. It is symmetrical with respect to its mean value. Its mean, median, and mode are all equal. For values farther and farther from the mean, the curve is closer and closer to the horizontal axis.

- **Normal distribution**: Normal distribution is the ideal bell-shaped distribution.

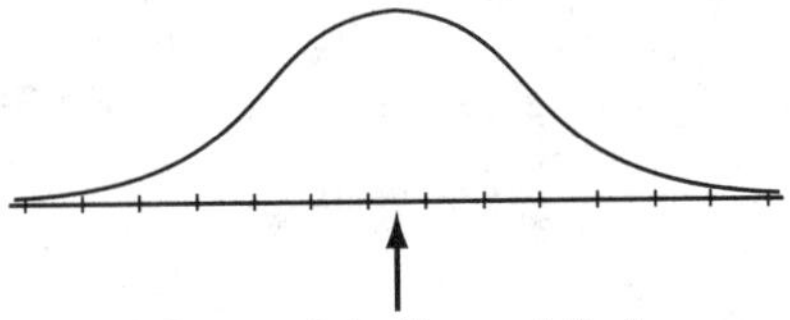

Example 1 Consider the histogram shown below. Classify the shape of the histogram as Uniform, U-shaped, Bell-shaped, Skewed Right, or Skewed Left.

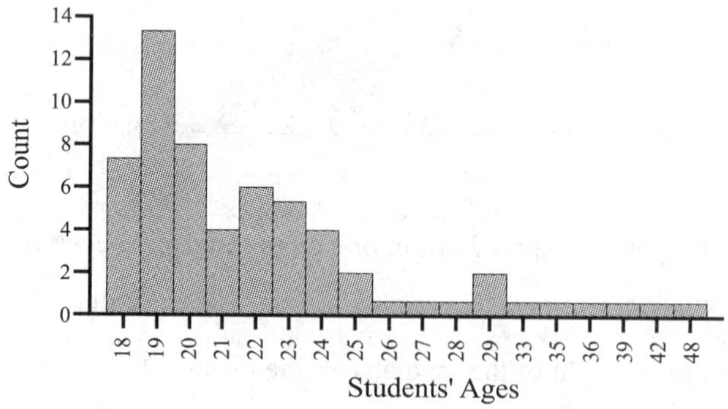

Step 1: The data "spreads out" from the center toward the right, so the distribution is skewed right.

Answer: This is a Skewed Right distribution.

Example 2 Consider the histogram shown below. What is the modal class of the histogram?

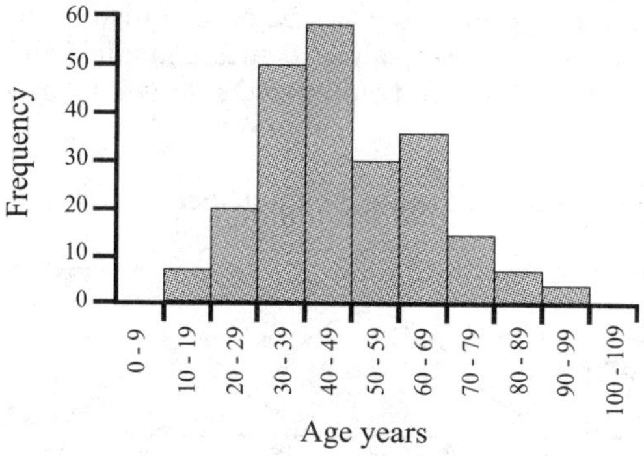

The interval with the largest frequency is 40 – 49.

Step 1: Determine the interval with the largest frequency. In this case, the interval is represented by the tallest bar, which has a frequency of 60.

Answer: The modal class is 40 - 49.

Example 3 Use the histogram shown below to create a frequency table for the data represented.

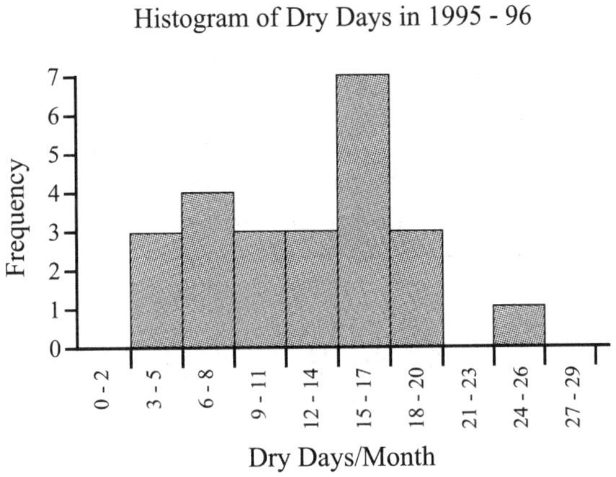

Step 1: The intervals along the bottom of the histogram form the first column in the table.

Interval	Frequency
0 – 2	
3 – 5	
6 – 8	
9 – 11	
12 – 14	
15 – 17	
18 – 20	
21 – 23	
24 – 26	
27 – 29	

Step 2: The heights of the bars along the vertical axis show the frequency for each interval.

Interval	Frequency
0 – 2	0
3 – 5	3
6 – 8	4
9 – 11	3
12 – 14	3
15 – 17	7
18 – 20	3
21 – 23	0
24 – 26	1
27 – 29	0

Answer: A frequency table for this histogram is given by:

Interval	Frequency
0 – 2	0
3 – 5	3
6 – 8	4
9 – 11	3
12 – 14	3
15 – 17	7
18 – 20	3
21 – 23	0
24 – 26	1
27 – 29	0

Example 4 Consider the histogram shown below. What is the median of the histogram?

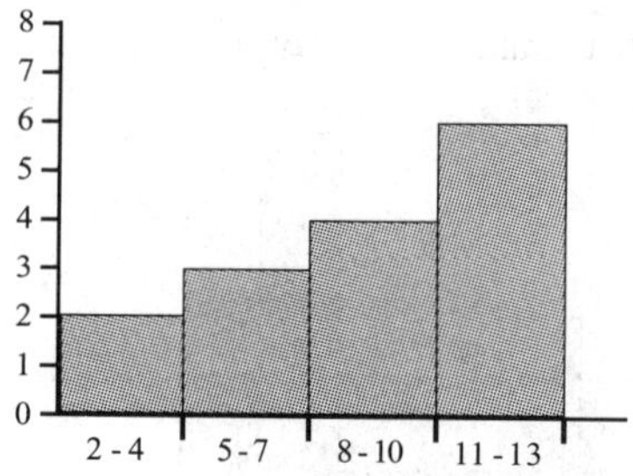

Interval	Frequency
2 – 4	2
5 – 7	3
8 – 10	4
11– 13	6
	15

Step 1: Create a frequency distribution table for the data.

$$\frac{2+4}{2} = \frac{6}{2} = 3$$

$$\frac{5+7}{2} = \frac{12}{2} = 6$$

$$\frac{8+10}{2} = \frac{18}{2} = 9$$

$$\frac{11+13}{2} = \frac{24}{2} = 12$$

Step 2: Find the midpoints for each interval and place them in the table.

Interval	Midpoint	Frequency
2 – 4	3	2
5 – 7	6	3
8 – 10	9	4
11– 13	12	6
		15

3, 3, 6, 6, 6, 9, 9, 9, 9, 12, 12, 12, 12, 12, 12

Step 3: Since the midpoints form the values in the data set, list them in order to find the middle value.

3, 3, 6, 6, 6, 9, 9, **9**, 9, 12, 12, 12, 12, 12, 12
The middle value is 9.

Step 4: Find the middle value.

Answer: The median of the histogram is 9.

Example 5 What is the mean of the histogram shown in Example 3?

Step 1: Since we have the frequency table, we first compute the midpoints of each of the intervals, and place them in the table.

$$\frac{0+2}{2} = 1 \qquad \frac{15+17}{2} = \frac{32}{2} = 16$$

$$\frac{3+5}{2} = \frac{8}{2} = 4 \qquad \frac{18+20}{2} = \frac{38}{2} = 19$$

$$\frac{6+8}{2} = \frac{14}{2} = 7 \qquad \frac{21+23}{2} = \frac{44}{2} = 22$$

$$\frac{9+11}{2} = \frac{20}{2} = 10 \qquad \frac{24+26}{2} = \frac{50}{2} = 25$$

$$\frac{12+14}{2} = \frac{26}{2} = 13 \qquad \frac{27+29}{2} = \frac{56}{2} = 28$$

Interval	Midpoint	Frequency
0 – 2	1	0
3 – 5	4	3
6 – 8	7	4
9 – 11	10	3
12 – 14	13	3
15 – 17	16	7
18 – 20	19	3
21 – 23	22	0
24 – 26	25	1
27 – 29	28	0

Step 2: Next we compute the total of all the frequencies and place them in the table.

$$0+3+4+3+3+7+3+0+1+0 = 24$$

Interval	Midpoint	Frequency
0 – 2	1	0
3 – 5	4	3
6 – 8	7	4
9 – 11	10	3
12 – 14	13	3
15 – 17	16	7
18 – 20	19	3
21 – 23	22	0
24 – 26	25	1
27 – 29	28	0
		24

Step 3: We then multiply the midpoint times the frequency for each interval, and find the sum of the products.

Interval	Midpoint	Frequency	Product
0 – 2	1	0	0
3 – 5	4	3	12
6 – 8	7	4	28
9 – 11	10	3	30
12 – 14	13	3	39
15 – 17	16	7	112
18 – 20	19	3	57
21 – 23	22	0	0
24 – 26	25	1	25
27 – 29	28	0	0
		24	303

Step 4: Finally, divide the sum of the products by the total frequency.

$$\frac{303}{24} = 12.625$$

Answer: The mean of the histogram is 12.625.

Problem Set

1. Classify the following histogram by its shape:

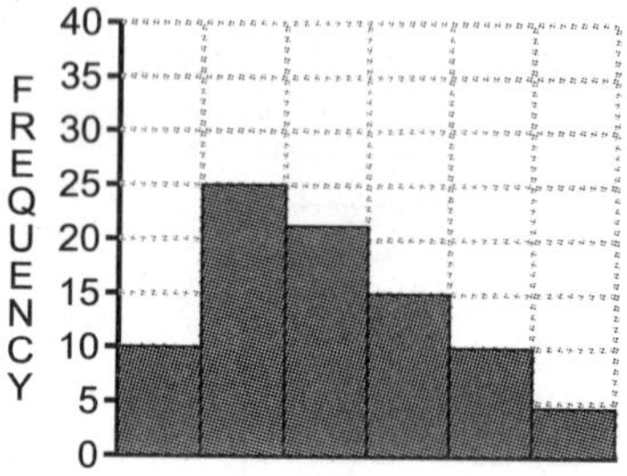

2. Classify the following histogram by its shape:

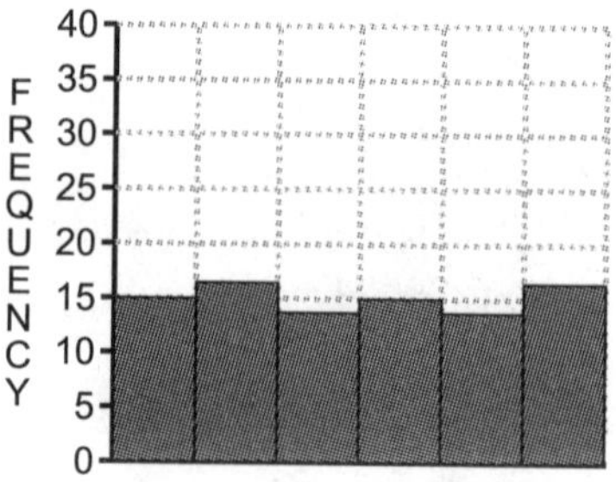

3. Classify the following histogram by its shape:

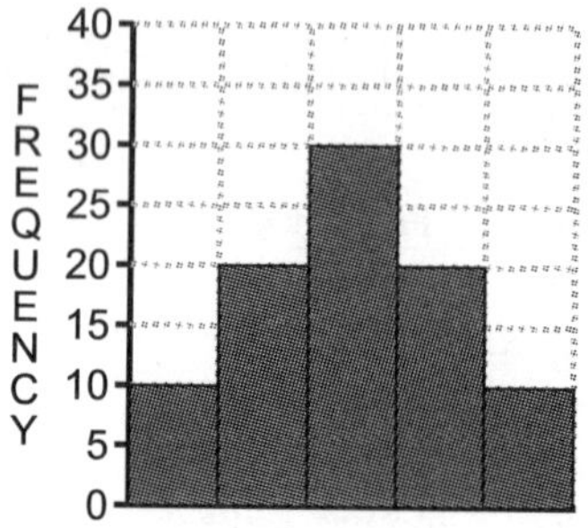

4. Classify the following histogram by its shape:

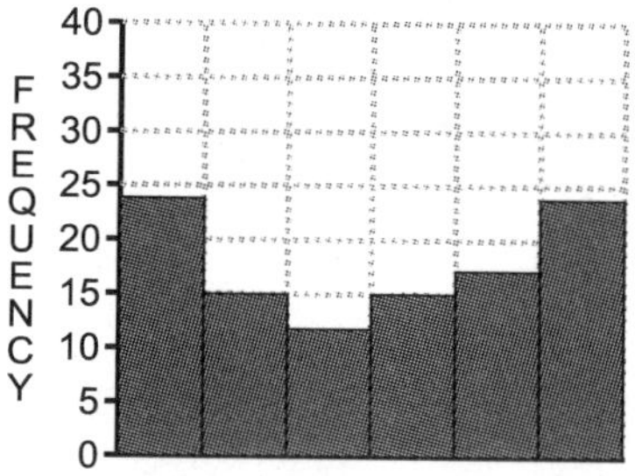

5. Classify the following histogram by its shape:

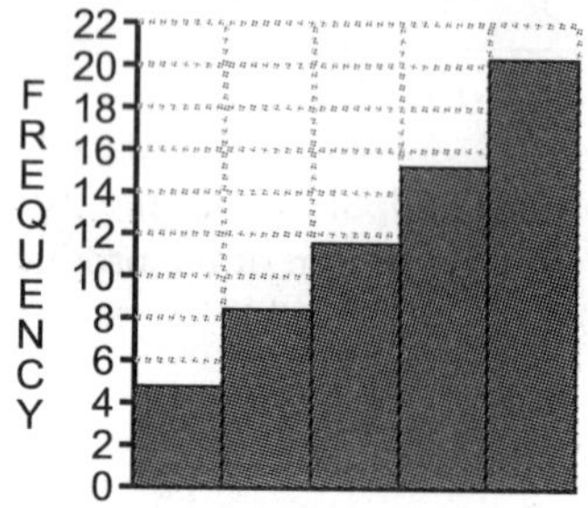

6. Classify the following histogram by its shape:

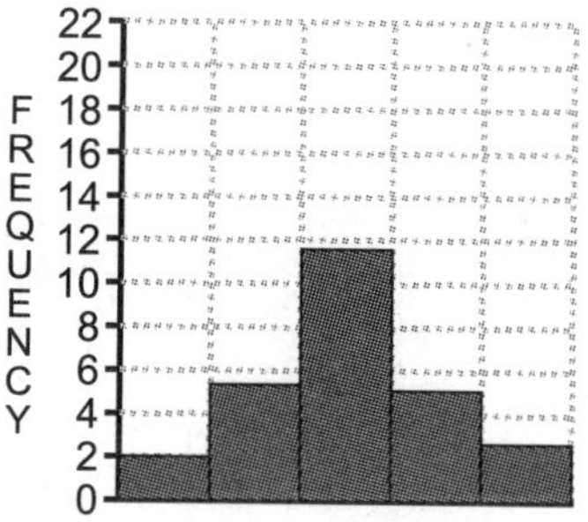

7. Classify the following histogram by its shape:

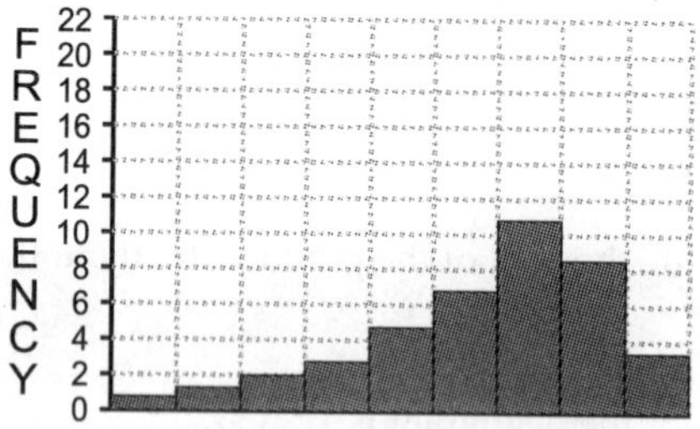

8. Which weight category includes the largest number of dogs?

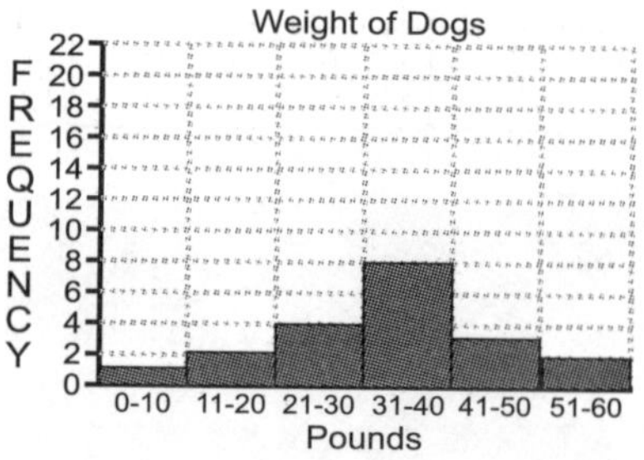

9. A survey was conducted to determine how many hours of TV students watch per week. Which range contains the largest number of students?

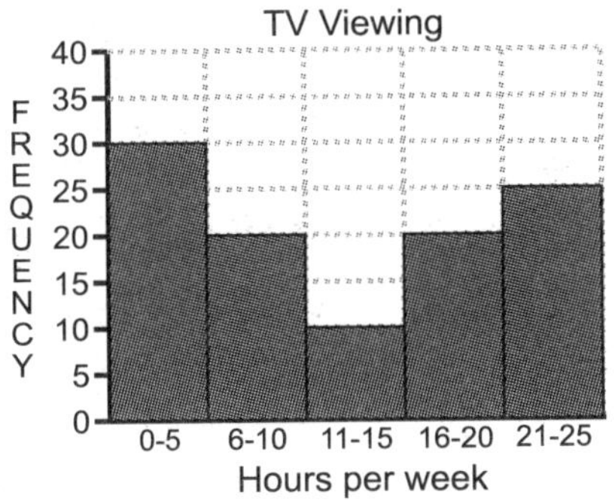

10. What range of grade averages includes the largest number of science students?

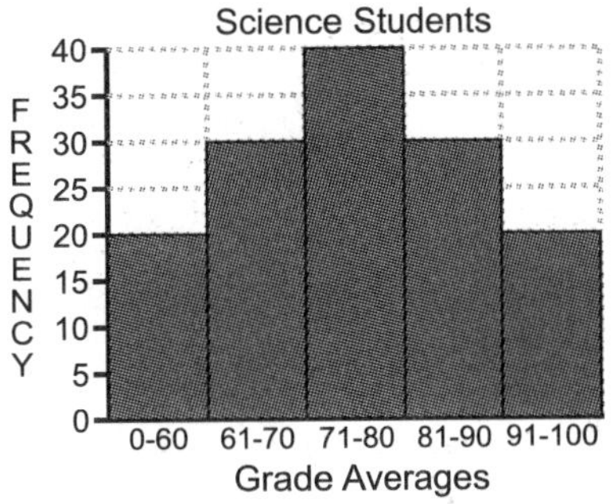

11. A company conducted a survey to determine the number of miles its employees traveled to get to work. Which mileage range represents the largest number of employees?

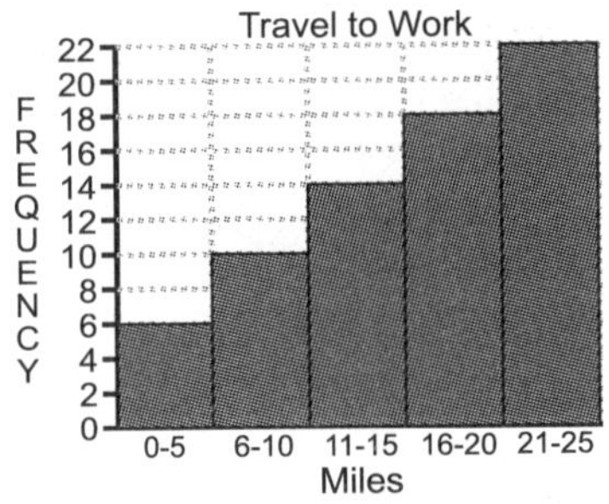

12. A school conducted a survey to determine how many minutes students study per day. Which range represents the study time of the largest number of students?

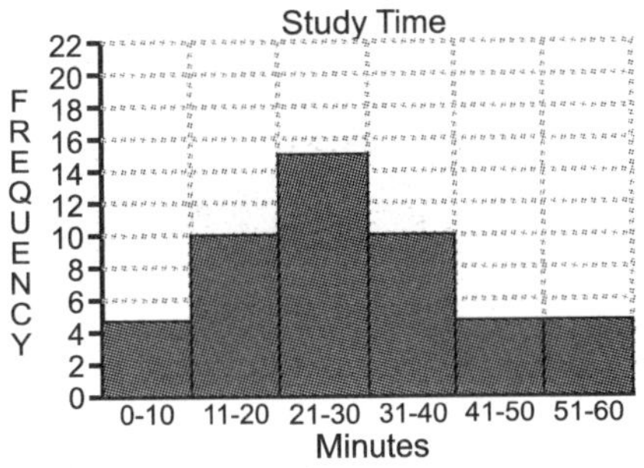

13. A restaurant conducted a survey to determine how long customers at a restaurant had to wait to be seated. What length of time represents the wait time of the largest number of customers?

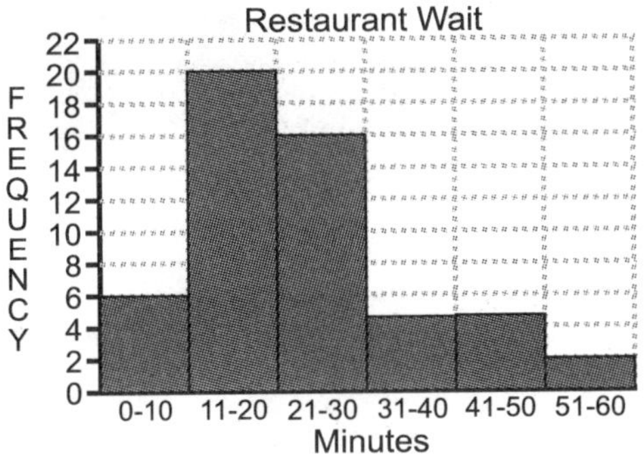

14. Use the histogram to find the mode of the grouped data.

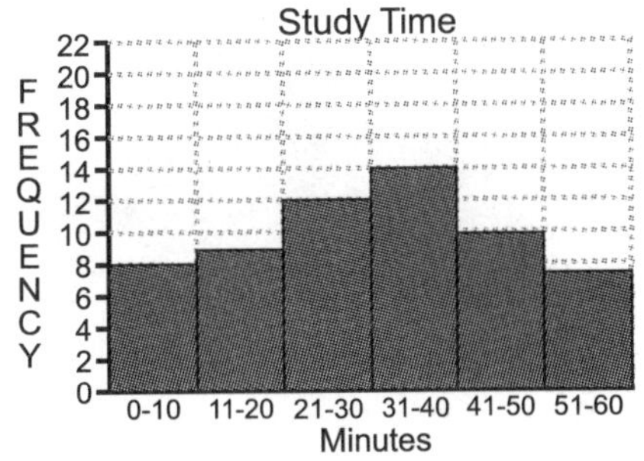

15. Use the histogram to find the mode of the grouped data.

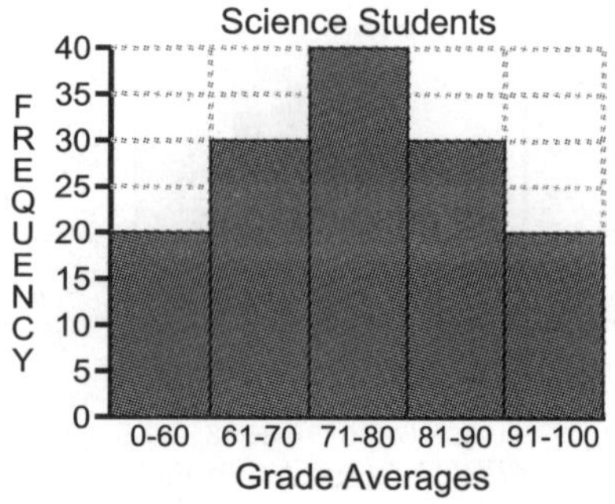

16. Use the histogram to find the mode of the grouped data.

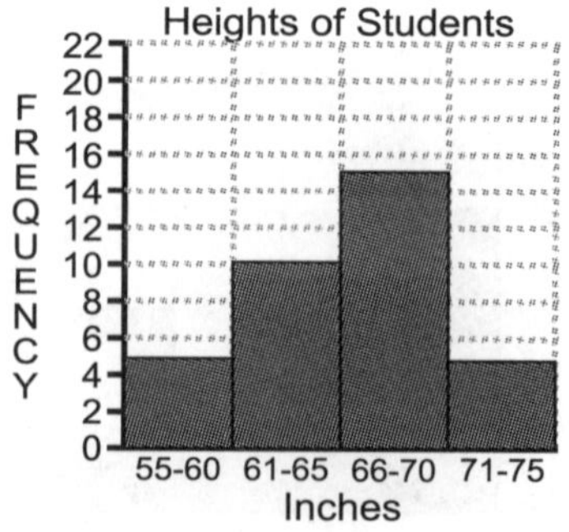

17. Use the histogram to find the mode of the grouped data.

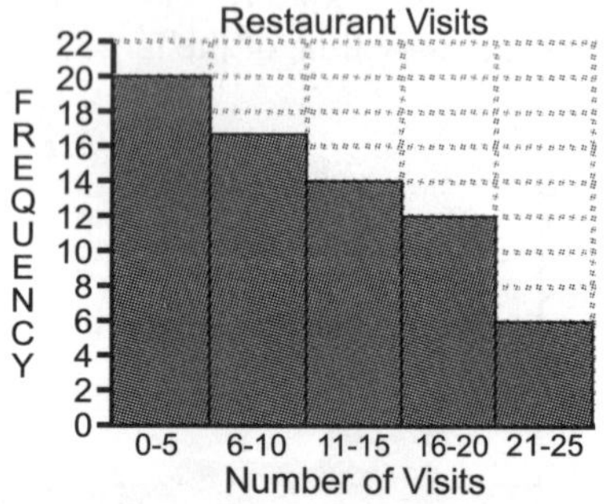

18. Use the histogram to find the mode of the grouped data.

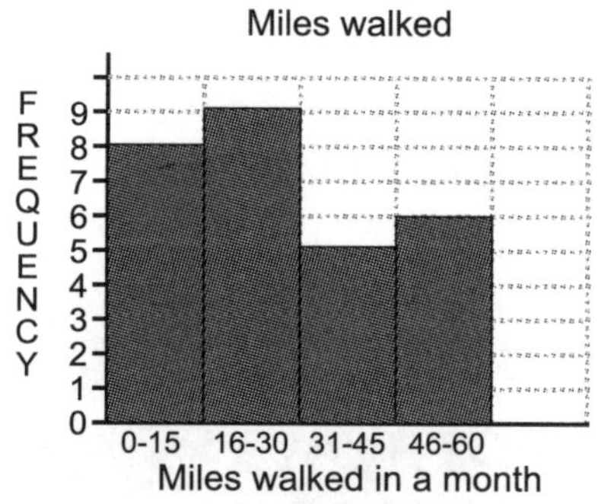

19. Use the histogram to find the mode of the grouped data.

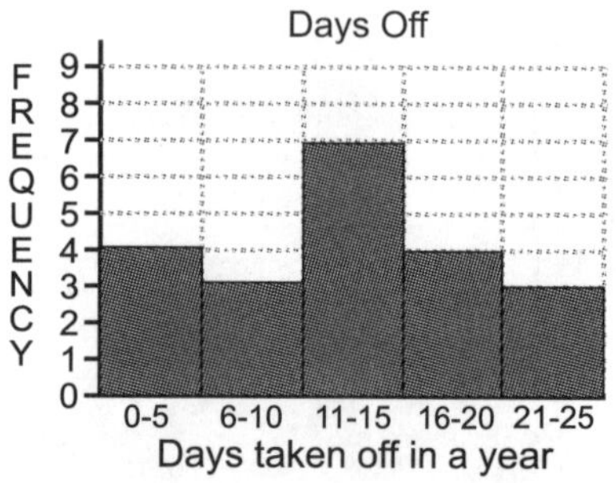

20. Use the histogram to find the mode of the grouped data.

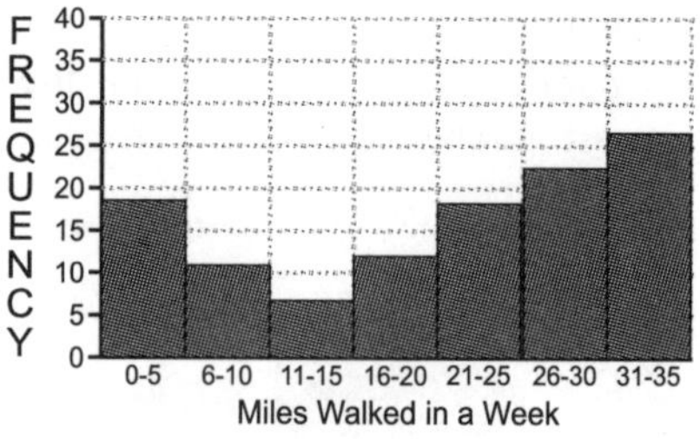

21. Draw a histogram using the following data: 1, 0, 19, 13, 12, 6, 7, 8, 17, 8, 4, 3, 1.

22. Draw a histogram using the following data: 35, 22, 5, 9, 19, 30, 40, 45, 6, 18, 29, 26, 38, 42, 12, 13.

23. Draw a histogram using the following data: 11, 20, 82, 12, 6, 8, 43, 74, 72, 22, 39, 90, 63, 88, 98, 36, 84.

24. Draw a histogram using the following data: 15, 18, 41, 50, 13, 22, 43, 55, 2, 24, 33, 59, 8, 6, 5, 3, 3, 28, 29, 30, 17, 19, 25, 35, 38, 52, 51, 55.

25. Draw a histogram using the following data: 8, 15, 22, 38, 42, 7, 19, 28, 37, 48, 5, 20, 27, 39, 36, 49, 41, 43, 43, 16, 33, 34, 26, 25.

26. Draw a histogram using the following data: 33, 19, 19, 16, 4, 2, 7, 9, 12, 13, 14, 13, 15, 17, 24, 25, 1, 8, 8, 18, 19.

27. Draw a histogram using the following data:
5, 22, 42, 17, 19, 31, 33, 94, 62, 70, 80, 10, 28, 50, 15, 30, 56, 60.

28. Create a frequency table to match the histogram.

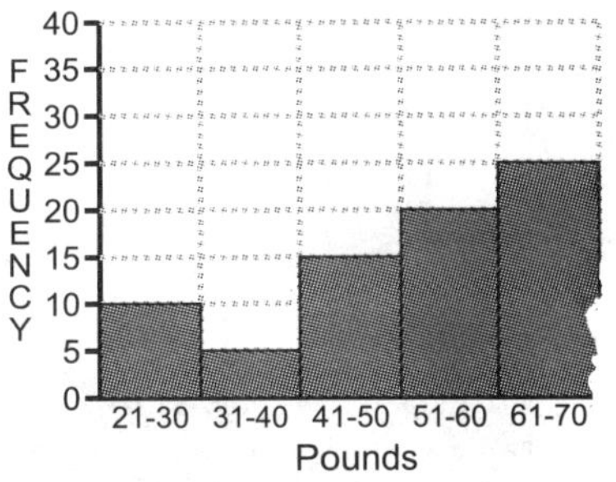

29. Create a frequency table to match the histogram.

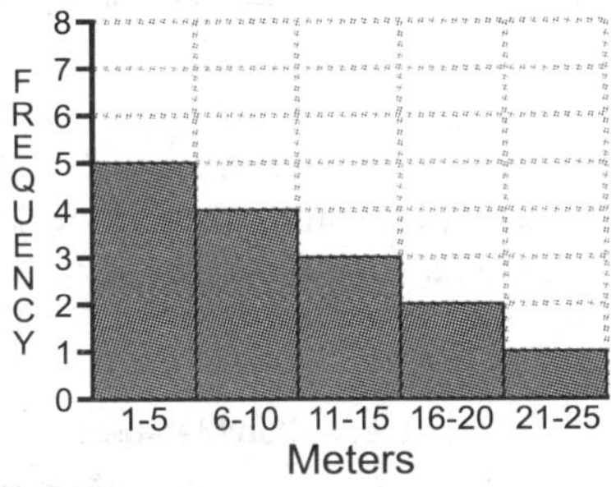

30. Create a frequency table to match the histogram.

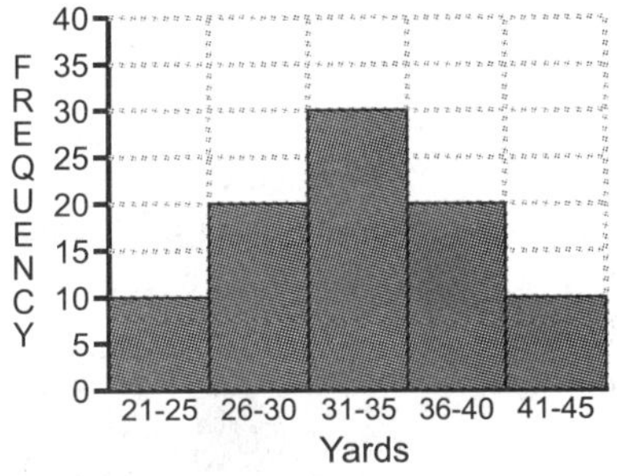

31. Create a frequency table to match the histogram.

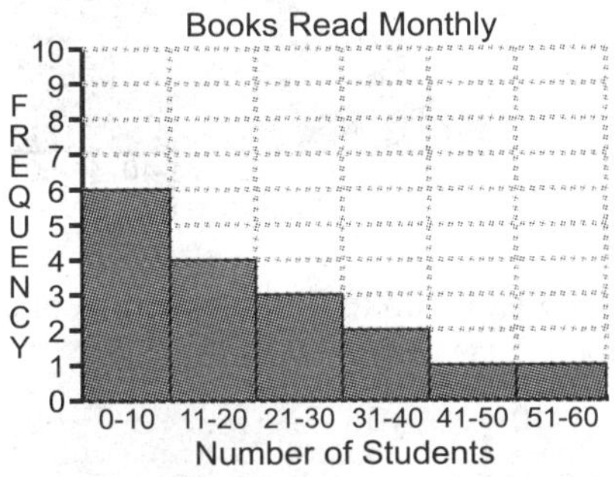

32. Create a frequency table to match the histogram.

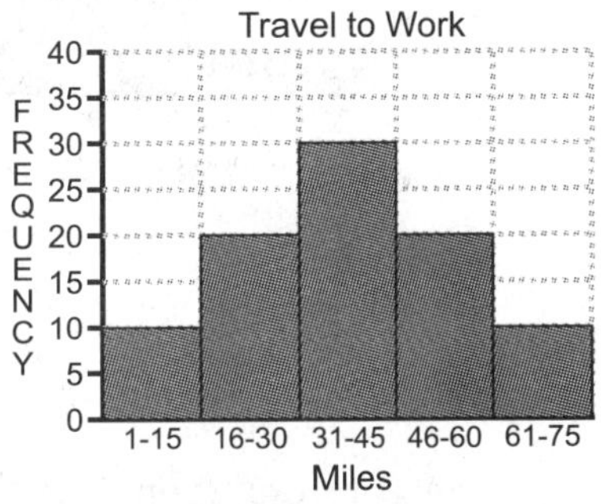

33. Create a frequency table to match the histogram.

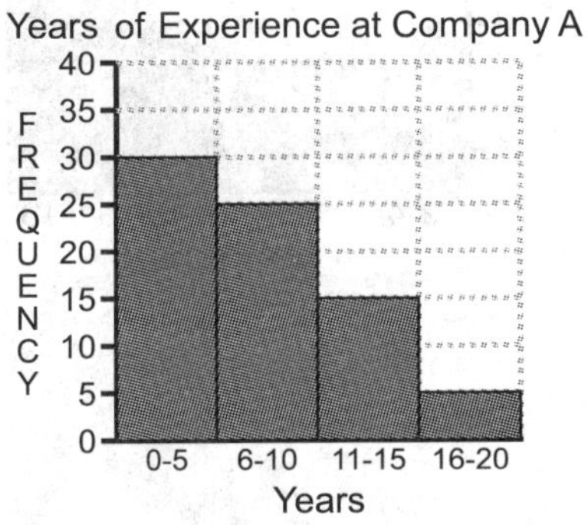

34. Create a frequency table to match the histogram.

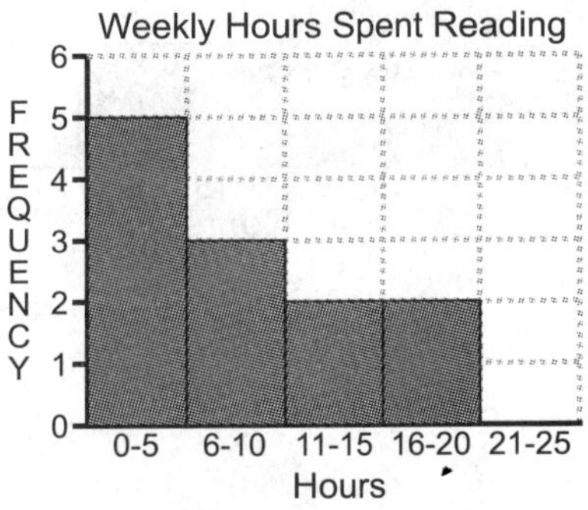

35. Create a frequency table to match the histogram.

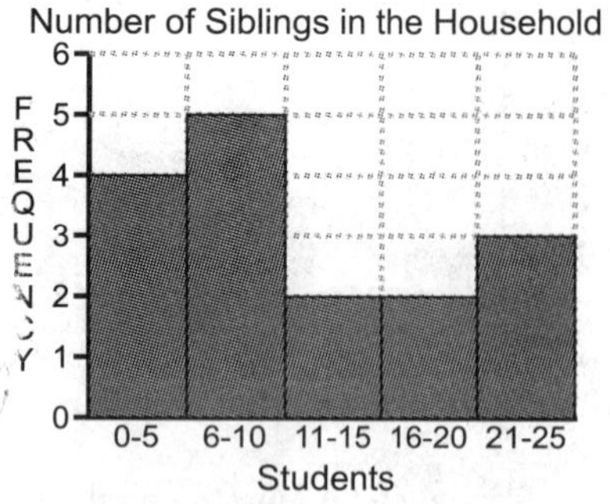

36. The histogram represents the number of movies rented by students in a certain year. Determine the order of the mean, median, and mode based on the shape of the histogram.

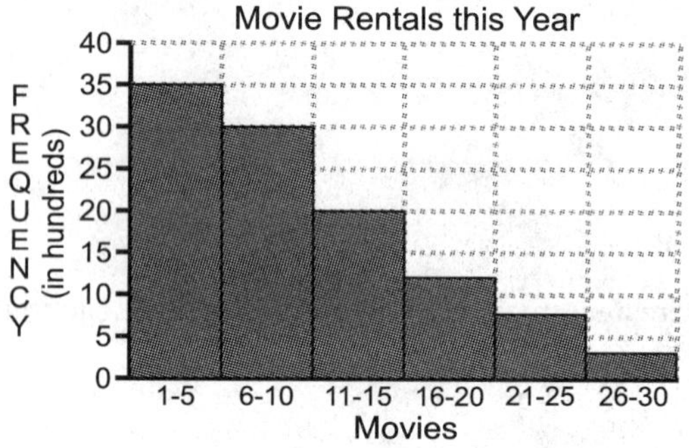

37. The histogram represents the test grades for science students. Determine the order of the mean, median and mode based on the shape of the histogram.

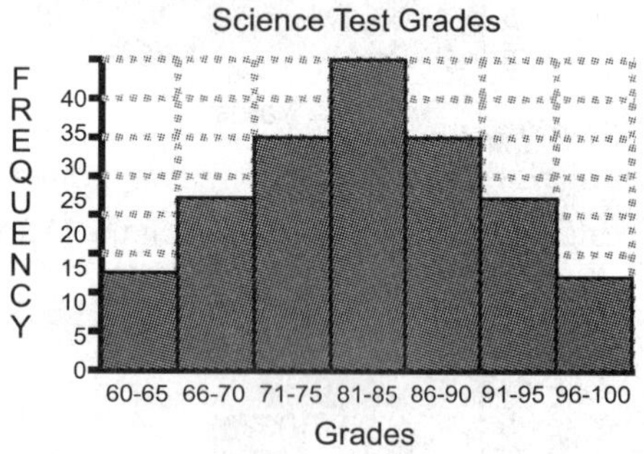

38. The histogram represents miles walked by fitness club members per month. Determine the order of the mean, median and mode based on the shape of the histogram.

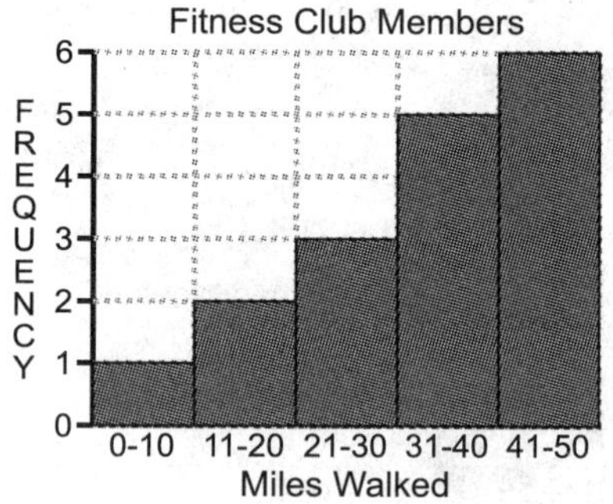

39. Use the histogram to find the mean of the grouped data.

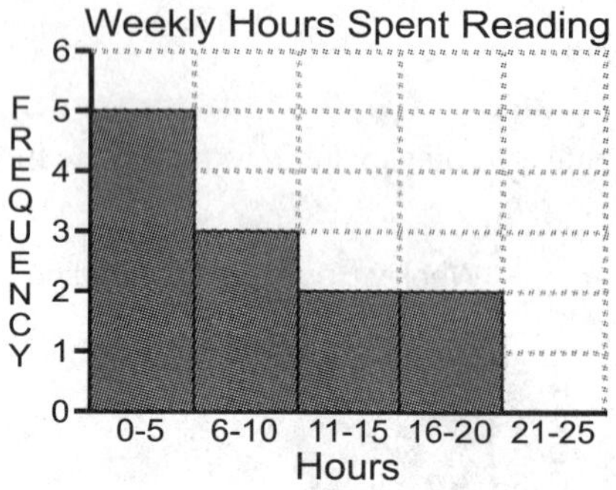

40. Use the histogram to find the median of the grouped data.

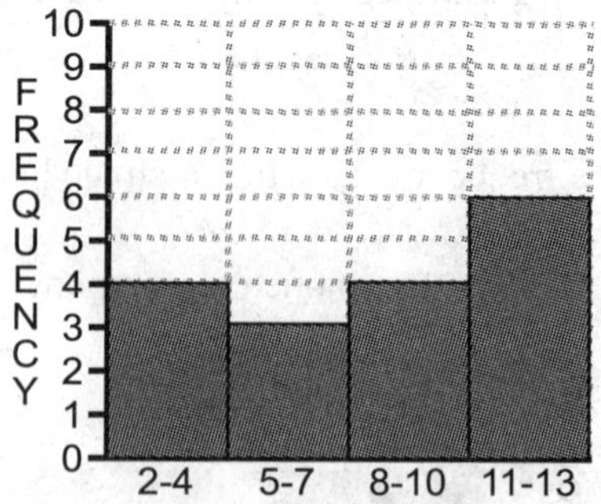

HA1-886: Unions and Intersections of Sets Using Venn Diagrams

In this lesson, we will examine unions and intersections of sets. If A and B are two sets, we will define the union of the two sets (denoted by $A \cup B$) as "the set of all elements that are contained in one or both of the sets." The operation of union is "putting the sets together."

Union of Sets	The union of two or more sets of objects consists of all elements that are contained in each of the sets.

We define the intersection of the two sets (denoted by $A \cap B$) as "the set of all objects that are contained in both sets simultaneously." The operation of intersection is "finding the common elements."

Intersection of Sets	The intersection of two or more sets of objects is the set of elements that is contained in all of the sets simultaneously.

In many cases a **Venn Diagram** can be useful when computing the union and the intersection of sets. Let's take a specific example. Suppose that A = {n, u, m, b, e, r} and B = {p, r, i, m, e}. The union of these sets is the set of all items that are in one or the other set or both:

$A \cup B$ = {n, u, m, b, e, r, p, i}

The intersection of these sets is the set of items they have in common:

$A \cap B$ = {m, e, r}

Now we consider a Venn Diagram to represent these sets and their unions and intersections. In a Venn Diagram we draw the sets as circles inside a larger rectangle.

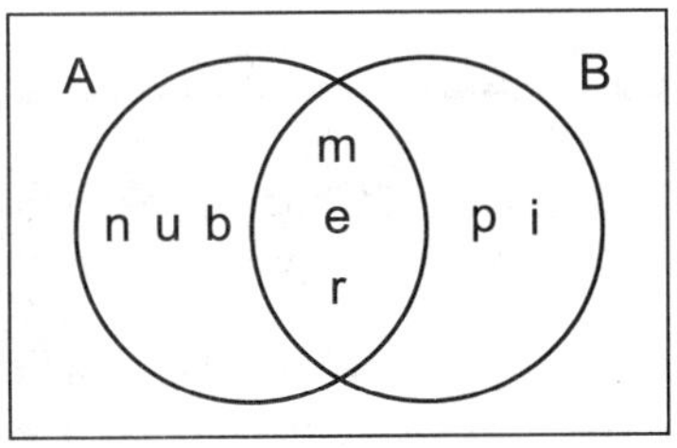

First notice that the only items inside the circle representing set A are the elements of set A. Similarly, the only items inside the circle representing set B are the elements of set B.

Notice that the union of A and B is the set of all items that appear in either of the circles representing sets A and B. The intersection of A and B is the part in the middle that is common to both sets.

The union can be represented by all points contained in the shaded region shown below.

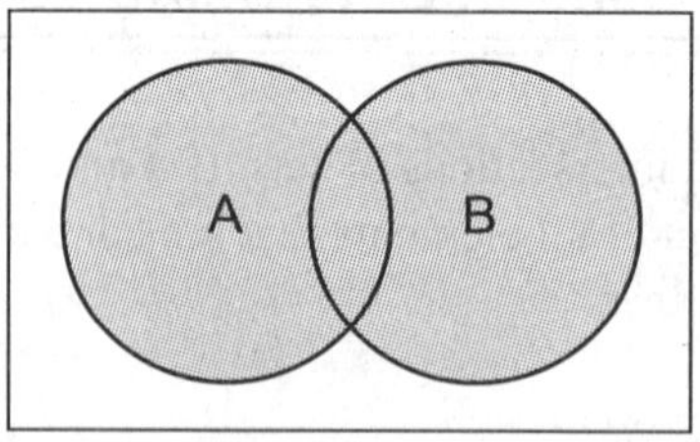

The intersection is the set of points in common, and in the picture it is represented by the shaded region shown below.

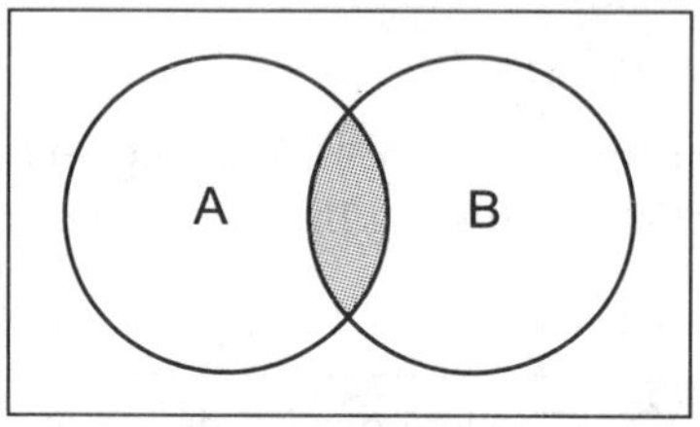

Example 1 Find $A \cap B$ in the Venn Diagram shown here:

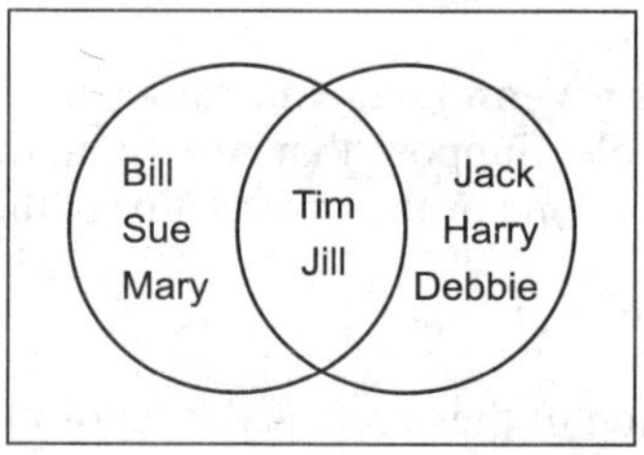

Step 1: The intersection is the set of items the two sets have in common. In this case, that is the set {Tim, Jill}.

Answer: $A \cap B = \{\text{Tim, Jill}\}$

Example 2 There are 28 students in a 9th grade Algebra class. Fourteen plan to study French, ten plan to study Spanish, and three plan to study both. Use a Venn Diagram to find out how many students plan to study either only French, only Spanish, or both?

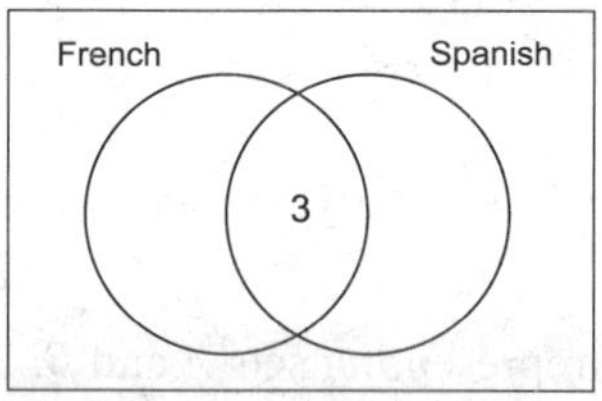

Step 1: Draw a Venn Diagram and label each set. Since we know three students plan to study both French and Spanish, we place 3 in the center.

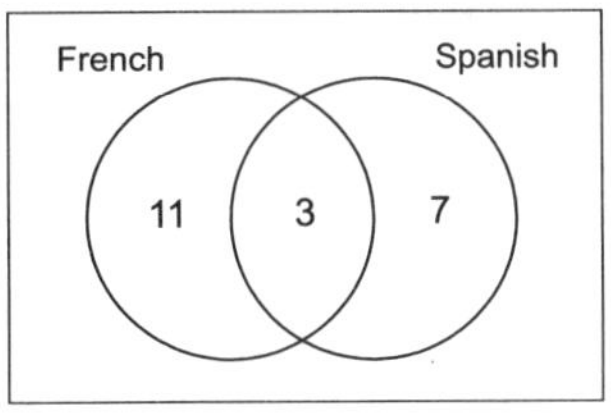

Step 2: We have accounted for 3 students who plan to take both. That leaves $14 - 3 = 11$ students who plan to study only French and $10 - 3 = 7$ students who plan to study only Spanish.

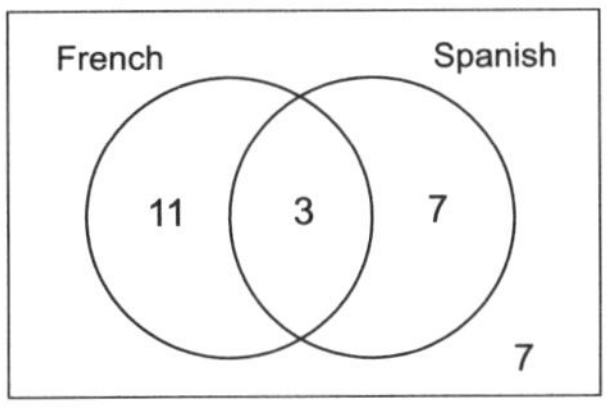

Step 3: We have accounted for $11 + 3 + 7 = 21$ students. That leaves $28 - 21 = 7$ students who do not plan to take French or Spanish.

Answer: A total of 21 students plan to study either French or Spanish or both.

Example 3 Consider the Venn Diagram shown indicating the number of elements in each region of the diagram. How many elements are there in $A \cup C$?

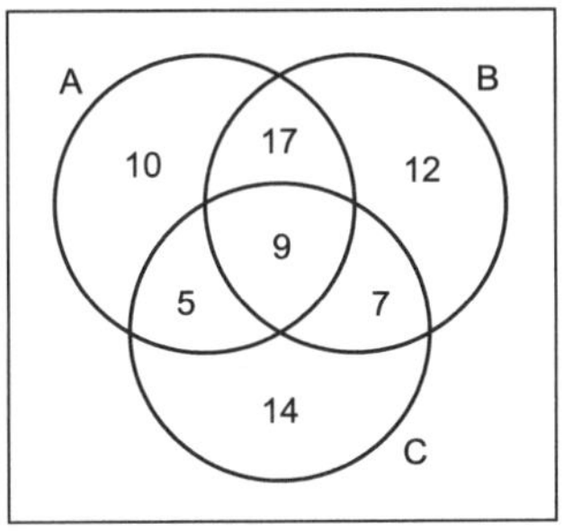

$$A \cup C = 10 + 17 + 5 + 9 + 7 + 14$$
$$= 62$$

Step 1: To find the number of elements of $A \cup C$, we add all of the regions that are in either A or C or both. (Be careful not to count an element twice.)

Answer: The number of elements in $A \cup C$ is 62.

Example 4 The sophomore class at a local high school has 284 students. In a recent survey, 189 said they like to go to the movies, and 112 said they like to roller skate. If 47 people are in both groups, complete a Venn Diagram to show how many people only like to go to the movies.

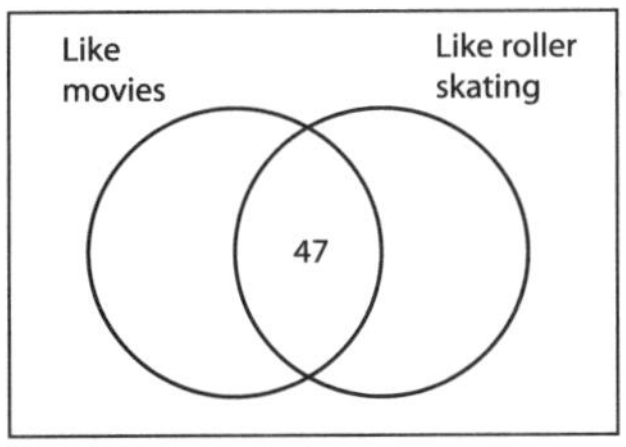

Step 1: Draw a Venn Diagram and label each set. Since we know 47 students are in both groups, we place 47 in the center.

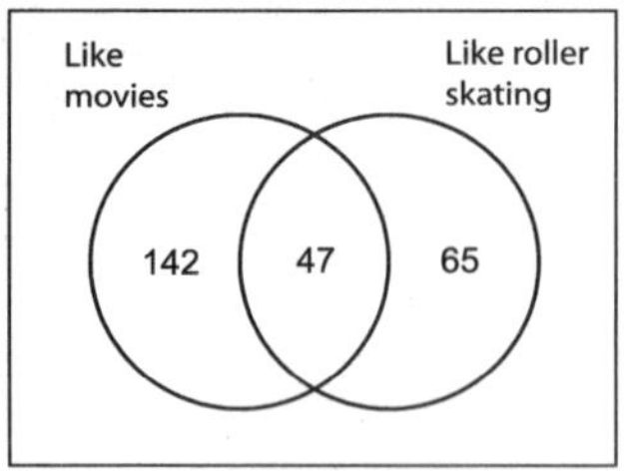

Step 2: We have accounted for 47 students who like both. That leaves $189 - 47 = 142$ students who only like to go to the movies and $112 - 47 = 65$ students who only like to go roller skating.

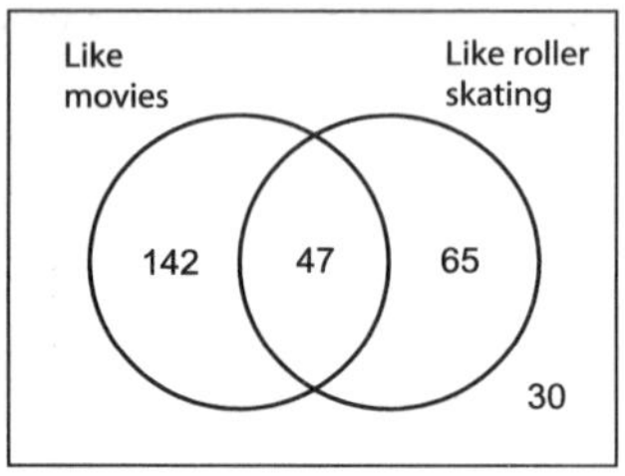

Step 3: We have accounted for $142 + 47 + 65 = 254$ students. That leaves $284 - 254 = 30$ students who like neither.

Answer: A total of 142 students only like to go to the movies.

Example 5 Four hundred twenty seniors at a local high school were surveyed about their plans after graduation, and the results are shown in the table. Use a Venn Diagram to find out how many students only plan to go to college.

Plan	Number
Go to college	304
Get a job	225
Get married	100
Job and college	145
College and married	70
Job and married	88
College, job, married	60

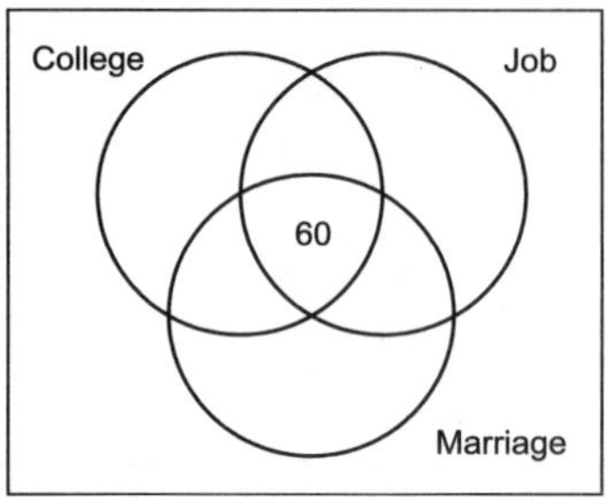

Step 1: Draw a Venn Diagram and label each set. Since we know 60 people will do all three things, we place 60 in the center.

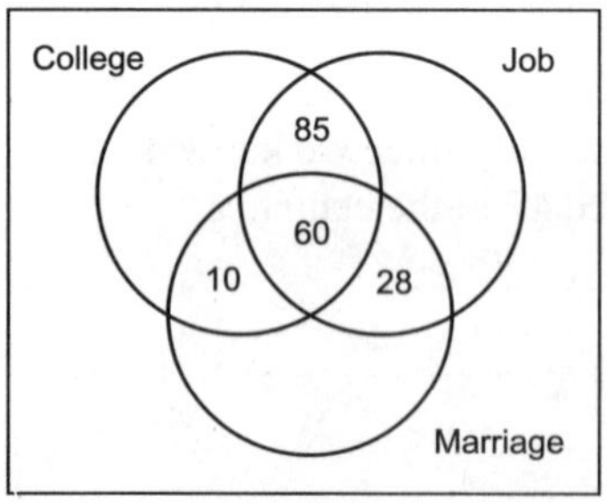

Step 2: We can see that there are $145 - 60 = 85$ people who will go to college and get a job. Likewise, there are $88 - 60 = 28$ people who will get a job and get married, and there are $70 - 60 = 10$ people who will go to college and get married.

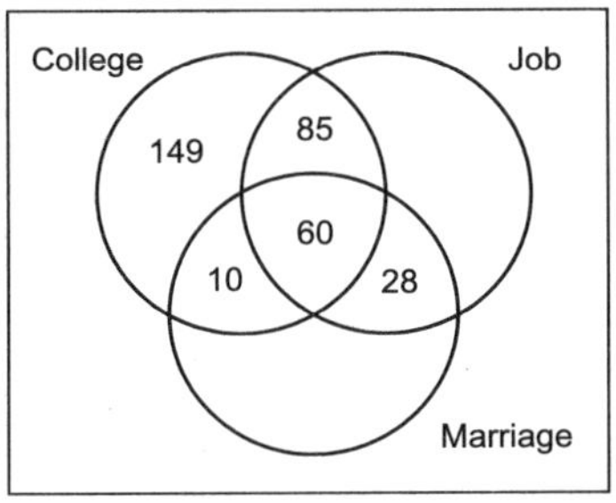

Step 3: So far $85 + 60 + 10 = 155$ people have been placed in the set of people going to college. Since we know from the table that a total of 304 people are going to college, that leaves $304 - 155 = 149$ people who only plan to go to college.

Answer: A total of 149 seniors only plan to go to college.

Problem Set

Solve:

1. Use the Venn Diagram to find the elements in the set $C \cap D$.

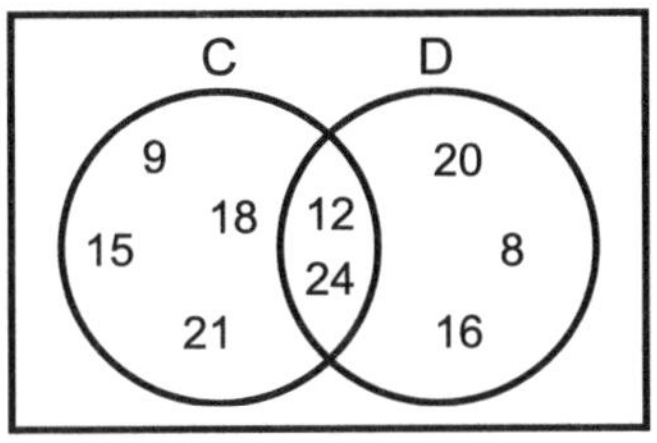

2. Use the Venn Diagram to find the elements in the set $F \cap G$.

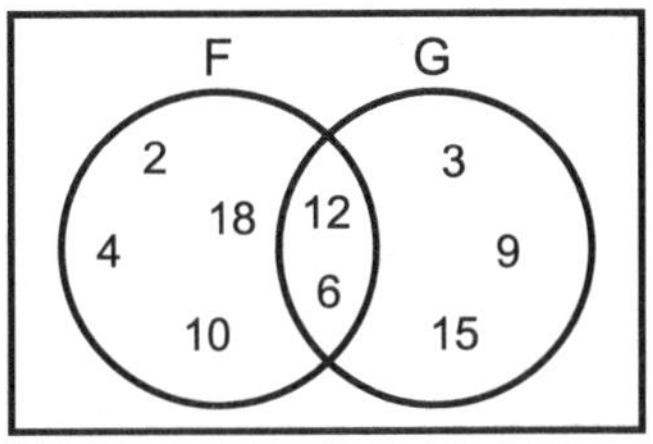

3. Use the Venn Diagram to find the elements in the set $M \cap N$.

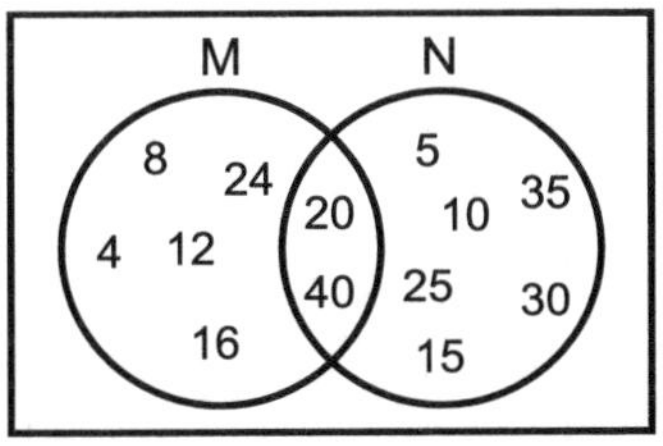

4. Use the Venn Diagram to find the elements in the set $A \cap B$.

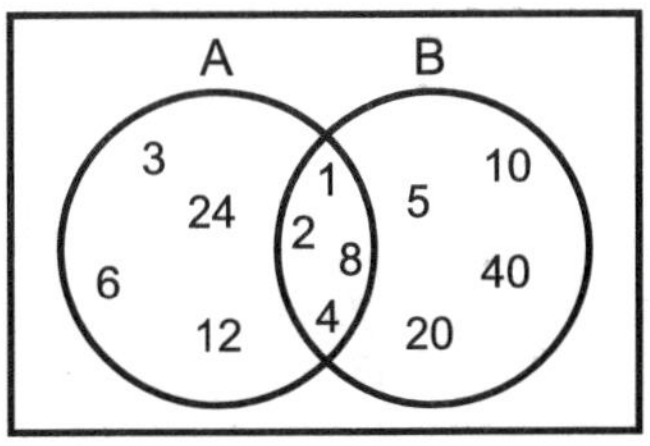

5. Use the Venn Diagram to find the elements in the set $X \cap Y$.

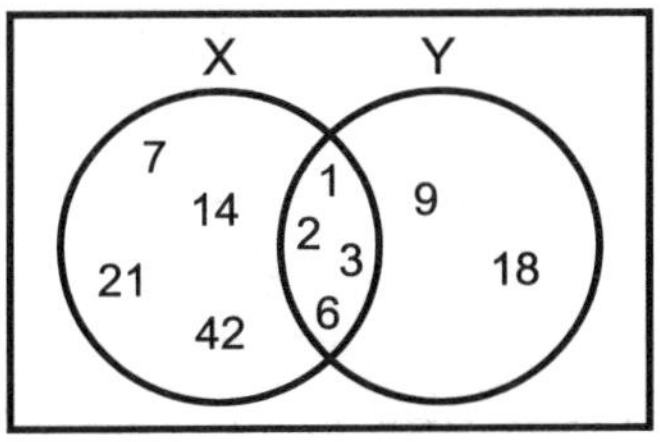

6. Use the Venn Diagram to find the elements in the set $H \cap J$.

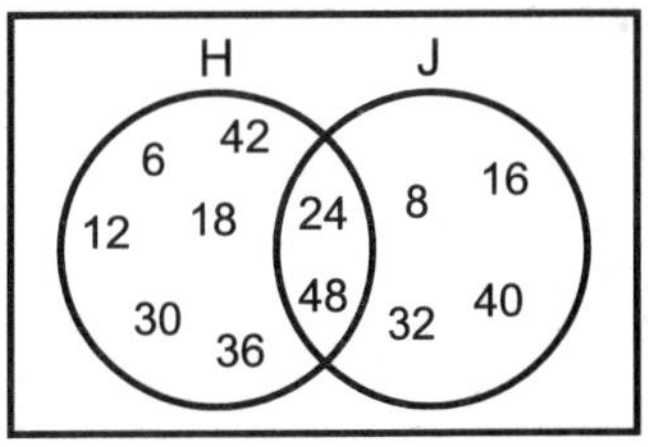

7. Use the Venn Diagram to find the elements in the set $B \cap C$.

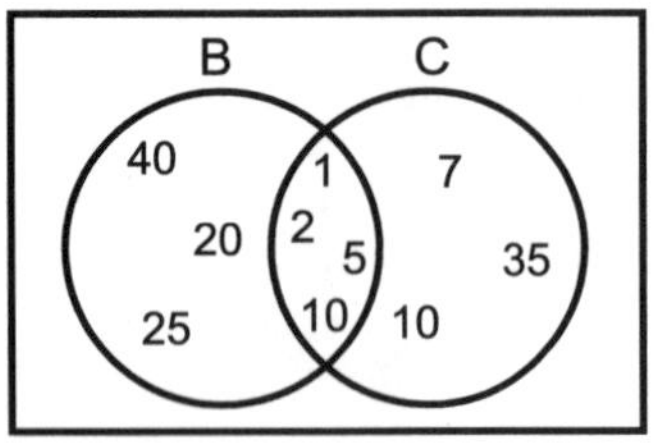

8. Use the Venn Diagram to find the elements in the set $F \cap G$.

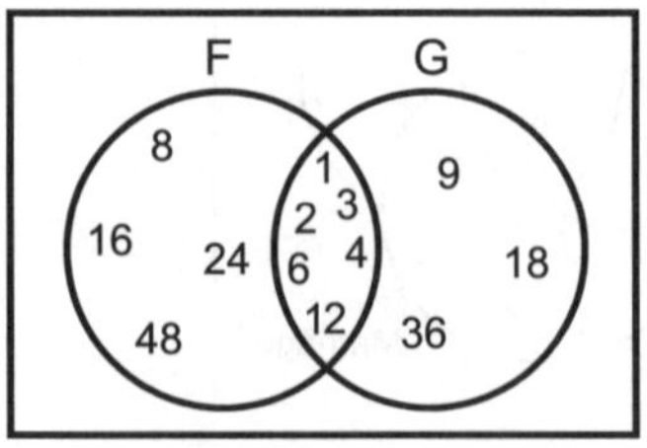

9. Use the Venn Diagram to find the elements in the set $X \cup Y$.

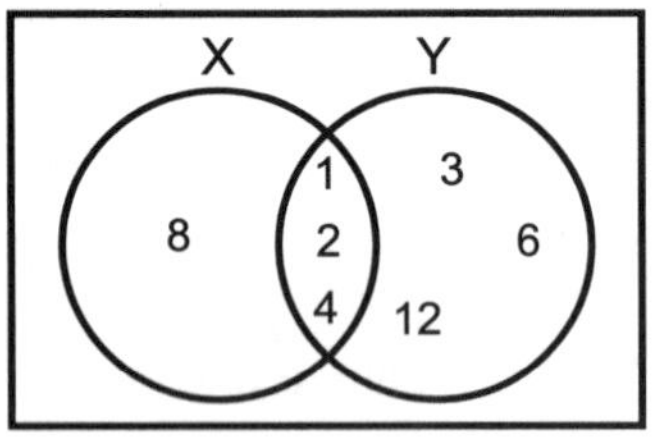

10. Use the Venn Diagram to find the elements in the set $C \cup D$.

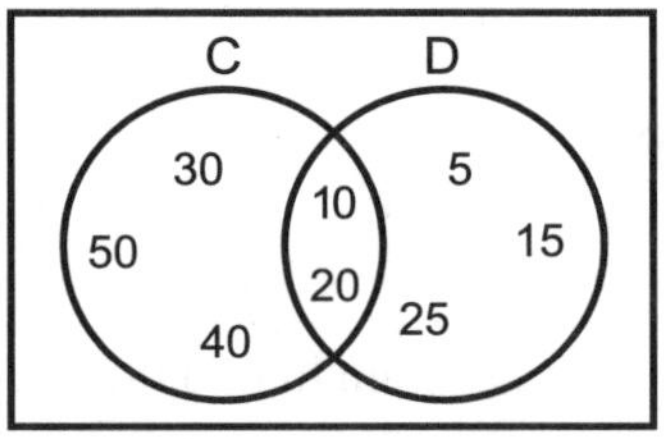

11. Use the Venn Diagram to find the elements in the set $E \cup F$.

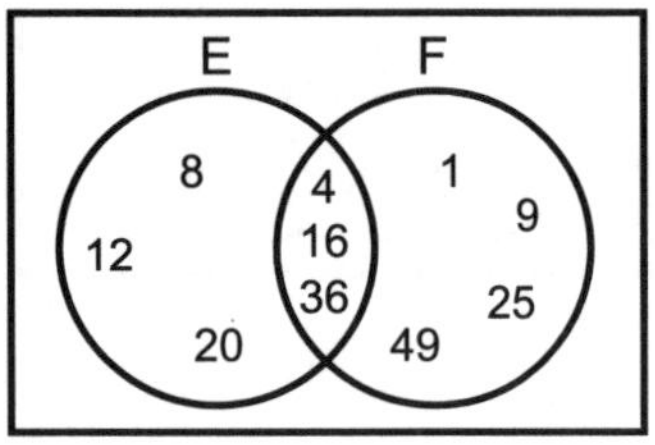

12. Use the Venn Diagram to find the elements in the set $P \cup Q$.

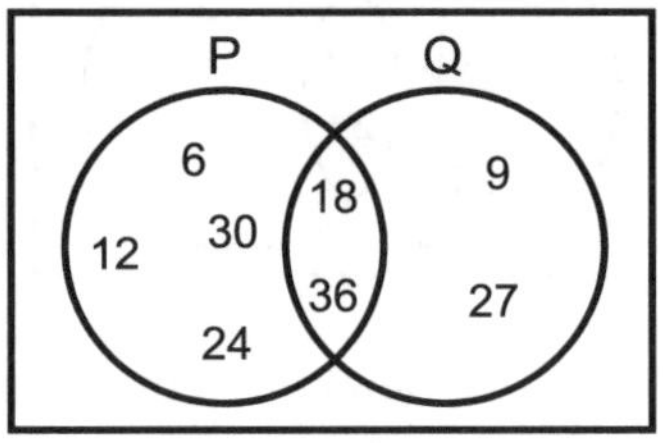

13. Use the Venn Diagram to find the elements in the set $S \cup T$.

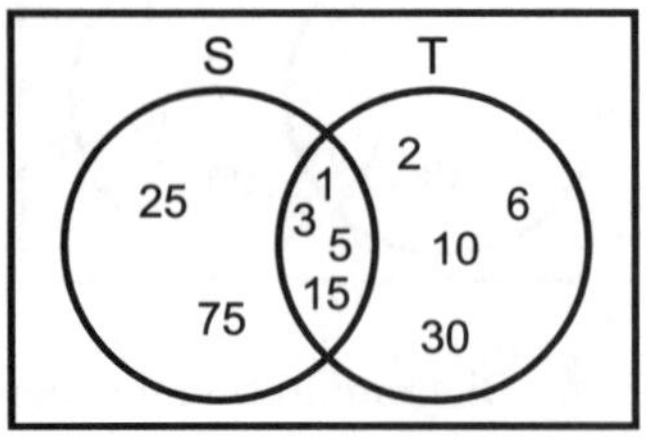

14. Use the Venn Diagram to find the elements in the set $X \cup Y$.

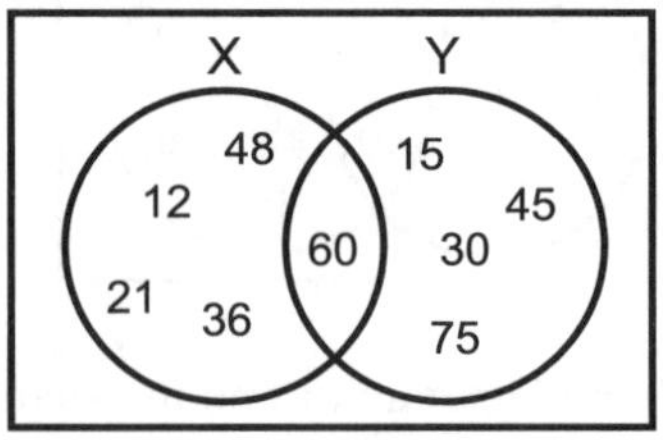

15. Use the Venn Diagram to find the elements in the set $E \cup F$.

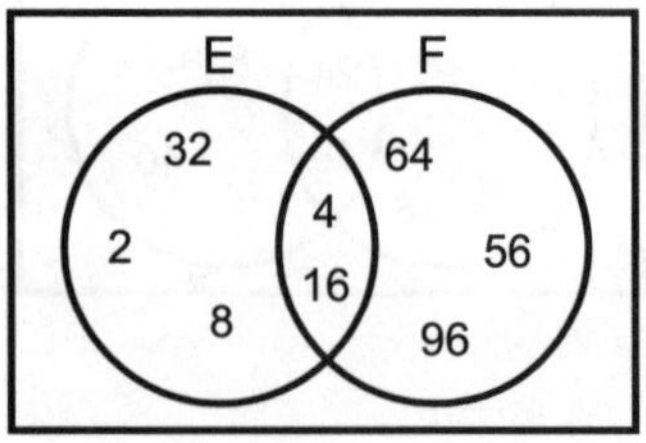

16. Find the number of elements in the set $A \cup B$.

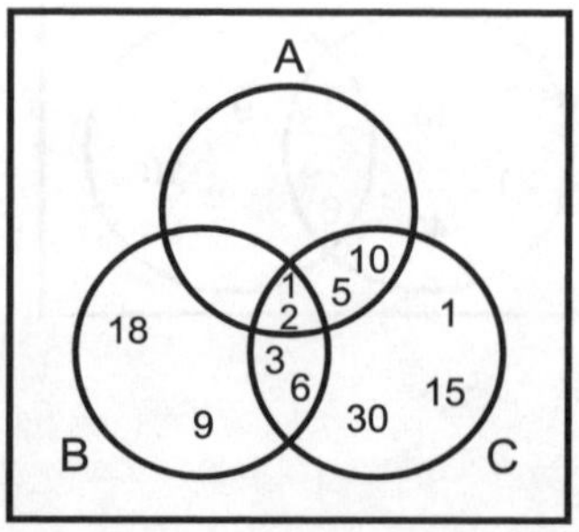

17. Find the number of elements in the set $A \cup C$.

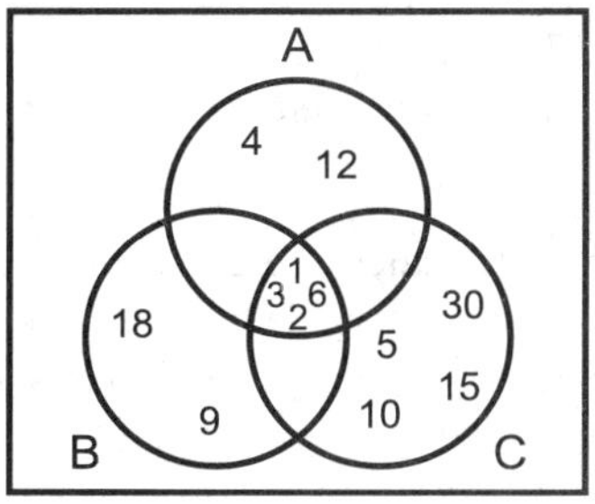

18. Find the number of elements in the set $B \cup C$.

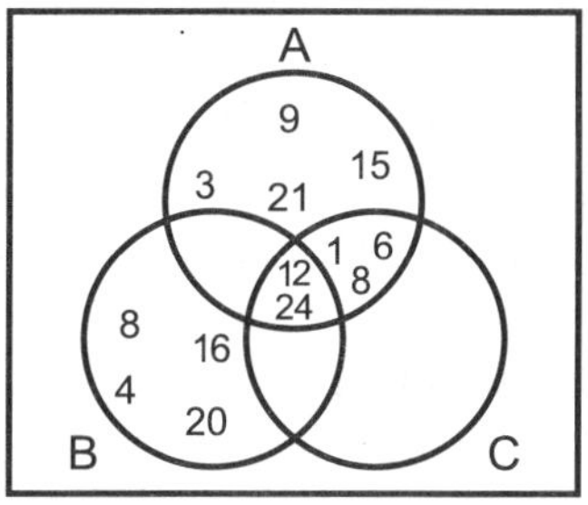

19. Find the number of elements in the set $X \cup Z$.

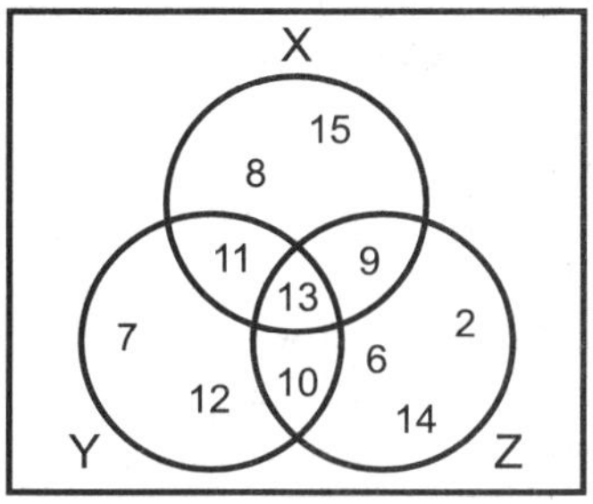

20. Find the number of elements in the set $C \cap D$.

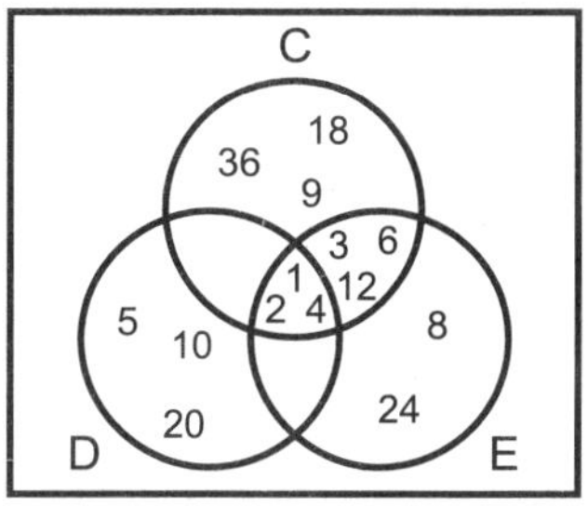

21. Find the number of elements in the set $X \cap Y$.

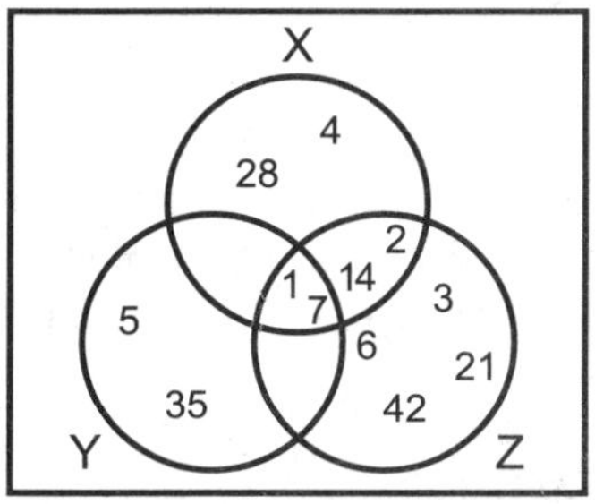

22. Find the number of elements in the set $A \cap B$.

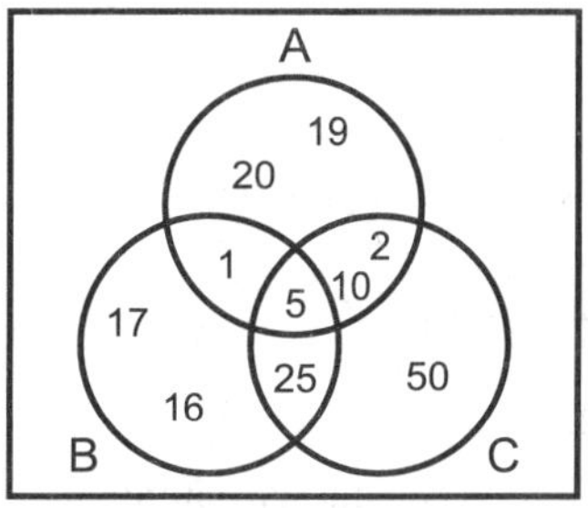

23. Find the number of elements in the set $Q \cap P$.

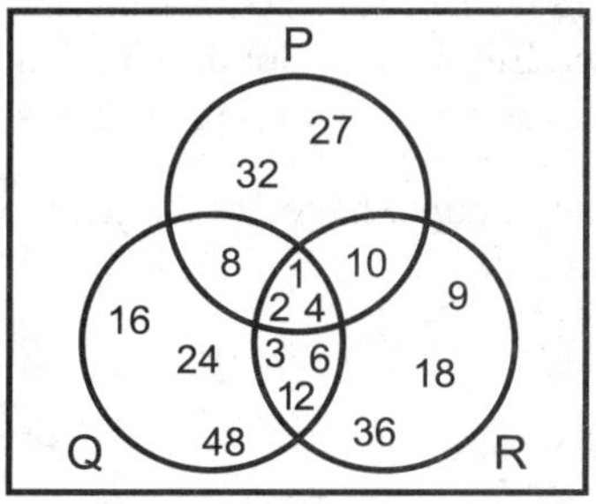

24. One hundred twenty-five customers bought pizza at Pizza Place on their "Special Price" night. The special includes toppings of pepperoni and sausage. Eighty-five customers had pepperoni, and 40 had sausage. Of these customers, 32 had both. How many customers had only pepperoni pizza?

25. One hundred seventy-five students attended summer school at a high school. Ninety-five students were enrolled in English, and 80 students were enrolled in Math. Of these, 61 students were enrolled in both classes. How many students were enrolled only in English?

26. At a Community College, 38 students signed up for Accounting and 49 students signed up for History. Of these, 21 students signed up for both classes. How many students signed up only for Accounting?

27. One hundred fifty people attended the annual Fourth of July picnic. Seventy-four people ate fruit salad and 91 ate potato salad. Of these, 53 ate both. How many people ate neither salad?

28. Of 300 fans at a baseball game, 123 fans were wearing hats and 216 were wearing sunglasses. Of these, 77 were wearing both. How many fans were wearing neither sunglasses nor a hat?

29. At a local hospital, 65 nurses will work the day shift and 72 will work the night shift. Of these, 39 will work both shifts. How many nurses will work only the day shift?

30. During a recent sales event at a clothing store, 91 customers purchased shirts and 64 purchased pants. Of these, 48 bought both pants and shirts. How many customers bought only shirts?

31. Five hundred people were surveyed to determine the types of transportation they use. The results were as follows:
200 use a car
100 use the bus
90 use the bicycle
40 use car and bicycle
30 use bus and bicycle
50 use car and bus
10 use all three

How many participants did not use any of the above transportation types?

32. Angela was checking her closet for dresses containing the primary colors: red, yellow, and blue. She came up with the following results:
30 dresses total
8 contain red
9 contain yellow
9 contain blue
5 contain red and yellow
4 contain yellow and blue
3 contain red and blue
1 contains all three colors

How many dresses contain more than one of these colors?

33. Eight hundred people were surveyed to determine the hobbies they prefer. The survey yielded the following results:
400 prefer sports
350 prefer gardening
300 prefer reading
200 prefer sports and gardening
45 prefer gardening and reading
100 prefer sports and reading
25 prefer all three hobbies

How many people prefer exactly one of these three hobbies?

34. Three hundred students were surveyed regarding the countries that they have visited. The countries they had visited were grouped into three regions: Europe, Canada, and South America. These were the results:
30 visited Europe
100 visited Canada
40 visited South America
10 visited Europe and Canada
20 visited Canada and South America
15 visited Europe and South America
5 visited countries in all three regions

How many people have not visited any of these regions?

35. Seven hundred people were surveyed to determine their tastes in snack foods. The following results were obtained:
250 like cookies
195 like chips
375 like candy
120 like cookies and chips
25 like chips and candy
16 like cookies and candy
10 like all three kinds of snack foods

How many people like only candy?

36. Two thousand, three hundred people were surveyed to determine the types of pets they own. The following results were obtained:
1,600 own a dog
1,400 own a cat
520 own a bird
930 own a dog and a cat
410 own a dog and a bird
50 own a cat and a bird
6 people own all three

How many people own more than one of these pets?

37. Judy surveyed the 35 kids in her class to determine the color of their pens. Her results were as follows:
15 have black
20 have blue
10 have red
6 have black and blue
3 have blue and red
2 have black and red
1 have all three

How many kids have only black pens?

38. An ice-cream parlor conducted a survey of 400 people to determine which toppings people prefer on their ice cream. The results were as follows:
100 like chocolate syrup
200 like whipped cream
150 like cherries
50 like chocolate syrup and whipped cream
20 like chocolate syrup and cherries
15 like whipped cream and cherries
6 like all three

How many like neither chocolate syrup, whipped cream, nor cherries on their ice cream?

39. Fifteen hundred people were surveyed to determine what colors they would prefer when shopping for a new car. The results were as follows:
800 like black
600 like silver
700 like red
100 like black and silver
250 like black and red
300 like silver and red
10 like all three

How many people like at least one of these colors?

40. John surveyed his class of 34 students to determine the kinds of books they enjoy. The results were as follows:
20 enjoy fiction
15 enjoy adventure
16 enjoy biography
10 enjoy fiction and adventure
8 enjoy fiction and biography
5 enjoy adventure and biography
4 enjoy all three

How many students in John's class don't enjoy any of these kinds of books?

HA1-889: Complementary and Supplementary Angles

Angles are given specific names when they have certain properties. Pairs of angles that have sums of measures of 90° or 180° are particularly important.

Complementary Angles	Two angles are complementary if the sum of their measures is 90°.

Supplementary Angles	Two angles are supplementary if the sum of their measures is 180°.

When the sum of the measures of two angles is known to be 90° or 180°, there is an algebraic relationship between the measures. Suppose we know that two angles are complementary, and we know the measure of one of them. We can find the measure of the other angle using algebra.

Example 1 $\angle A$ and $\angle B$ are complementary angles. If $m\angle A = 57°$, find $m\angle B$.

Let $x = m\angle B$.	**Step 1:**	Assign a variable to the unknown quantity.
$m\angle A + m\angle B = 90°$	**Step 2:**	The definition of complementary angles relates the known quantity and the unknown quantity.
$57 + x = 90$	**Step 3:**	Substitute $m\angle A = 57°$ and $m\angle B = x°$ into the equation.
$57 - 57 + x = 90 - 57$	**Step 4:**	Subtract 57 from both sides.
$x = 33°$	**Step 5:**	Simplify.
	Answer:	$m\angle B = 33°$

Similarly, suppose we know that two angles are supplementary and we know the measure of one of them. We can find the measure of the other angle algebraically.

Example 2 $\angle M$ and $\angle L$ are supplementary angles. If $m\angle M = 122°$, find $m\angle L$.

Let $x = m\angle L$.	**Step 1:**	Assign a variable to the unknown quantity.
$m\angle M + m\angle L = 180°$	**Step 2:**	The definition of supplementary angles relates the known quantity and the unknown quantity.
$122 + x = 180$	**Step 3:**	Substitute $m\angle M = 122°$ and $m\angle L = x°$ into the equation.
$122 - 122 + x = 180 - 122$	**Step 4:**	Subtract 122 from both sides.
$x = 58°$	**Step 5:**	Simplify to find the value of the variable.
	Answer:	$m\angle L = 58°$

Sometimes the measures of complementary and supplementary angles are given as algebraic expressions. We can use the definitions of complementary and supplementary angles to write an equation. Then, we can solve the equation for the unknown variable. We use this value to find the measures of the angles.

Example 3 $\angle P$ and $\angle Q$ are complementary angles. If $m\angle P = (2x + 11)°$ and $m\angle Q = (3x - 36)°$, find the measures of angles P and Q.

$$m\angle P + m\angle Q = 90°$$

Step 1: The definition of complementary angles relates the unknown quantities.

$$(2x + 11) + (3x - 36) = 90$$

Step 2: Substitute the algebraic expressions $m\angle P = (2x + 11)°$ and $m\angle Q = (3x - 36)°$ into the equation.

$$5x - 25 = 90$$

Step 3: Combine like terms on the left-hand side of the equation.

$$5x - 25 + 25 = 90 + 25$$
$$5x = 115$$

Step 4: Add 25 to both sides of the equation and simplify.

$$\frac{5x}{5} = \frac{115}{5}$$
$$x = 23$$

Step 5: Divide both sides of the equation by 5 and simplify.

$$m\angle P = 2(23) + 11$$
$$m\angle P = 46 + 11 = 57°$$

Step 6: Substitute the value of x to find $m\angle P$. .

$$m\angle Q = 3(23) - 36$$
$$m\angle Q = 69 - 36 = 33°$$

Step 7: Substitute the value of x to find $m\angle Q$. .

Answer: $m\angle P = 57°$ and $m\angle Q = 33°$

Example 4 $\angle U$ and $\angle V$ are supplementary angles. If $m\angle U = (8x)°$ and $m\angle V = (5x - 41)°$, find the measures of angles U and V.

$$m\angle U + m\angle V = 180°$$

Step 1: The definition of supplementary angles relates the unknown quantities.

$$8x + (5x - 41) = 180$$

Step 2: Substitute the algebraic expressions $m\angle U = (8x)°$ and $m\angle V = (5x - 41)°$ into the equation.

$$13x - 41 = 180$$

Step 3: Combine like terms on the left-hand side of the equation.

$$13x - 41 + 41 = 180 + 41$$
$$13x = 221$$

Step 4: Add 41 to both sides of the equation and simplify.

$$\frac{13x}{13} = \frac{221}{13}$$
$$x = 17$$

Step 5: Divide both sides of the equation by 13 and simplify.

$$m\angle U = 8(17) = 136°$$

Step 6: Substitute the value of x to find $m\angle U$.

$m\angle V = 5(17) - 41$ $m\angle V = 85 - 41 = 44°$	**Step 7:** Substitute the value of x to find $m\angle V$.
	Answer: $m\angle U = 136°$ and $m\angle V = 44°$

Problems involving complementary or supplementary pairs of angles may also be given in words. In such problems, we introduce a variable. Once we specify a variable, we can write all of the unknown quantities in terms of the variable. We can then set up an equation and solve it.

Example 5 The supplement of an angle is twelve degrees greater than five times the measure of the angle itself. Find the measure of the angle.

Let x = the measure of the angle. Then, $5x + 12$ = the measure of the supplement.	**Step 1:** Assign a variable to the unknown quantity and use it to write algebraic expressions for the measure of the angle and its supplement.
$x + (5x + 12) = 180$	**Step 2:** Substitute the algebraic expressions into the equation.
$6x + 12 = 180$	**Step 3:** Combine like terms on the left-hand side of the equation.
$6x + 12 - 12 = 180 - 12$ $6x = 168$	**Step 4:** Subtract 12 from both sides of the equation and simplify.
$\frac{6x}{6} = \frac{168}{6}$ $x = 28°$	**Step 5:** Divide both sides of the equation by 6 and simplify.
$28°$ = the measure of the angle	**Step 6:** Evaluate the necessary expression using the value found to answer the question.
	Answer: The angle has a measure of 28°.

Problem Set

1. T and V are complementary angles. The measure of angle V is 85°. Find the measure of angle T.

2. N and P are complementary angles. The measure of angle N is 69°. Find the measure of angle P.

3. X and Y are complementary angles. The measure of angle X is 41°. Find the measure of angle Y.

4. F and G are complementary angles. The measure of angle G is 36°. Find the measure of angle F.

5. J and K are complementary angles. The measure of angle J is 55°. Find the measure of angle K.

6. L and M are complementary angles. The measure of angle M is 45°. Find the measure of angle L.

7. P and Q are complementary angles. The measure of angle P is 79°. Find the measure of angle Q.

8. R and S are complementary angles. The measure of angle S is 19°. Find the measure of angle R.

9. T and V are supplementary angles. The measure of angle V is 58°. Find the measure of angle T.

10. X and Y are supplementary angles. The measure of angle X is 172°. Find the measure of angle Y.

11. X and Y are supplementary angles. The measure of angle X is $41°$. Find the measure of angle Y.

12. F and G are supplementary angles. The measure of angle G is $30°$. Find the measure of angle F.

13. J and K are supplementary angles. The measure of angle J is $150°$. Find the measure of angle K.

14. L and M are supplementary angles. The measure of angle M is $45°$. Find the measure of angle L.

15. P and Q are supplementary angles. The measure of angle P is $89°$. Find the measure of angle Q.

16. Angles A and B are complementary. If $m\angle A = x°$ and $m\angle B = (4x - 35)°$, find the measures of angles A and B.

17. Angles C and D are complementary. If $m\angle C = (x - 18)°$ and $m\angle D = (3x)°$, find the measures of angles C and D.

18. Angles P and Q are complementary. If $m\angle P = (2x - 10)°$ and $m\angle Q = (3x)°$, find the measures of angles P and Q.

19. Angles R and S are complementary. If $m\angle R = (5x)°$ and $m\angle S = (x + 12)°$, find the measures of angles R and S.

20. Angles T and V are complimentary. If $m\angle T = (2x + 3)°$ and $m\angle V = (2x - 13)°$, find the measures of angles T and V.

21. Angles J and K are complimentary. If $m\angle J = (x - 15)°$ and $m\angle K = (x - 1)°$, find the measures of angles J and K.

22. Angles L and M are complimentary. If $m\angle L = (6x)°$ and $m\angle M = (x + 27)°$, find the measures of angles L and M.

23. Angles F and G are complimentary. If $m\angle F = (x + 26)°$ and $m\angle G = (3x)°$, find the measures of angles F and G.

24. Angles A and B are supplementary. If $m\angle A = (x)°$ and $m\angle B = (3x - 36)°$, find the measures of angles A and B.

25. Angles C and D are supplementary. If $m\angle C = (2x - 15)°$ and $m\angle D = (3x)°$, find the measures of angles C and D.

26. Angles P and Q are supplementary. If $m\angle P = (2x + 10)°$ and $m\angle Q = (3x)°$, find the measures of angles P and Q.

27. Angles R and S are supplementary. If $m\angle \mathrm{R} = (5x)°$ and $m\angle \mathrm{S} = (x + 12)°$, find the measures of angles R and S.

28. Angles T and V are supplementary. If $m\angle T = (2x + 3)°$ and $m\angle V = (2x - 13)°$, find the measures of angles T and V.

29. Angles J and K are supplementary. If $m\angle J = (x - 15)°$ and $m\angle K = (x - 1)°$, find the measures of angles J and K.

30. Angles L and M are supplementary. If $m\angle L = (6x)°$ and $m\angle M = (x + 33)°$, find the measures of angles L and M.

31. The measure of an angle is eighteen less than one-third of its complement. Find the measure of the angle.

32. The measure of an angle is fifteen more than four times its complement. Find the measure of the angle.

33. The measure of an angle is six more than five times its complement. Find the measure of the angle.

34. The measure of an angle is forty-six more than the measure of its complement. Find the measure of the angle.

35. The measure of an angle is twenty more than one-fourth the measure of its complement. Find the measure of the angle.

36. The measure of an angle is eight less than its supplement. Find the measure of the angle.

37. The measure of an angle is fifteen less than twice its supplement. Find the measure of the angle.

38. The measure of an angle is eighteen more than five times its supplement. Find the measure of the angle.

39. The measure of an angle is eighteen more than half of its supplement. Find the measure of the angle.

40. The measure of an angle is four less than three times its supplement. Find the measure of the angle.

HA1-890: Using Models to Derive Formulas for Two-Dimensional Geometric Figures

A complex **two-dimensional (2-D) geometric figure** is created using basic geometric figures. Two-dimensional geometric objects are flat objects with length and width, but no thickness. Circles, triangles, and squares are examples of two-dimensional objects located in planes. A plane is a flat surface that extends infinitely in all directions.

For example, an **annulus** is the region between two circles with the same center, as shown below:

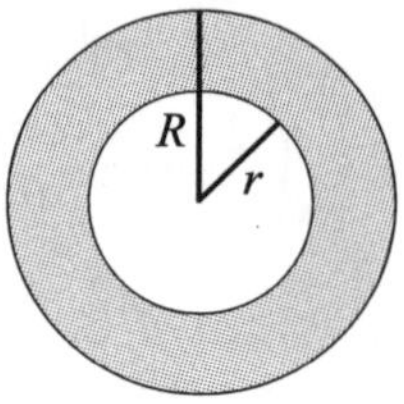

We can find the area and perimeter of this figure using what we know about the area and perimeter of circles. Below are some basic facts and formulas from plane geometry that are useful when working with complex figures.

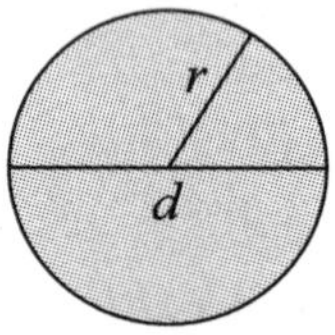

CIRCLE

radius: r
diameter: $d = 2r$

Area: $A = \pi r^2$
Circumference: $C = 2\pi r$
$\pi \approx 3.14$

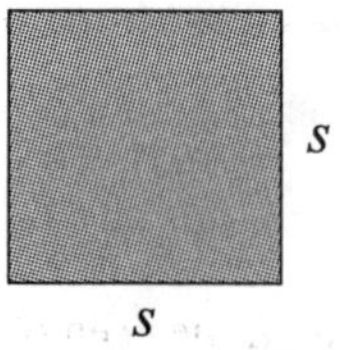

SQUARE

side: s

Area: $A = s^2$
Perimeter: $P = 4s$

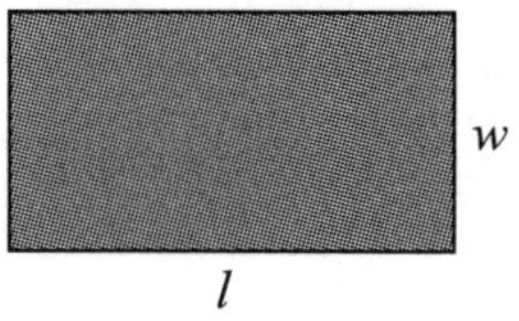

RECTANGLE

length: l
width: w

Area: $A = lw$
Perimeter: $P = 2l + 2w$

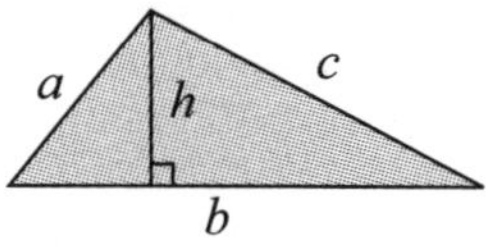

TRIANGLE

sides: a, b, c
height: h

Area: $A = \frac{1}{2}bh$
Perimeter: $P = a + b + c$

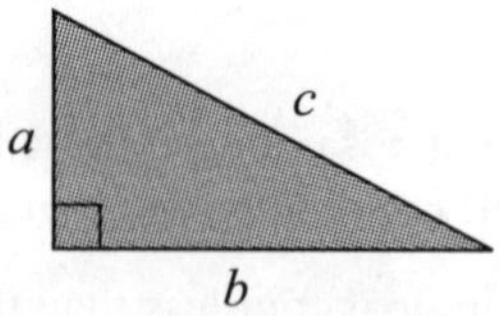

RIGHT TRIANGLE

legs: a, b
hypotenuse: c
Pythagorean Theorem:
$a^2 + b^2 = c^2$

Area: $A = \frac{1}{2}ab$
Perimeter: $P = a + b + c$

Example 1 Find the area and circumference of the annulus shown below:

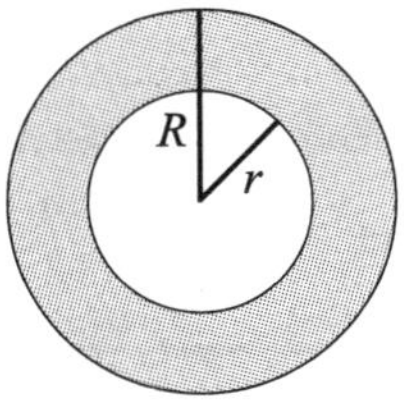

Step 1: The area of the annular region is the area of the shaded region.

Area of annulus $= \pi R^2 - \pi r^2$

Step 2: The area of the larger circle is πR^2. The area of the smaller circle is πr^2. So, the area of the annulus is $\pi R^2 - \pi r^2$.

Step 3: The circumference of the annular region is the sum of the circumferences of the larger circle with radius R and of the smaller circle with the radius r.

Circumference of the annulus $= 2\pi R + 2\pi r$

Step 4: The circumference of the larger circle is $2\pi R$. The circumference of the smaller circle is $2\pi r$.

Answer: The area of the annulus is $\pi R^2 - \pi r^2$ and the circumference of the annulus is $2\pi R + 2\pi r$.

Example 2 We want to make a concrete patio in the shape of a square of dimension x with a circle of radius $\frac{x}{10}$ removed from the center. Find the area of the patio as a function of x.

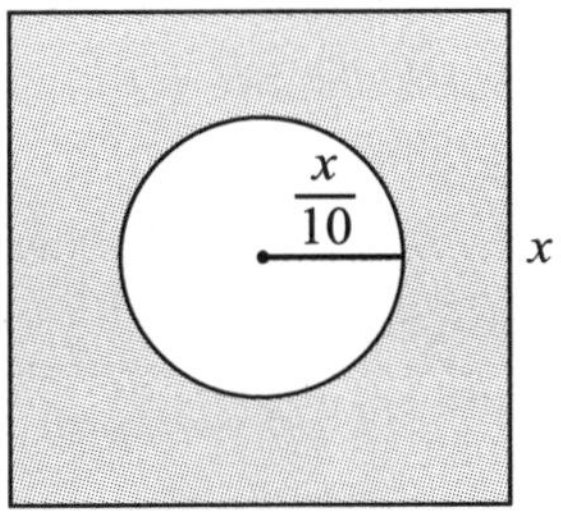

Step 1: To find the area of the patio we start with the area of a square and remove the area of the circle.

$A = s^2$

$A = x^2$

Step 2: Find the area of the square.

$A = \pi r^2$

$= \pi\left(\frac{x}{10}\right)^2$

Step 3: Find the area of the circle.

Area of patio $= x^2 - \pi\left(\frac{x}{10}\right)^2$

Step 4: Find the area of the patio by subtracting the two areas.

Answer: The area of the patio is $x^2 - \pi\left(\frac{x}{10}\right)^2$.

Example 3 Find the area and perimeter of the figure shown below.

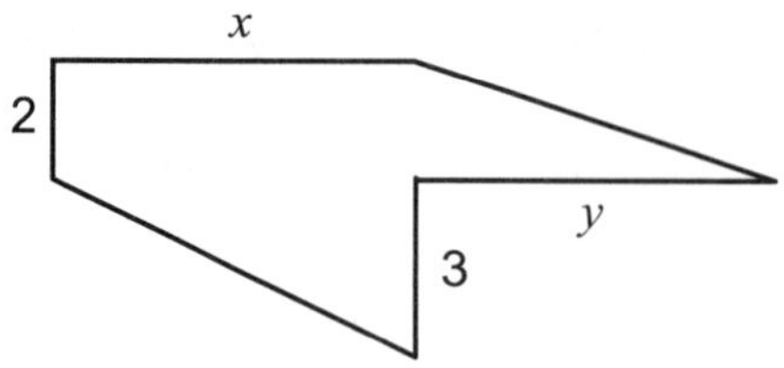

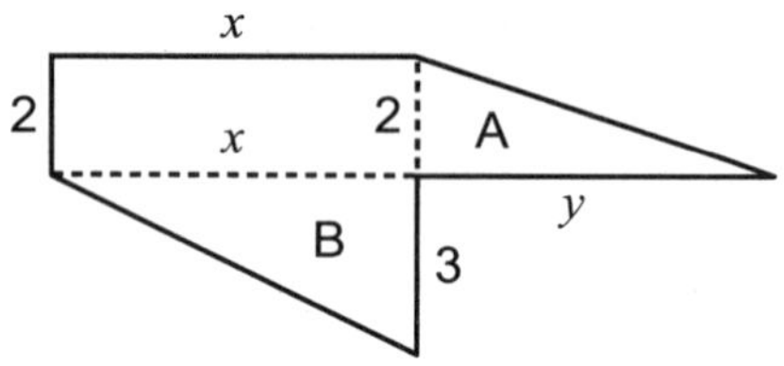

Step 1: To find the area and perimeter we will divide the figure into a rectangle and two triangles.

Area of a rectangle: $= lw$

$= 2x$

Area of triangle B: $= \frac{1}{2}bh$

$= \frac{1}{2}(3)x$

$= \frac{3}{2}x$

Area of triangle A: $= \frac{1}{2}bh$

$= \frac{1}{2}(2)y$

$= y$

Step 2: To find the area of the figure we must find the areas of the rectangle and of the two triangles.

Total area $= 2x + \frac{3}{2}x + y$

$= \frac{7}{2}x + y$

Step 3: Add the individual areas.

$a^2 + b^2 = c^2$

Step 4: Use the Pythagorean Theorem, $a^2 + b^2 = c^2$, to determine the hypotenuses of the two triangles.

Hypotenuse of triangle B: $a^2 + b^2 = c^2$

$x^2 + 3^2 = c^2$

$x^2 + 9 = c^2$

$\sqrt{x^2 + 9} = c$

Step 5: Determine the hypotenuse of triangle B, given the two legs, x and 3.

Hypotenuse of triangle A: $a^2 + b^2 = c^2$

$y^2 + 2^2 = c^2$

$y^2 + 4 = c^2$

$\sqrt{y^2 + 4} = c$

Step 6: Determine the hypotenuse of triangle A, given the two legs, y and 2.

$$P = x + \sqrt{y^2 + 4} + y + 3 + \sqrt{x^2 + 9} + 2$$
$$= x + y + 5 + \sqrt{y^2 + 4} + \sqrt{x^2 + 9}$$

Step 7: Find the perimeter, *P*, of the figure by adding all of the outer edges.

Answer: The area of the figure is $\frac{7}{2}x + y$ and the perimeter of the figure is $x + y + 5 + \sqrt{y^2 + 4} + \sqrt{x^2 + 9}$.

Example 4 If the area of the triangle shown below is 32 square units, find the value of x.

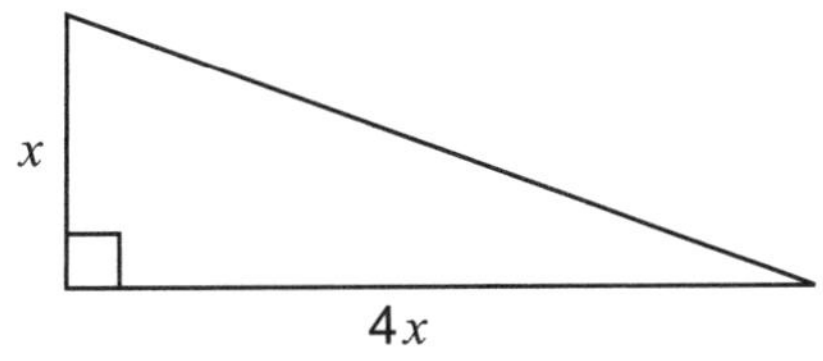

Area of a triangle $= \frac{1}{2}bh$

$= \frac{1}{2}(4x)(x)$

Step 1: Substitute the values for the base and height into the formula for the area of a triangle.

$32 = \frac{1}{2}(4x)(x)$

Step 2: Since the area of the triangle is 32, set this equation equal to 32.

$32 = \frac{1}{\cancel{2}}(\cancel{4}^{2}x)(x)$

$32 = 2x^2$

$16 = x^2$

$4 = x$

Step 3: Simplify and solve for x.

Answer: The value of x is 4 units.

Example 5 The sides of a rectangular pool are given below. What are all possible values of x?

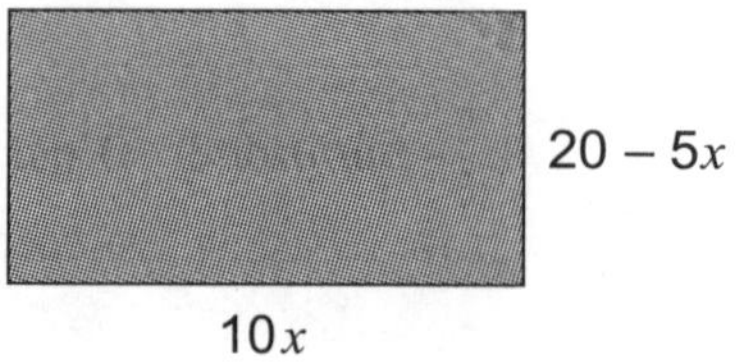

$20-5x>0$ and $10x>0$	**Step 1:** Length and width must be positive. Therefore, $20-5x>0$ and $10x>0$.
$20-5x>0$ $20-5x+5x>0+5x$ $20>5x$ $4>x$	**Step 2:** Solve the inequality: $20-5x>0$.
$10x>0$ $x>0$	**Step 3:** Solve the inequality: $10x>0$.
$x>0$ and $x<4$ $0<x<4$	**Step 4:** Put both inequality statements together.

Answer: The possible values of x are $0<x<4$.

Problem Set

1. Find the area of the given figure.

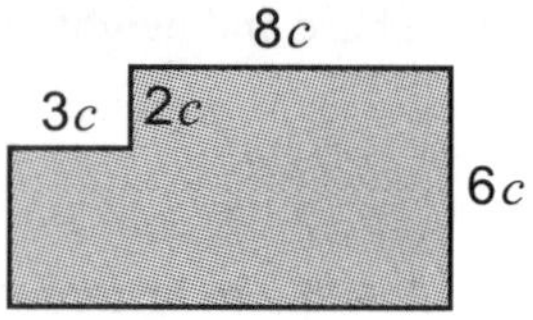

2. Find the perimeter of the window.

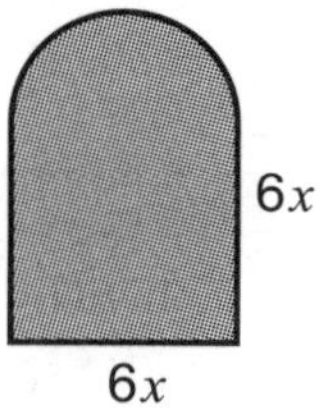

3. Cindy wants to build a flower bed designed as shown. In order to purchase the correct amount of border material, she needs to determine its perimeter. Find the perimeter of the flower bed using the figure.

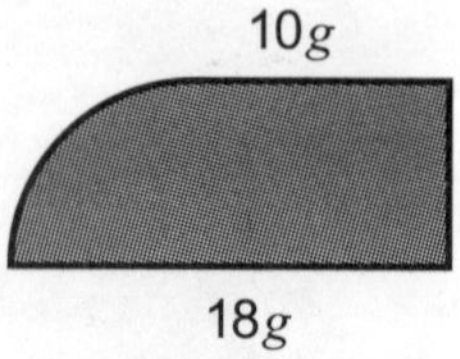

4. A garden has the following shape. Find the area of the garden.

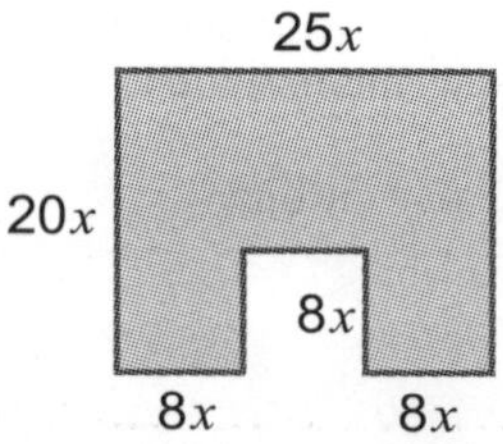

5. An ice-cream cone filled with strawberry ice cream has the shape of an equilateral triangle (all sides equal) with a semicircle top having a radius of $6x$. Find the perimeter.

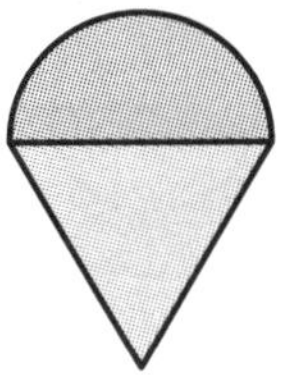

6. Find the area of the window given by the following shape.

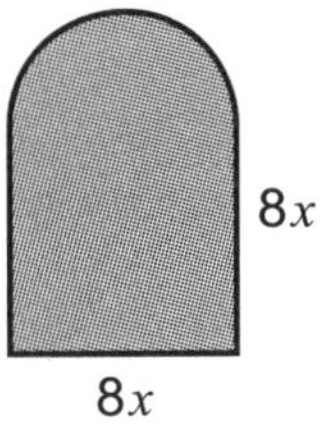

7. A play area is made up of a square and an isosceles (two sides equal) right triangle. Find the perimeter of the play area.

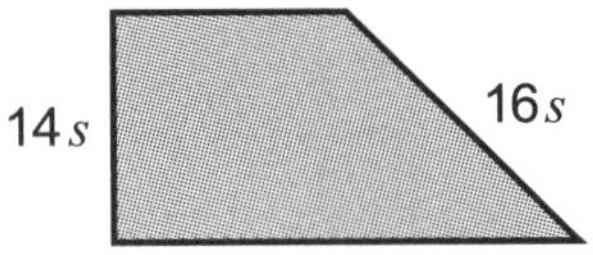

8. The side of a house has the shape of a rectangle with a triangle on top. Find its area.

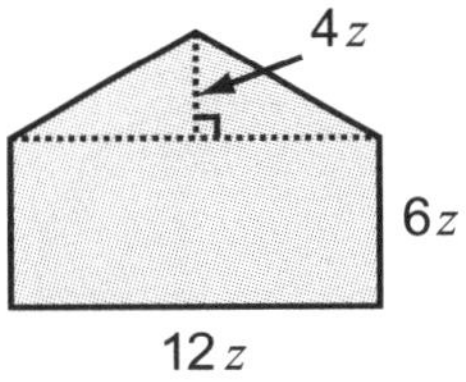

9. The side of a house has the shape of a rectangle with an isosceles triangle (two sides equal) on top. Find its perimeter.

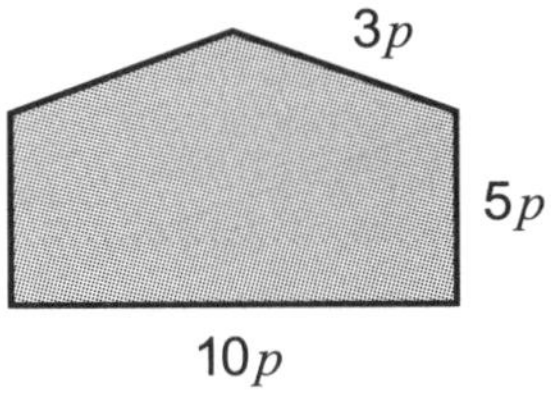

10. Find the area of the figure.

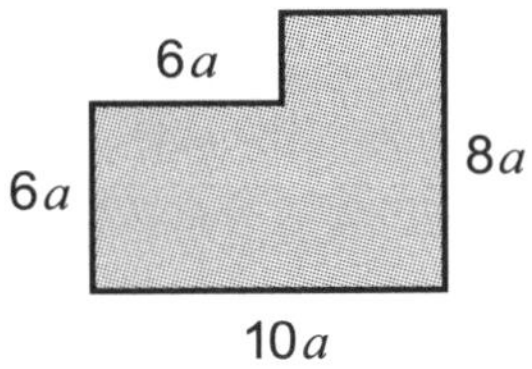

11. Cynthia has to seat several guests at a dinner party. She wants to make sure that the table is large enough to seat each guest. Find the perimeter of the table.

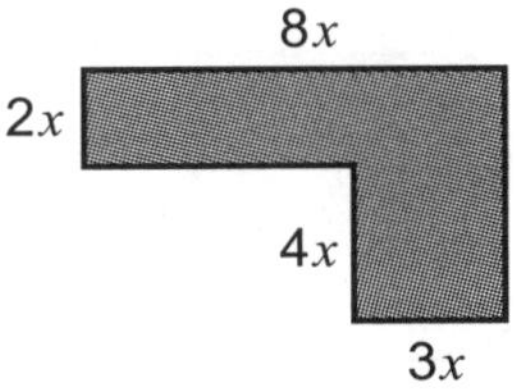

12. A walkway is to have the shape of a right triangle attached to a rectangle as shown. The legs of the right triangle are equal in length. Find the area of the walkway.

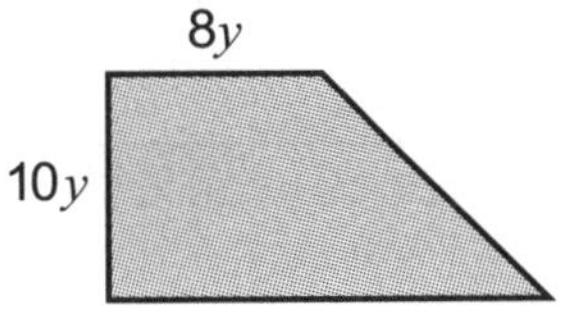

13. A walkway is to have the shape of a rectangle attached to a semicircle as shown.
Find its perimeter.

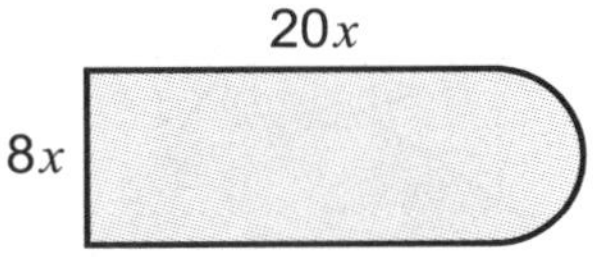

14. An ice-cream cone filled with chocolate ice cream has the shape of an isosceles triangle (two sides equal) with a semicircle on top. The semicircle has a radius of $5x$. Find the area.

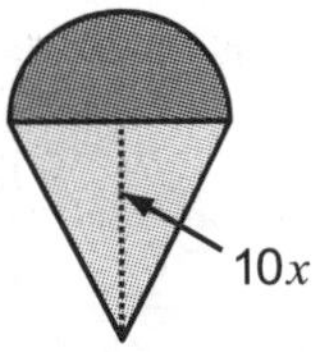

15. A concrete patio with the shape of the figure below is to be built in a space having dimensions $16x$ by $24x$. Find the perimeter of the patio.

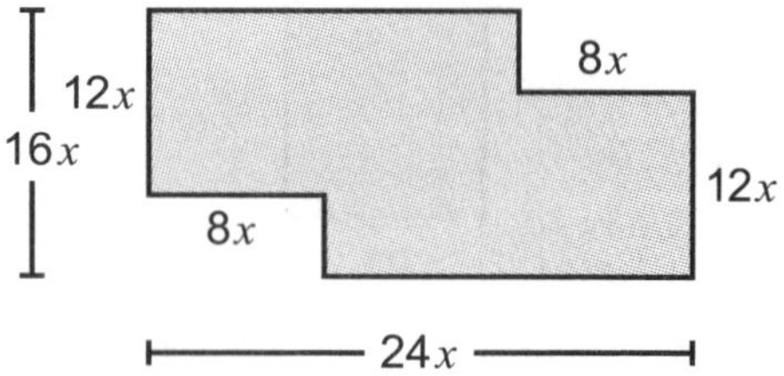

16. The shaded figure below is called an annulus. It is the region between two circles with the same center. Find the area of the given annulus.

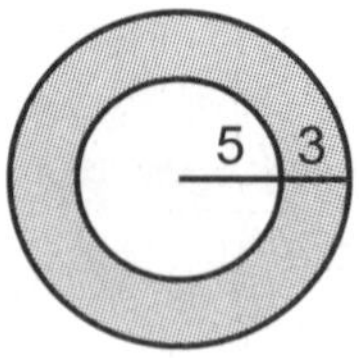

17. The circumference of the given circle is 24π and its diameter is $6x + 12$. Find the value of x.

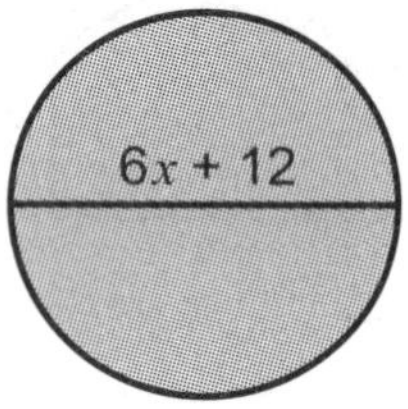

18. The area of the given rectangle is 108 square inches. Find the value of x.

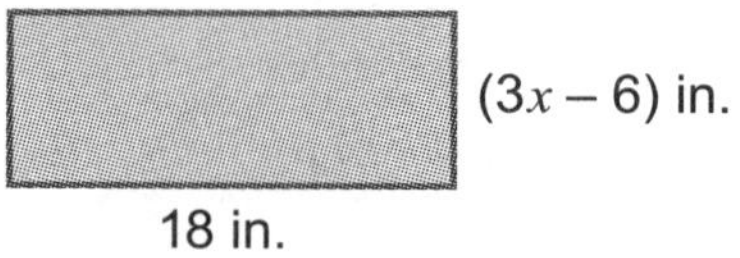

19. A tile contains the pattern shown. The side of the square has a length of $18x$. The diameter of the circle is equal to the side of the square. Find the area of the shaded region.

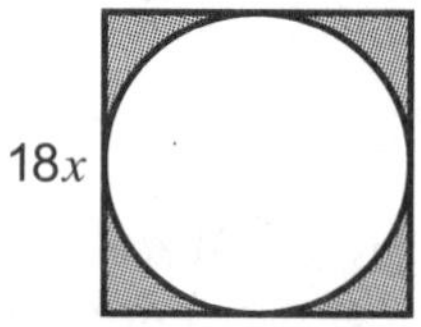

20. The area of the given right triangle is 90 square feet. Find the value of w.

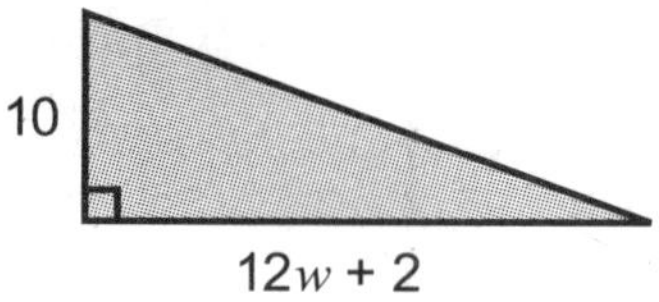

21. Four congruent squares with sides of length x have been cut from a rectangle that has a length of 16 and a width of 12. Find the area of the shaded region.

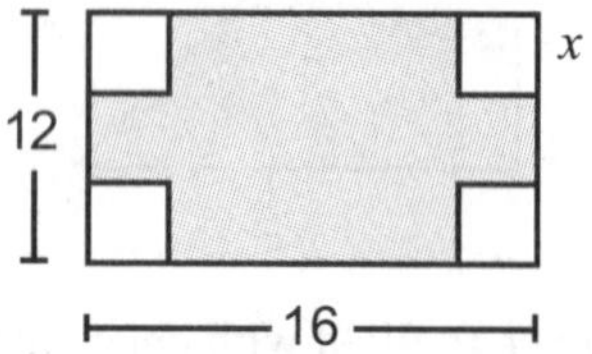

22. The perimeter of the given square is 96 cm. Find the value of x.

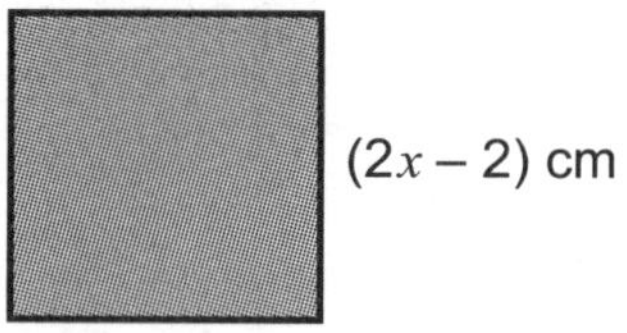

23. The matting of a picture frame has a width of $4x$ as shown in the figure. Find its area if the inside rectangular picture is 8 inches wide and 10 inches long.

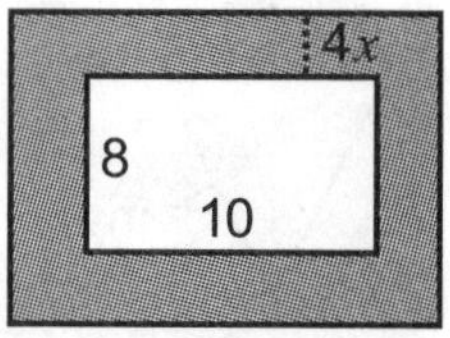

24. The area of the given triangle is 144 square meters. Find the value of x.

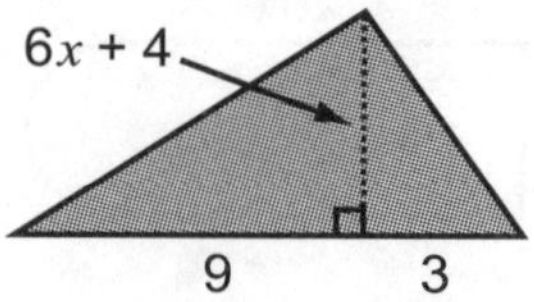

25. The figure below is a circle with a square hole in its center. The side of the square hole is one-fourth of the diameter of the circle which is $20x$. Find the area of the given figure.

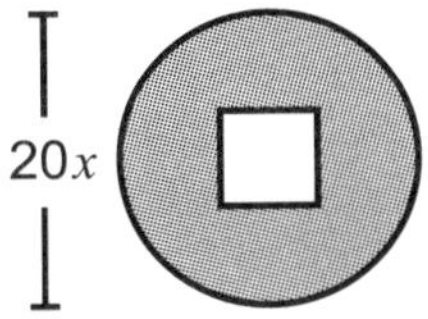

26. The perimeter of the given triangle is 60 inches. Find the value of x.

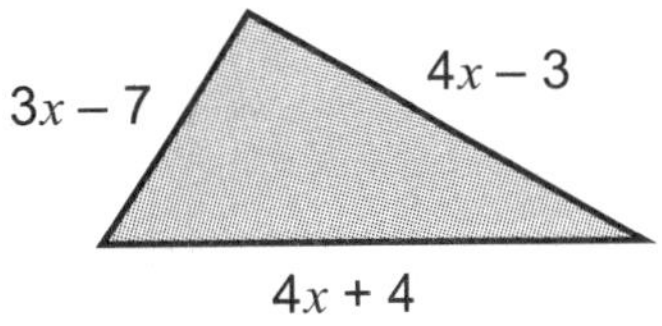

27. A large triangle has a smaller similar triangle cut out of its center as shown in the figure. The base and altitude of the smaller triangle are one-half of the base and altitude of the larger triangle, respectively. Find the area of the shaded region.

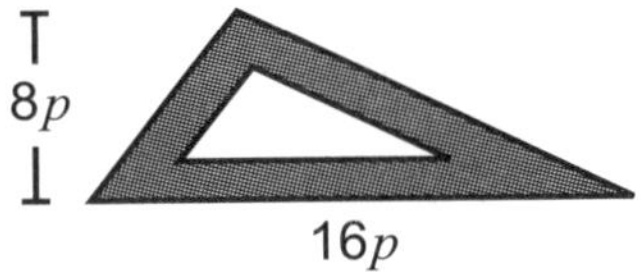

28. The perimeter of the given rectangle is 68. Find the value of q.

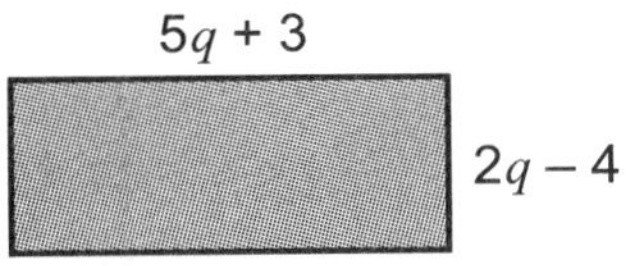

29. A right triangle is removed from a circle as shown in the figure. The legs of the right triangle are 3x and 4x. The hypotenuse of the right triangle is the diameter of the circle. Find the area of the shaded section.

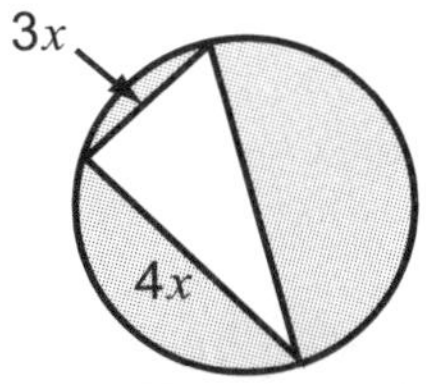

30. The perimeter of the given right triangle is 72. Find the value of x.

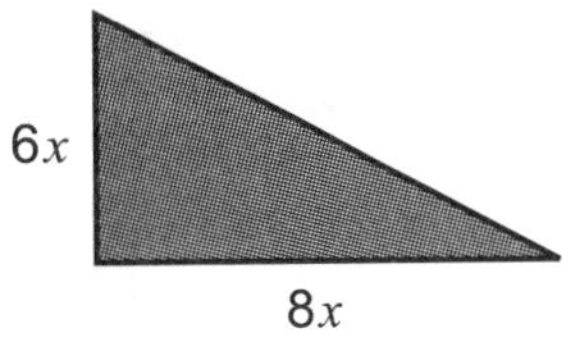

31. Teresa has a rectangular garden with length and width as shown. Find all possible values of p.

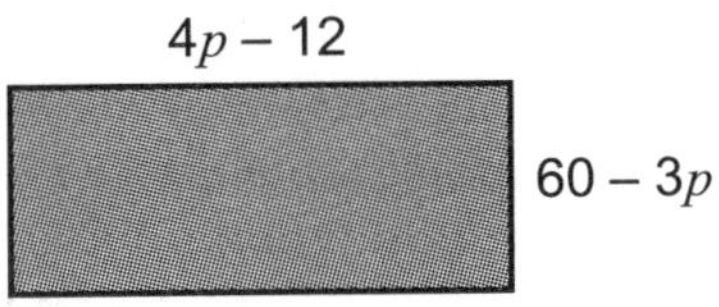

32. A square has sides of length $4x - 68$. Find all possible values of x.

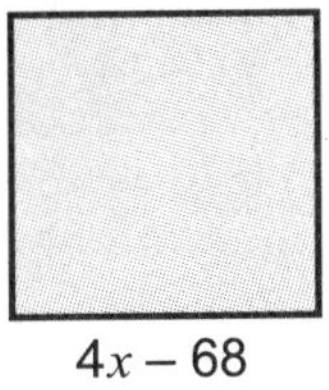

33. A triangle has sides with lengths shown in the figure. What are all possible values of x?

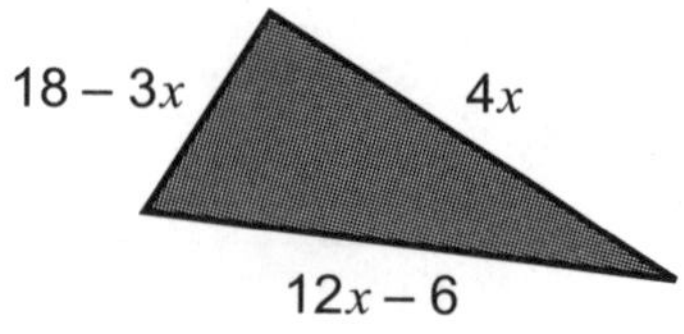

34. A picture frame has a width of $2x$. Its outer dimensions are 30 inches by 20 inches. What are the possible values of x so that an inside rectangle exists?

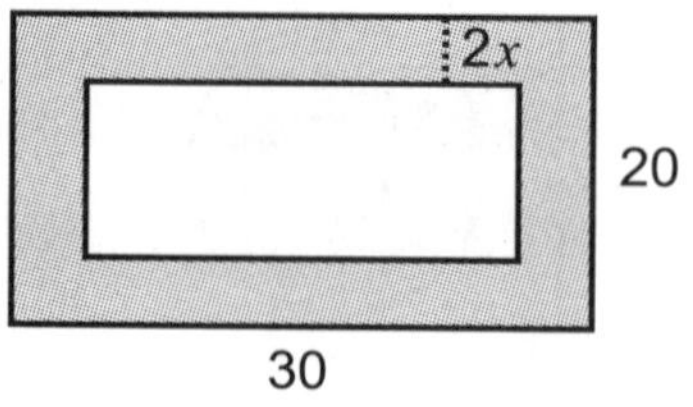

35. The base and altitude of a triangle are shown in the figure. What are all possible values of v?

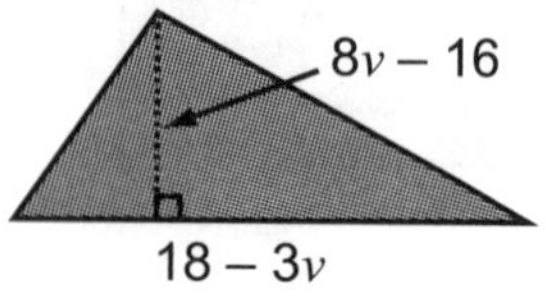

36. The sides of a rectangle are given in the figure. What are all possible values of x?

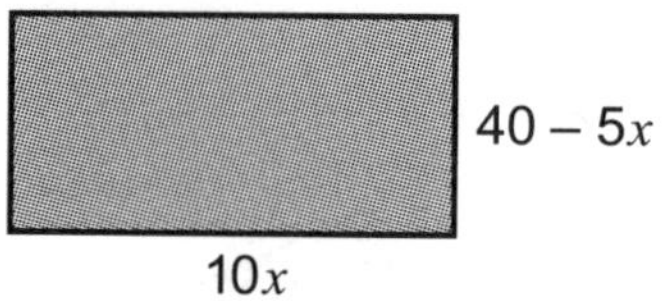

37. The sides of a triangle are as shown in the figure. What are all possible values of x?

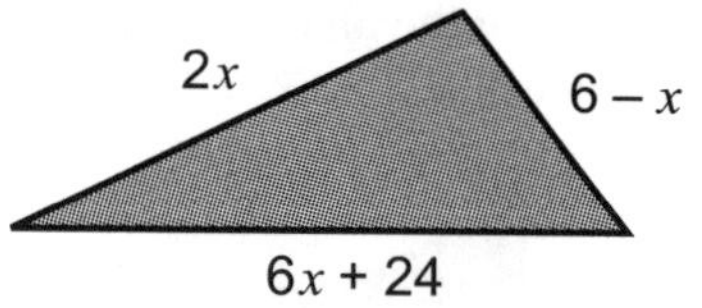

38. The altitude that is shown in the triangle separates its base into two segments whose lengths are shown. What are all possible values of x?

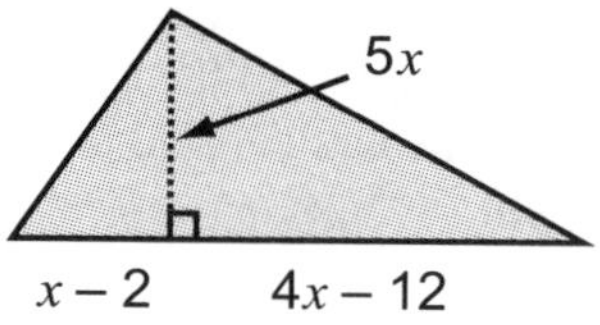

39. The side of a house is a triangle sitting on a square, as shown in the figure. What are all possible values of x?

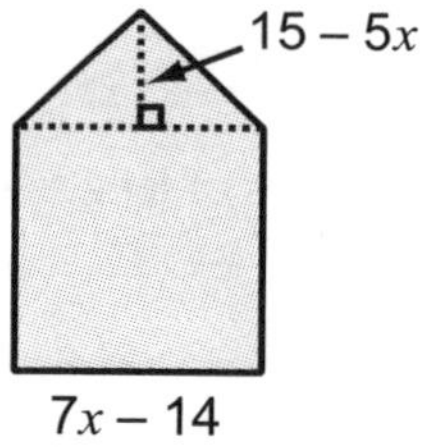

40. A rectangle has the dimensions shown in the figure. What are the possible values of y?

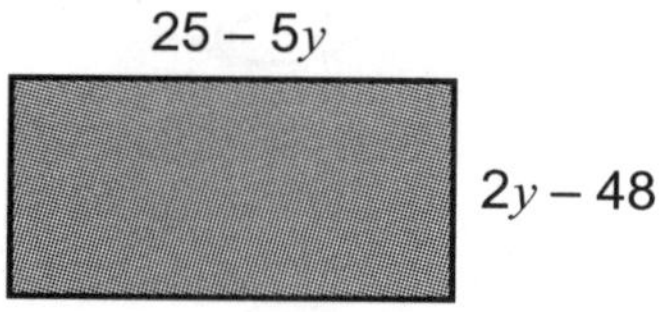

HA1-891: Using Models to Derive Formulas for Three-Dimensional Geometric Solids

Three-dimensional (3-D) geometric objects are also often called solids. They have length, width, and height, and they take up space, or volume. Boxes, cylinders, and spheres are examples of 3-D objects. The space in which we live and move is three-dimensional space, and most objects around us are three-dimensional objects.

A complex three-dimensional solid is one that is created using simple solids. For example, the frustum of a cone is the solid formed by cutting off the top of a cone.

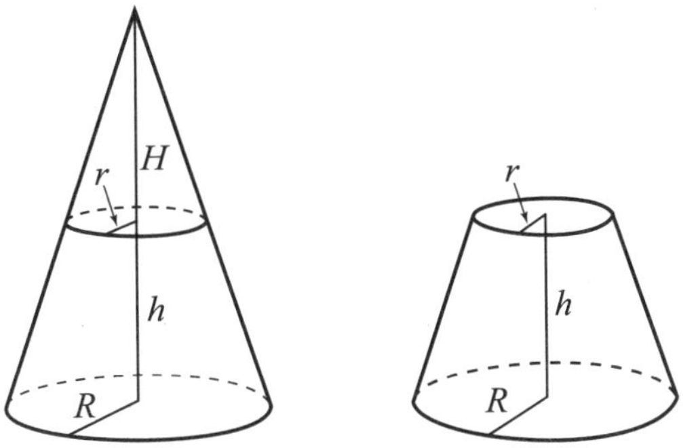

Below are some basic facts and formulas from solid geometry that will be useful when working with more complex solids.

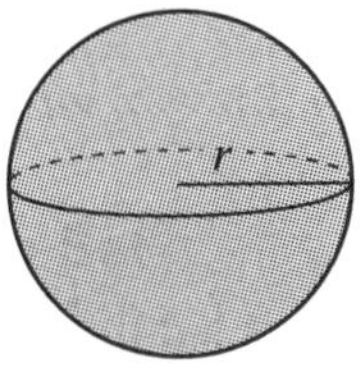

SPHERE

Radius = r

Volume: $A = \frac{4}{3}\pi r^3$

Surface Area: $S = 4\pi r^2$

$\pi \approx 3.14$

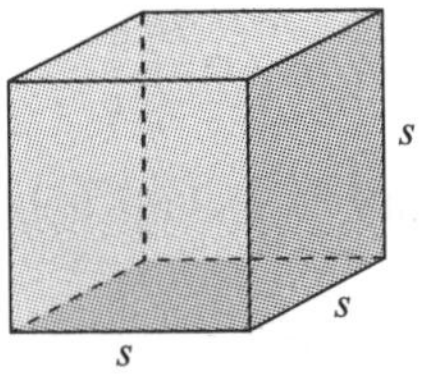

CUBE

Side = s

Volume: $V = s^3$

Surface Area: $S = 6s^2$

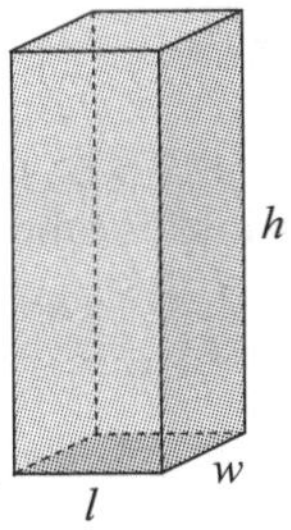

RECTANGULAR PRISM

Length = l
Width = w
Height = h

Volume: $V = lwh$

Surface Area:
$S = 2(lw + lh + wh)$

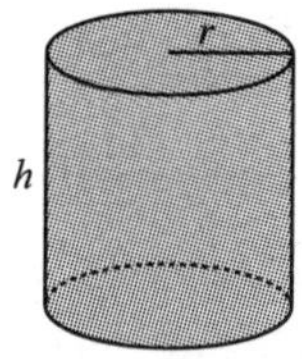

CYLINDER

Base radius = r
Height = h

Volume: $V = \pi r^2 h$

Surface Area (lateral):
$S = 2\pi rh$

Total Surface Area:
$S = 2\pi r^2 + 2\pi rh$

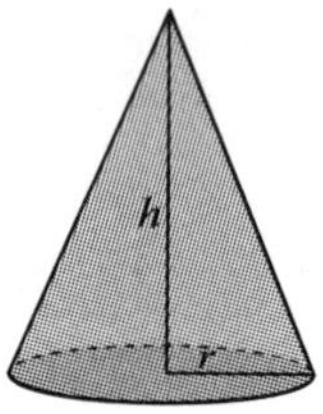

CIRCULAR CONE

Base Radius = r
Height = h

Volume: $V = \frac{1}{3}\pi r^2 h$

Surface Area (lateral):
$S = \pi r\sqrt{r^2 + h^2}$

Total Surface Area:
$S = \pi r^2 + \pi r\sqrt{r^2 + h^2}$

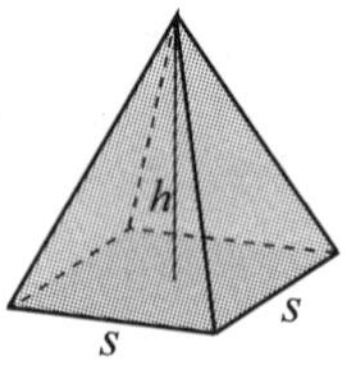

PYRAMID

Side = s
Height = h

Volume: $V = \frac{1}{3}s^2 h$

Example 1 Find an expression to represent the surface area of the rectangular prism.

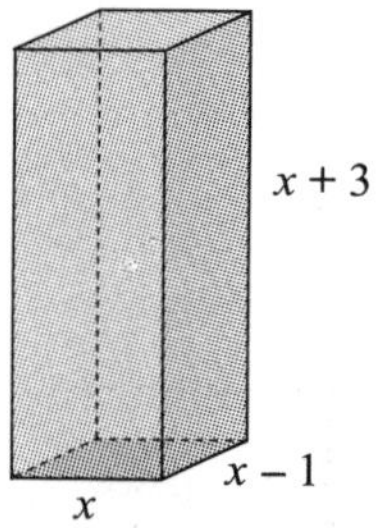

$l = x$ $w = x - 1$ $h = x + 3$	**Step 1:** Identify the length, *l*, the width, *w*, and the height, *h*, of the rectangular prism.
$S = 2(lw + lh + wh)$	**Step 2:** The surface area of a rectangular prism is $2(lw + lh + wh)$.
$S = 2[x(x-1) + x(x+3) + (x-1)(x+3)]$	**Step 3:** Replace *l*, *w*, and *h* with the expressions for the length, width and height.
$S = 2(x^2 - x + x^2 + 3x + x^2 + 3x - x - 3)$ $= 2(3x^2 + 4x - 3)$ $= 6x^2 + 8x - 6$	**Step 4:** Simplify.

Answer: The expression to represent the surface area is

$$S = 6x^2 + 8x - 6.$$

Example 2 Find the volume of a rectangular prism with width that is 3 units longer than its length and with height that is double its length.

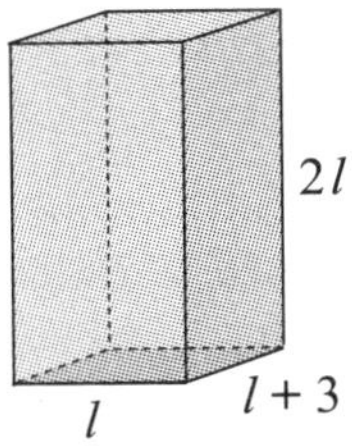

Step 1: Let l = length. Therefore, the width, is $w = l+3$ and the height is $h = 2l$.

$V = lwh$

Step 2: The volume of a rectangular prism is $V = lwh$.

$V = l(l+3)(2l)$

Step 3: Replace l, w, and h with the expressions for the length, width, and height.

$V = (l^2 + 3l)(2l)$
$= 2l^3 + 6l^2$

Step 4: Simplify.

Answer: The volume of the rectangular prism is $V = 2l^3 + 6l^2$.

Example 3 A cylindrical tank is $\frac{4}{5}$ full of water. The base diameter is equal to the height of the tank. Find the volume of the water in the tank in terms of h, the height of the tank.

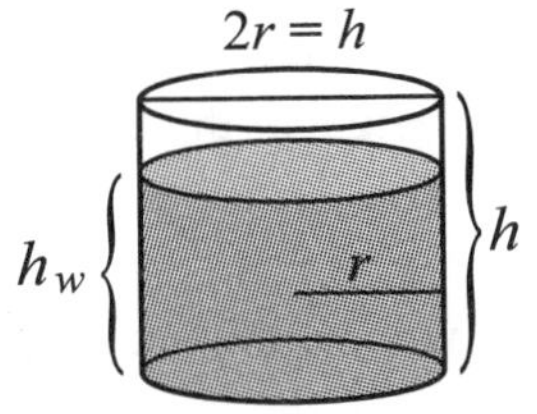

$d = h$
$r = \frac{1}{2}d$
$r = \frac{1}{2}h$

Step 1: Since the diameter is equal to the height of the tank, the radius is one-half the height.

$h_w = \frac{4}{5}h$

Step 2: Since the tank is $\frac{4}{5}$ full of water, the height of the water, h_w, is $\frac{4}{5}$ times the height of the tank.

$$V = \pi r^2 h$$
$$= \pi\left(\frac{1}{2}h\right)^2\left(\frac{4}{5}h\right)$$
$$= \pi\frac{1}{4}h^2\frac{4}{5}h$$
$$= \frac{1}{5}\pi h^3$$

Step 3: Find the volume of a cylindrical tank, $V = \pi r^2 h$, where r is the radius of the tank and h is the height of the tank.

Answer: The volume of the water in the tank is $\frac{1}{5}\pi h^3$.

Example 4 A rocket is made in the form of a cone surmounted on a cylinder. The height of the cylinder is equal to two times the diameter of the base of the cone. The height of the cone is equal to the base diameter of the cone. Find the volume of the rocket in terms of the radius, r.

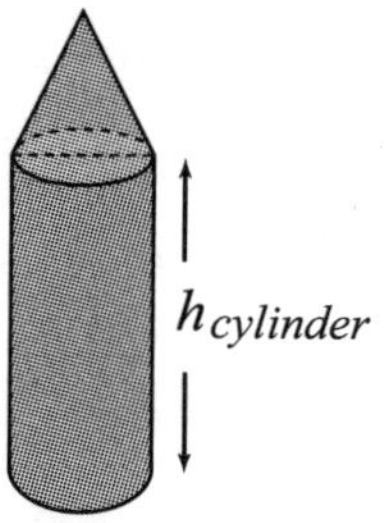

$$d = 2r$$
$$h_{cylinder} = 2d$$
$$= 2(2r)$$
$$= 4r$$

Step 1: Since the diameter of the base of the cone is two times the radius, and the height of the cylinder is two times the diameter, then the height of the cylinder is $4r$.

$$h_{cone} = d$$
$$= 2r$$

Step 2: Since the height of the cone is equal to the base diameter of the cone, the height of the cone is $2r$.

$$V = \pi r^2 h_{cylinder}$$
$$= \pi r^2(4r)$$
$$= 4\pi r^3$$

Step 3: Find the volume of the cylinder, $V = \pi r^2 h_{cylinder}$.

$$V = \frac{1}{3}\pi r^2 h_{cone}$$
$$= \frac{1}{3}\pi r^2(2r)$$
$$= \frac{2}{3}\pi r^3$$

Step 4: Find the volume of the cone, $V = \frac{1}{3}\pi r^2 h_{cone}$.

$$V_{rocket} = V_{cylinder} + V_{cone}$$
$$= 4\pi r^3 + \frac{2}{3}\pi r^3$$
$$= \frac{14}{3}\pi r^3$$

Step 5: The volume of the rocket is equal to the sum of the volume of the cylinder and the volume of the cone.

Answer: The volume of the rocket in terms of r is $\frac{14}{3}\pi r^3$.

Example 5 Find the height in terms of x of a cylinder with a base radius of $2x$ and with a surface area that is the same as the surface area of a sphere with a radius of $2x$.

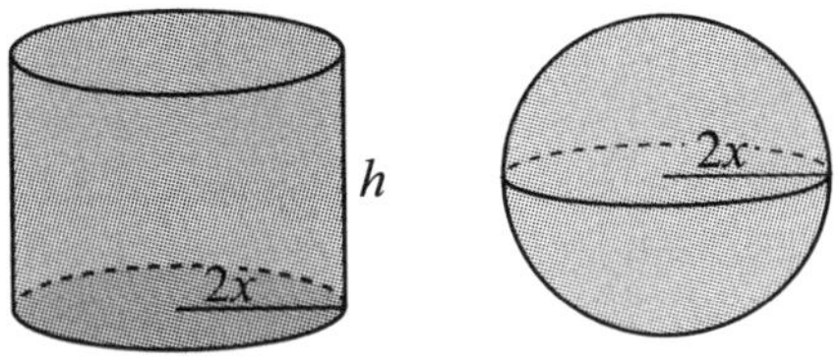

$$r_{sphere} = r_{cylinder} = 2x$$
$$S_{sphere} = S_{cylinder}$$

Step 1: The radii of the sphere and the cylinder are the same. The surface areas of the sphere and cylinder are the same.

$$S_{sphere} = 4\pi r^2$$
$$= 4\pi(2x)^2$$
$$= 4\pi(4x^2)$$
$$= 16\pi x^2$$

Step 2: Find the surface area of the sphere, $S_{sphere} = 4\pi r^2$.

$$S_{cylinder} = 2\pi rh + 2\pi r^2$$
$$= 2\pi(2x)h + 2\pi(2x)^2$$
$$= 4\pi xh + 2\pi(4x^2)$$
$$= 4\pi xh + 8\pi x^2$$

Step 3: Find the surface area of the cylinder, $S_{cylinder} = 2\pi rh + 2\pi r^2$.

$$S_{sphere} = S_{cylinder}$$
$$16\pi x^2 = 4\pi xh + 8\pi x^2$$

Step 4: Set the surface areas of the sphere and the cylinder equal to each other.

$$16\pi x^2 = 4\pi xh + 8\pi x^2$$
$$8\pi x^2 = 4\pi xh$$
$$\frac{8\pi x^2}{4\pi x} = \frac{4\pi xh}{4\pi x}$$
$$2x = h$$

Step 5: Solve for h to find the height of the cylinder.

Answer: The height of the cylinder is $2x$.

Problem Set

Perform the indicated operation:

1. Find the expression to represent the volume of the rectangular prism shown.

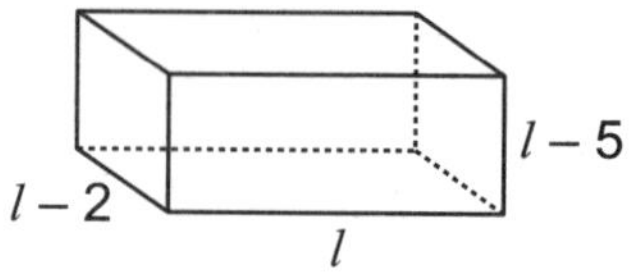

2. Find the expression to represent the volume of the rectangular prism shown.

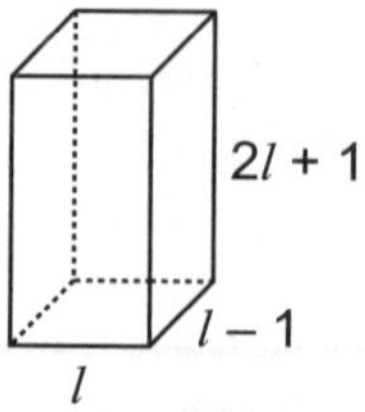

3. Find the expression to represent the volume of the rectangular prism shown.

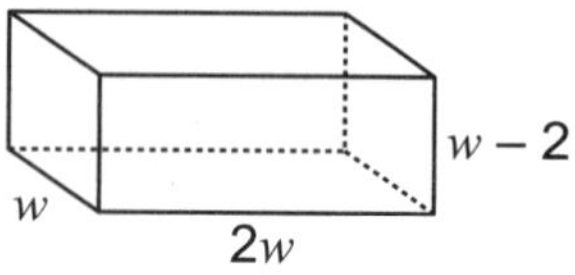

4. Find the expression to represent the volume of the rectangular prism shown.

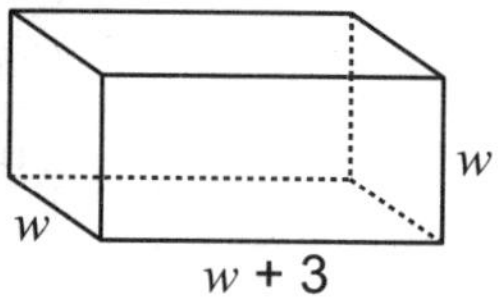

5. Find the expression to represent the volume of the rectangular prism shown.

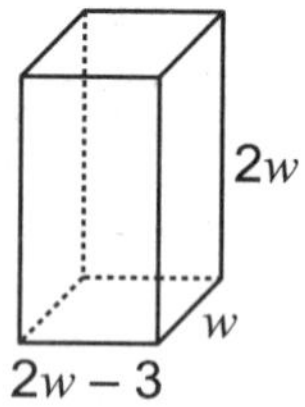

6. Find the expression to represent the volume of the rectangular prism shown.

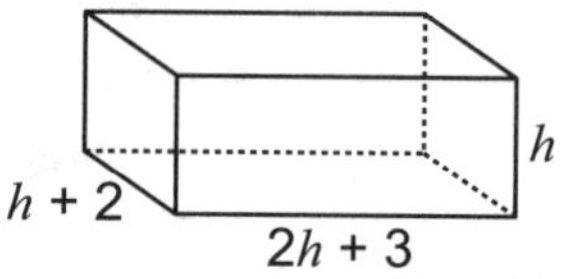

7. Find the expression to represent the volume of the rectangular prism shown.

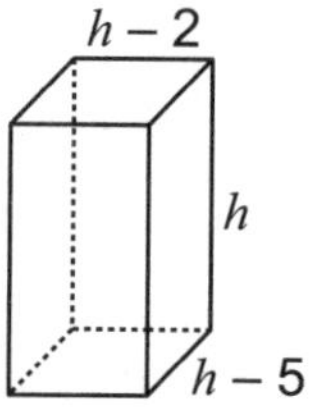

8. Find the total surface area of a can if the height of the can is double its radius, r.

9. Find the total surface area of a can if the height of the can is five times its radius, r.

10. Find the total surface area of a can if the height of the can is twice its radius, r.

11. Find the total surface area of a can if the height of the can is triple its radius, r.

12. Find the total surface area of a can if the height of the can is four times its radius, r.

13. Find the total surface area of a can if the height of the can is half its radius, r.

14. Find the total surface area of a can if the height of the can is ten times its radius, r.

15. Find the total surface area of a can if the height of the can is one-fourth its radius, r.

16. A cylindrical water tank with the base radius r and a height that is twice the diameter of the base is filled with water up to three-fourths of its height. Find the volume of the water in the water tank.

17. A cylindrical water tank with the base radius r and a height that is three times the diameter of the base, is filled with water up to half of its height. Find the volume of the water in the water tank.

18. A cylindrical water tank with the base radius r and a height that is twice the diameter of the base, is filled with water up to half of its height. Find the volume of the water in the water tank.

19. A cylindrical tank has a base diameter that is equal to twice its height. If the tank is filled with water up to two-thirds of its height, find the volume of the water in the tank in terms of the height of the tank, h.

20. Cylinder A has a height that is equal to twice its base radius. Cylinder B has a height equal to three times the base radius of Cylinder A and a radius equal to the height of Cylinder A. Express the volume of Cylinder B in terms of the radius, r, of Cylinder A.

21. A cylindrical tank has a height that is 3 times its base radius. A new larger cylindrical tank is built such that its height and base radius are each 4 times as large as the smaller tank. Find the volume of the larger tank in terms of the base radius, r, of the smaller tank.

22. The base radius of a cylinder is equal to its height, h. A larger cylinder has a base radius and a height that are each four times as large as those of the smaller cylinder. Find the volume of the larger cylinder in terms of the height of the smaller cylinder.

23. A silo is shaped like a cylinder with a hemisphere on top. The height, h, of the cylinder is equal to 2 times the base radius of the cylinder. The radius of the hemisphere is equal to the radius of the base of the cylinder. Find the volume of the silo.

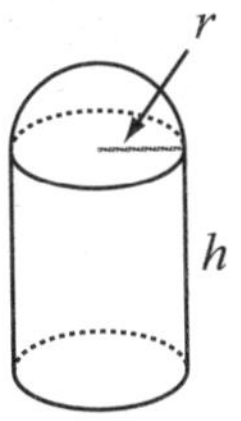

24. A capsule has a shape of a cylinder with a hemisphere at the top and at the bottom. The radii of the base of the cylinder and the hemispheres are x, while the height of the cylinder is $3x$. Find the expression for the surface area of the capsule.

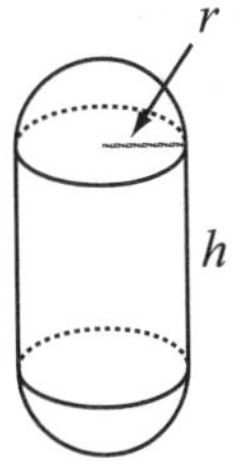

25. A capsule has a shape of a cylinder with a hemisphere at the top and at the bottom. The radii of the base of the cylinder and the hemispheres are x, while the height of the cylinder is $5x$. Find the expression for the surface area of the capsule.

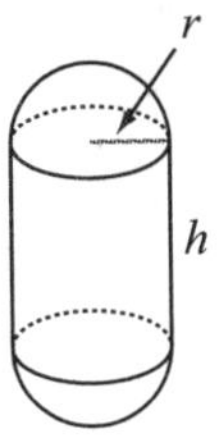

26. A house has the shape of a cube with a square-based pyramid as its roof. The sides of the cube and of the pyramid are x, while the height of the pyramid is $2x$. Find the expression for the volume of the house.

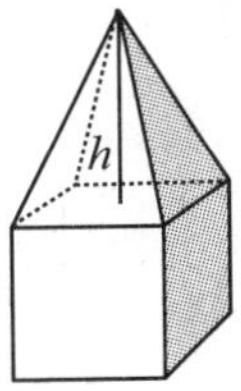

27. A silo has the shape of a cylinder with a hemisphere at the top. The radius of the base of the cylinder and of the hemisphere is x, while the height of the cylinder is $4x$. Find the expression for the surface area of the silo.

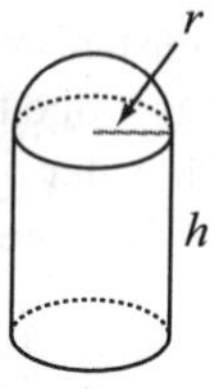

28. A silo has the shape of a cylinder with a hemisphere at the top. The radius of the base of the cylinder and of the hemisphere is x, while the height of the cylinder is $3x$. Find the expression for the surface area of the silo.

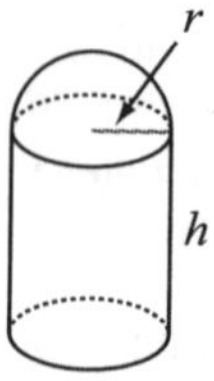

29. A pencil is shaped like a cylinder with a cone sitting on the top. The height of the cylinder is equal to 7 times the diameter of the base of the cone. The height of the cone is equal to the base radius of the cone. The base radii of the cone and the cylinder are equal. Find the volume of the pencil in terms of r, the radius of the base of the cylinder.

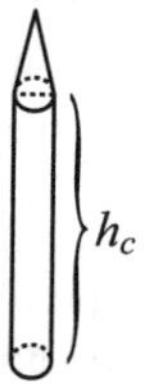

30. An ice-cream cone is full of ice cream, and there is an additional hemisphere of ice cream sitting on the top. Find the total volume of the ice cream. The height of the cone is 6 times the radius. The radii of the cone and the hemisphere are both equal to r.

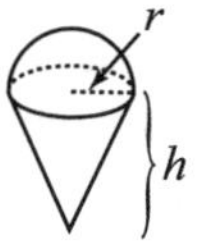

31. Find the height in terms of x of a rectangular prism whose length and width is $2x$, and whose volume is the same as the volume of a cube whose side is $4x$.

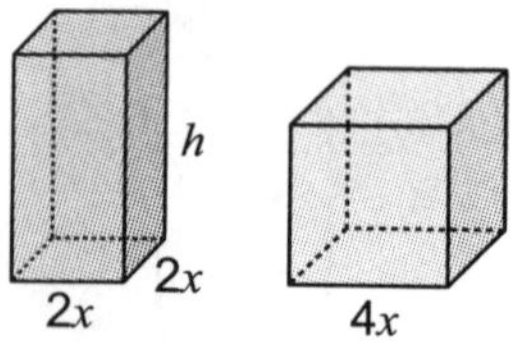

32. Find the height in terms of x of a rectangular prism whose width is $3x$, whose length is twice its width, and whose volume is the same as the volume of a cube whose side is $3x$.

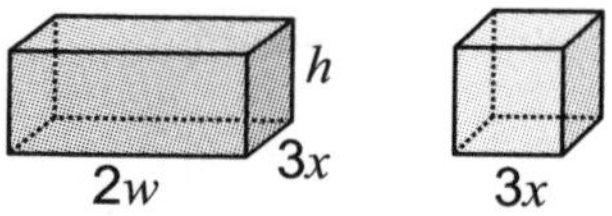

33. Find the height in terms of x of a rectangular prism whose length and width are $\frac{1}{4}x$, and whose surface area is the same as the surface area of a cube whose side is x.

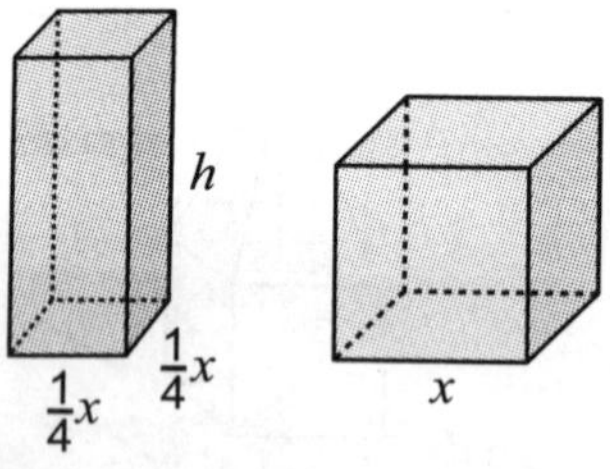

34. Find the height in terms of x of a cone whose radius is $2x$, and whose volume is the same as the volume of a sphere whose radius is $3x$.

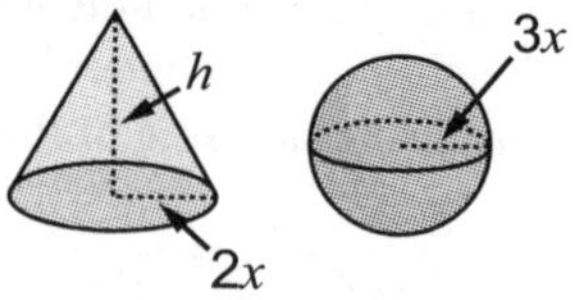

35. Find the height in terms of x of a cone whose radius is $3x$, and whose volume is the same as the volume of a sphere whose radius is $3x$.

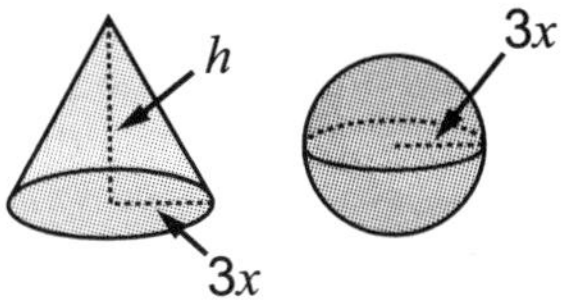

36. Find the height in terms of x of a cylinder whose base radius is $3x$, and whose surface area is the same as the surface area of a sphere whose radius is $3x$.

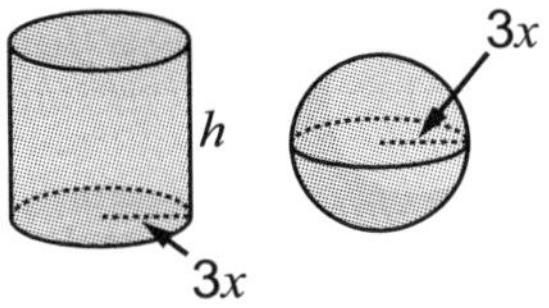

37. Find the height in terms of x of a cylinder whose base radius is $6x$, and whose volume is the same as the volume of a sphere whose radius is $2x$.

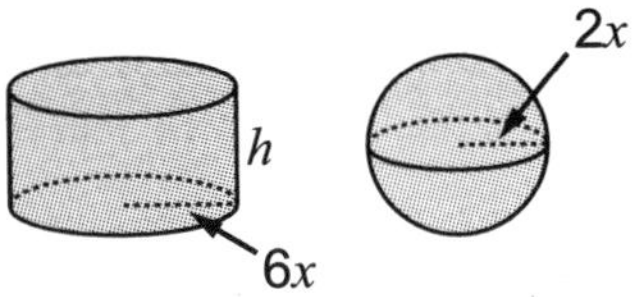

38. Find the height in terms of x of a cylinder whose base radius is $3x$, and whose volume is the same as the volume of a sphere whose radius is x.

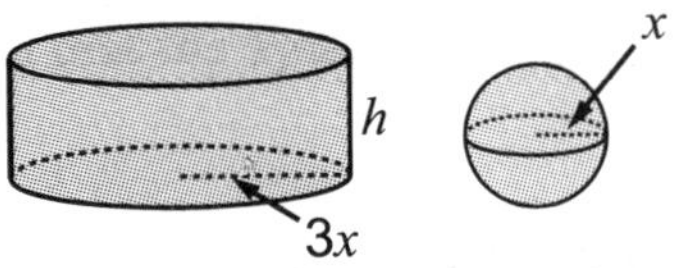

39. Find the height in terms of x of a rectangular prism whose length and width is x, and whose volume is the same as the volume of a cube whose side is x.

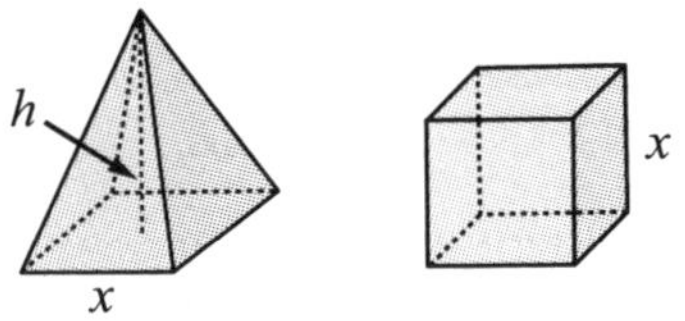

40. Find the height in terms of x of a rectangular prism whose length and width is $5x$, and whose surface area is the same as the surface area of a cube whose side is $4x$.

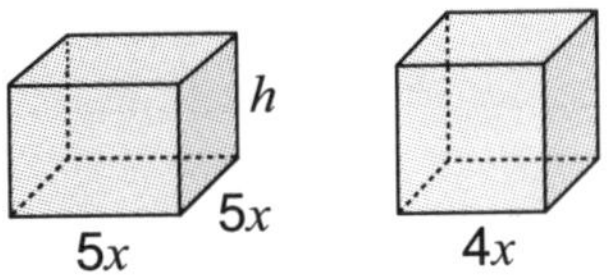

HA1-893: Constructing Solids from Different Perspectives

In this lesson we will see how to match the drawing of a three-dimensional object with its two-dimensional views from the top, front, and side. In the following examples, we will show three views of three-dimensional objects: the view from the top of the object, the view from the viewer's left side, and the view of the front of the object.

We will start with a cube. The vertices are labeled to make it easy to identify the top, front, and side faces. Here we see the view of the cube from one corner that is called the isometric drawing of the cube.

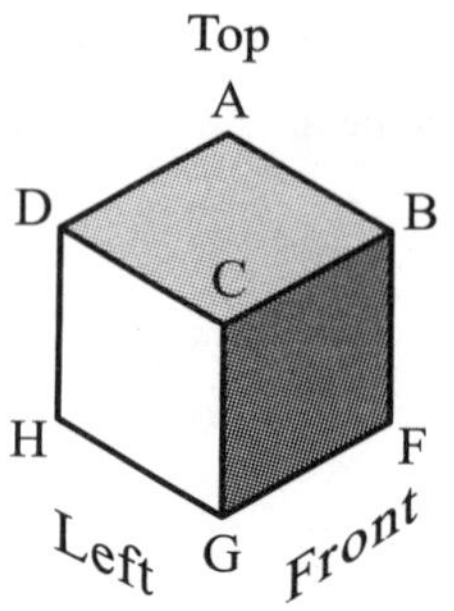

When we view this cube from the top, we see the square ABCD.

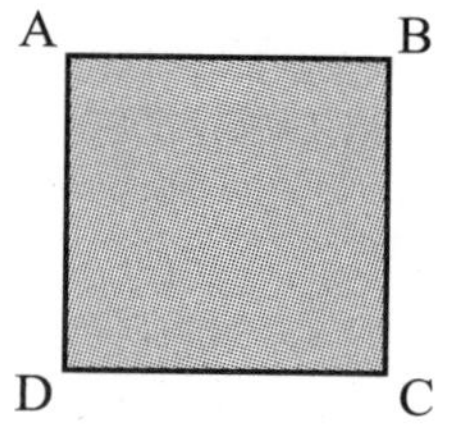

When we view it from the front, we see the square CBFG.

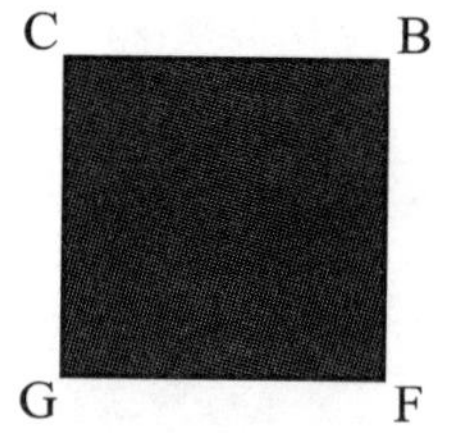

When we view the cube from the left side of this diagram, we see the square DCGH.

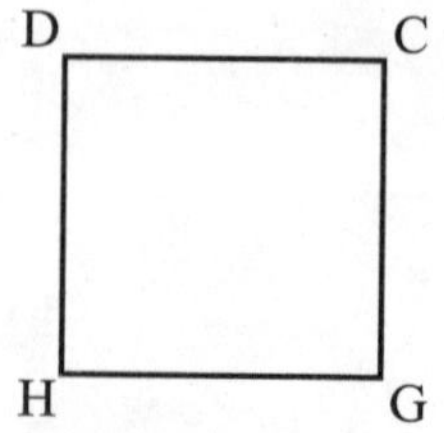

Example 1 Consider the following isometric drawing of a cube. Which of the following is the top view of the cube?

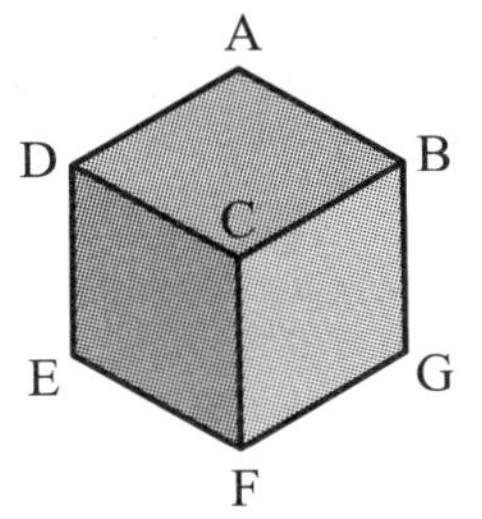

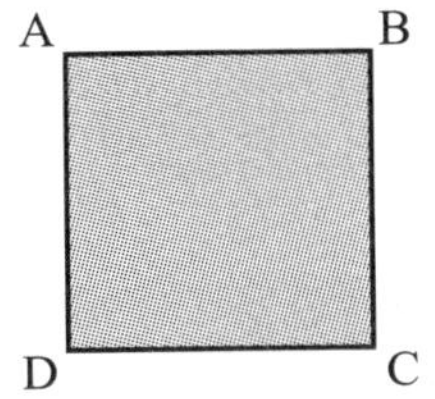

Step 1: When we view the cube from the top we see the square ABCD.

Answer: The top view of the cube is ABCD.

Example 2 Consider the following isometric drawing of a triangular pyramid. Which of the following is the left view of the pyramid?

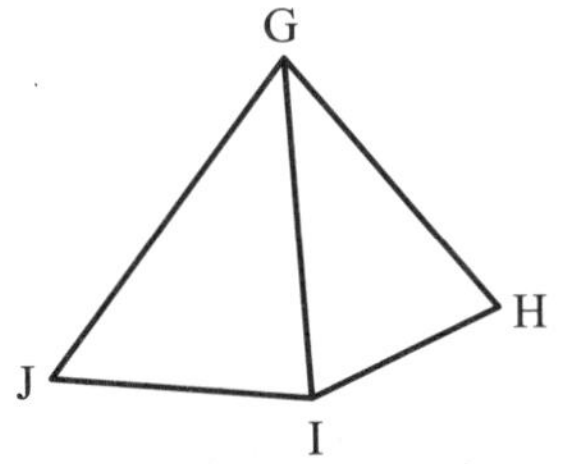

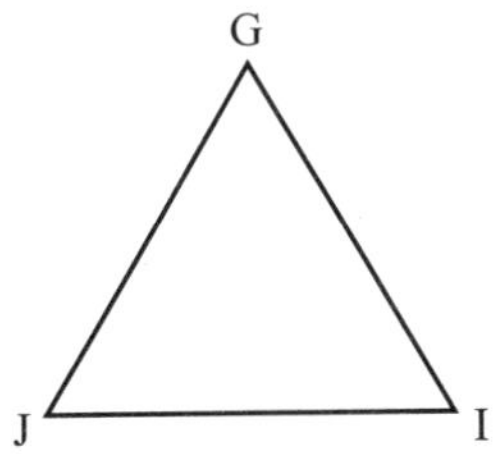

Step 1: When we view the pyramid from the left we see the triangle GIJ.

Answer: The left view of the triangular pyramid is GIJ.

Example 3 What figure could we have if the view from the top is a circle and the front and left views show squares?

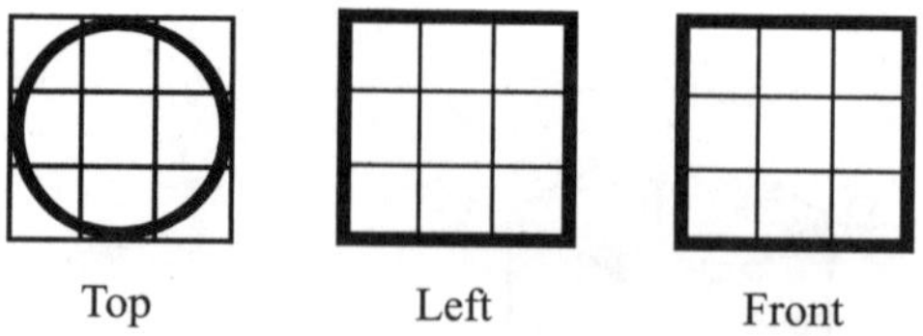

Step 1: A cylinder is a 3-dimensional geometric figure that can have a circular top view and square views from the front and left.

Answer: These views form a cylinder.

Let's look at the following solid figure. The top, left, and front views are shown.

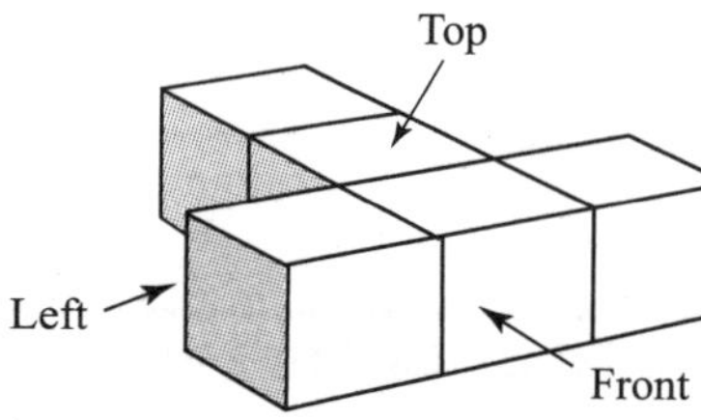

When we view the figure from the top, we see an inverted letter "T."

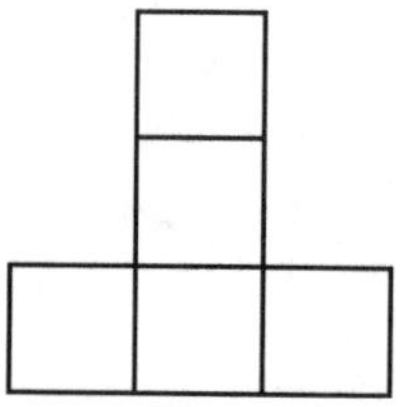

Top

When we view the figure from the left, we see three blocks.

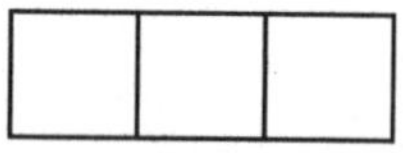

Left

When we view the figure from the front, we also see three blocks.

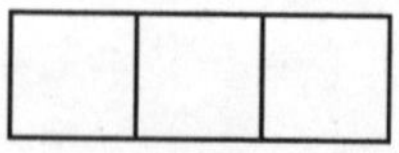

Front

Example 4 For the solid figure given, draw the top, front, and left views.

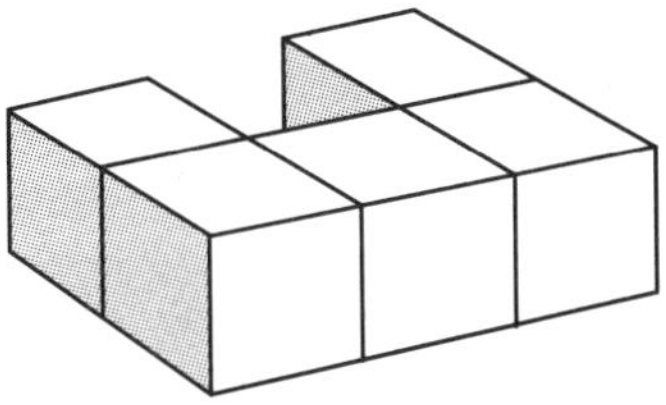

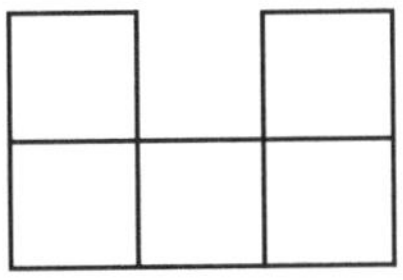

Top

Step 1: The top view will show a "U" shape.

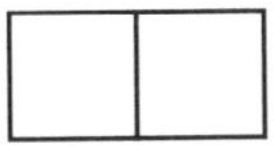

Left

Step 2: The left view will show two blocks.

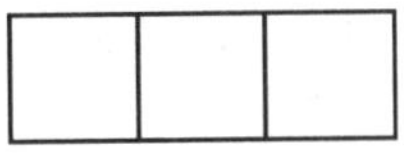

Front

Step 3: The front view will show three blocks.

Answer: See the diagrams above.

Example 5 Draw the figure that has the following top, front, and left views.

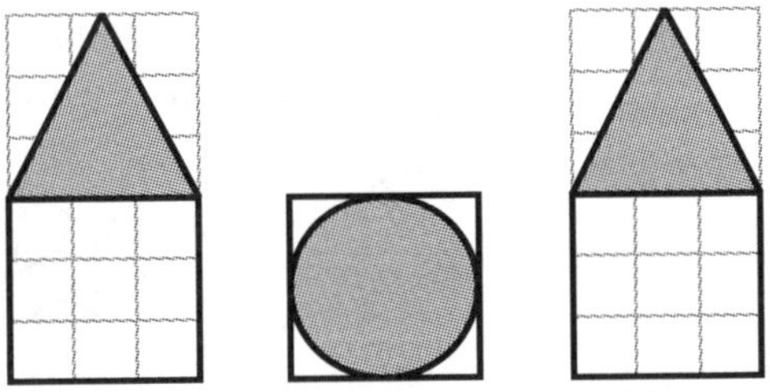

Left Top Front

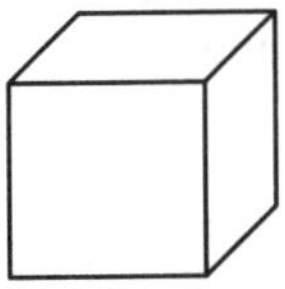

Step 1: The two different shades (gray and white) indicate two different shapes. The white figure has a top, left, and front view of a square. Therefore, we conclude that the white figure on the bottom is a cube.

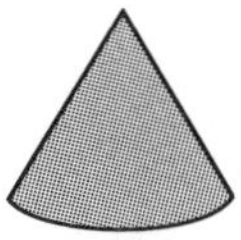

Step 2: The front and left views of the top figure are triangles. Since the top view of the top figure is a circle, we conclude that the top figure is a cone.

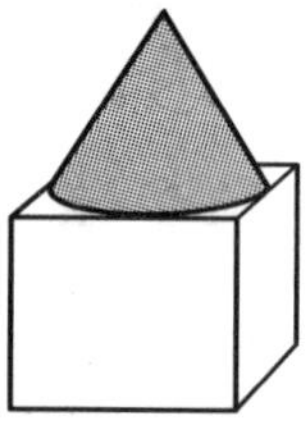

Step 3: Combining the two figures gives a solid with a bottom figure of a cube and a top figure of a cone.

Answer: See the figure in Step 3.

Problem Set

1. Consider the following isometric drawing of a cube. Name the top view of the cube.

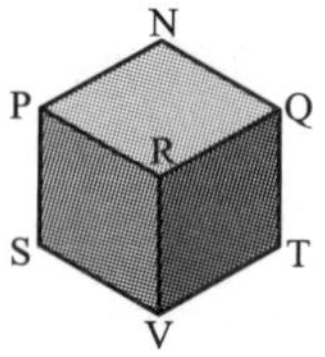

2. Consider the following isometric drawing of a cube. Name the left view of the cube.

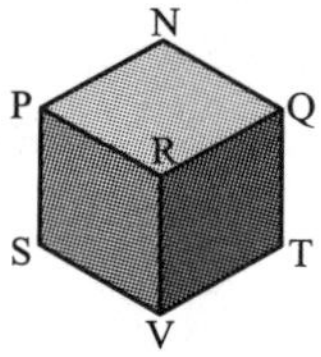

3. Consider the following isometric drawing of a cube. Name the front view of the cube.

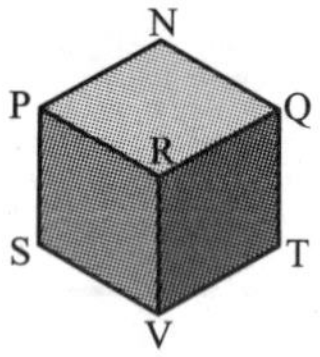

4. Consider the following isometric drawing of a cube. Name the top view of the cube.

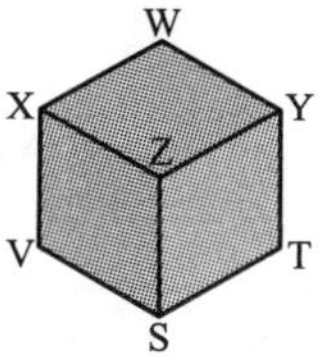

5. Consider the following isometric drawing of a cube. Name the left view of the cube.

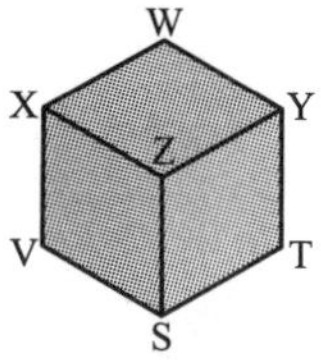

6. Consider the following isometric drawing of a cube. Name the front view of the cube.

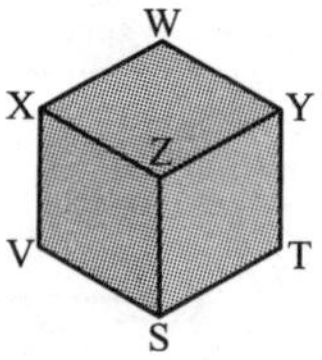

7. Consider the following isometric drawing of a cube. Name the left view of the cube.

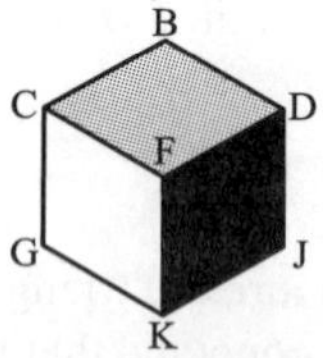

8. Consider the following isometric drawing of a cube. Name the front view of the cube.

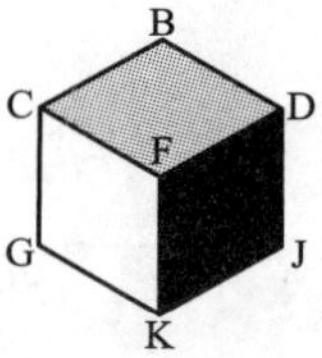

9. Consider the following isometric drawing of a pyramid. Name the left view of the pyramid.

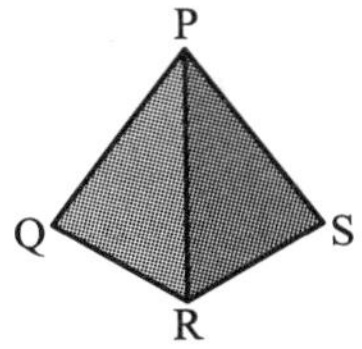

10. Consider the following isometric drawing of a pyramid. Name the front view of the pyramid.

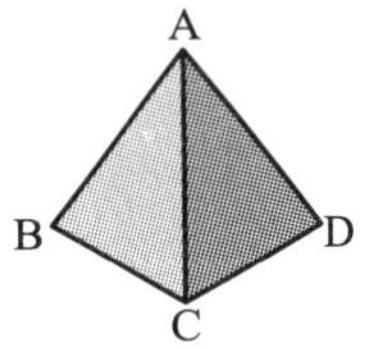

11. Consider the following isometric drawing of a pyramid. Name the left view of the pyramid.

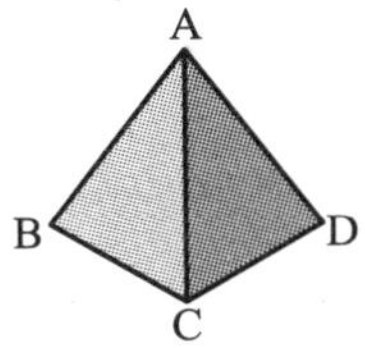

12. Consider the following isometric drawing of a pyramid. Name the front view of the pyramid.

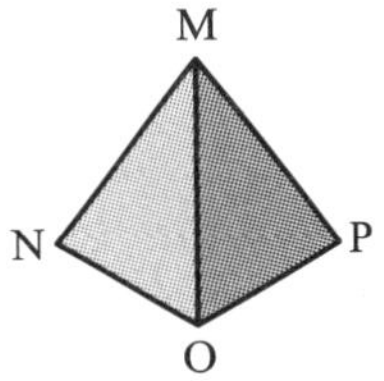

13. Consider the following isometric drawing of a pyramid. Name the left view of the pyramid.

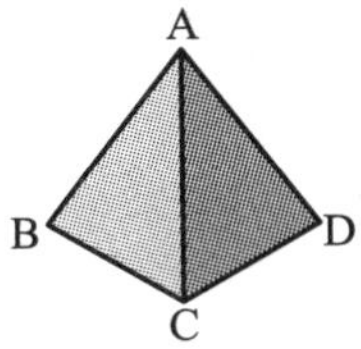

14. Consider the following isometric drawing of a pyramid. Name the front view of the pyramid.

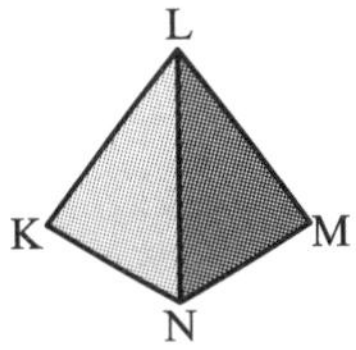

15. Consider the following isometric drawing of a pyramid. Name the left view of the pyramid.

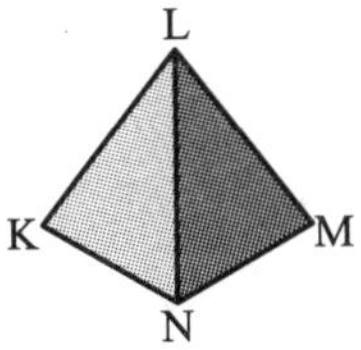

16. What geometric figure could we have if the view from the top is a circle and the front and left views are triangles as shown?

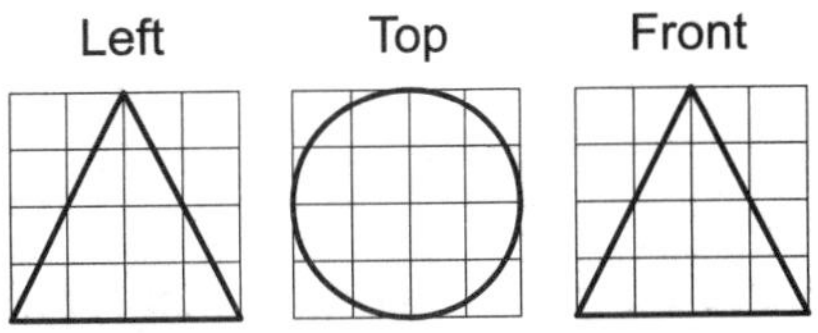

17. What geometric figure could we have if the view from the top is a square and the front and left views are congruent squares as shown?

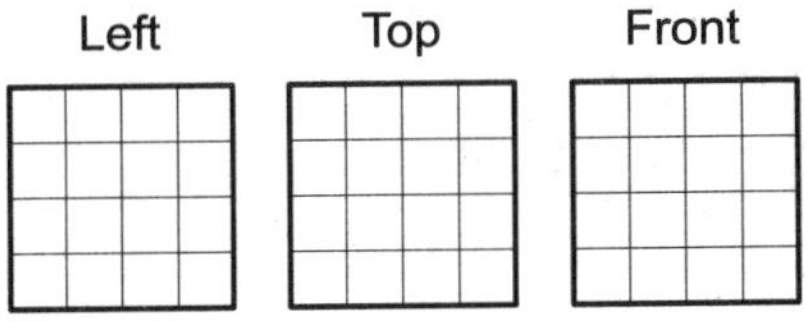

18. What figure could we have if the view from the top is a circle and the front and left views are rectangles as shown?

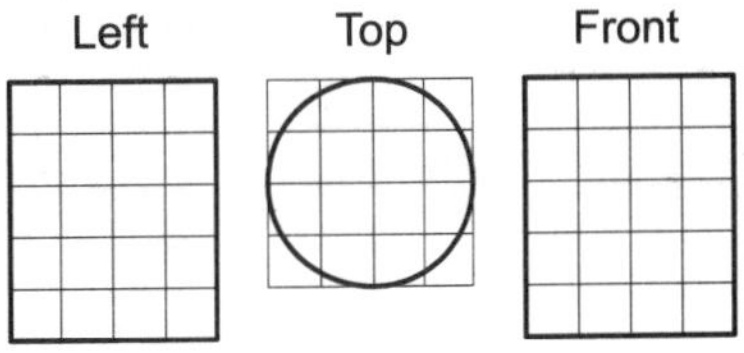

19. What figure could we have if the view from the top is a square and the front and left views are triangles shown?

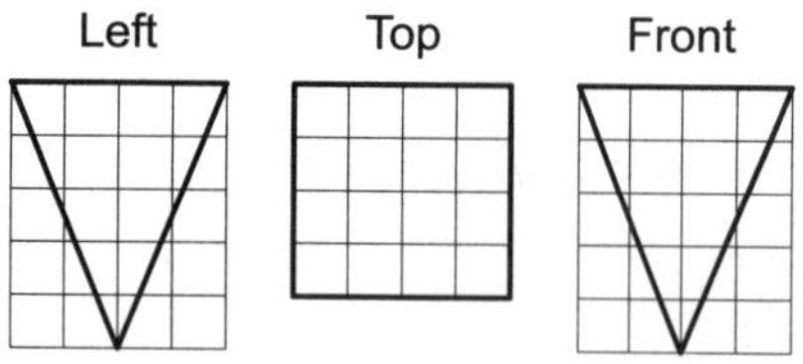

20. What figure could we have if the view from the top is a square and the views from the left and front are rectangles as shown?

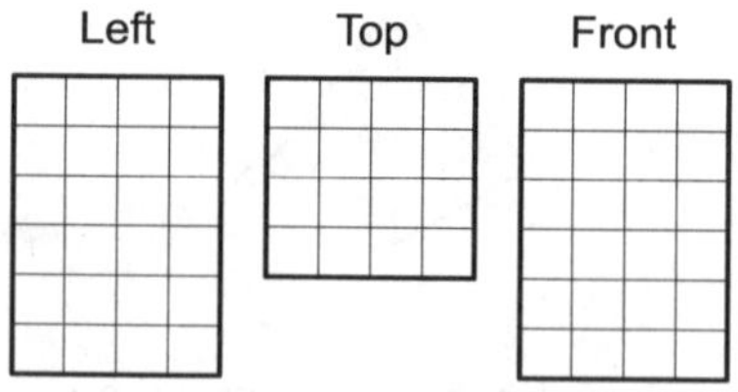

21. What figure could we have if the view from the top is a circle and the front and left views are squares?

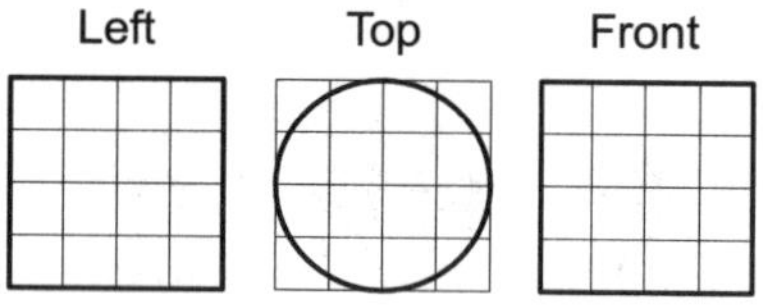

22. What figure could we have if the view from the top is a triangle and the views from the left and front are squares as shown?

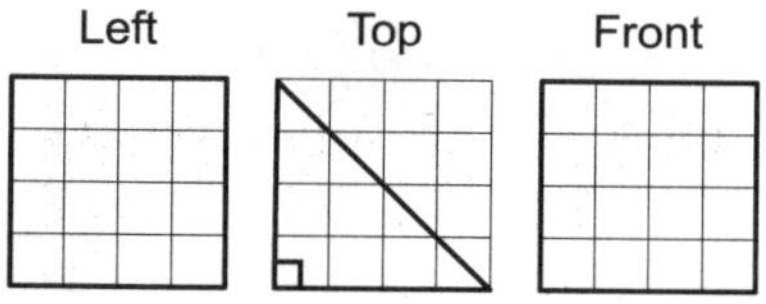

23. What figure could we have if the view from the top is a circle and the views from the front and left are triangles as shown?

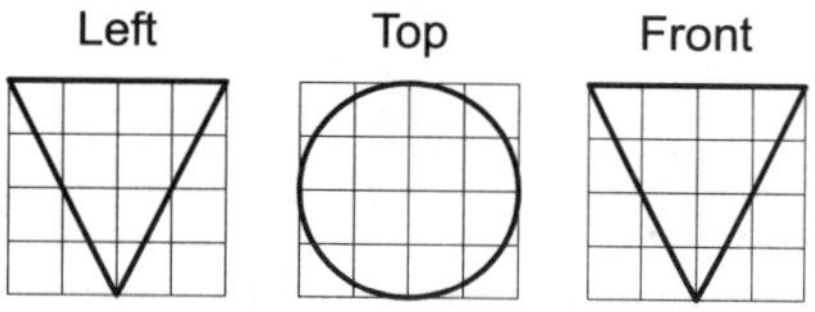

24. Construct the isometric drawing of the solid figure that is represented by the three views that are given.

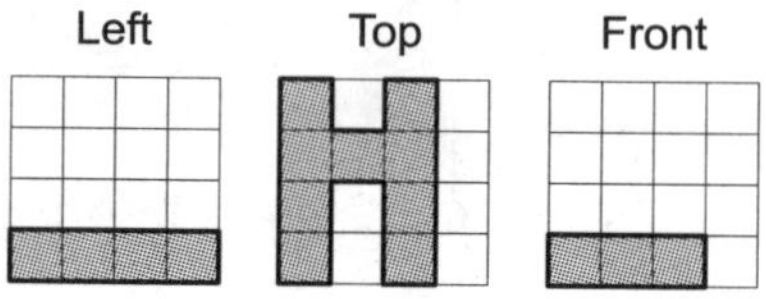

25. Construct the isometric drawing of the solid figure that is represented by the three views that are given.

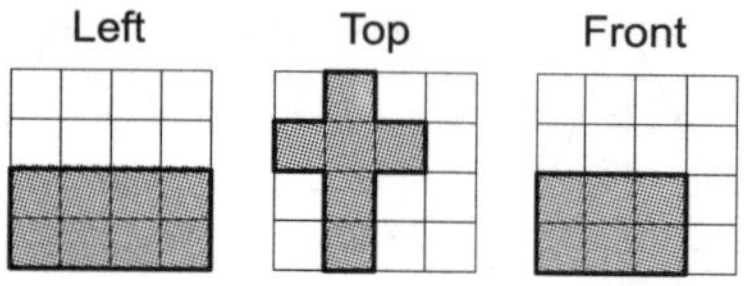

26. Construct the isometric drawing of the solid figure that is represented by the three views that are given.

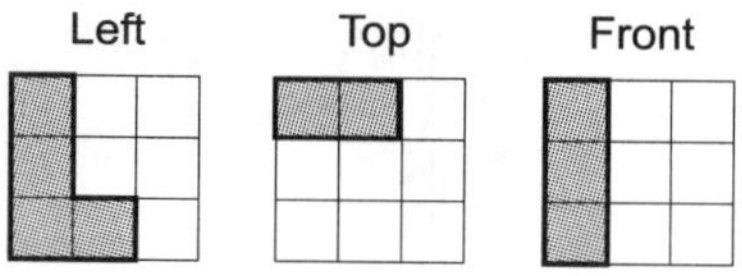

27. Construct the isometric drawing of the solid figure that is represented by the three views that are given.

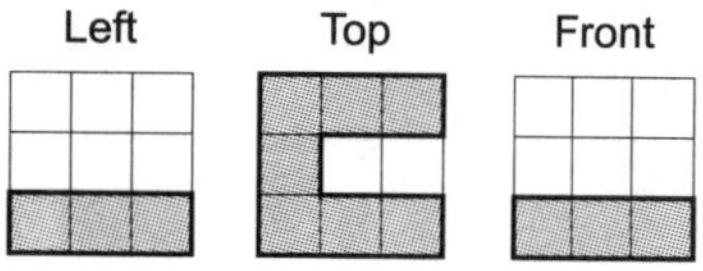

28. Construct the isometric drawing of the solid figure that is represented by the three views that are given.

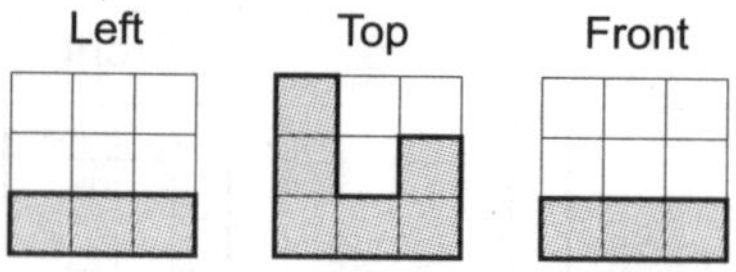

29. Construct the isometric drawing of the solid figure that is represented by the three views that are given.

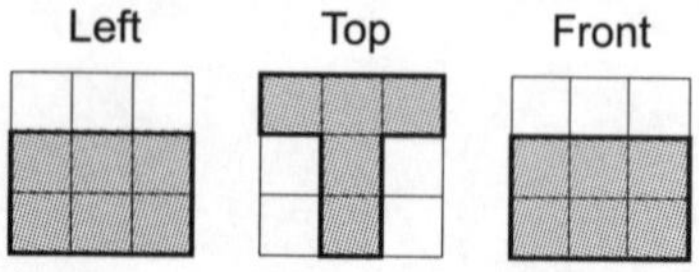

30. Construct the isometric drawing of the solid figure that is represented by the three views that are given.

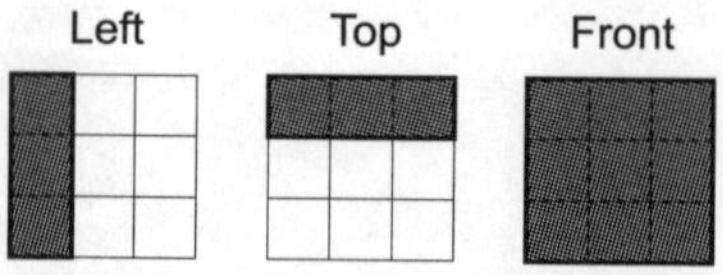

31. Construct the isometric drawing of the object that could be represented by the three views that are given.

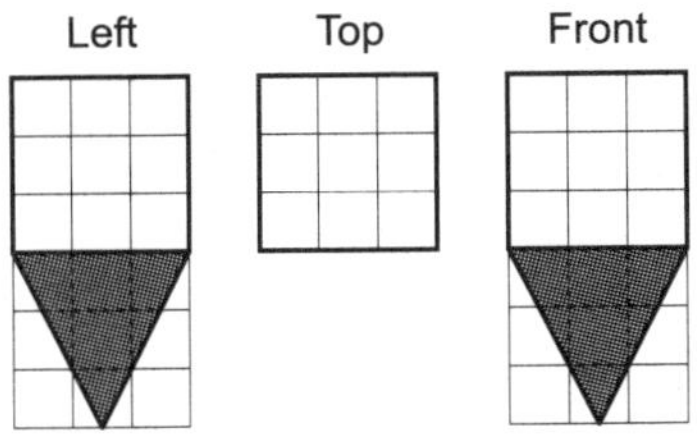

32. Construct the isometric drawing of the object that could be represented by the three views that are given.

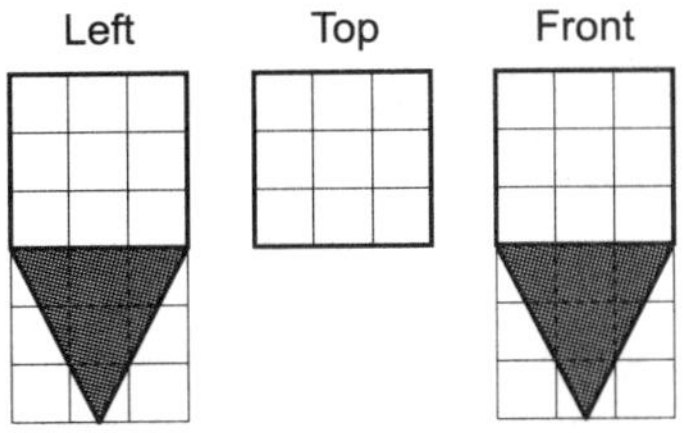

33. Construct the isometric drawing of the solid figure that is represented by the three views that are given.

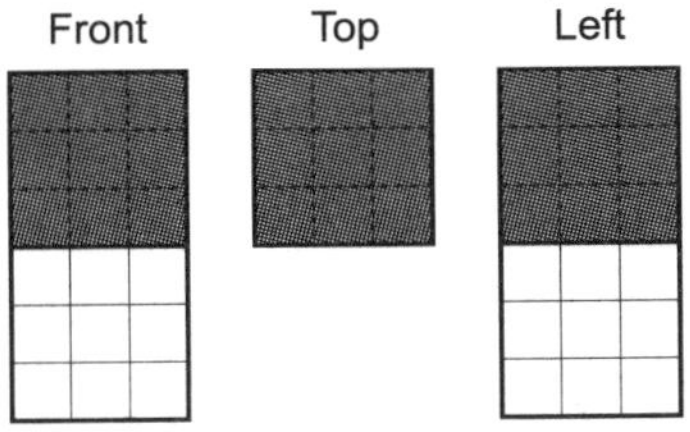

34. Construct the isometric drawing of the solid figure that is represented by the three views that are given.

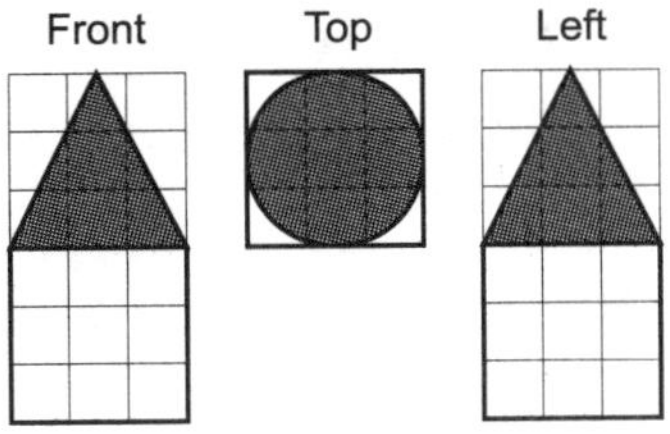

35. Construct the isometric drawing of the solid figure that is represented by the three views that are given.

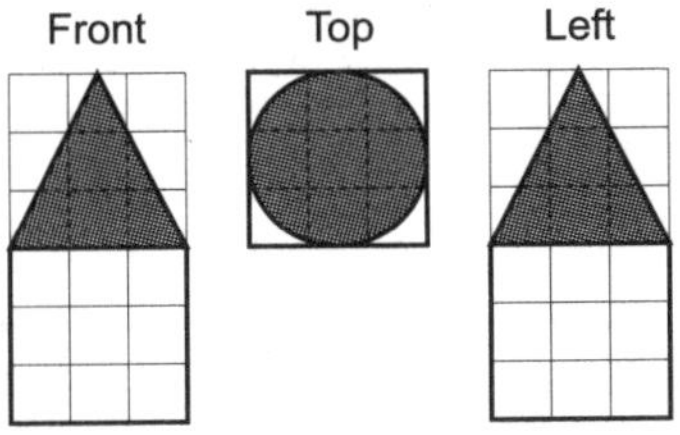

36. Construct the isometric drawing of the solid figure that is represented by the three views that are given.

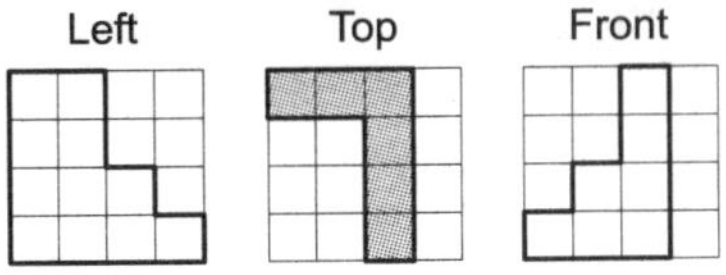

37. Construct the isometric drawing of the solid figure that is represented by the three views that are given.

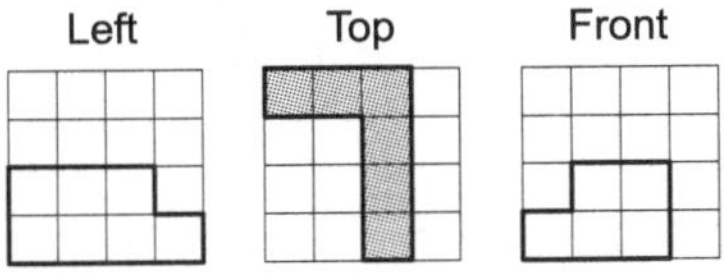

38. Construct the isometric drawing of the solid figure that is represented by the three views that are given.

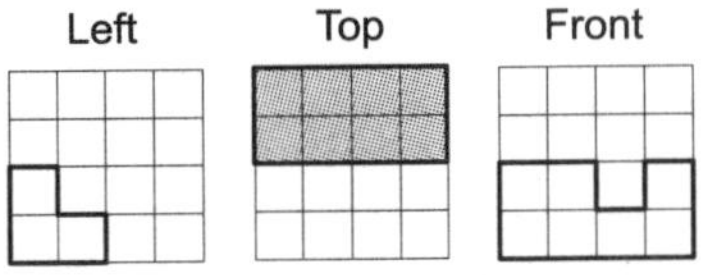

39. Construct the isometric drawing of the solid figure that is represented by the three views that are given.

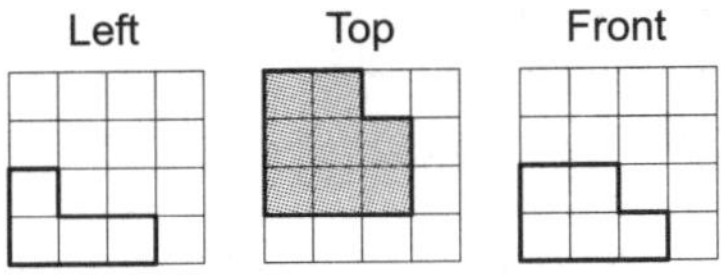

40. Construct the isometric drawing of the solid figure that is represented by the three views that are given.

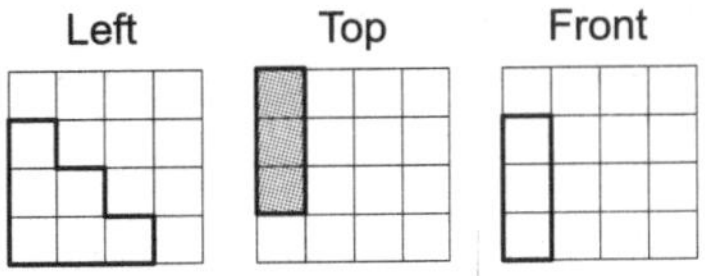

HA1-920: Simplifying Algebraic Expressions Using the Distributive Property

In this lesson we will learn how to simplify algebraic expressions using the Distributive Property. To use this property, we multiply the term outside the parentheses by each term inside the parentheses. The same procedure can be applied when multiplying a monomial by a polynomial or when multiplying two binomials. Remember when we multiply two binomials, or a binomial and a trinomial, we use the Distributive Property of multiplication repeatedly.

Distributive Property of Multiplication	For real numbers a, b, and c: $a(b+c) = ab+ac$.
Product of Two Binomials	$(a+b)(c+d)$ $= ((a+b)c+(a+b)d)$ $= ac+bc+ad+bd$
Product of a Binomial and a Trinomial	$(a+b)(c+d+e)$ $= (a+b)c+(a+b)d+(a+b)e$ $= ac+bc+ad+bd+ae+be$

Example 1 Simplify $4x^2(2x^3+3x^2)$.

$4x^2(2x^3+3x^2)$
$= (4x^2)(2x^3)+(4x^2)(3x^2)$

Step 1: In order to solve this equation, we must multiply the term outside the parentheses by each term inside the parentheses. We will do this by using the Distributive Property of multiplication. Therefore, we get $(4x^2)(2x^3)+(4x^2)(3x^2)$.

$= (4\cdot 2\cdot x^2\cdot x^3)+(4\cdot 3\cdot x^2\cdot x^2)$
$= 8x^5+12x^4$

Step 2: Now we must simplify the equation. Remember when multiplying terms with the same base we need to add the exponents. So $(4)(2) = 8$ and $(x^2)(x^3) = x^5$; $(4)(3) = 12$ and $(x^2)(x^2) = x^4$.

Answer: The result is $8x^5+12x^4$.

Example 2 Simplify $x^2(x-2x^4)$.

$x^2(x-2x^4) = (x^2)(x)-(x^2)(2x^4)$

Step 1: In order to solve this equation we must multiply the term outside the parentheses by each term inside the parentheses.

$= (x^2)(x) - (x^2)(2x^4)$
$= x^3 - 2x^6$

Step 2: Now we will simplify the equation by multiplying the terms in parentheses.

Answer: The result is $x^3 - 2x^6$.

Example 3 Simplify $x^2y^2(2xy + 3x^2y^2 + 4xy^3)$.

$x^2y^2(2xy + 3x^2y^2 + 4xy^3)$
$=(x^2y^2)(2xy) + (x^2y^2)(3x^2y^2) + (x^2y^2)(4xy^3)$

Step 1: In order to solve this equation we must multiply the term outside the parentheses by each term inside the parentheses.

$=(x^2y^2)(2xy) + (x^2y^2)(3x^2y^2) + (x^2y^2)(4xy^3)$
$=2x^3y^3 + 3x^4y^4 + 4x^3y^5$

Step 2: We now need to multiply all the terms in parentheses and add their results.

Answer: Since there are no like terms to add, the answer is $2x^3y^3 + 3x^4y^4 + 4x^3y^5$.

Example 4 Simplify $(2x^3 + x^2)(4x^2 + 3x^2)$.

$(2x^3 + x^2)(4x^2 + 3x^2)$
$=(2x^3 + x^2)(4x^2) + (2x^3 + x^2)(3x^2)$

Step 1: We will use the Distributive Property of multiplication to solve this problem. In this case we will multiply the expression in the first set of parentheses by each term in the second set of parentheses.

$= (2x^3)(4x^2) + (2x^3)(3x^2) + (x^2)(4x^2) + (x^2)(3x^2)$
$= 8x^5 + 6x^5 + 4x^4 + 3x^4$
$= 14x^5 + 7x^4$

Step 2: We now need to multiply all the terms in parentheses, add their results and combine like terms.

Answer: The answer is $14x^5 + 7x^4$.

Example 5 Simplify $(3x^3y^2 + 4x^2y)(5x^2y^3 + xy)$

$(3x^3y^2 + 4x^2y)(5x^2y^3 + xy)$
$= (3x^3y^2 + 4x^2y)(5x^2y^3) + (3x^3y^2 + 4x^2y)(xy)$

Step 1: To solve this equation we need to use the Distributive Property to multiply the first expression by each term in the second expression.

$=(3x^3y^2)(5x^2y^3) + (4x^2y)(5x^2y^3) + (3x^3y^2)(xy) + (4x^2y)(xy)$
$= 15x^5y^5 + 20x^4y^4 + 3x^4y^3 + 4x^3y^2$

Step 2: We now need to multiply all the terms in parentheses and add their results. Since there are no like terms the answer is stated here.

Answer: Since there are no like terms the answer is
$15x^5y^5 + 20x^4y^4 + 3x^4y^3 + 4x^3y^2$

Example 6 Simplify $(xy^2)(x^3y^2 + xy^2 + x)$

$$(xy^2)(x^3y^2 + xy^2 + x)$$
$$= (xy^2)(x^3y^2) + (xy^2)(xy^2) + (xy^2)(x)$$

Step 1: To solve this equation we need to use the Distributive Property to multiply the first expression by each term in the second expression.

$$= (xy^2)(x^3y^2) + (xy^2)(xy^2) + (xy^2)(x)$$
$$= x^4y^4 + x^2y^4 + x^2y^2$$

Step 2: We now need to multiply all the terms in parentheses and add their results.

Answer: The answer is $x^4y^4 + x^2y^4 + x^2y^2$.

Problem Set

Simplify completely:

1. $3x^3(7x^5 - 3x^2)$

2. $2x^2(7x^7 - 3x^4)$

3. $x^3(7x^7 - 3x^4)$

4. $4x^3(7x^7 - 3x^4)$

5. $4x^3(7x^8 - 2x^4)$

6. $x^3y^2(4x^3y - 3xy^2 - 3x^3y)$

7. $x^3y^2(4xy + 3x^2y - 3x^3y)$

8. $x^3y^2(4xy^2 + 3x^2y^2 - 3x^3y)$

9. $x^3y^2(4xy - 3xy + 3x^3y)$

10. $x^3y^2(4xy + 3xy + 8xy)$

11. $(2x^3 + x^2)(3x^2 - 4x)$

12. $(2x^3 - x^2)(3x^2 - 4x)$

13. $(3x^3 + x^2)(3x^2 + 4x)$

14. $(3x^3 + x^2)(x^2 + 4x)$

15. $(5x^3 + x^2)(x^2 + 4x)$

16. $(2x^3y^2 - 3x^2y)(2x^2y^3 + 4xy)$

17. $(2x^3y^2 + 3x^2y)(2x^2y^3 - 4xy)$

18. $(2x^3y^2 + 3x^2z)(2x^2y^3 - 4xz)$

19. $(2x^3y^2 - 3x^2z)(2x^2y^3 - 4xz)$

20. $(2x^3y^2 - 3x^2z)(2x^2y^3 - 5xz)$

21. $(2x^3y^2 - 3x^2y)(2x^2y^3 - 5xy)$

22. $(2x^2 + x)(3x^2 - 4x - 9)$

23. $(2x^2 + x)(3x^2 + 4x + 9)$

24. $(3x^2 - x)(3x^2 + x + 9)$

25. $(3x^2 + x)(3x^2 - 4x + 9)$

26. $(3x^2 - x)(3x^2 + 4x + 9)$

27. $(3x^2 + x)(3x^2 - x + 9)$

28. $(3x^2 + 2)^2$

29. $(3x^2 + 2y)^2$

30. $(3x^2 - 2y)^2$

31. $(3x^2 + 4y)^2$

32. $(4x^2 + 4y)^2$

33. $(4x^2 - 4y)^2$

34. $(3x^2 - 3y)^2$

35. $(4x^2 + 3y)^2$

36. $(2xy^2 + 3x^2y)(3x^3y^2 + 4xy^2 - 6x)$

37. $(2xy^2 + 3x^2y)(3x^3y^2 + 4xy^2 + 6x)$

38. $(2xy^2 - 3x^2y)(3x^3y^2 + 4xy^2 + 6x)$

39. $(2xy^2 + 3x^2y)(-3x^3y^2 + 4xy^2 + 6x)$

40. $(4xy^2 - 3x^2y)(3x^3y^2 - 4xy^2 + 6x)$

Answers to Selected Exercises

HA1-003: Order of Operations

1. 23 **3.** 53 **5.** 87 **7.** 34 **9.** 20
11. 15 **13.** 925 **15.** 6 **17.** 30 **19.** 12

HA1-005: Evaluating Algebraic Expressions

1. 18 **3.** 4 **5.** 18 **7.** 40 **9.** 11
11. 22 **13.** 60 **15.** 240 **17.** 48 **19.** 2

HA1-015: Graphing Real Numbers Using a Number Line

1. –2 **3.** H **5.** L **7.** –2 **9.** –6
11. 1.25 **13.** 0 **15.** C **17.** $\frac{1}{2}$ **19.** 0

HA1-020: Classifying Numbers into Subsets of Real Numbers

1. Integers and Real **3.** 0, 4 **5.** Whole **7.** Irrational
9. Integers **11.** 2, 14 **13.** Rational **15.** Real
17. Rational **19.** π

HA1-025: Comparing and Ordering Real Numbers

1. $>$ **3.** $=$ **5.** $>$ **7.** True **9.** $1 > 2$
11. –1 **13.** 2 **15.** $<$ **17.** $=$ **19.** 3.25

HA1-030: Using Opposites and Absolute Values

1. 14 **3.** –29 **5.** –3.65 **7.** 39 **9.** 84
11. –1.8 **13.** –3.7 **15.** 256 **17.** –12.8 **19.** –7.1

HA1-035: Adding Real Numbers Using a Number Line

1. $5 + (-2) = 3$ **3.** $-3 + 2 = -1$ **5.** $-2 + (-1) = -3$
7. $-2 + 4 = 2$ **9.** $-3 + (-3) = -6$ **11.** $-2 + (-3) + 6 = 1$
13. $-2 + (-3) + (-3) = -8$ **15.** $-8 + 3 + (-3) = -8$ **17.** $-3 + (-6) + 5 = -4$
19. $-1.50 + 0.75 = -0.75$

HA1-040: The Addition Rule for Real Numbers

1. –8	**3.** 13	**5.** Negative	**7.** Negative	**9.** Negative
11. –28	**13.** –8	**15.** 21	**17.** –39	**19.** 7

HA1-045 Subtracting Real Numbers

1. –5	**3.** –4	**5.** 15	**7.** –97	**9.** –55
11. –6	**13.** –15.9	**15.** –4.5	**17.** 0.5	**19.** $-\frac{1}{6}$

HA1-050: Multiplying Real Numbers

1. 120	**3.** Negative	**5.** Negative	**7.** Zero	**9.** Negative
11. 0	**13.** 0	**15.** 0	**17.** –160	**19.** $\frac{80}{3}$

HA1-055: Dividing Real Numbers

1. 0	**3.** Negative	**5.** Negative	**7.** Negative	**9.** 3
11. Undefined	**13.** 6	**15.** 14	**17.** 20	**19.** $\frac{8}{9}$

HA1-060: Evaluating Expressions Using the Order of Operations

1. 26	**3.** –4	**5.** –23	**7.** 47	**9.** –15
11. 41	**13.** –3	**15.** –2	**17.** 2	**19.** 5

HA1-062: Adding, Subtracting, Multiplying, and Dividing Real Numbers

1. –15	**3.** –13	**5.** –15	**7.** –15	**9.** –20
11. –18	**13.** –12	**15.** –20	**17.** 16	**19.** 3
21. 4	**23.** 18	**25.** –7	**27.** –7	**29.** 11
31. –2	**33.** –14	**35.** 1	**37.** 4	**39.** 6

HA1-065: Evaluating Expressions Containing Exponents

1. –25	**3.** –100	**5.** 7	**7.** $a^2 \cdot b^2$
9. $2 \cdot 2 \cdot 2 \cdot 5 \cdot 5 \cdot 5 \cdot 5$	**11.** 98	**13.** 36	**15.** –64
17. $\frac{4}{9}$	**19.** 27		

HA1-070: Evaluating Formulas for Given Values of the Variables

1. 8.64 **3.** 45 **5.** 7,000 **7.** –46 **9.** –51

11. 14.13 **13.** 240 **15.** 7 **17.** 5,024 **19.** 395,718.5 square feet

HA1-075: Simplifying Algebraic Expressions by Combining Like Terms

1. Yes **3.** Yes **5.** 1 **7.** No **9.** $-2a$

11. $5d+4$ **13.** $3r-8$ **15.** $11-2h$ **17.** $-8+7k$ **19.** $22x+8y-7$

HA1-076: Basic Distributive Property

1. $5a+20$ **3.** $4c+28$ **5.** $7g+77$ **7.** $10c+10a$ **9.** $15c-45$

11. $30-10p$ **13.** $12n-72$ **15.** $7a-7b$ **17.** $15x-3$ **19.** $20x+15$

21. $4x-3$ **23.** $2x+16$ **25.** $3x+21$ **27.** $8x+40$ **29.** $-8x-32$

31. $\frac{2}{3}x+8$ **33.** $\frac{4}{5}x+20$ **35.** $\frac{3}{7}x+12$ **37.** $\frac{5}{8}x+30$ **39.** $\frac{3}{4}x+33$

HA1-079: Using a Concrete Model to Simplify Algebraic Expressions

1. $(2x+1)+(x+2)$ **3.** $(-3x+2)+(x-1)$ **5.** $(3x-2)+(-x+1)$ **7.** $(2x-3)+(-3x-2)$

9. 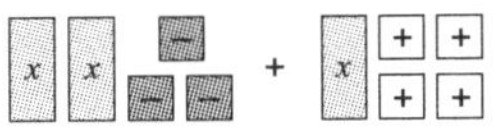 **11.** 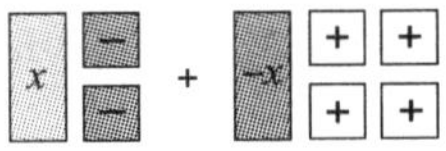 **13.**  **15.** 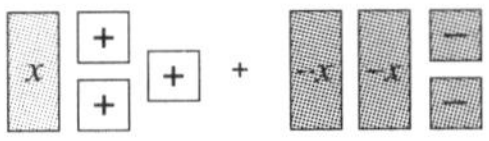

17. 2 **19.** $-4x+1$ **21.** $5x+1$ **23.** $x-1$ **25.** –2 **27.** $x-1$

29. $-3x+1$ **31.** $3x+4$ **33.** $x+10$ **35.** $-x-1$ **37.** $4x+2$ **39.** $-x+2$

HA1-080: Simplifying and Evaluating Algebraic Expressions Containing Grouping Symbols

1. $h+6$ **3.** $-5-h$ **5.** $-w^2-8$ **7.** $5+7h$ **9.** $3+9v^2$

11. $-h^2-7$ **13.** $5f-7$ **15.** –4 **17.** 90 **19.** $5m+76$

HA1-085: Simplifying Expressions Using the Properties of Real Numbers

1. 71 **3.** 77 **5.** No **7.** Yes

9. No **11.** Yes **13.** Transitive Property of Equality **15.** Commutative Property for Addition

17. $5(-2)+5(6)$ **19.** $-12 = 3n+9$

HA1-090: Simplifying Expressions Using the Property of –1

1. True **3.** False **5.** False **7.** False **9.** True

11. True **13.** True **15.** False **17.** False **19.** $8 - c$

HA1-095: Translating Word Phrases Into Algebraic Expressions

1. $k + m$ **3.** bd **5.** $k - t$ **7.** $n - 25$

9. $10n^2$ **11.** x multiplied by y **13.** k increased by c **15.** $8(n + 12)$

17. $2n + 28$ **19.** $p - 21$

HA1-100: Finding Solution Sets of Open Sentences from Given Replacement Sets

1. $\{1\}$ **3.** $\{-1, 0, 1, 2\}$ **5.** $\{0, 4\}$ **7.** $\{0, 7\}$ **9.** $\{0, 1.5\}$

11. $\{-2, -1\}$ **13.** $\{4\}$ **15.** $\{3, 4\}$ **17.** $\{15, 20\}$ **19.** $\{-1\}$

HA1-104: Translating Word Statements into Equations

1. $x - 9$ **3.** $x + 5$ **5.** $6b$ **7.** $\frac{75}{a}$

9. $3m = 45$ **11.** $7 - q = 30$ **13.** $n - 17 = 52$ **15.** $2g = 42$

17. $4y + 7$ **19.** $\frac{p}{15} - 3$ **21.** $3n - 19$ **23.** $77 - 4w = 22$

25. $42 = 8y - 12$ **27.** $5(z + 4) = 25$ **29.** $21 + \frac{4}{r} = 65$ **31.** $3\left(\frac{4}{n} + 7\right) = 50$

33. $\frac{b}{12} - 7(3) = 34$ **35.** $\frac{7c}{10} + 12 = 19$ **37.** $22 + (12 - 4g) = 30$ **39.** $35 + \frac{8}{3n} = 48$

HA1-105: Translating Word Statements into Inequalities

1. $n + 1 > 10$ **3.** $\frac{n}{2} < 5$ **5.** $n + 9 \leq 20$ **7.** $2n \leq 7$ **9.** $13 - n \leq 3$

11. $n + 2 > 6$ **13.** $\frac{x + 5}{7} \leq 13$ **15.** $\frac{x}{5} + 1 \leq 10$ **17.** $230 - x < 210$ **19.** $7x > 150$

21. $425 - x > 300$ **23.** $4x < 55$ **25.** $5.50x \leq 35$ **27.** $8 + 0.35x \leq 12$ **29.** $34 + 2x \geq 90$

31. $8x \geq 110$ **33.** $8x \geq 74$ **35.** $0.8x \geq 34$ **37.** $2.5x \geq 48$ **39.** $6x \geq 54$

HA1-115: Using the Addition and Subtraction Properties for Equations

1. 6 **3.** –22 **5.** –79 **7.** 76 **9.** –115

11. –59 **13.** 42 **15.** 43 **17.** –135 **19.** –15.5

HA1-120: Using the Multiplication and Division Properties for Equations

1. 2 **3.** Multiply by $\frac{5}{2}$ **5.** Multiply by $\frac{-2}{5}$ **7.** Multiply by $\frac{11}{9}$ **9.** Multiply by $\frac{5}{3}$

11. 28 **13.** 147 **15.** 56 **17.** –4 **19.** 28

HA1-124: Using a Concrete Model to Solve One- and Two-Step Equations

1. 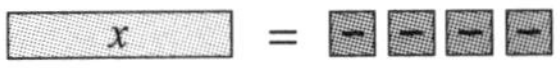

$x = -4$

3. 

$x = -2$

5. 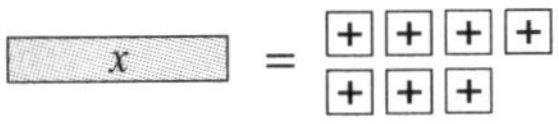

$x = 7$

7. 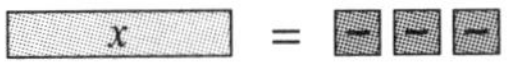

$x = -3$

9. 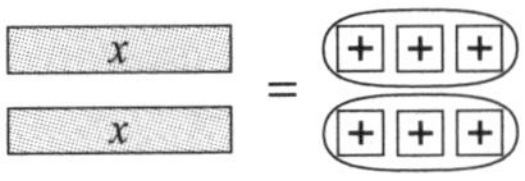

$x = 3$

11. 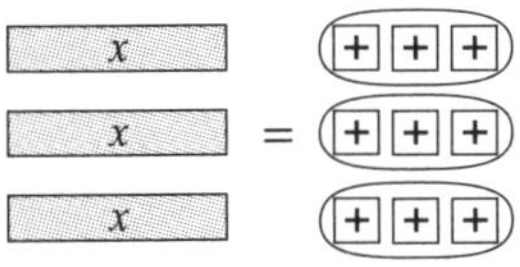

$x = 3$

13. 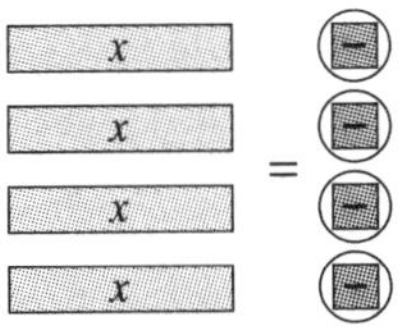

$x = -1$

15. 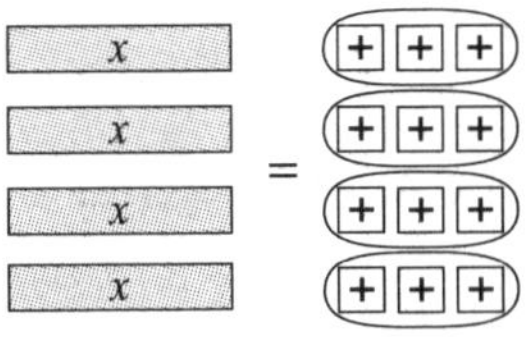

$x = 3$

17. 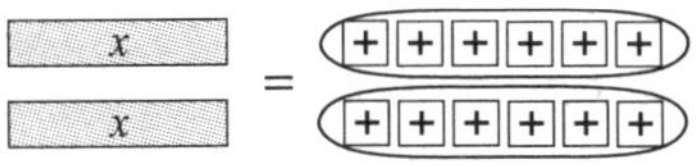

$x = 6$

Danielle's brother is 6 years old.

19. 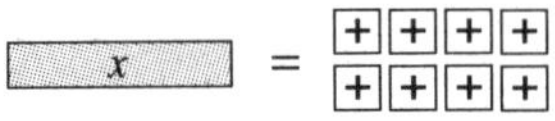

$x = 8$

Brenton worked 8 hours this week.

21. 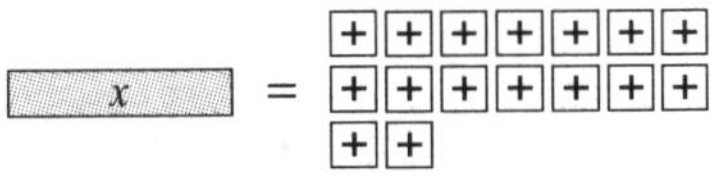

$x = 16$

Monique originally had $16.00.

23. 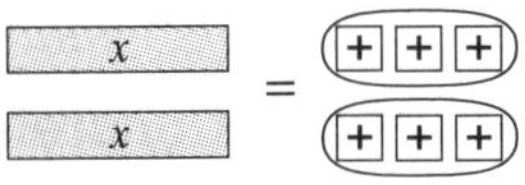

$x = 3$

25. 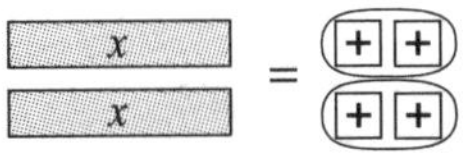

$x = 2$

27. 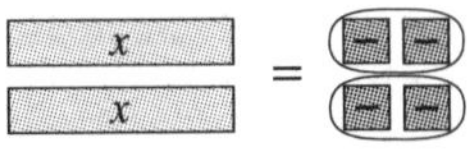

$x = -2$

29. 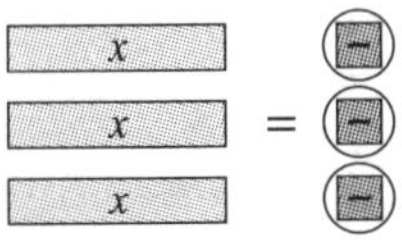

$x = -1$

31. 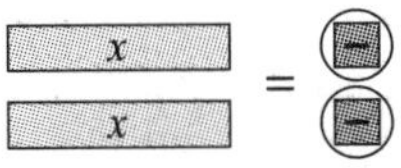

$x = -1$

33. 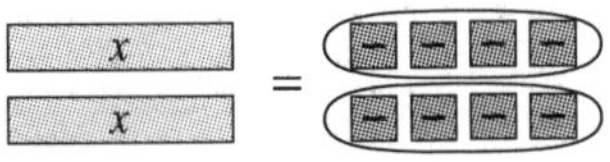

$x = -4$

35. 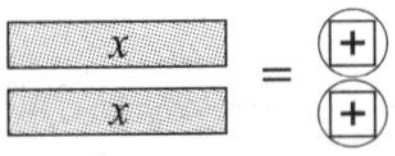

$x = 1$

37. 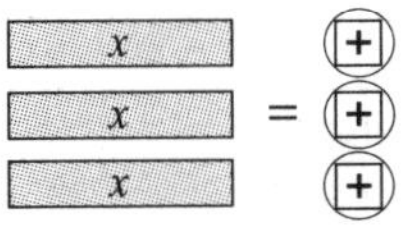

$x = 1$

39. 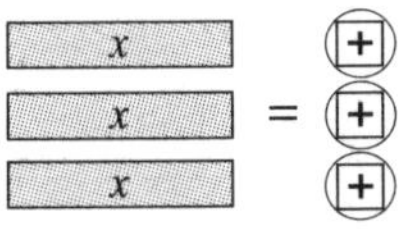

$x = 1$

HA1-125: Solving Equations Using More Than One Property

1. 13	**3.** –9	**5.** 10	**7.** 11	**9.** 22
11. –3	**13.** 13	**15.** 0	**17.** 42	**19.** –2.6

HA1-130: Identifying Postulates, Theorems, and Properties

1. Additive Identity

3. Commutative Property of Addition

5. Associative Property of Addition

7. Multiplicative Inverse Property

9. Multiplicative Identity

11. False

13. True

15. True

17. False

19. Commutative Property of Addition

HA1-135: Evaluating Formulas

1. 36	**3.** 135	**5.** 70	**7.** 20	**9.** 46
11. 16	**13.** 8 feet	**15.** 200	**17.** 200	**19.** 295 minutes

HA1-140: Solving Equations by Combining Like Terms

1. 4	**3.** –7	**5.** 7	**7.** –10	**9.** –12
11. 0	**13.** 14	**15.** $\frac{21}{5}$	**17.** –1	**19.** 14

HA1-144: Using a Concrete Model to Solve Equations with Variables on Both Sides

1. $x = 1$ **3.** $x = -3$ **5.** $x = -2$ **7.** $x = 3$

9. 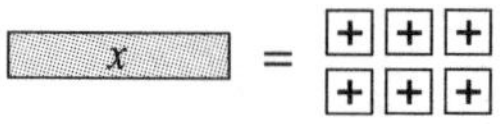

$x = 6$

11. 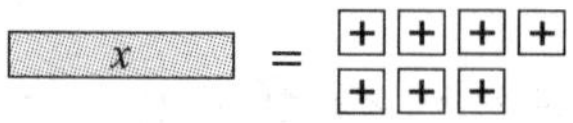

$x = 7$

13. 

$x = 1$

15. 

$x = -1$

17. 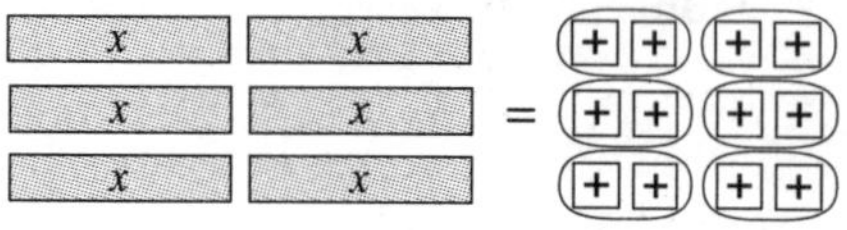

$x = 2$

19. 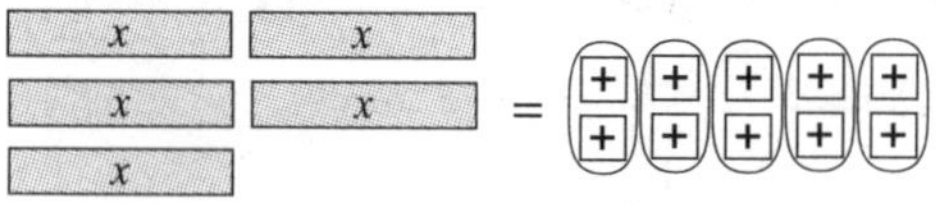

$x = 2$

21. 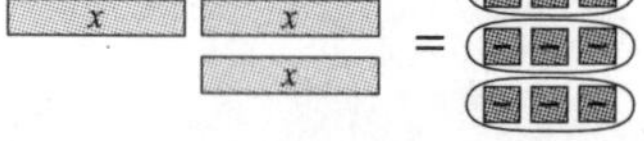

$x = -3$

23. 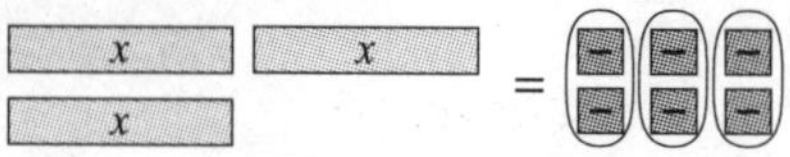

$x = -2$

25. 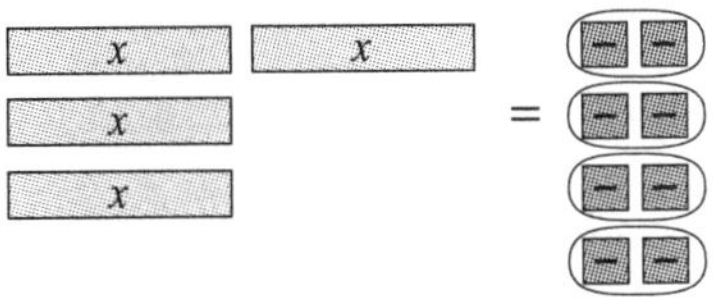

$x = -2$

27. 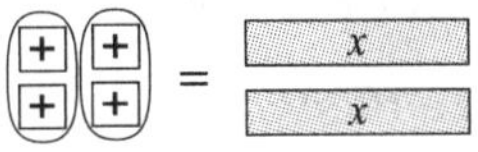

$x = 2$

29. 

$x = 1$

31. 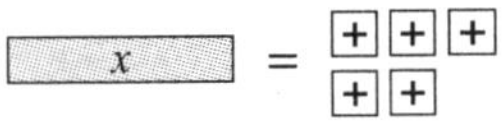

$x = 5$

33. 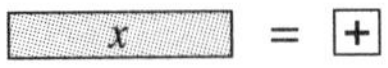

$x = 1$

35. 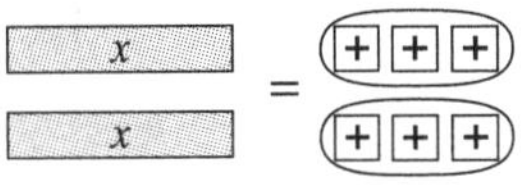

$x = 3$

37. 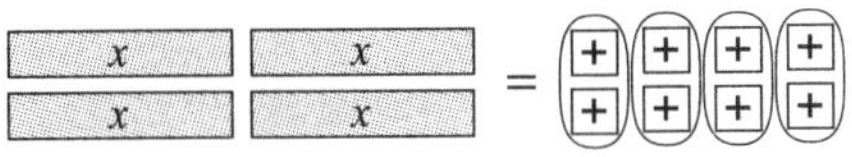

$x = 2$

39. 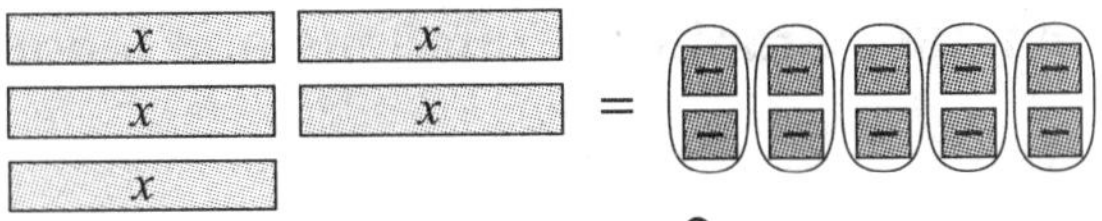

$x = -2$

HA1-145: Solving Equations with Variables on Both Sides

1. –6 **3.** –5 **5.** 6 **7.** 8 **9.** –8

11. 15 **13.** 6 **15.** 7 **17.** –3 **19.** –7

HA1-150: Writing an Equation to Solve Word Problems

1. 24 **3.** 23 **5.** –11 **7.** 11 **9.** –9

11. 230 pounds **13.** $1,839.25 **15.** 14 years old **17.** 60 months **19.** 21 inches

HA1-155: Writing an Equation to Solve Consecutive Integer Problems

1. 12, 13 **3.** 3, 4 **5.** –5, –4 **7.** 1, 2, 3 **9.** –5, –4, –3

11. 2, 3, 4, 5 **13.** 6, 8, 10 **15.** 41, 43 **17.** 1, 3, 5, 7 **19.** 22

HA1-160: Writing an Equation to Solve Distance, Rate, and Time Problems

1. 1.5 hours **3.** 396 miles **5.** 2.5 hours **7.** 9 hours **9.** 120 mph

11. 4 hours **13.** $450y + 475y = 1,850$ **15.** $60(c + 1) = 65c$

17. $2t + 2(t - 5) = 250$ **19.** 12 hours

HA1-165: Using Equations to Solve Percent Problems

1. 3 **3.** 3,000 **5.** 2,613 **7.** 2.3 **9.** 10.89

11. 300 **13.** 56 **15.** 6,500 **17.** 820 **19.** 50%

HA1-170: Solving Percent of Change Problems

1. \$384.00 **3.** \$22.10 **5.** \$16.20 **7.** \$14.40 **9.** \$18.75

11. 10% **13.** 30% **15.** 4% **17.** 20% **19.** 20%

HA1-175: Solving Literal Equations

1. $d = e - 12$

3. $a - h = b$ or $b = a - h$

5. $m = g - 7$

7. $w = \frac{V}{lh}$

9. $j - n + c = m$ or $m = j - n + c$

11. $w - a + d = v$ or $v = w - a + d$

13. $d = \frac{3e}{4}$

15. $f - np = a$ or $a = f - np$

17. $\frac{2A}{h} = b$ or $b = \frac{2A}{h}$

19. $\frac{5}{2}a - b = h$ or $h = \frac{5}{2}a - b$

HA1-180: Graphing Equations and Inequalities on the Number Line

1. $n \leq -1$

3. $x > 0$

5. {all real numbers less than –5}

7. {All real numbers not equal to 3}

9.

11.
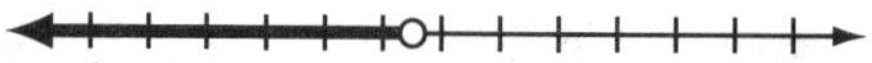

13. $n \geq -2$

15. $y \neq -2$

17.

19.

HA1-185: Solving Inequalities Using the Addition and Subtraction Properties

1. $-32 \leq m$ or $m \geq -32$

3. $a > 8$

5. $f < -32$

7. $-40 < a$ or $a > -40$

9. $-11 > y$ or $y < -11$

11. $y \leq -2$
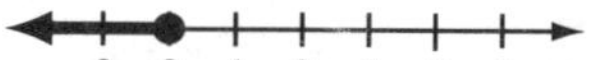

13. $h \geq 2$
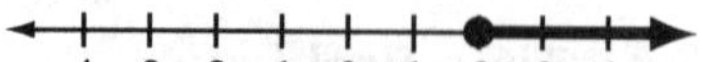

15. $g > -4$

17. $x > 3$
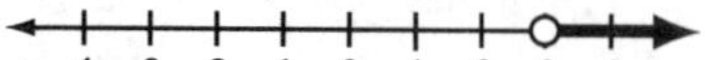

19. $y > -75$

HA1-190: Solving Inequalities Using the Multiplication and Division Properties

1. $x > -2$

3. $a > 4$

5. $m > 64$

7. $b > -14$

9. $n > -64$

11. $x > -13$

13. $n \geq 12$

15. $x > -3$

17. $m \le -3$

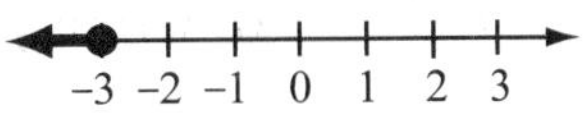

19. $y \ge 3$

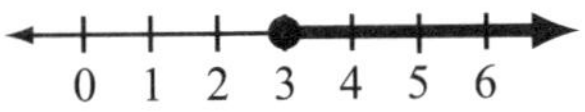

HA1-195: Solving Inequalities Using More Than One Property

1. $x < 12$
3. $x < 4$
5. $x > -5$
7. $x \ge -2$
9. $x \ge -20$
11. $x > -20$
13. $x \ge 20$
15.

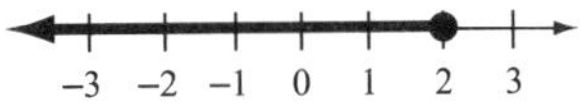

17.

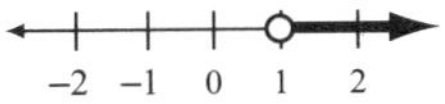

19.

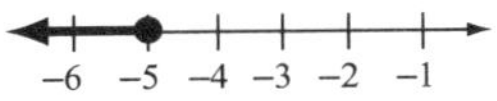

HA1-200: Combined Inequalities

1.

3.

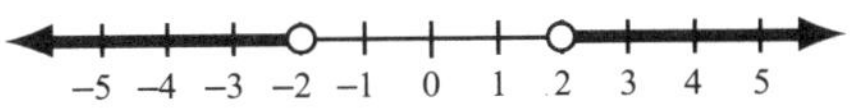

5.

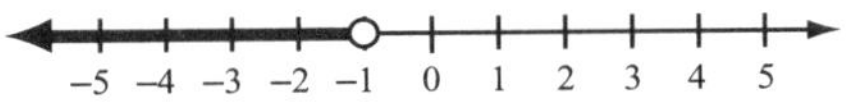

7.

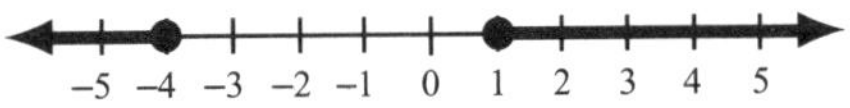

9.

11. $-1 \le x \le 4$

13. $-2 < x < 1$

15. $-3 \le x \le 2$

17. $x < -4$ or $x > -2$

19.

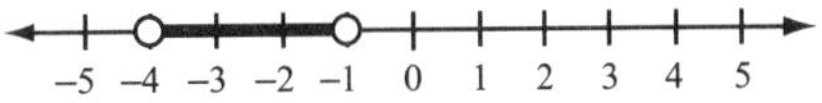

HA1-205: Solving Combined Inequalities

1. $x > 4$ or $x < 1$
3. $x < 2$ or $x > 3$
5. $-9 < x < 2$
7. $x > 4$ or $x < 10$
9. $-3 \le x \le 1$

11. $x < 10$ or $x > 5$

13. $x > -6$ or $x < -1$

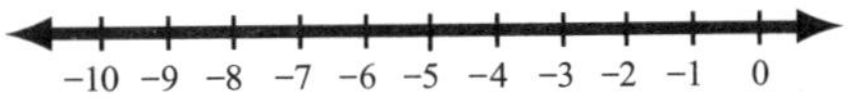

15. $x > -2$

17. $x < -7$

19. $-1 < x < 6$

HA1-220: Identifying and Multiplying Monomials

1. Yes
3. Yes
5. Yes
7. x^{11}
9. $x^{12}y^{8}$
11. $-14c^{6}d^{5}$
13. $12x^{5}y^{8}$
15. $-15x^{12}y^{8}$
17. $-21x^{20}y^{11}$
19. $-24x^{13}y^{13}$

21. $21xy$ square inches

HA1-225: Dividing Monomials and Simplifying Expressions Having an Exponent of Zero

1. $\frac{1}{x^2}$ **3.** $\frac{1}{y^4}$ **5.** $\frac{1}{x}$ **7.** a^3 **9.** 1

11. $\frac{b^4}{a}$ **13.** $\frac{y^3}{x^6}$ **15.** -5 **17.** $2x$ **19.** $-6a^4$

HA1-230: Raising a Monomial or Quotient of Monomials to a Power

1. y^{72} **3.** m^{48} **5.** $\frac{x^{10}}{y^{15}}$ **7.** $\frac{16}{x^2}$ **9.** $\frac{y^2}{36}$

11. $81x^4y^6$ **13.** $8x^9y^{12}z^{15}$ **15.** $9a^{12}b^{24}c^2$ **17.** $48x^9$ **19.** $\frac{-27a^3}{8b^9c^6}$

HA1-235: Writing, Multiplying, and Dividing Numbers Written in Scientific Notation

1. No **3.** Yes **5.** 7.32×10^4 **7.** 1.5×10^7 **9.** 8.0×10^{-3}

11. 8.4×10^5 **13.** 4.4×10^{12} **15.** 1.9×10^{-3} **17.** 2.4×10^{18} **19.** 8.0×10^9

HA1-240: Identifying the Degree of Polynomials and Simplifying by Combining Like Terms

1. 10 **3.** 9 **5.** Monomial **7.** Trinomial **9.** 18

11. $-8mx^3 + 10m^3x^2 + 7mx$ **13.** $-6mx^3 + 10mx^2 + 2mx$ **15.** $5x^4 + 7x^3 + 6x^2 + 10$

17. $4x^4 + 4x^3 + 8x$ **19.** $-6x^3y^2 + 8x^2y^3$

HA1-245: Adding and Subtracting Polynomials

1. $12b^2 + 3b - 8$ **3.** $-7m^2 + 8m - 35$ **5.** $11m^2 - 2m + 5$ **7.** $12m - 9x - 2z$

9. $2n^2 - 11n - 1$ **11.** $14n + 3m + 4a$ **13.** $5n^2 - 7n + 6$ **15.** $-19m^2 + 8m - 19$

17. $21x - 4y$ dollars **19.** $12x + 22xy - 6y$ inches

HA1-255: Multiplying Two Binomials Using the FOIL Method

1. $x^2 + 11x + 30$ **3.** $y^2 + 13y + 22$ **5.** $d^2 - 9d + 18$ **7.** $k^2 - 7k + 6$

9. $s^2 - 3s - 40$ **11.** $n^2 - 49$ **13.** $2x^2 - 9x + 9$ **15.** $2x^2 - 3x - 27$

17. $5y^2 - 13y - 28$ 19. $3x^2 + 5xy - 12y^2$ 21. $x^2 - 2x - 15$ square feet

HA1-260: Squaring a Binomial and Finding the Product of a Sum and Difference

1. $a^2 - 1$ 3. $x^2 - 9$ 5. $d^2 - 144$ 7. $b^2 - 2b + 1$

9. $d^2 - 14d + 49$ 11. $25y^2 - 70y + 49$ 13. $9m^2 - 48m + 64$ 15. $9 + 6k + k^2$

17. $36m^2 - 48m + 16$ 19. $9x^2 - 24xy + 16y^2$

HA1-265: Writing a Number in Prime Factorization and Finding the Greatest Common Factor

1. $2^2 \cdot 3^2$ 3. $2^2 \cdot 3 \cdot 7$ 5. composite 7. prime 9. $2^2 \cdot 3 \cdot 11$

11. 14 13. 22 15. 81 17. 9 19. 21

HA1-270: Factoring the Greatest Common Monomial Factor from a Polynomial

1. $5y^2$ 3. p^2 5. $2w^4$ 7. $7a^2b^2$

9. $2w^4y$ 11. 2 13. $4xy$ 15. $3x^2z^3$

17. $7b^4(1 + 2b^6)$ 19. $-5x^3(1 + 4x^2)$ 21. $8x^3y^4(1 + 2xy)$ 23. $5bc(bc^2 - 1)$

25. $2x^2(4x^2 - 1)$ 27. $x^4(1 - 2x^2)$ 29. $4d^8(1 + 4d^2)$ 31. $y^2(5z^3 - 2yz^4 + 1)$

33. $x^2(16x^2 + 8x + 1)$ 35. $9p^2q(2pq^2 + 1 - p^2q)$ 37. $-2x^3(xy^3 + 1 + 8xy)$ 39. $7a^2b(ab^2 - 2b - 1)$

HA1-271: Factoring Trinomials and Differences of Squares Using Algebra Tiles

1. $3x^2 + 7x + 2$ 3. $2x^2 - x - 1$ 5. $(x + 1)(x + 2)$

7. $(x - 2)(2x - 1)$

9.
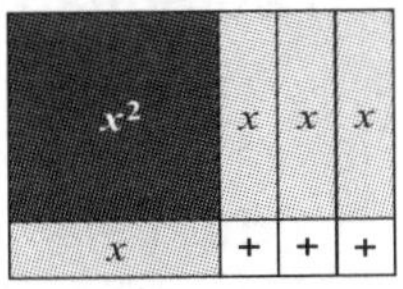
$(x - 1)(x + 3)$

11.
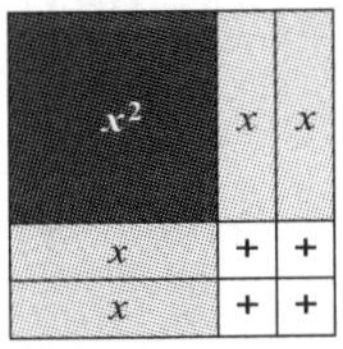
$(x + 2)(x + 2)$

13.
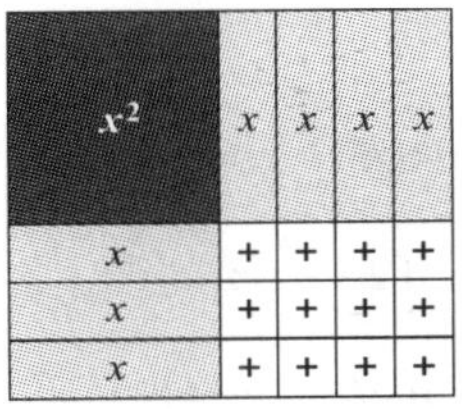
$(x + 3)(x + 4)$

15.
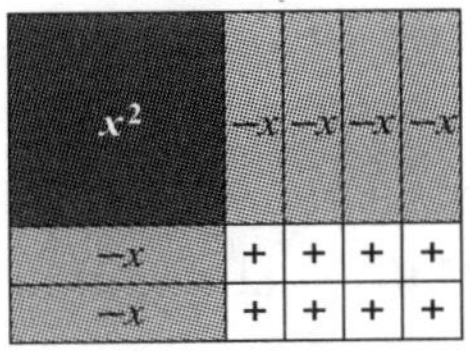
$(x - 2)(x - 4)$

17.
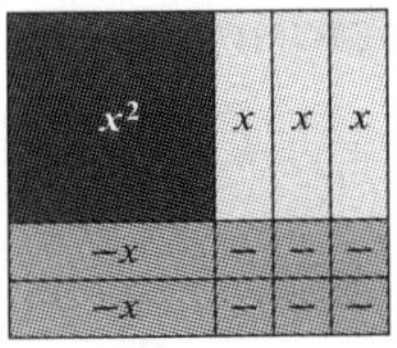
$(x - 2)(x + 3)$

19.

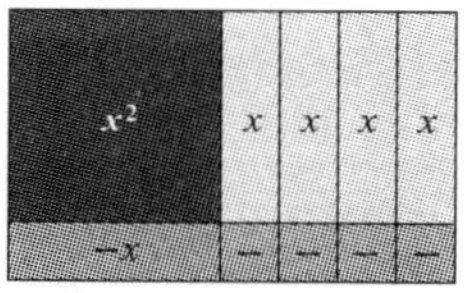

$(x - 1)(x + 4)$

21.

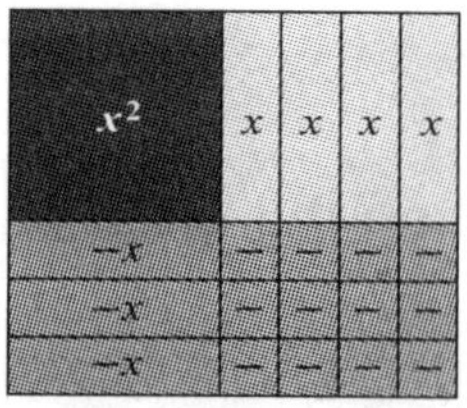

$(x - 3)(x + 4)$

23.

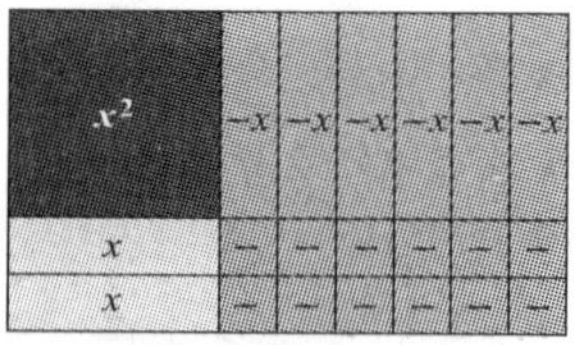

$(x + 2)(x - 6)$

25.

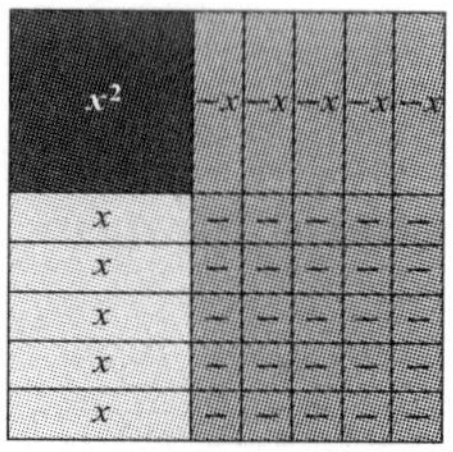

$(x + 5)(x - 5)$

27.

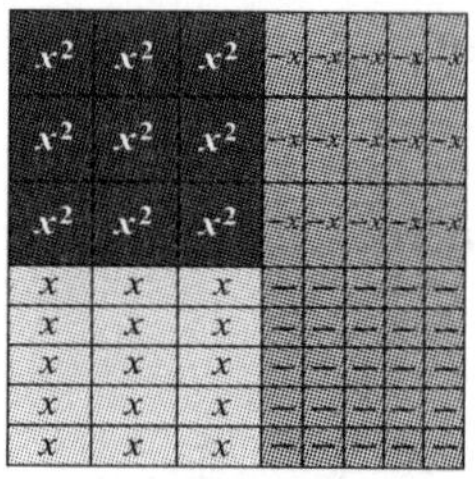

$(3x + 5)(3x - 5)$

29.

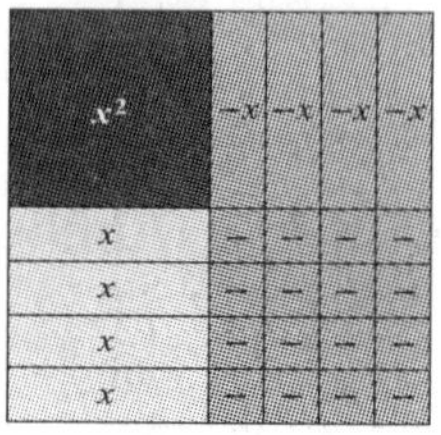

$(x + 4)(x - 4)$

31.

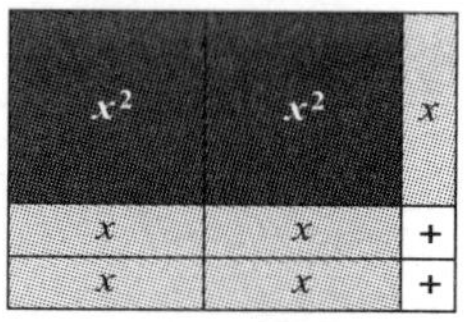

$(x + 2)(2x + 1)$

33.

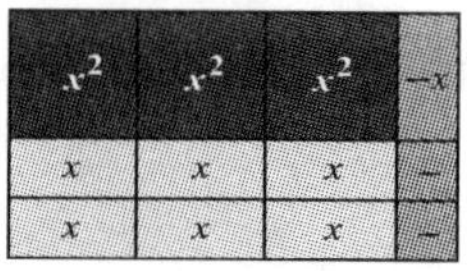

$(x + 2)(3x - 1)$

35.

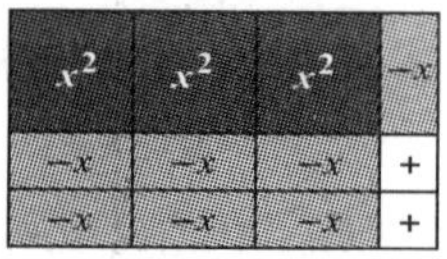

$(x - 2)(3x - 1)$

37.

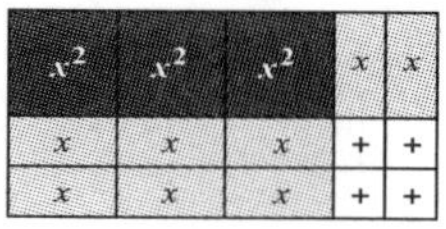

$(x + 2)(3x + 2)$

39.

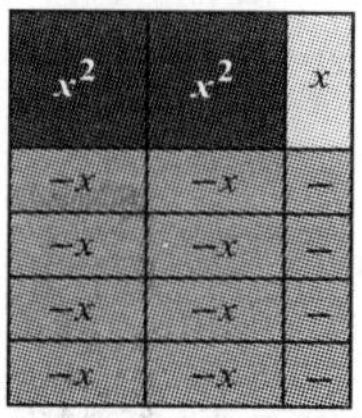

$(x - 4)(2x + 1)$

HA1-275: Factoring the Difference Between Two Squares and Perfect Square Trinomials

1. Yes **3.** Yes **5.** Yes **7.** $(x-3)(x+3)$

9. $(m-12)(m+12)$ **11.** $(2r-7)(2r+7)$ **13.** $(5c-9)(5c+9)$ **15.** Yes

17. Yes **19.** $(d+2)^2$

HA1-276: Factoring Sums and Differences of Cubes

1. $(y-z)(y^2+yz+z^2)$ **3.** $(m+p)(m^2-mp+p^2)$ **5.** $(k-m)(k^2+km+m^2)$

7. $(s+v)(s^2-sv+v^2)$ **9.** $(c+11)(c^2-11c+121)$ **11.** $(5+n)(25-5n+n^2)$

13. $(w-1)(w^2+w+1)$ **15.** $(r-13)(r^2+13r+169)$ **17.** $(e+7)(e^2-7e+49)$

19. $(h+8)(h^2-8h+64)$ **21.** $(n-10)(n^2+10n+100)$ **23.** $(s-12)(s^2+12s+144)$

25. $(9-2x)(81+18x+4x^2)$ **27.** $(11x+8)(121x^2-88x+64)$ **29.** $(3f-10)(9f^2+30f+100)$

31. $(d+2e)(d-2e)(d^2-2de+4e^2)(d^2+2de+4e^2)$

33. $(b+4m)(b-4m)(b^2-4bm+16m^2)(b^2+4bm+16m^2)$

35. $(d+7h)(d-7h)(d^2-7dh+49h^2)(d^2+7dh+49h^2)$

37. $(4y+9x)(4y-9x)(16y^2-36xy+81x^2)(16y^2+36xy+81x^2)$

39. $(10s+8r)(10s-8r)(100s^2-80rs+64r^2)(100s^2+80rs+64r^2)$

HA1-280: Factoring $x^2 + bx + c$ When c is Greater Than Zero

1. $z-5$ **3.** $x+10$ **5.** $x-5y$ **7.** $x+4$ **9.** $x+4$

11. $(x+4)(x+7)$ **13.** $(x+5)(x+7)$ **15.** $(x+5)(x+6)$ **17.** $(x+3)(x+7)$ **19.** $(x-3)(x-5)$

HA1-285: Factoring $x^2 + bx + c$ When c is Less Than Zero

1. $x+1$ **3.** $x+4$ **5.** $x-4$ **7.** $y+10$

9. $x-6$ **11.** $(w+12)(w-3)$ **13.** $(w-20)(w+2)$ **15.** $(w+36)(w-2)$

17. $(x-5)(x+9)$ **19.** $(w-48y)(w+2y)$

HA1-290: Factoring $ax^2 + bx + c$

1. $3x+1$ **3.** $2x-3y$ **5.** $3x-2$ **7.** $4y+1$ **9.** $2x+5$

11. $(3x-1)(3x+2)$ **13.** $(2x-1)(x-4)$ **15.** $(3x+4)(2x-1)$

17. $(4x+1)(2x-1)$ **19.** $(3x+5y)(5x+3y)$

HA1-291: Factoring Quadratic Equations Using the Graphing Calculator

1. $(x-2)(x-9)$ **3.** $-(x+7)(x-2)$ **5.** $(y+4)(y+17)$

7. $-(4x-3)(x-4)$ **9.** $-(3x+1)(2x-5)$

HA1-295: Factoring by Removing a Common Factor and Grouping

1. $2x+4$ **3.** $4y-3$ **5.** $5a-1$ **7.** $x-5$ **9.** $a-3b$

HA1-300: Factoring a Polynomial Completely

1. $5(x-1)(x-1)$ **3.** $a(1+6ab-b)$ **5.** $5(x+6)(x-2)$ **7.** $9x(x+5)$

9. $8a(b-8)$ **11.** $b^2(b+3)(b-3)$ **13.** $-4x(x+4)^2$ **15.** $8a(x+2)(x-2)$

17. $(x^2 - 5)(x - 1)$ **19.** $9x(3x - 5)(x + 1)$

HA1-305: Solving Polynomial Equations by Factoring

1. {3,–3} **3.** {13,–13} **5.** {–3,–9} **7.** {6,–9} **9.** {0,–2}

11. {–8} **13.** {2,–12} **15.** {6,–6} **17.** {5,–2} **19.** {–1}

HA1-310: The Practical Use of Polynomial Equations

1. 2, 3 **3.** 11, 12 **5.** 16, 18 **7.** 12, 14 **9.** 4 sides

11. 7 feet **13.** 4 feet **15.** 6 feet **17.** 4 feet **19.** 5, 6

HA1-355: Dividing Polynomials

1. $3m + 2$ **3.** $4y^2 - 1$ **5.** $c^3 + 2c^2 - \frac{4}{5}$ **7.** $\frac{b^2}{2} + b - 2$

9. $10m^6 - 5m^5 - \frac{1}{2}$ **11.** $y + 2 + \frac{18}{y - 5}$ **13.** $c + 5 + \frac{38}{c - 7}$ **15.** $x + 7$

17. $y - 4$ **19.** $3x^2 - 4x + 5 - \frac{7}{2x + 3}$

HA1-362: Solving Work Problems

1. $\frac{4}{7}$ **3.** $\frac{1}{3}$ **5.** $\frac{3}{8}$ **7.** $\frac{3}{20}$ **9.** 30 minutes

11. $\frac{75}{4}$ minutes **13.** 12 hours **15.** 40 minutes **17.** 10 p.m. **19.** 2:30 p.m.

HA1- 370: Graphing Ordered Pairs on a Coordinate Plane

1. (0, 6) **3.** (–4, 4) **5.** (–6,–6) **7.** $\left(6\frac{1}{2}, -2\frac{1}{2}\right)$ **9.** on the *x*-axis

HA1-375: Identifying Solutions of Equations in Two Variables

1. Not a solution **3.** Solution **5.** Not a solution **7.** Solution **9.** Not a solution

11. (–4, 7) **13.** (3, –2) **15.** (1, 0) **17.** (–1, 2) **19.** No

HA1-380: Graphing Linear Equations

1.

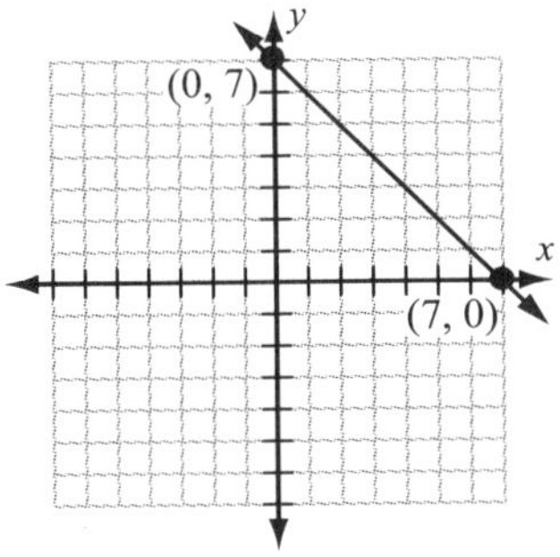

3.

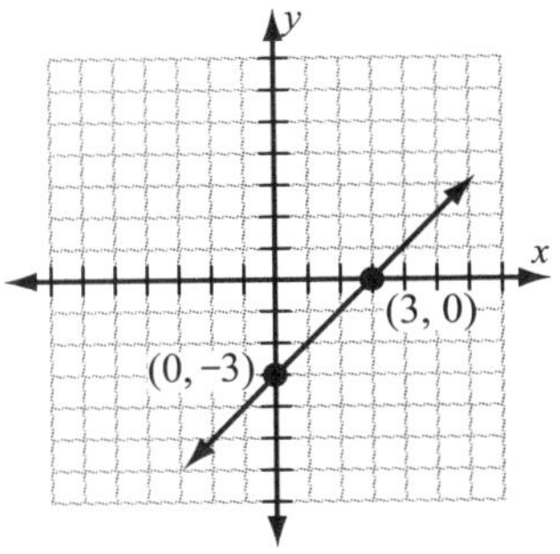

5.

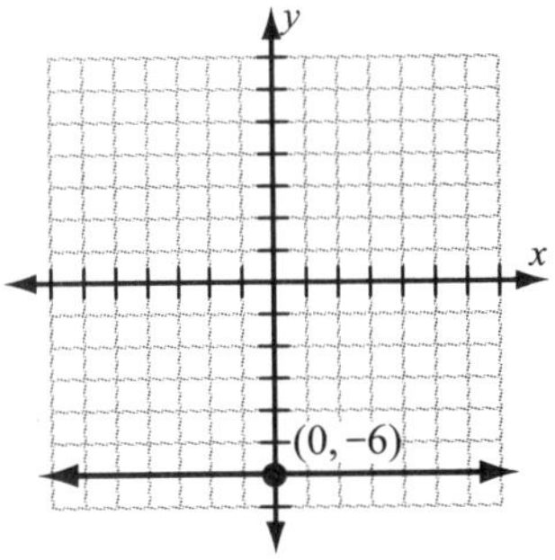

7.

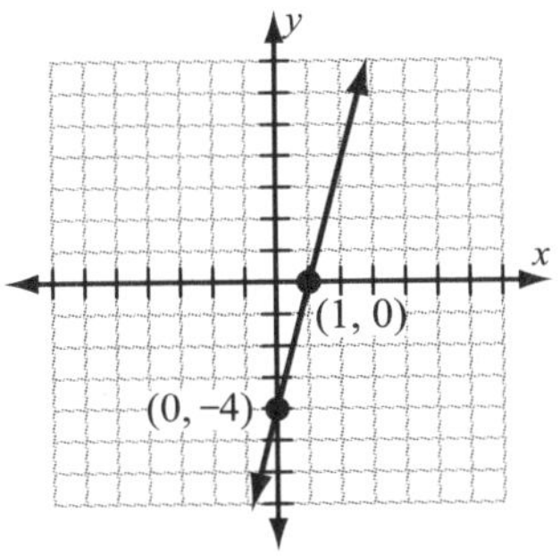

9.

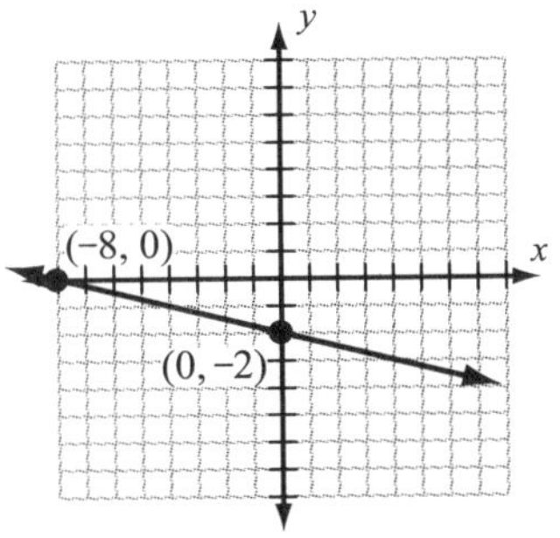

HA1-382: Solving Linear Equations Using the Graphing Calculator

1. 2 **3.** 11 **5.** 114 **7.** $-\frac{2}{9}$ **9.** $\frac{1}{2}$

HA1-385: Finding the Slope of a Line from its Graph or from the Coordinates of Two Points

1. 1 **3.** 2 **5.** 5 **7.** 0 **9.** 6

11. $-\frac{1}{3}$ **13.** $-\frac{1}{5}$ **15.** –1 **17.** $\frac{2}{3}$ **19.** No slope

HA1-394: Interchanging Linear Equations between Standard Form and Slope-Intercept Form

1. $y = -2x + \frac{7}{3}$ **3.** $y = -4x + 8$ **5.** $y = -5x + 7$ **7.** $y = 3x - 8$ **9.** $12x + y = 7$

11. $7x + y = -8$ **13.** $10x - y = 8$ **15** $9x - y = 1$ **17.** $y = \frac{7}{3}x - 4$ **19.** $y = -\frac{5}{4}x + 4$

21. $y = -\frac{3}{4}x + 4$ **23.** $3x + 5y = -40$ **25.** $5x + 6y = -24$ **27.** $8x + 3y = 48$ **29.** $4x + 7y = 63$

31. $5x - 15y = -3$ **33.** $2x - 9y = -3$ **35.** $9x - 24y = -4$ **37.** $72x - 45y = -20$ **39.** $4x - 6y = -1$

HA1-395: Finding the Equation of a Line Parallel or Perpendicular to a Given Line

1. $y = 8x + 5$ **3.** $y = \frac{1}{2}x - 10$ **5.** $y = -\frac{3}{7}x - 9$ **7.** $y = -\frac{9}{2}x + 2$

9. 4 **11.** $\frac{1}{2}$ **13.** $\frac{5}{4}$ **15.** 1

17. Neither **19.** Parallel **21.** Perpendicular **23.** Parallel

25. $y = \frac{3}{4}x + 7$ **27.** $y = -\frac{1}{4}x - 3$ **29.** $y = 2x + 20$ **31.** $y = -\frac{1}{3}x + 7$

33. $y = -x + 3$ **35.** $y = 5x + 40$ **37.** $y = -\frac{3}{2}x + 14$ **39.** $y = \frac{1}{2}x - 7$

HA1-398: Graphing Linear Equations Using Slope and y-intercept or Slope and a Point

1.

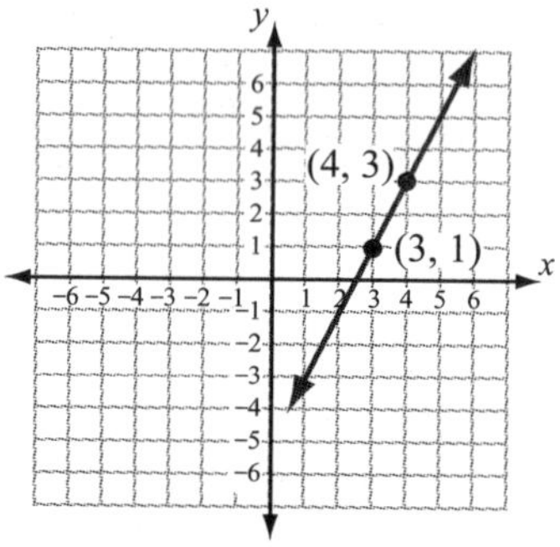

3.

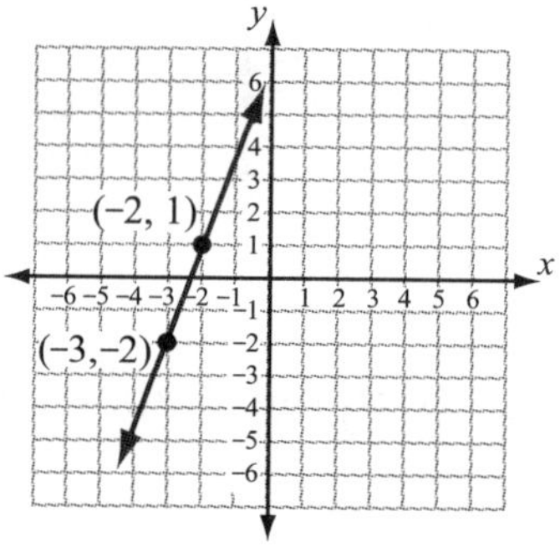

5.

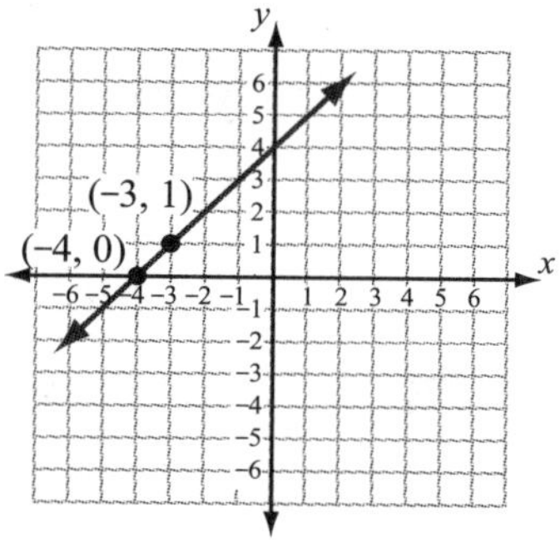

7.

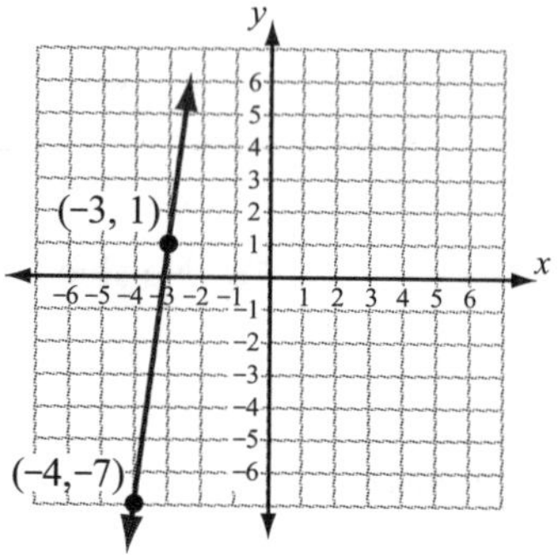

9.

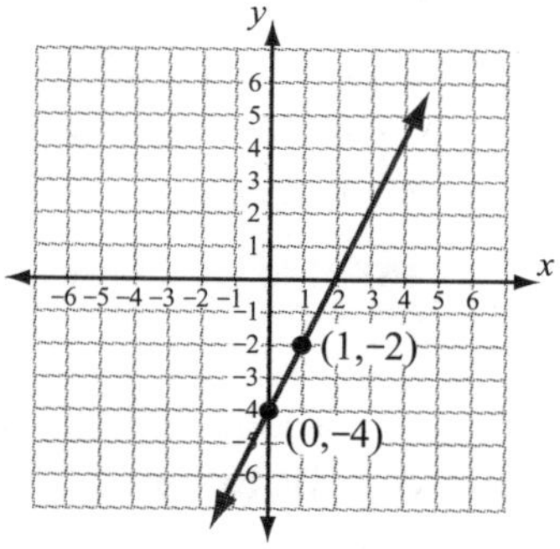

11.

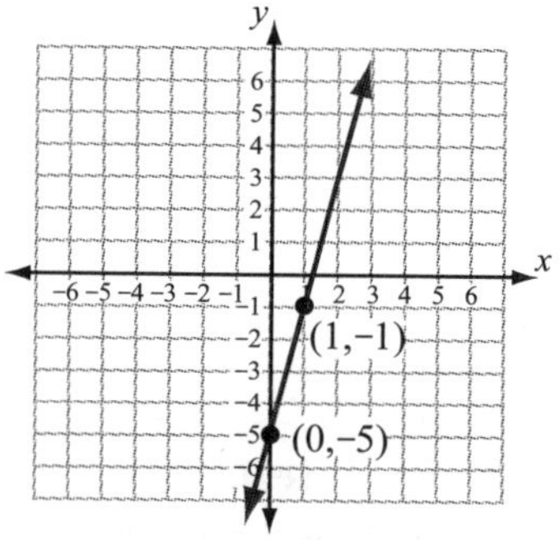

13.

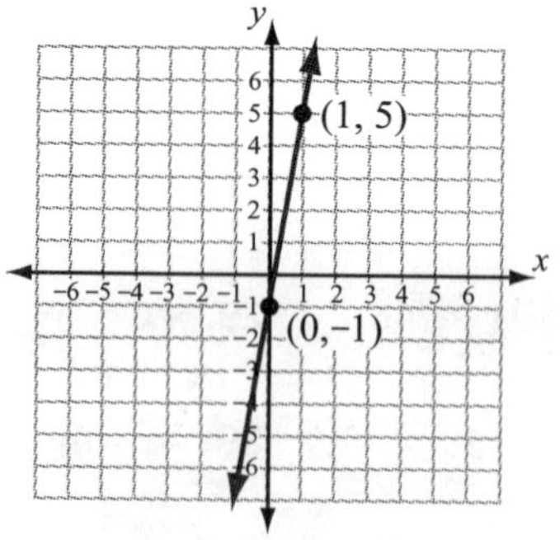

15.

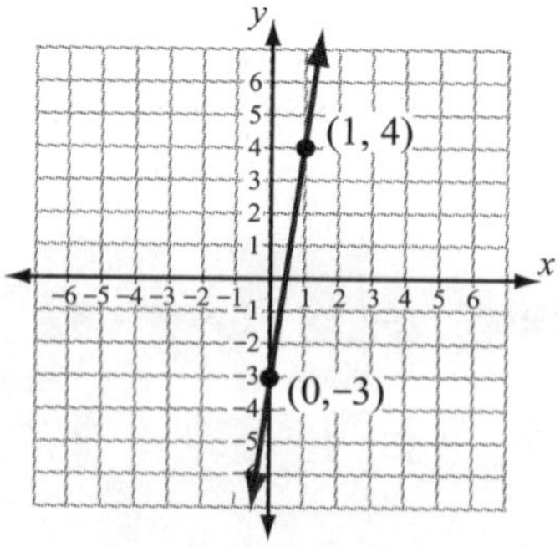

17.

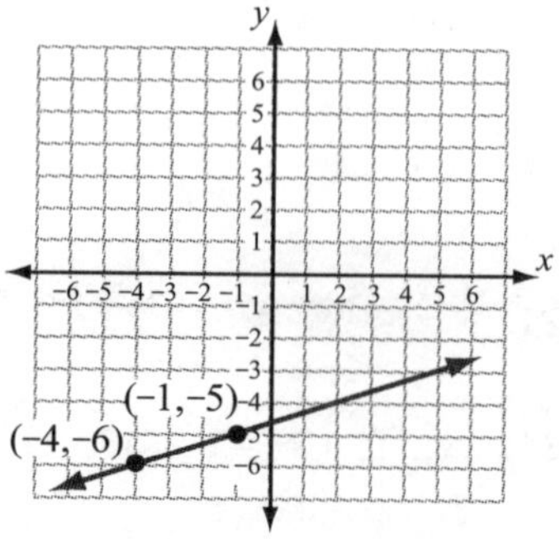

19.

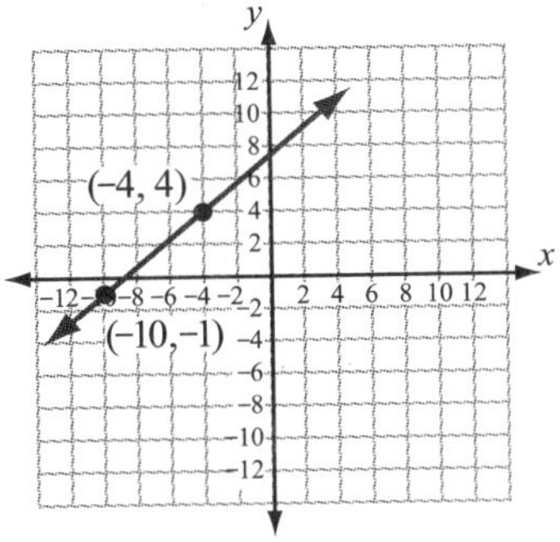

21.

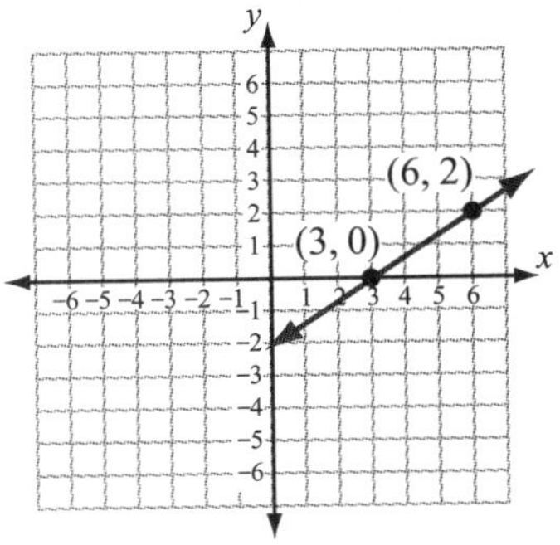

23.

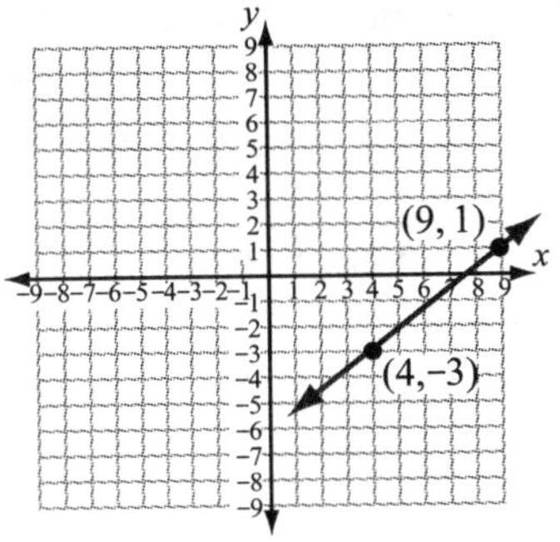

25.

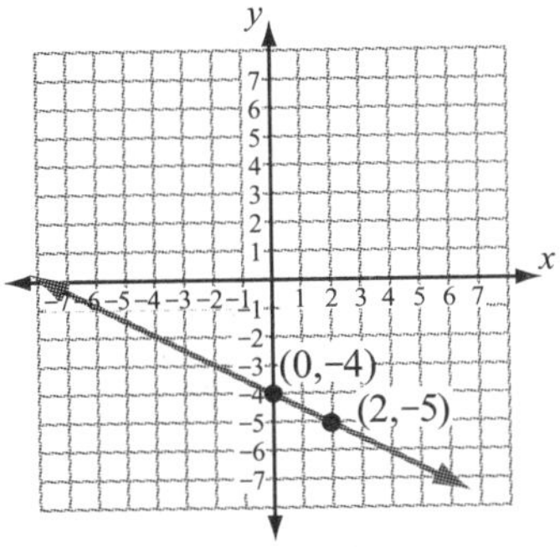

27.

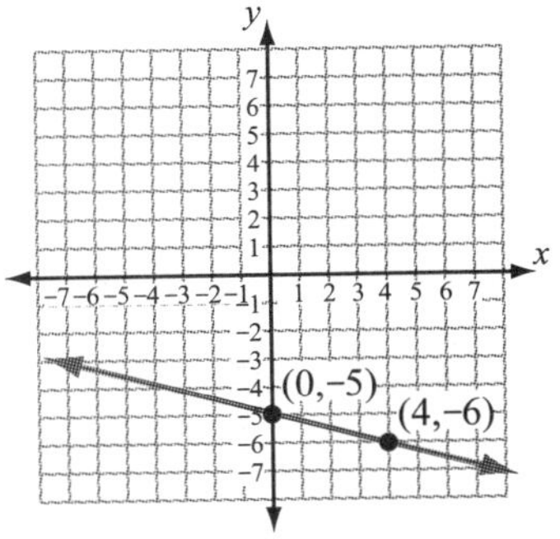

29.

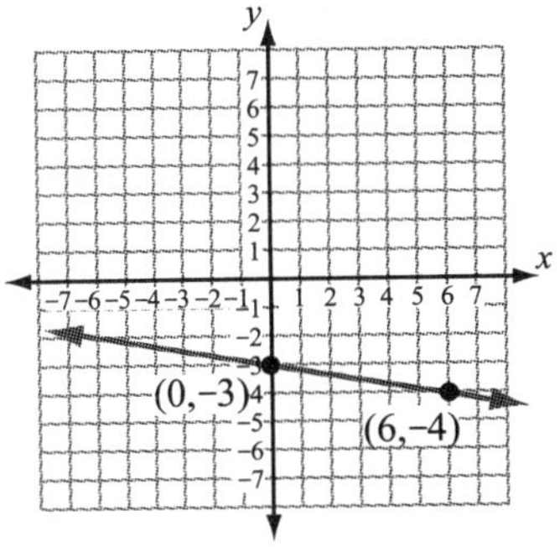

31.

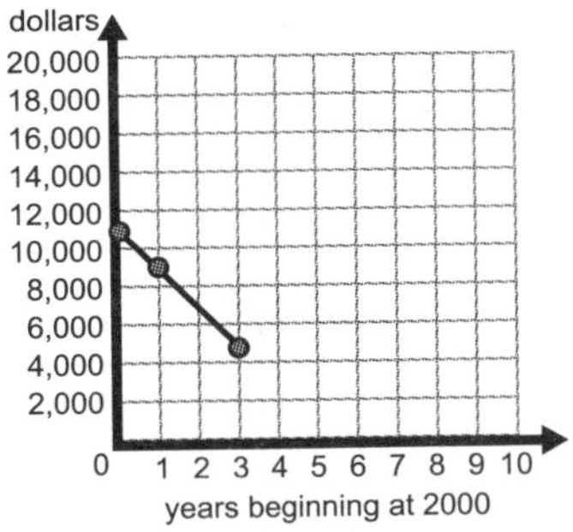

33.

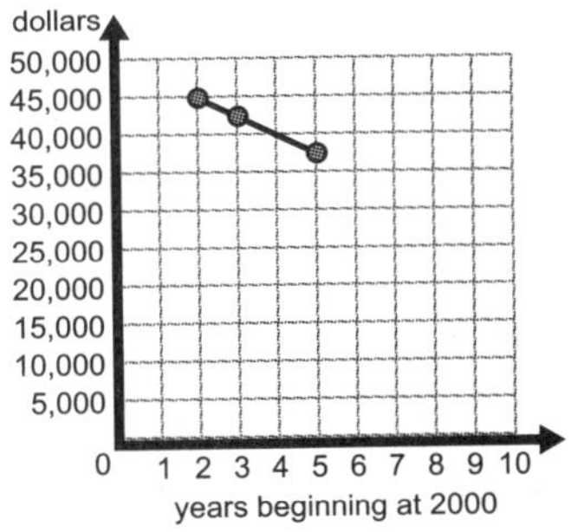

35.

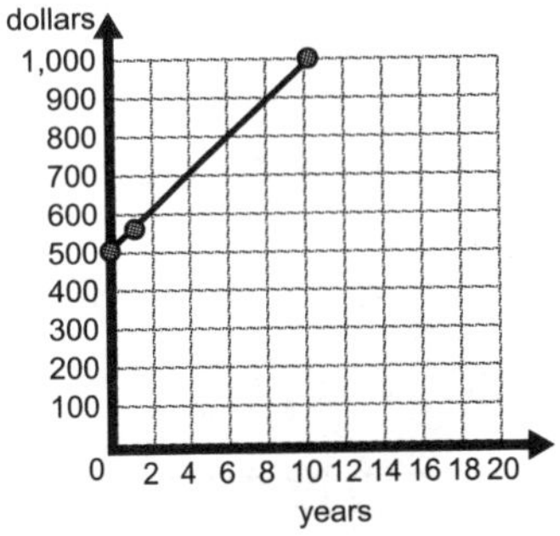

37.

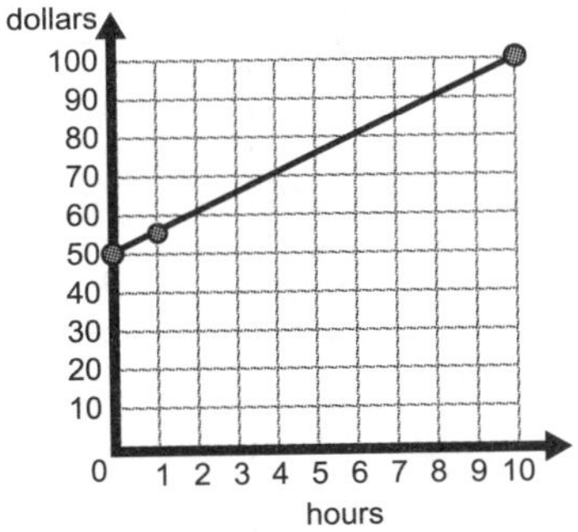

39.

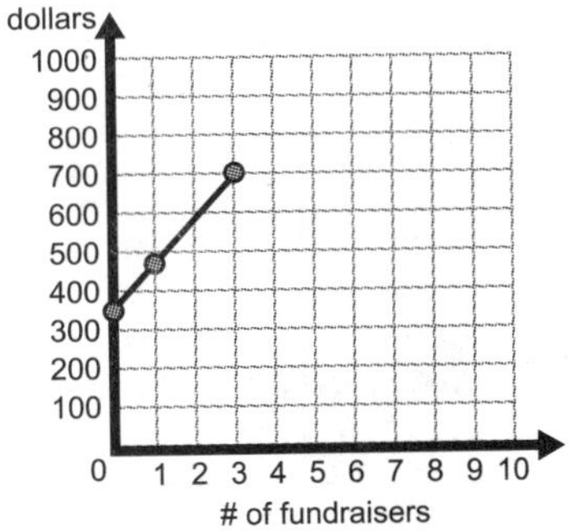

HA1-401: How Variations of "m" and "b" Affect the Graph of $y = mx + b$

1. $y = 10x$ **3.** $y = -8x$ **5.** $y = \frac{3}{7}x$ **7.** $y = -7x$

9. $y = x$ **11.** $y = -\frac{2}{7}x$ **13.** $y = 9x$ **15.** $y = 0$

17. The angle is less than 45°.

19. The angle is equal to 45°.

21. The angle is greater than 45°.

23. The angle is greater than 45°.

25. The angle is equal to 45°.

27. The angle is less than 45°.

29. The angle is greater than 45°.

31. $m_A > m_B > 0$
$b_A > b_B$

33. $|m_A| > |m_B|$ $m_B < 0$
$m_A < 0$ $b_A = b_B = 3$

35. $m_A > 0 > m_B$ $b_A = b_B = 2$
$|m_A| > |m_B|$

37. $|m_B| > |m_A|$ $m_B < 0$
$m_A < 0$ $b_B > b_A$

39. $m_A < 0$ $|m_A| > |m_B|$
$m_B < 0$ $b_A > b_B$

HA1-405: Determining an Equation of a Line Given the Slope and Coordinates of One Point

1. $y = -2x - 6$ **3.** $y = 6x - 26$ **5.** $y = 2x + 2$ **7.** $y = -3x + 13$ **9.** $y = -\frac{1}{4}x + \frac{7}{4}$

11. $2x - y = 9$ **13.** $3x - y = 1$ **15.** $4x + y = 0$ **17.** $x + 2y = 8$ **19.** $y = 5x + 10$

HA1-410: Determining the Equation of a Line Given the Coordinates of Two Points

1. $y = 2x + 3$ **3.** $y = -x + 1$ **5.** $y = 2x - 2$ **7.** $y = \frac{1}{2}x$ **9.** $y = \frac{1}{6}x + \frac{8}{3}$

11. $2x - 3y = -17$ **13.** $5x - 2y = 11$ **15.** $3x + y = 16$ **17.** $3x - y = 12$ **19.** $y = 0.1x$

HA1-415: Graphing Linear Inequalities with Two Variables

1.
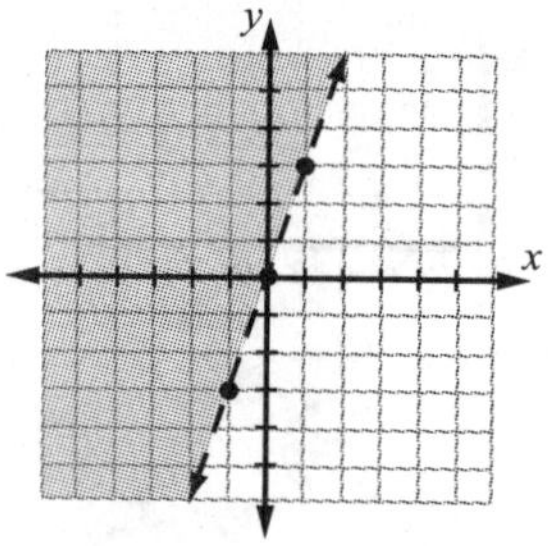

3.
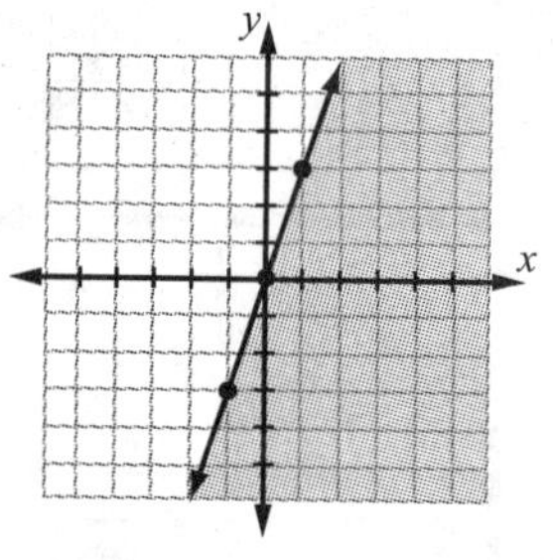

5.
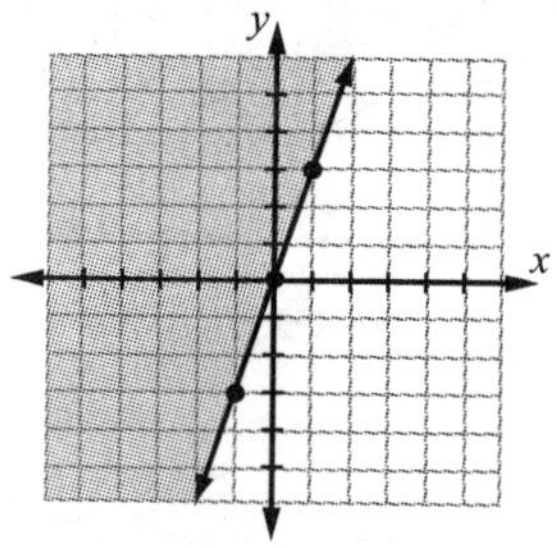

7.
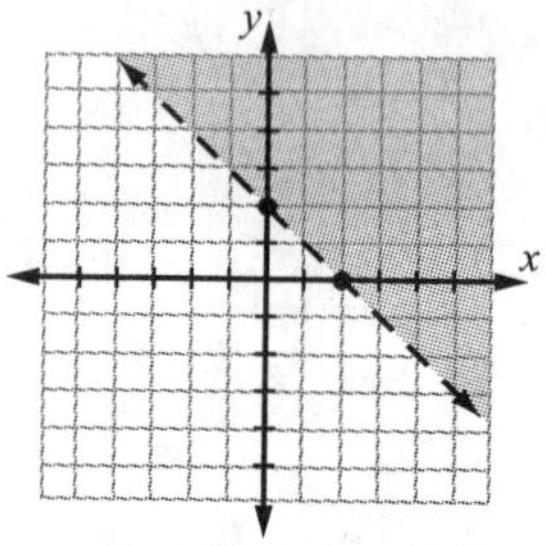

9.
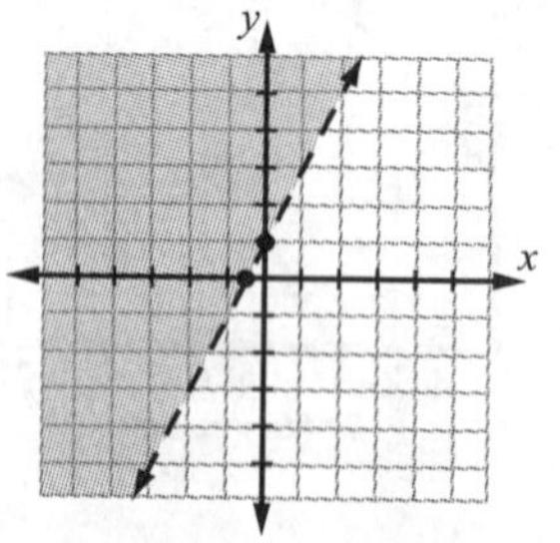

HA1-416: Graphing Linear Inequalities with Two Variables Using the Graphing Calculator

1. (6, 3) **3.** (2, 5) **5.** (3, 7) **7.** (3, 2)

9.

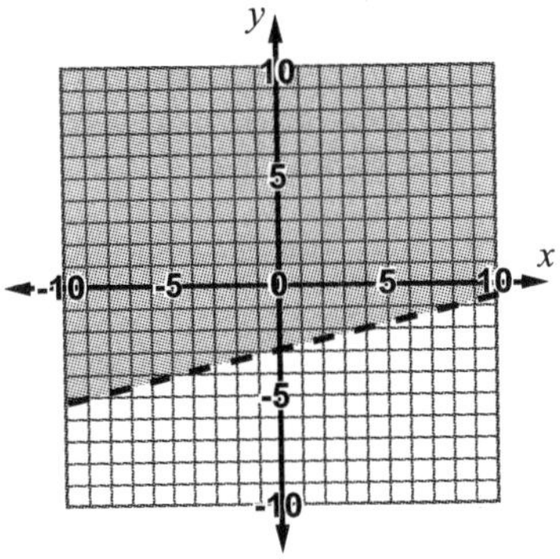

HA1-442: Interpreting Graphs of Functions in Real-Life Situations

1. *f* is decreasing. **3.** *f* is increasing. **5.** *f* is increasing and decreasing. **7.** *f* is increasing.

9. $x > 5$ **11.** $x < -4$ **13.** $x < 0$ **15.** $-3 < x < 3$

17. The number of cars sold was greatest in month 3.

19. The population was the smallest in 1988.

21. The increase in profit was the greatest between July and August.

23. Maria had the most money in her savings account after 4 months.

25.

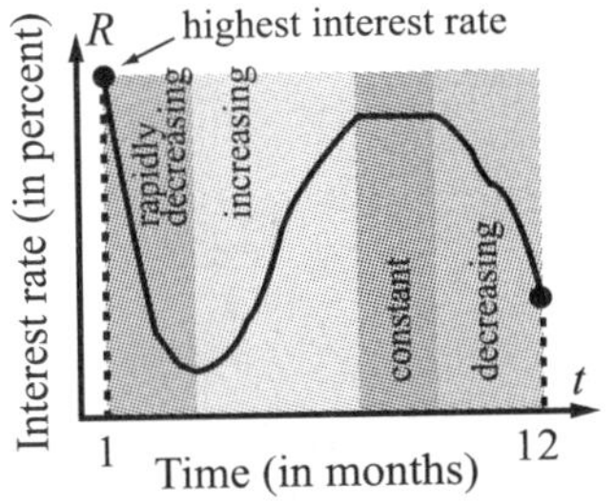

27.

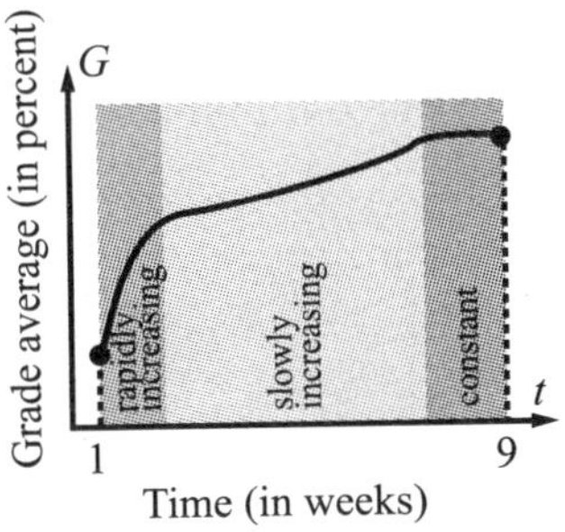

29.

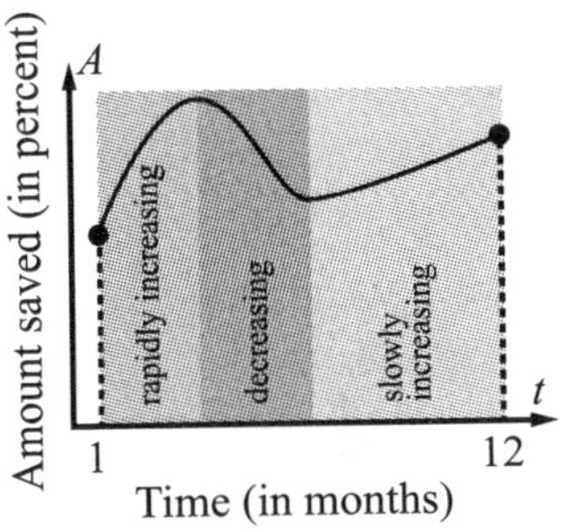

31. Although the sales increased for the first 3 months, they increased rapidly between months 2 and 3. Sales were then constant for a month, and then decreased slowly.

33. The population decreased for the first 3 years, but decreased rapidly between 1993 and 1994. Then, it increased for 2 years but decreased again for the next 4 years.

35. The cost of a tank of gasoline decreased for 3 months to $24 where it remained for 4 months. It rapidly increased between months 8 and 9, then increased more.

37. The interest rate decreased for the first 2 months and then remained constant at 4.2% for 4 months. Then the rate decreased again for the final 3 months.

39. The water level in the lake fell from 5 inches to 3 inches above normal and stayed there for 2 months. Then it increased to its highest level of 8 inches above normal.

HA1-449: Applying Inductive and Deductive Reasoning

1. If the storm is a hurricane, then the winds are more than 73 miles per hour.

3. If you do not wake up on time, then you will be late for school.

5. If you exceed the speed limit, then you will get a speeding ticket.

7. If n is an even integer, then $n + 2$ is an even integer.

9. Washing machine

11. Octopus

13. 1

15. A person who lives in Baton Rouge, Louisiana

17. You noticed that 3^2 is an odd number, 7^2 is an odd number, and 11^2 is an odd number. You conclude that x^2 is always an odd number.

19. Inductive reasoning

21. Deductive reasoning

23. Deductive reasoning

25. If you do not live in Texas, then you do not live in the United States.

27. If $x + 2$ is greater than 0, then x is greater than 0.

29. If the animal is a fish, then the animal lives in the sea.

31. Jamie made more baskets than Joe.

33. Florida has warmer weather than Michigan.

35. If $-(x+3) = 2(x+6)$, then $x = -5$.

37. You do not get hungry in the morning.

39. The sun is not shining.

HA1-492: Simplifying Square and Cube Roots

1. -8 **3.** -10 **5.** 14 **7.** 8 **9.** a^{12}

11. z^{10} **13.** y^8 **15.** c^9 **17.** $9x^4y^7$ **19.** $5p^8q^7r^4$

21. $3a^6b^2$ **23.** $6x^7y^8z^6$ **25.** $\frac{6a^8}{11b^6}$ **27.** $\frac{12x^3y^2}{13z^8}$ **29.** $\frac{2a^7b^5}{5c^9}$

31. $4x^5$ **33.** $3p^{10}q^6$ **35.** $9x^{15}$ **37.** $11c^9$ **39.** $6x^4y^5z^6$

HA1-540: Finding the Mean, Median, and Mode from Data and Frequency Distribution Tables

1. 90° **3.** 3 **5.** \$17.25 **7.** \$15.50 **9.** 3

11. 81 **13.** 1,980 **15.** 3 **17.** Hal **19.** 49.5

HA1-541: Analyzing Data Using the Measures of Central Tendency and the Range

1. 4 **3.** 2 **5.** \$0.93 **7.** No mode

9. Brown **11.** 73.5 inches **13.** 8 inches **15.** Black and white

17. The median is greater than the mean.

19. The median, mode, and range are all 17 phone calls.

21. Mode

23. Mode

25. Mean **27.** Median **29.** Mode **31.** Mean and range

33. Range **35.** Mean, median, and range **37.** Mean and mode **39.** Mean and median

HA1-545: Making a Frequency Distribution Table

1. No **3.** Yes **5.** 10 **7.** 900 – 949 **9.** 5

11.

Interval	Tally	Frequency
40 – 44	II	2
45 – 49	III	3
50 – 54	I	1
55 – 59	III	3
60 – 64	II	2
65 – 69	IIII	4
	Total	15

13.

Interval	Tally	Frequency
15 – 19	IIII	4
20 – 24	IIII	4
25 – 29	卌	5
	Total	13

15.

Interval	Tally	Frequency
60 – 69	III	3
70 – 79	III	3
80 – 89	卌	5
90 – 99	IIII	4
	Total	15

17.

Interval	Tally	Frequency
30 – 39	卌 I	6
40 – 49	IIII	4
50 – 59	II	2
	Total	12

19.

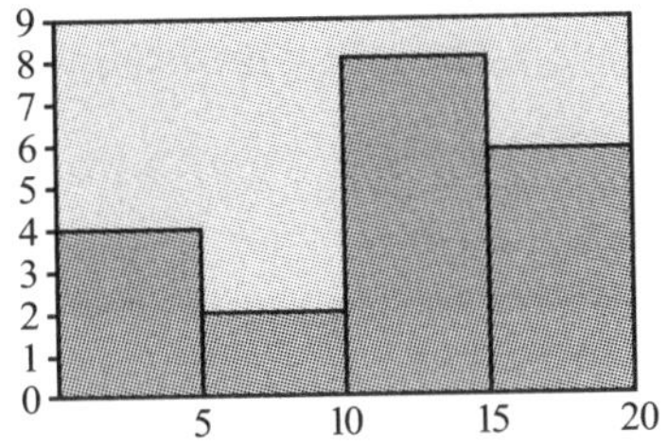

HA1-555: Computing the Range, Variance, and Standard Deviation of a Set of Data

1. 20.2 **3.** 90.14 **5.** 31.58 **7.** 24 **9.** 8

11. 39.16 **13.** 3.84 **15.** 52.58 **17.** 17.27 **19.** 4.16

HA1-560: Determining Probability of an Event and Complementary Event from a Random Experiment

1. $\frac{20}{67}$ **3.** $\frac{5}{28}$ **5.** $\frac{6}{25}$ **7.** $\frac{4}{25}$ **9.** $\frac{7}{30}$

11. $\frac{7}{13}$ **13.** $\frac{6}{13}$ **15.** $\frac{3}{14}$ **17.** $\frac{45}{67}$ **19.** $\frac{2}{17}$

HA1-565: Solving Problems Involving Independent, Dependent, and Mutually Exclusive and Inclusive Events

1. $\frac{1}{3}$ **3.** $\frac{2}{5}$ **5.** $\frac{3}{16}$ **7.** $\frac{1}{12}$ **9.** $\frac{17}{62}$ **11.** $\frac{5}{6}$

HA1-800: Solving Equations

1. 4 **3.** 12 **5.** 12 **7.** 252 **9.** 4

11. –7 **13.** –32 **15.** 140 **17.** $\frac{23}{2}$ **19.** 18

HA1-801: Solving Inequalities

1. $a < 7$ **3.** $b < -2$ **5.** $h < -59$

7. $h > 12$ 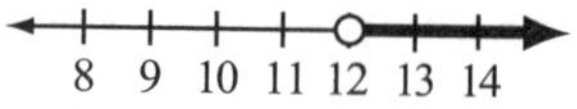

9. $f \leq 5$ 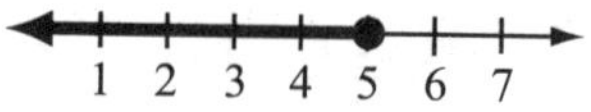

11. $h > 15$ **13.** $f < 43$ **15.** $x > -4$

17. $k \leq 4$ **19.** $w \geq -11$

HA1-802: Factoring

1. $(x+2)(x+2)$ **3.** $(x+2)(x+9)$ **5.** $(x+4)(x+8)$ **7.** $(x+5)(x+8)$ **9.** $(x+2)(x+10)$

11. $(x-2)(x-4)$ **13.** $(y-4)(y-7)$ **15.** $(a-3)(a-6)$ **17.** $(z-5)(z-7)$ **19.** $(x+8)(x-12)$

HA1-803: Solving Systems of Equations

1. $(1, 1)$ **3.** $(7, 0)$ **5.** $(3, 7)$ **7.** $(-1, -1)$

9. $(-2, -3)$ **11.** $(-1, -2)$ **13.** $(-4, 3)$ **15.** $(-1, 1)$

17. $(3, 0)$ **19.** Infinite number of solutions

HA1-804: Solving Systems of Equations by Graphing

1. Yes **3.** No **5.** $(4, -2)$ **7.** $(-4, 1)$

9. Since the graphs are the same, the system has an infinite number of solutions.

HA1-805: Applying Algebra Concepts

1. 10 feet **3.** 22.63 inches **5.** 11.83 meters **7.** 7.6 feet

9. 35 adult tickets and 45 children tickets

HA1-815: Simplifying Expressions with Negative and Zero Exponents

1. 121 **3.** 1 **5.** 5^{-2} **7.** 8^{-2} **9.** 1

11. $\frac{1}{25}$ **13.** $\frac{1}{81y^4}$ **15.** $\frac{27x^3}{16y^4}$ **17.** \$2,411 **19.** 100

HA1-818: Simplifying Expressions Using the Division Properties of Exponents

1. $\frac{1}{a^2}$ **3.** 7^{10} **5.** $\frac{64}{125}$ **7.** $\frac{125}{216}$ **9.** $\frac{4a^8}{25b^{14}}$

11. $\frac{81a^{20}}{16b^{32}}$ 13. $\frac{49}{4x^4}$ 15. $\frac{64}{729a^9}$ 17. $\frac{1}{x^8y^7}$ 19. $\frac{27b^3}{8a^5}$

HA1-840: Introduction to Matrices

1. 4×5
3. 5×4
5. 2×4
7. 4×4
9. row 2, column 2
11. row 2, column 4
13. row 4, column 2
15. row 1, column 3

17.

	Rochelle	Myra
A	2	2
B	2	1
C	1	2

19.

	Daryl	Phoebe
Heritage	3	2
Breast cancer	1	1
American flag	1	2

21.

	Kelly	Michelle
SUV	2	1
truck	2	2
car	1	2

23.

	Computer A	Computer B
On	2	4
Off	3	1

25.

	Republican	Democrat
for	2	4
against	4	2

27.

	Batch A	Batch B
Good	3	2
Bad	3	4

29.

	Customer A	Customer B	Customer C
approved	4	2	3
denied	1	3	2

31. Leslie bought 71 stocks and Michelle bought 76 stocks.

33. Keisha earned \$373 in week 1 and \$402 in week 2.

35. Judy saved \$39.55 and Jasmine saved \$46.95.

37. 19 wins and 11 losses

39. The average high temperature is 84° and the average low temperature is 66°.

HA1-845: Operations with Matrices

1. $\begin{bmatrix} 6 & 13 \\ 13 & 13 \end{bmatrix}$

3. $\begin{bmatrix} 8 & 7 \\ 11 & 12 \end{bmatrix}$

5. $\begin{bmatrix} 3 & 3 & 3 & 3 \\ 3 & 3 & 3 & 3 \\ 3 & 3 & 3 & 3 \\ 3 & 3 & 3 & 3 \end{bmatrix}$

7. $\begin{bmatrix} 4 & 4 & 4 \\ 5 & 7 & 9 \\ 9 & 12 & 11 \end{bmatrix}$

9. $\begin{bmatrix} 9 & 9 \\ 3 & 3 \\ 3 & 9 \\ 12 & 3 \\ 9 & 9 \\ 3 & 6 \end{bmatrix}$

11. $\begin{bmatrix} 8 & 2 \\ 6 & 4 \\ 8 & 2 \\ 4 & 10 \end{bmatrix}$

13. $\begin{bmatrix} 12 & 36 & 0 \\ 28 & 20 & 8 \\ 0 & 24 & 16 \end{bmatrix}$

15. $\begin{bmatrix} 15 & 12 \\ 9 & 6 \end{bmatrix}$

17. $\begin{bmatrix} 14 & 14 & 12 \\ 26 & 8 & 12 \\ 28 & 16 & 10 \end{bmatrix}$

19. $\begin{bmatrix} -11 & -6 \\ 14 & -13 \end{bmatrix}$

21. $\begin{bmatrix} 8 & 14 & 6 \\ 16 & 8 & 0 \\ 20 & 6 & 10 \end{bmatrix}$

23. $\begin{bmatrix} -15 & 14 & 10 \\ 12 & -3 & 17 \end{bmatrix}$

25. $\begin{bmatrix} -3 & -2 \\ -6 & -5 \end{bmatrix}$

27. $\begin{bmatrix} 13 & 19 \\ 18 & 21 \end{bmatrix}$

29. $\begin{bmatrix} 13 & 7 & 1 \\ 2 & 0 & 2 \\ 16 & 3 & 0 \end{bmatrix}$

31. $\begin{bmatrix} 266 & 245 \\ 175 & 210 \end{bmatrix}$

33. $\begin{bmatrix} 432 & 585 \\ 558 & 441 \end{bmatrix}$

35. $\begin{bmatrix} 424 & 496 & 232 \\ 424 & 508 & 244 \\ 420 & 504 & 252 \\ 432 & 504 & 260 \\ 436 & 536 & 264 \end{bmatrix}$

37. $\begin{bmatrix} 588 & 432 \\ 930 & 762 \\ 408 & 492 \end{bmatrix}$

39. $\begin{bmatrix} 110 & 130 \\ 182 & 142 \end{bmatrix}$

HA1-860: Using the Laws of Exponents

1. 1,000 **3.** 10 **5.** 625 **7.** 1 **9.** d^{15}

11. b^{10} **13.** x^4 **15.** e^{75} **17.** x^8 **19.** m^{30}

HA1-861: Simplifying Expressions with Negative and Zero Exponents

1. 1 **3.** -1 **5.** $-\frac{1}{25}$ **7.** $\frac{1}{4}$ **9.** $\frac{1}{x^5}$

11. x^4 **13.** x^8 **15.** y^2 **17.** $\frac{x^8}{y^6}$ **19.** $\frac{y^{12}}{4096x^9}$

HA1-862: Dividing Polynomials Using Factoring

1. x **3.** x **5.** $\frac{x}{2}$ **7.** $-\frac{1}{5x}$ **9.** $8x^2$

11. $5k^2 - k^3$ **13.** $42x^4 + 50x$ **15.** $2b^2 - 3$ **17.** $4y^2 - 3$ **19.** $x + 7$

HA1-863: Dividing Polynomials Using Long Division

1. $y - 8 + \frac{51}{y+6}$ **3.** $k + 8 + \frac{2}{k-4}$ **5.** $x + 1 - \frac{14}{x+2}$ **7.** $x + 2 + \frac{2}{x+3}$

9. $y + 8 + \frac{48}{y-5}$ **11.** $3y + 15 + \frac{88}{y-6}$ **13.** $4x + 4 - \frac{12}{x-2}$ **15.** $4y - 2 + \frac{5}{y+1}$

17. $a + 10 + \frac{19}{a - 10}$

19. $a^2 + 4a + 15 + \frac{72}{a - 4}$

HA1-864: Dividing Polynomials Using Synthetic Division

1. $6x^3 - 3x^2 + 2x + 6$
3. No
5. –2
7. Yes
9. $x - 3$
11. –4
13. 0
15. $2x^2 - 6x - 2$
17. $2x^3 - 2x^2 - x + 6$
19. $9x^2 + 6$

HA1-866: Drawing a Line Using Slope-Intercept Form and Determining if Two Lines are Parallel or Perpendicular

1. $m = -\frac{2}{5}$ and $b = \frac{1}{5}$
3. $m = -3$ and $b = 6$
5. $m = 2$ and $b = -3$
7. $m = -12$ and $b = -14$
9. $m = \frac{5}{3}$ and $b = 3$

11.

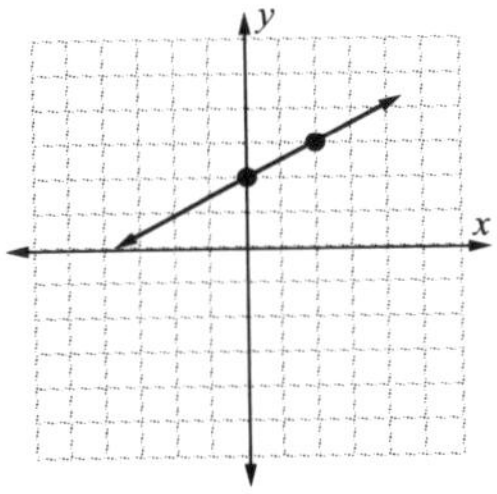

13.

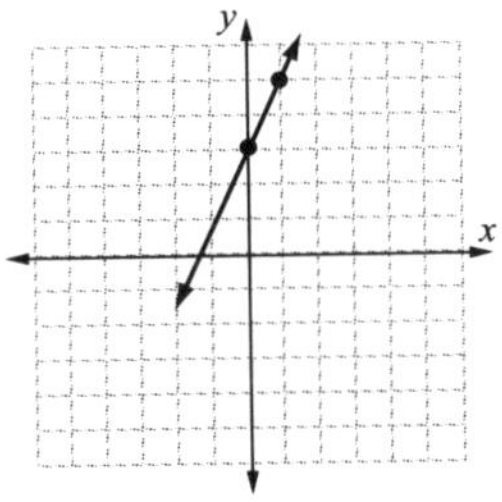

15.

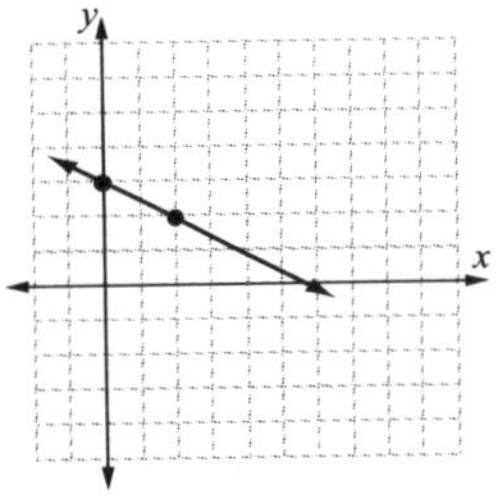

17.

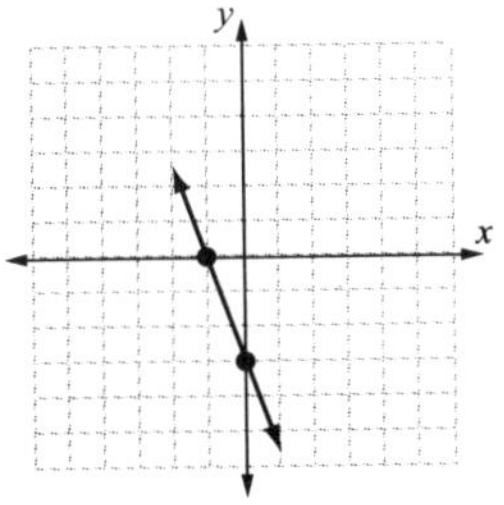

19. Parallel

HA1-877: Drawing Inferences and Making Predictions from Tables and Graphs

1. 17 cars
3. 1 student
5. $8.00
7. Monday
9. 15 minutes
11. 163 cars
13. $y = 4x$
15. 15°
17. $y = 2.5x + 5$
19. 30

HA1-879: Applying Counting Techniques to Permutations and Combinations

1. 6
3. 5,040
5. 6
7. 6
9. 120
11. 132
13. 60
15. 756
17. 336
19. 120

HA1-885: Histograms and the Normal Distribution

1. skewed
3. bell-shaped
5. skewed
7. skewed
9. 0-5
11. 21-25
13. 11-20
15. 75.5
17. 2.5
19. 13

21.

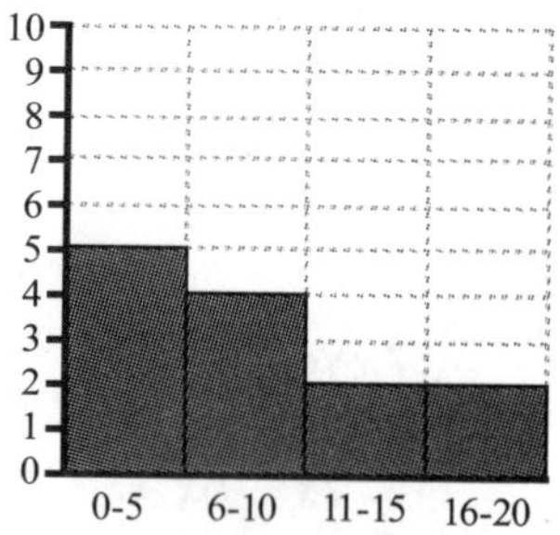

23.

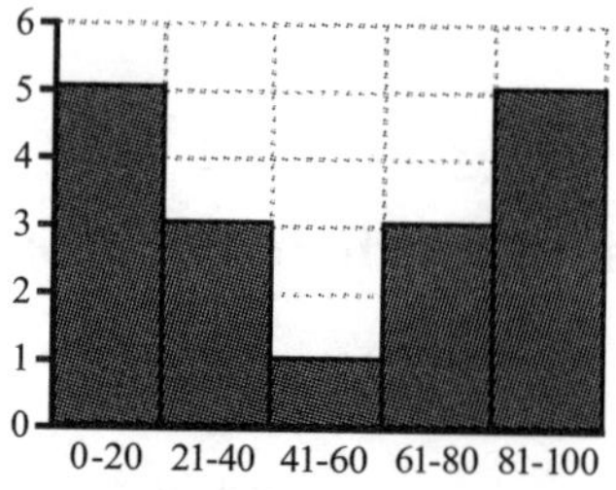

25.

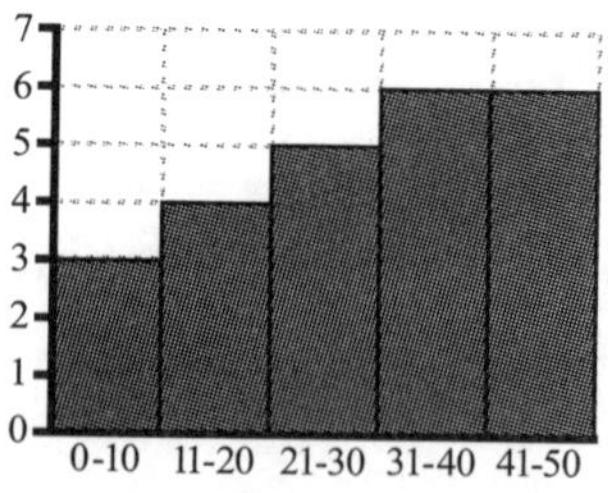

27.

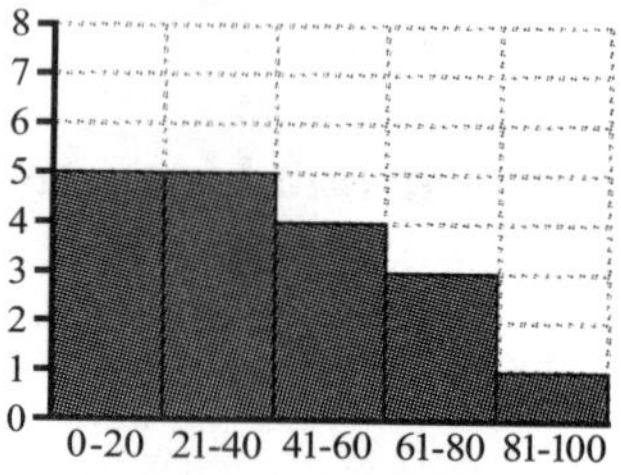

29.

Meters	Frequency
1-5	5
6-10	4
11-15	3
16-20	2
21-25	1

31.

Students	Frequency
1-10	6
11-20	4
21-15	3
31-40	2
41-50	1
51-60	1

33.

Years	Frequency
0-5	30
6-10	25
11-15	15
16-20	5

35.

Students	Frequency
0-5	4
6-10	5
11-15	2
16-20	2
21-25	3

37. mean = median = mode

39. 8.21

HA1-886: Unions and Intersections of Sets Using Venn Diagrams

1. {12, 24} **3.** {20, 40} **5.** {1, 2, 3, 6}

7. {1, 2, 5, 10} **9.** {1, 2, 3, 4, 6, 8, 12} **11.** {1, 4, 8, 9, 12, 16, 20, 25, 36, 49}

13. {1, 2, 3, 5, 6, 10, 15, 25, 30, 75} **15.** {2, 4, 8, 16, 32, 56, 64, 96} **17.** 10 **19.** 9

21. 2 **23.** 4 **25.** 34 students **27.** 38 **29.** 26 nurses **31.** 220

33. 435 **35.** 344 **37.** 8 **39.** 1,460

HA1-889: Complimentary and Supplementary Angles

1. 5° **3.** 49° **5.** 35° **7.** 11° **9.** 122°

11. 139° **13.** 30° **15.** 91° **17.** $m\angle C = 9°$, $m\angle D = 81°$ **19.** $m\angle R = 65°$, $m\angle S = 25°$

21. $m\angle J = 38°$, $m\angle K = 52°$ **23.** $m\angle F = 42°$, $m\angle G = 48°$ **25.** $m\angle C = 63°$, $m\angle D = 117°$ **27.** $m\angle R = 140°$, $m\angle S = 40°$ **29.** $m\angle J = 83°$, $m\angle K = 97°$

31. 9° **33.** 76° **35.** 34° **37.** 115° **39.** 72°

HA1-890: Using Models to Derive Formulas for Two-Dimensional Geometric Figures

1. $60c^2$ **3.** $(36 + 4\pi)g$ **5.** $(24 + 6\pi)x$ **7.** $72s$ **9.** $26p$

11. $28x$ **13.** $(48 + 4\pi)x$ **15.** $80x$ **17.** 2 **19.** $(324 - 81\pi)x^2$

21. $192 - 4x^2$ **23.** $144x + 64x^2$ **25.** $(100\pi - 25)x^2$ **27.** $48p^2$ **29.** $\left(\frac{25}{4}\pi - 6\right)x^2$

31. $3 < p < 20$ **33.** $\frac{1}{2} < x < 6$ **35.** $2 < v < 6$ **37.** $0 < x < 6$ **39.** $2 < x < 3$

HA1-891: Using Models to Derive Formulas for Three-Dimensional Solids

1. $V = l^3 - 7l^2 + 10l$ **3.** $V = 2w^3 - 4w^2$ **5.** $V = 4w^3 - 6w^2$ **7.** $V = h^3 - 7h^2 + 10h$

9. $S = 12\pi r^2$ **11.** $S = 8\pi r^2$ **13.** $S = 3\pi r^2$ **15.** $S = \frac{5}{2}\pi r^2$

17. $3\pi r^3$ **19.** $\frac{2}{3}\pi h^3$ **21.** $192\pi r^3$ **23.** $\frac{8}{3}\pi r^3$

25. $14\pi x^2$ **27.** $11\pi x^2$ **29.** $\frac{43}{3}\pi r^3$ **31.** $16x$

33. $\frac{47}{8}x$ **35.** $12x$ **37.** $\frac{8}{27}x$ **39.** $3x$

HA1-893: Constructing Solids from Different Perspectives

1. PNQR **3.** RQTV **5.** XZSV **7.** CFKG

9. PRQ **11.** ABC **13.** CBA **15.** KLN

17. cube **19.** square pyramid **21.** cylinder **23.** cone

25.
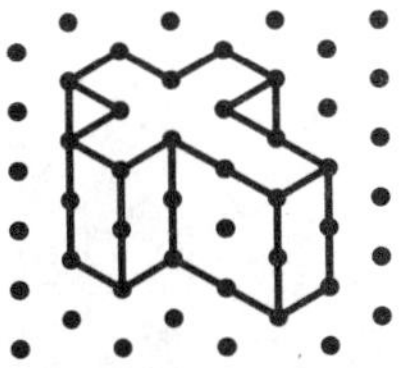

27.

29.
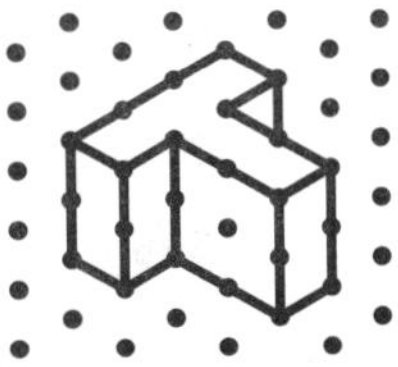

31.
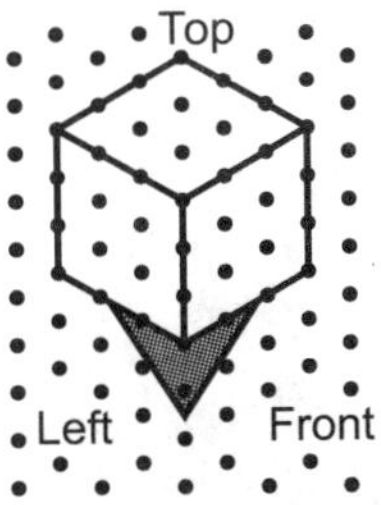

33.
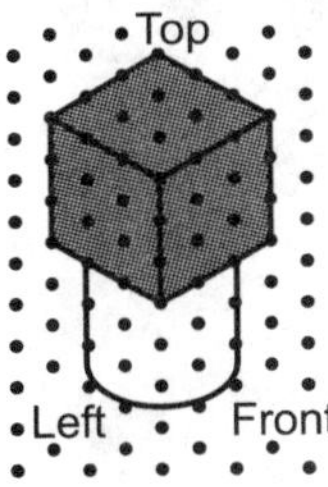

35.
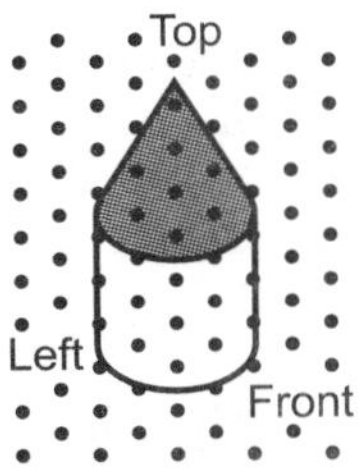

37.
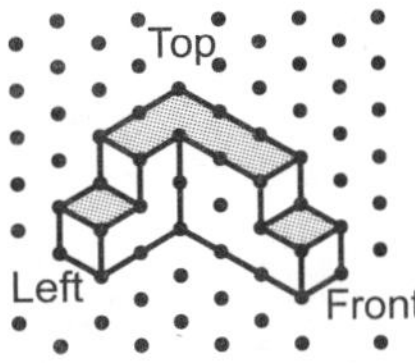

39.
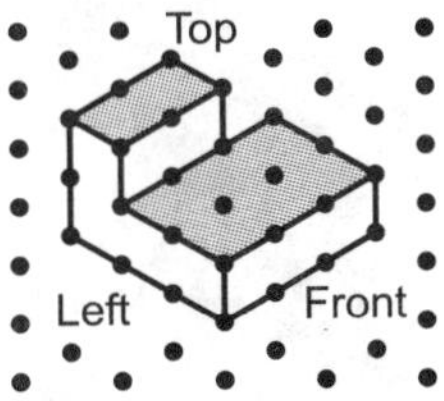

HA1-920: Simplifying Algebraic Expressions Using the Distributive Property

1. $21x^8 - 9x^5$ **3.** $7x^{10} - 3x^7$ **5.** $28x^{11} - 8x^7$

7. $4x^4y^3 + 3x^5y^3 - 3x^6y^3$

9. $x^4y^3 + 3x^6y^3$

11. $6x^5 - 5x^4 - 4x^3$

13. $9x^5 + 15x^4 + 4x^3$

15. $5x^5 + 21x^4 + 4x^3$

17. $4x^5y^5 + 6x^4y^4 - 8x^4y^3 - 12x^3y^2$

19. $4x^5y^5 - 6x^4y^3z - 8x^4y^2z + 12x^3z^2$

21. $4x^5y^5 - 6x^4y^4 - 10x^4y^3 + 15x^3y^2$

23. $6x^4 + 11x^3 + 22x^2 + 9x$

25. $9x^4 - 9x^3 + 23x^2 + 9x$

27. $9x^4 + 26x^2 + 9x$

29. $9x^4 + 12x^2y + 4y^2$

31. $9x^4 + 24x^2y + 16y^2$

33. $16x^4 - 32x^2y + 16y^2$

35. $16x^4 + 24x^2y + 9y^2$

37. $6x^4y^4 + 9x^5y^3 + 8x^2y^4 + 12x^3y^3 + 12x^2y^2 + 18x^3y$

39. $-6x^4y^4 - 9x^5y^3 + 8x^2y^4 + 12x^3y^3 + 12x^2y^2 + 18x^3y$

Index

F

G

H

I